D0145945

Structural Geology

Lavishly illustrated in color, this textbook takes an applied approach to introduce undergraduate students to the basic principles of structural geology. The book provides unique links to industry applications in the upper crust, including petroleum and groundwater geology, which highlight the importance of structural geology in exploration and exploitation of petroleum and water resources. Topics range from faults and fractures forming near the surface to shear zones and folds of the deep crust. Students are engaged through examples and parallels drawn from practical everyday situations, enabling them to connect theory with practice. Containing numerous end-of-chapter problems, e-learning modules, and with stunning field photos and illustrations, this book provides the ultimate learning experience for all students of structural geology.

Haakon Fossen is Professor of Structural Geology at the University of Bergen, Norway, where he is affiliated with the Department of Earth Science, the Natural History Collections, and the Centre for Integrated Petroleum Research (CIPR). His professional career has also involved work as an exploration and production geologist/geophysicist for Statoil and periods of geologic mapping and mineral exploration in Norway. His research ranges from hard to soft rocks and includes studies of folds, shear zones, formation and collapse of the Caledonian Orogen, numerical modeling of deformation (transpression), the evolution of the North Sea rift, and studies of deformed sandstones in the western United States. He has conducted extensive field work in various parts of the world, notably Norway, Utah/Colorado and Sinai, and his research is based on field mapping, microscopy, physical and numerical modeling, geochronology and seismic interpretation. Professor Fossen has been involved in editing several international geology journals, has authored over 90 scientific publications, and has written two books and several book chapters. He has taught undergraduate structural geology courses for over ten years and has a keen interest in developing electronic teaching resources to aid student visualization and understanding of geologic structures.

Structural Geology

Haakon Fossen

UNIVERSITY OF BERGEN, NORWAY

CAMBRIDGE UNIVERSITY PRESS

Cambridge, New York, Melbourne, Madrid, Cape Town, Singapore,
São Paulo, Delhi, Dubai, Tokyo

Cambridge University Press
The Edinburgh Building, Cambridge CB2 8RU, UK

Published in the United States of America by Cambridge University Press, New York

www.cambridge.org
Information on this title: www.cambridge.org/9780521516648

© Haakon Fossen 2010

This publication is in copyright. Subject to statutory exception
and to the provisions of relevant collective licensing agreements,
no reproduction of any part may take place without
the written permission of Cambridge University Press.

First published 2010

Printed in the United Kingdom at the University Press, Cambridge

A catalogue record for this publication is available from the British Library

Library of Congress Cataloging-in-Publication Data

Fossen, Haakon, 1961–
Structural geology / Haakon Fossen.
    p.   cm.
ISBN 978-0-521-51664-8 (Hardback)
1. Geology, Structural.   I. Title.
QE601.F687 2010
551.8–dc22   2010011781

ISBN 978-0-521-51664-8 Hardback

Additional resources for this publication at www.cambridge.org/9780521516648

Cambridge University Press has no responsibility for the persistence or
accuracy of URLs for external or third-party internet websites referred to
in this publication, and does not guarantee that any content on such
websites is, or will remain, accurate or appropriate.

Contents

HOW TO USE THIS BOOK

Each chapter starts with a general **introduction**, which presents a context for the topic within structural geology as a whole. These introductions provide a roadmap for the chapter and will help you to navigate through the book.

BOX 4.2 | VECTORS, MATRICES AND TEN

A scalar is a real number, reflecting temperature, mass no direction. A vector has both magnitude (length) an or velocity. A matrix is a two-dimensional array of nu meaning that they have 9 or 4 components). Matrice

The term tensor is, in rock mechanics, applied to v scalars as tensors of order zero, vectors as first-orde Hence, for our purposes, the terms matrix and seco cases where numbers are arranged in matrices that

The main text contains **highlighted terms** and **key expressions** that you will need to understand and become familiar with. Many of these terms are listed in the **Glossary** at the back of the book. The Glossary allows you to easily look up terms whenever needed and can also be used to review important topics and key facts. Each chapter also contains a series of **highlighted statements** to encourage you to pause and review your understanding of what you have read.

Most chapters have one or more **boxes** containing in-depth information about a particular subject, helpful examples or relevant background information. Other important points are brought together in the **chapter summaries. Review questions** should be used to test your understanding of the chapter before moving on to the next topic. **Answers** to these questions are given on the book's web-page.

Review questions

1. When is it appropriate to use the term pressure in geology?

2. How can we graphically visualize the state of stress in two and th

3. Where could we expect to find tensile stress in the crust?

4. How will the shape and orientation of the stress ellipsoid change system?

5. Will the stress tensor (matrix) look different if we choose a differe

6. A diagonal tensor has numbers on the diagonal running from the with all other entries being zero. What does a diagonal stress tens

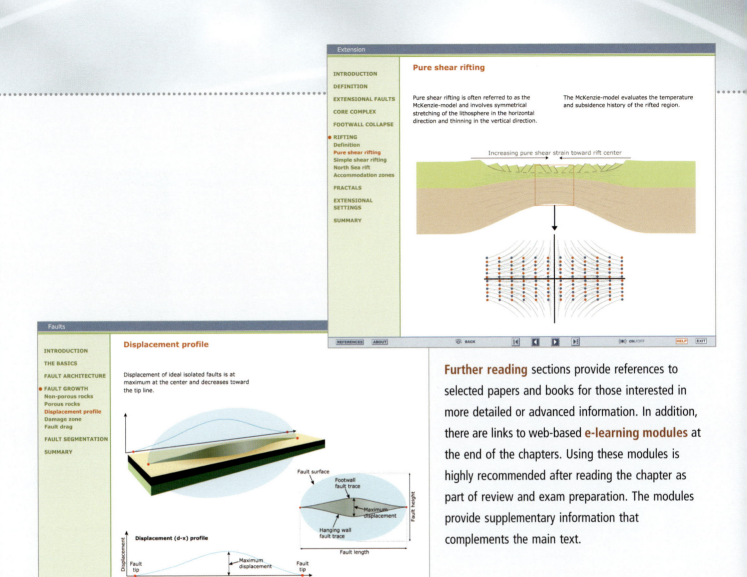

Pure shear rifting

INTRODUCTION

DEFINITION

EXTENSIONAL FAULTS

CORE COMPLEX

FOOTWALL COLLAPSE

● RIFTING
  Definition
  Pure shear rifting
  Simple shear rifting
  North Sea rift
  Accommodation zones

FRACTALS

EXTENSIONAL
SETTINGS

SUMMARY

Pure shear rifting is often referred to as the McKenzie-model and involves symmetrical stretching of the lithosphere in the horizontal direction and thinning in the vertical direction.

The McKenzie-model evaluates the temperature and subsidence history of the rifted region.

Increasing pure shear strain toward rift center

REFERENCES   ABOUT   BACK   ON /OFF   HELP   EXIT

Faults

Displacement profile

INTRODUCTION

THE BASICS

FAULT ARCHITECTURE

● FAULT GROWTH
  Non-porous rocks
  Porous rocks
  Displacement profile
  Damage zone
  Fault drag

FAULT SEGMENTATION

SUMMARY

Displacement of ideal isolated faults is at maximum at the center and decreases toward the tip line.

Fault surface
Footwall fault trace
Maximum displacement
Hanging wall fault trace
Fault length
Fault height

Displacement (d-x) profile

Displacement
Fault tip
Maximum displacement
Fault tip
Length

REFERENCES   ABOUT   BACK   ON /OFF   HELP   EXIT

Further reading sections provide references to selected papers and books for those interested in more detailed or advanced information. In addition, there are links to web-based **e-learning modules** at the end of the chapters. Using these modules is highly recommended after reading the chapter as part of review and exam preparation. The modules provide supplementary information that complements the main text.

Web-based resources

Specially prepared resources, unique to this book, are available from the book's web-page: www.cambridge.org/fossen. These are:

- Flash based e-learning modules that combine animations, text, illustrations and photographs. These present key aspects of structural geology in a highly visual and interactive environment.

- All of the figures for each chapter as jpeg files for use by instructors and readers.

- Supplementary figures illustrating additional geologic structures and field examples.

- Answers to the review questions presented at the end of each chapter.

- Additional exercises and solutions.

- A repository for further images, animations, videos, exercises and other resources provided by readers and instructors as a community resource.

Preface

This textbook is written to introduce undergraduate students, and others with a general geologic background, to basic principles, aspects and methods of structural geology. It is mainly concerned with the structural geology of the crust, although the processes and structures described are relevant also for deformation that occurs at deeper levels within our planet. Further, remote data from Mars and other planets indicate that many aspects of terrestrial structural geology are relevant also beyond our own planet.

The field of structural geology is very broad, and the content of this book presents a selection of important subjects within this field. Making the selection has not been easy, knowing that lecturers tend to prefer their own favorite aspects of, and approaches to, structural geology, or make selections according to their local departmental course curriculum. Existing textbooks in structural geology tend to emphasize the ductile or plastic deformation that occurs in the middle and lower crust. In this book I have tried to treat the frictional regime in the upper crust more extensively so that it better balances that of the deeper parts of the crust, which makes some chapters particularly relevant to courses where petroleum geology and brittle deformation in general are emphasized.

Obtaining this balance was one of several motivating factors for writing this book, and is perhaps related to my mixed petroleum geology and hard-rock structural geology experience. Other motivating factors include the desire to make a book where I could draw or redraw all of the illustrations and be able to present the first full-color book in structural geology. I also thought that a fundamental structural geology text of the twenty-first century should come with specially prepared e-learning resources, so the package of e-learning material that is presented with this book should be regarded as part of the present book concept.

Book structure

The structure of the book is in many ways traditional, going from strain (Chapters 2 and 3) to stress (Chapters 4 and 5) and via rheology (Chapter 6) to brittle deformation (Chapters 7 and 8). Of these, Chapter 2 contains material that would be too detailed and advanced for some students and classes, but selective reading is possible. Then, after a short introduction to the microscale structures and processes that distinguish crystal-plastic from brittle deformation (Chapter 10), ductile deformation structures such as folding, boudinage, foliations and shear zones are discussed (Chapters 11–15). Three consecutive chapters then follow that are founded on the three principal tectonic regimes (Chapters 16–18) before salt tectonics and restoration principles are presented (Chapters 19 and 20). A final chapter, where links to metamorphic petrology as well as stratigraphy are drawn, rounds off the book, and suggests that structural geology and tectonics largely rely on other disciplines. The chapters do not have to be read in numerical order, and most chapters can be used individually.

Emphasis and examples

The book seeks to cover a wide ground within the field of structural geology, and examples presented in the text are from different parts of the world. However, pictures and illustrations from a few geographic areas reappear. One of those is the North Sea rift system, notably the Gullfaks oil field, which I know quite well from my years with the Norwegian oil company Statoil. Another is the Colorado Plateau (mostly Utah), which over the last two decades has become one of my favorite places to do field work. A third, and much wetter and greener one, is the Scandinavian Caledonides. From this ancient orogen I have chosen a number of examples to illustrate structures typical of the plastic regime.

Acknowledgments

During the writing of this textbook I have built on experience and knowledge achieved through my entire career, from early days as a student, via various industrial and academic positions, to the time I have spent writing the manuscript. In this respect I want to thank fellow students, geologists and professors with whom I have interacted during my time at the Universities of Bergen, Oslo, Minnesota and Utah, at Utah State University, in Statoil and at the Geological Survey of Norway. In particular, my advisers and friends Tim Holst, Peter Hudleston and Christian Teyssier deserve thanks for sharing their knowledge during my three years in Minnesota, and among the many fellow PhD students there special thanks are due to Jim Dunlap, Eric Heatherington, David Kirschner, Labao Lan and, particularly, Basil Tikoff for valuable discussions and exchange of ideas as we were exploring various aspects of structural geology. Among coworkers and colleagues I wish to extend special thanks to Roy Gabrielsen, who contributed to the Norwegian book on which this book builds, Jonny Hesthammer for good company in Statoil and intense field discussions, Egil Rundhovde for co-leading multiple field trips to the Colorado Plateau, and to Rich Schultz who is always keen on intricate discussions on fracture mechanics and deformation bands in Utah and elsewhere.

Special thanks also go to Wallace Bothner, Rob Butler, Nestor Cardozo, Declan DePaor, Jim Evans, James Kirkpatrick, Stephen Lippard, Christophe Pascal, Atle Rotevatn, Zoe Shipton, Holger Stunitz and Bruce Trudgill for reading and commenting on earlier versions of the text. I am also thankful to colleagues and companies who assisted in finding appropriate figures and seismic examples of structures, each of which is acknowledged in connection with the appearance of the illustration in the book, and to readers who will send their comments to me so that improvements can be made for the next edition.

Symbols

a	long axis of ellipse representing a microcrack
A	area;
	empirically determined constant in flow laws
c	short axis of ellipse representing a microcrack
C	cohesion or cohesional strength of a rock
C_f	cohesive strength of a fault
d	offset
d_{cl}	thickness of clay layer
D	displacement;
	fractal dimension
D_{max}	maximum displacement along a fault trace or on a fault surface
$\mathbf{D}$	deformation (gradient) matrix
$e = \varepsilon$	elongation
$\dot{e} = \dot{\varepsilon}$	elongation rate (de/dt)
$\dot{e}_x$ and $\dot{e}_y$	elongation rates in the x and y directions (s^{-1})
$\mathbf{e}_1$, $\mathbf{e}_2$ and $\mathbf{e}_3$	eigenvectors of deformation matrix, identical to the three axes of strain ellipsoid
$\bar{e}$	logarithmic (natural) elongation
$\bar{e}_s$	natural octahedral unit shear
E	Young's modulus;
	activation energy for migration of vacancies through a crystal ($J\,mol^{-1}\,K^{-1}$)
E^*	activation energy
$\mathbf{F}$	force vector ($kg\,m\,s^{-2}$, N)
F_n	normal component of the force vector
F_s	shear component of the force vector
g	acceleration due to gravity (m/s^2)
h	layer thickness
h_0	initial layer thickness
h_T	layer thickness at onset of folding (buckling)
ISA_{1-3}	instantaneous stretching axes
K	bulk modulus
K_i	stress intensity factor
K_c	fracture toughness
k	parameter describing the shape of the strain ellipsoid (lines in the Flinn diagram)
k_x and k_y	pure shear components, diagonal elements in the pure shear and simple shear matrices
l	line length (m)

l_0	line length prior to deformation (m)
L	velocity tensor (matrix)
L	fault length;
	wavelength
L_d	dominant wavelength
L_T	actual length of a folded layer over the distance of one wavelength
n	exponent of displacement-length scaling law
p_f	fluid pressure
P	pressure (Pa)
Q	activation energy
R	ellipticity or aspect ratio of ellipse (long over short axis);
	gas constant ($J\,kg^{-1}\,K^{-1}$)
R_f	final ellipticity of an object that was non-circular prior to deformation
R_i	initial ellipticity of an object (prior to deformation)
R_s	same as R, used in connection with the R_f/ϕ-method to distinguish it from R_f
R_{xy}	X/Y
R_{yz}	Y/Z
s	stretching
$\dot{\mathbf{S}}$	stretching tensor, symmetric part of **L**
t	time (s)
T	temperature (K or °C);
	uniaxial tensile strength (bar);
	local displacement or throw of a fault when calculating SGR and SSF
v	velocity vector (m/s)
V	volume (m^3)
V_0	volume prior to deformation
V_P	velocity of P-waves
V_s	velocity of S-waves
w	vorticity vector
w	vorticity
W	vorticity (or spin) tensor, which is the skew-symmetric component of **L**
W_k	kinematic vorticity number
x	vector or point in a coordinate system prior to deformation
x′	vector or point in a coordinate system after deformation
x, y, z	coordinate axes, z being vertical
X, Y, Z	principal strain axes; $X \geq Y \geq Z$
Z	crustal depth (m)
α	thermal expansion factor (K^{-1});
	Biot poroelastic parameter;
	angle between passive marker and shear direction at onset of non-coaxial deformation (Chapter 15);
	angle between flow apophyses (Chapter 2)
α'	angle between passive marker and shear direction after a non-coaxial deformation
β	stretching factor, equal to s
Δ	volume change factor
$\Delta\sigma$	change in stress

γ	shear strain
$\bar{\gamma}_{\text{oct}}$	octahedral shear strain
$\dot{\gamma}$	shear strain rate
Γ	non-diagonal entry in deformation matrix for subsimple shear
η	viscosity constant (N s m^{-2})
λ	quadratic elongation
λ_1, λ_2 and λ_3	eigenvalues of deformation matrix
$\sqrt{\lambda_1}$, $\sqrt{\lambda_2}$ and $\sqrt{\lambda_3}$	length of strain ellipse axes
μ	shear modulus; viscosity
μ_{f}	coefficient of sliding friction
μ_{L}	viscosity of buckling competent layer
μ_{M}	viscosity of matrix to buckling competent layer
ν	Poisson's ratio; Lode's parameter
θ	angle between the normal to a fracture and σ_1; angle between ISA_1 and the shear plane
θ'	angle between X and the shear plane
ρ	density (g/cm^3)
σ	stress $(\Delta F / \Delta A)$ (bar: $1\ \text{bar} = 1.0197\ \text{kg/cm}^2 = 10^5\ \text{Pa} = 10^6\ \text{dyne/cm}^2$)
$\boldsymbol{\sigma}$	stress vector (traction vector)
$\sigma_1 > \sigma_2 > \sigma_3$	principal stresses
$\bar{\sigma}$	effective stress
σ_{a}	axial stress
σ_{dev}	deviatoric stress
σ_{diff}	differential stress $(\sigma_1 - \sigma_3)$
σ_{H}	max horizontal stress
σ_{h}	min horizontal stress
σ_{h}^*	average horizontal stress in thinned part of the lithosphere (constant-horizontal-stress model)
σ_{m}	mean stress $(\sigma_1 + \sigma_2 + \sigma_3)/3$
σ_{n}	normal stress
σ_{r}	remote stress
σ_{s}	shear stress
σ_{t}	tectonic stress
σ_{tip}	stress at tip of fracture or point of max curvature along pore margin
σ_{tot}	total stress $(\sigma_{\text{m}} + \sigma_{\text{dev}})$
σ_{v}	vertical stress
$\sigma_{\text{n}}^{\text{g}}$	normal stress at grain–grain or grain–wall contact areas in porous medium
$\sigma_{\text{n}}^{\text{w}}$	average normal stress exerted on wall by grains in porous medium
ϕ	internal friction (rock mechanics); angle between X and a reference line at onset of deformation (R_f/ϕ-method)
ϕ'	angle between X and a reference line after a deformation (R_f/ϕ-method)
Φ	porosity
ψ	angular shear
$\boldsymbol{\omega}$	angular velocity vector

Chapter 1

Structural geology and structural analysis

Structural geology is about folds, faults and other deformation structures in the lithosphere – how they appear and how and why they formed. Ranging from features hundreds of kilometers long down to microscopic details, structures occur in many different settings and have experienced exciting changes in stress and strain – information that can be ours if we learn how to read the code. The story told by structures in rocks is beautiful, fascinating and interesting, and it can also be very useful to society. Exploration, mapping and exploitation of resources such as slate and schist (building stone), ores, groundwater, and oil and gas depend on structural geologists who understand what they observe so that they can present reasonable interpretations and predictions. In this first chapter we will set the stage for the following chapters by defining and discussing fundamental concepts and some of the different data sets and methods that structural geology and structural analysis rely on. Depending on your background in structural geology, it may be useful to return to this chapter after going through other chapters in this book.

1.1 Approaching structural geology

For us to understand structural geology we need to observe deformed rocks and find an explanation for how and why they ended up in their present state. Our main methods are field observations, laboratory experiments and numerical modeling. All of these methods have advantages and challenges. Field examples portray the final results of deformation processes, while the actual deformation history may be unknown. Progressive deformation can be observed in laboratory experiments, but how representative are such hour- or perhaps week-long observations of geologic histories that span thousands to millions of years in nature? Numerical modeling, where we use computers and mathematical equations to model deformation, is hampered by simplifications necessary for the models to be runable with today's codes and computers. However, by combining different approaches we are able to obtain realistic models of how structures form and what they mean. Field studies will always be important, as any modeling, numerical or physical, must be based directly or indirectly on accurate and objective field observations and descriptions. Objectivity during fieldwork is both important and challenging, and field studies in one form or another are the main reason why many geologists chose to become geoscientists!

1.2 Structural geology and tectonics

The word **structure** is derived from the Latin word *struere*, to build, and we could say:

A geologic structure is a geometric configuration of rocks, and structural geology deals with the geometry, distribution and formation of structures.

It should be added that **structural geology** only deals with structures created during rock deformation, not with primary structures formed by sedimentary or magmatic processes. However, deformation structures can form through the modification of primary structures, such as folding of bedding in a sedimentary rock.

The closely related word **tectonics** comes from the Greek word *tektos*, and both structural geology and tectonics relate to the building and resulting structure of the Earth's lithosphere, and to the motions that change and shape the outer parts of our planet. We could say that tectonics is more closely connected to the underlying processes that cause structures to form:

Tectonics is connected with external and often regional processes that generate a characteristic set of structures in an area or a region.

By external we mean external to the rock volume that we study. External processes or causes are in many cases plate motions, but can also be such things as forceful intrusion of magma, gravity-driven salt or mud diapirs, flowing glaciers and meteor impacts. Each of these "causes" can create characteristic structures that define a **tectonic style**, and the related tectonics can be given special names. **Plate tectonics** is the large-scale part of tectonics that directly involves the movement and interaction of lithospheric plates. Within the realm of plate tectonics, expressions such as subduction tectonics, collision tectonics and rift tectonics are applied for more specific purposes.

Glaciotectonics is the deformation of sediments and bedrock (generally sedimentary rocks) at the toe of an advancing ice sheet. In this case it is the pushing of the ice that creates the deformation, particularly where the base of the glacier is cold (frozen to the substrate).

Salt tectonics deals with the deformation caused by the (mostly) vertical movement of salt through its overburden (see Chapter 19). Both glaciotectonics and salt tectonics are primarily driven by gravity, although salt tectonics can also be closely related to plate tectonics. For example, tectonic strain can create fractures that enable salt to gravitationally penetrate its cover, as discussed in Chapter 19. The term **gravity tectonics** is generally restricted to the downward sliding of large portions of rocks and sediments, notably of continental margin deposits resting on weak salt or overpressured shale layers. Raft tectonics is a type of gravity tectonics occurring in such environments, as mentioned in Chapter 19. Smaller landslides and their structures are also considered examples of gravity tectonics by some, while others regard such surficial processes as **non-tectonic**. Typical non-tectonic deformation is the simple compaction of sediments and sedimentary rocks due to loading by younger sedimentary strata.

Neotectonics is concerned with recent and ongoing crustal motions and the contemporaneous stress field. Neotectonic structures are the surface expression of faults in the form of fault scarps, and important data sets stem from seismic information from earthquakes (such as focal mechanisms, Box 9.1) and changes in elevation of regions detected by repeated satellite measurements.

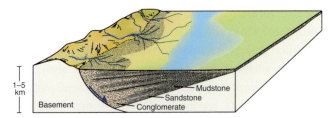

Figure 1.1 Illustration of the close relationship between sedimentary facies, layer thickness variations and syndepositional faulting (growth fault) along the margin of a sedimentary basin.

At smaller scales, **microtectonics** describes microscale deformation and deformation structures visible under the microscope.

Structural geology typically pertains to the observation, description and interpretation of structures that can be mapped in the field. How do we recognize deformation or **strain** in a rock? "Strained" means that something primary or preexisting has been geometrically modified, be it cross stratification, pebble shape, a primary magmatic texture or a preexisting deformation structure. Hence strain can be defined as a change in length or shape, and recognizing strain and deformation structures actually requires solid knowledge of undeformed rocks and their primary structures.

Being able to recognize tectonic deformation depends on our knowledge of primary structures.

The resulting deformation structure also depends on the initial material and its texture and structure. Deforming sandstone, clay, limestone or granite results in significantly different structures because they respond differently. Furthermore, there is often a close relationship between tectonics and the formation of rocks and their primary structures. Sedimentologists experience this as they study variations in thickness and grain size in the hanging wall (down-thrown side) of syndepositional faults. This is illustrated in Figure 1.1, where the gradual rotation and subsidence of the down-faulted block gives space for thicker strata near the fault than farther away, resulting in wedge-shaped strata and progressively steeper dips down section. There is also a facies variation, with the coarsest-grained deposits forming near the fault, which can be attributed to the fault-induced topography seen in Figure 1.1.

Another close relationship between tectonics and rock forming processes is shown in Figure 1.2, where forceful rising and perhaps inflating of magma deforms the outer and oldest part of the pluton and its country rock.

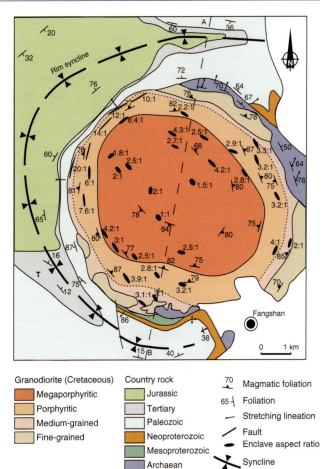

Granodiorite (Cretaceous)
- ▨ Megaporphyritic
- ▨ Porphyritic
- ▨ Medium-grained
- ▨ Fine-grained

Country rock
- ▨ Jurassic
- ▨ Tertiary
- ▨ Paleozoic
- ▨ Neoproterozoic
- ▨ Mesoproterozoic
- ▨ Archaean

- 70 Magmatic foliation
- 65 Foliation
- ― Stretching lineation
- / Fault
- ● Enclave aspect ratio
- ✕ Syncline

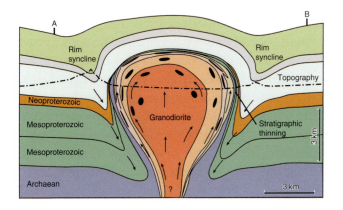

Figure 1.2 Structural geology can be linked to processes and mechanisms other than plate stresses. This map and profile from a granodioritic pluton southwest of Beijing, China, portray close connection between forceful intrusion of magma, strain and folds in the country rock. Black ellipses indicate strain, as discussed in Chapters 2 and 3. The strain (deformation) pattern within and around the pluton can be explained in terms of diapirism, where the intrusion ascends and squeezes and shears its outer part and the surrounding country rock to create space. Based on He *et al.* (2009).

Forceful intrusion of magma into the crust is characterized by deformation near the margin of the pluton, manifested by folding and shearing of the layers in Figure 1.2. Ellipses in this figure illustrate the shape of enclaves (inclusions), and it is clear that they become more and more elongated as we approach the margin of the pluton. Hence, the outer part of the pluton has been flattened during a forceful intrusion history.

Metamorphic growth of minerals before, during, and after deformation may also provide important information about the pressure–temperature conditions during deformation, and may contain textures and structures reflecting kinematics and deformation history. Hence, sedimentary, magmatic and metamorphic processes may all be closely associated with the structural geology of a locality or region.

These examples relate to strain, but structural geologists, especially those dealing with brittle structures of the upper crust, are also concerned with **stress**. Stress is a somewhat diffuse and abstract concept to most of us, since it is invisible. Nevertheless, there will be no strain without a stress field that exceeds the rock's resistance against deformation. We can create a stress by applying a force on a surface, but at a point in the lithosphere stress is felt from all directions, and a full description of such a state of stress considers stress from all directions and is therefore three-dimensional. There is always a relationship between stress and strain, and while it may be easy to establish from controlled laboratory experiments it may be difficult to extract from naturally formed deformation structures.

Structural geology covers deformation structures formed at or near the Earth's surface, in the cool, upper part of the crust where rocks have a tendency to fracture, in the hotter, lower crust where the deformation tends to be ductile, and in the underlying mantle. It embraces structures at the scale of hundreds of kilometers down to micro- or atomic-scale structures, structures that form almost instantaneously, and structures that form over tens of millions of years.

A large number of subdisciplines, approaches and methods therefore exist within the field of structural geology. The oil exploration geologist may be considering trap-forming structures formed during rifting or salt tectonics, while the production geologist worries about sub-seismic sealing faults (faults that stop fluid flow in porous reservoirs; Section 8.7). The engineering geologist may consider fracture orientations and densities in relation to a tunnel project, while the university professor uses structural mapping, physical modeling or computer modeling to understand mountain-building processes. The methods and approaches are many, but they serve to understand the structural or tectonic development of a region or to predict the structural pattern in an area. In most cases structural geology is founded on data and observations that must be analyzed and interpreted. Structural analysis is therefore an important part of the field of structural geology.

Structural data are analyzed in ways that lead to a tectonic model for an area. By **tectonic model** we mean a model that explains the structural observations and puts them into context with respect to a larger-scale process, such as rifting or salt movements. For example, if we map out a series of normal faults indicating E–W extension in an orogenic belt, we have to look for a model that can explain this extension. This could be a rift model, or it could be extensional collapse during the orogeny, or gravity-driven collapse after the orogeny. Age relations between structures and additional information (radiometric dating, evidence for magmatism, relative age relations and more) would be important to select a model that best fits the data. It may be that several models can explain a given data set, and we should always look for and critically evaluate alternative models. In general, a simple model is more attractive than a complicated one.

1.3 Structural data sets

Planet Earth represents an incredibly complex physical system, and the structures that result from natural deformation reflect this fact through their multitude of expressions and histories. There is thus a need to simplify and identify the one or few most important factors that describe or lead to the recognition of deformation structures that can be seen or mapped in naturally deformed rocks. **Field observations** of deformed rocks and their structures represent the most direct and important source of information on how rocks deform, and objective observations and careful descriptions of naturally deformed rocks are the key to understanding natural deformation. Indirect observations of geologic structures by means of various **remote sensing methods**, including satellite data and seismic surveying, are becoming increasingly important in our mapping and description of structures and tectonic deformation. **Experiments** performed in the laboratory give us valuable knowledge of how various physical conditions, including stress field, boundary condition, temperature or the physical properties of the deforming material, relate to deformation. **Numerical models**, where rock deformation is simulated on a computer, are also useful as they allow us to control the various parameters and properties that influence deformation.

Experiments and numerical models not only help us understand how external and internal physical conditions control or predict the deformation structures that form, but also give information on how deformation structures evolve, i.e. they provide insights into the deformation history. In contrast, naturally deformed rocks represent end-results of natural deformation histories, and the history may be difficult to read out of the rocks themselves. Numerical and experimental models allow one to control rock properties and boundary conditions and explore their effect on deformation and deformation history. Nevertheless, any deformed rock contains some information about the history of deformation. The challenge is to know what to look for and to interpret this information. Numerical and experimental work aids in completing this task, together with objective and accurate field observations.

Numerical, experimental and remotely acquired data sets are important, but should always be based on field observations.

1.4 Field data

It is hard to overemphasize the importance of traditional field observations of deformed rocks and their structures. Rocks contain more information than we will ever be able to extract from them, and the success of any physical or numerical model relies on the accuracy of observation of rock structures in the field. Direct contact with rocks and structures that have not been filtered or interpreted by people or computers is invaluable.

Unfortunately, our ability to make objective observations is limited. What we have learned and seen in the past strongly influences our visual impressions of deformed rocks. Any student of deformed rocks should therefore train himself or herself to be objective. Only then can we expect to discover the unexpected and make new interpretations that may contribute to our understanding of the structural development of a region and to the field of structural geology in general. Many structures are overlooked until the day that someone points out their existence and meaning, upon which they all of a sudden appear "everywhere". Shear bands in strongly deformed ductile rocks (mylonites) are one such example (Figure 15.25). They were either overlooked or considered as cleavage until the late 1970s, when they were properly described and interpreted. Since then, they have been described from almost every major shear zone or mylonite zone in the world.

Traditional fieldwork involves the use of simple tools such as a hammer, measuring device, topomaps, a hand lens and a compass, and the data collected are mainly structural orientations and samples for thin section studies. This type of data collection is still important, and is aided by modern global positioning system (GPS) units and high-resolution aerial and satellite photos. More advanced and detailed work may involve the use of a portable laser-scanning unit, where pulses of laser light strike the surface of the Earth and the time of return is recorded. This information can be used to build a detailed topographic or geometrical model of the outcrop, onto which one or more high-resolution field photographs can be draped. An example of such a model is shown in Figure 1.3, although the advantage of virtually moving around in the model cannot be demonstrated by a flat picture. Geologic observations such as the orientation of layering or fold axes can then be made on a computer.

In many cases, the most important way of recording field data is by use of careful field sketches, aided by photographs, orientation measurements and other measurements that can be related to the sketch. Sketching also forces the field geologist to observe features and details that may otherwise be overlooked. At the same time, sketches can be made so as to emphasize relevant information and neglect irrelevant details. Field sketching is, largely, a matter of practice.

1.5 Remote sensing and geodesy

Satellite images, such as those shown in Figure 1.4a, c, are now available at increasingly high resolutions and are a valuable tool for the mapping of map-scale structures. An increasing amount of such data is available on the World Wide Web, and may be combined with digital elevation data to create three-dimensional models. Ortho-rectified **aerial photos** (orthophotos) may give more or other details (Figure 1.4b), with resolutions down to a few tens of centimeters in some cases. Both ductile structures, such as folds and foliations, and brittle faults and fractures are mappable from satellite images and aerial photos.

In the field of neotectonics, **InSAR** (Interferometric Synthetic Aperture Radar) is a useful remote sensing technique that uses radar satellite images. Beams of radar waves are constantly sent toward the Earth, and an image is generated based on the returned information. The intensity of the reflected information reflects the composition of the ground, but the phase of the wave as it hits and becomes reflected is also recorded. Comparing phases enables us to monitor millimeter-scale changes in elevation and geometry of the surface, which may reflect active tectonic

Figure 1.3 Mediumfjellet, Svalbard, based on LIDAR (LIght Detection And Ranging) data (laser scanning from helicopter) and photos. This type of model, which actually is three dimensional, allows for geometric analysis on a computer and provides access to otherwise unreachable exposures. The lower figures are more detailed views. Modeling by Simon Buckley.

movements related to earthquakes. In addition, accurate digital elevation models (see next section) and topographic maps can be constructed from this type of data.

GPS data in general are an important source of data that can be retrieved from GPS satellites to measure plate movements (Figure 1.5). Such data can also be collected on the ground by means of stationary GPS units with down to millimeter-scale accuracy.

1.6 DEM, GIS and Google Earth

Conventional paper maps are still useful for many field mapping purposes, but rugged laptops, tablets and handheld devices now allow for direct digitizing of structural features on digital maps and images and are becoming more and more important. Field data in digital form can be combined with elevation data and other data by

Figure 1.4 (a) Satellite image of the Canyonlands National Park area, Utah. The image reveals graben systems on the east side of the Colorado River. An orthophoto (b) reveals that the grabens run parallel to fractures, and a high-resolution satellite image (c) shows an example of a graben stepover structure. *Source*: Utah AGRC.

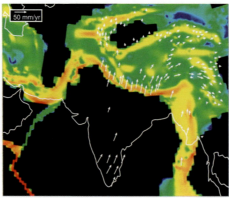

Figure 1.5 Use of GPS data from stationary GPS stations worldwide over time can be used to map relative plate motions and strain rates. (Left) White arrows (velocity vectors) indicating motions relative to Europe. The vectors clearly show how India is moving into Eurasia, causing deformation in the Himalaya–Tibetan Plateau region. (Right) Strain rate map based on GPS data. Calculated strain rates are generally less than 3×10^{-6} y^{-1} or 10^{-13} s^{-1}. Warm colors indicate high strain rates. Similar use of GPS data can be applied to much smaller areas where differential movements occur, for example across fault zones. From the project The Global Strain Rate Map (http://jules.unavco.org). See Kreemer *et al.* (2003) for more information.

means of a Geographical Information System (GIS). By means of GIS we can combine field observations, various geologic maps, aerial photos, satellite images, gravity data, magnetic data, typically together with a digital elevation model, and perform a variety of mathematical and statistical calculations. A **digital elevation model (DEM)** is a digital representation of the topography or shape of a surface, typically the surface of the Earth, but a DEM can be made for any geologic surface or interface that can be mapped in three dimensions. Surfaces mapped from cubes of seismic data are now routinely presented as DEMs and can easily be analyzed in terms of geometry and orientations.

Inexpensive or free access to geographic information exists, and this type of data was revolutionized by the development of Google Earth in the first decade of this century. The detailed data available from Google Earth and related sources of digital data have taken the mapping of faults, lithologic contacts, foliations and more to a new level, both in terms of efficiency and accuracy. Because of the rapid evolution of this field, further information and resources will be posted at the webpage of this book.

1.7 Seismic data

In the mapping of subsurface structures, seismic data are invaluable and since the 1960s have revolutionized our understanding of fault and fold geometry. Some seismic data are collected for purely academic purposes, but the vast majority of seismic data acquisition is motivated by exploration for petroleum and gas. Most seismic data are thus from rift basins and continental margins.

Acquisition of seismic data is, by its nature, a special type of remote sensing (acoustic), although always treated separately in the geo-community. Marine seismic reflection data (Figure 1.6) are collected by boat, where a sound source (air gun) generates sound waves that penetrate the crustal layers under the sea bottom. Microphones can also be put on the sea floor. This method is more cumbersome, but enables both seismic S- and P-waves to be recorded (S-waves do not travel through water). Seismic data can also be collected onshore, putting the sound source and microphones (geophones) on the ground. The onshore sound source would usually be an explosive device or a vibrating truck, but even a sledgehammer or specially designed gun can be used for very shallow and local targets.

The sound waves are reflected from layer boundaries where there is an increase in acoustic impedance, i.e. where there is an abrupt change in density and/or the velocity with which sound waves travel in the rock. A long line of microphones, onshore called geophones and offshore referred to as hydrophones, record the reflected sound signals and the time they appear at the surface. These data are collected in digital form and processed by computers to generate a seismic image of the underground.

Seismic data can be processed in a number of ways, depending on the focus of the study. Standard reflection seismic lines are displayed with two-way travel time as the vertical axis. Depth conversion is therefore necessary to create an ordinary geologic profile from those data. Depth conversion is done using a velocity model that depends on the lithology (sound moves faster in sandstone than in shale, and yet faster in limestone) and burial depth (lithification

BOX 1.1 | MARINE SEISMIC ACQUISITION

Offshore collection of seismic data is done by a vessel that travels at about 5 knots while towing arrays of air guns and streamers containing hydrophones a few meters below the surface of the water. The tail buoy helps the crew locate the end of the streamers. The air guns are activated periodically, such as every 25 m (about every 10 seconds), and the resulting sound wave that travels into the Earth is reflected back by the underlying rock layers to hydrophones on the streamer and then relayed to the recording vessel for further processing.

The few sound traces shown on the figure indicate how the sound waves are both refracted across and reflected from the interfaces between the water and Layer 1, between Layer 1 and 2, and between Layer 2 and 3. Reflection occurs if there is an increase in the product between velocity and density from one layer to the next. Such interfaces are called reflectors. Reflectors from a seismic line image the upper stratigraphy of the North Sea Basin (right). Note the upper, horizontal sea bed reflector, horizontal Quaternary reflectors and dipping Tertiary layers. Unconformities like this one typically indicate a tectonic event. Note that most seismic sections have seconds (two-way time) as vertical scale.

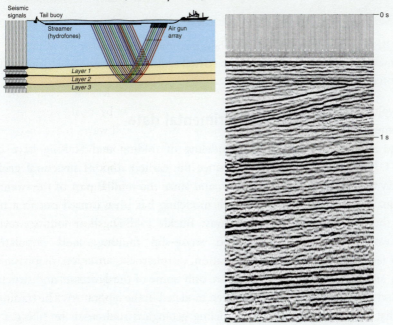

leads to increased velocity). In general it is the interpretation that is depth converted. However, the seismic data themselves can also be depth migrated, in which case the vertical axis of the seismic sections is depth, not time. This provides more realistic displays of faults and layers, and takes into account lateral changes in rock velocity that may cause visual or geometrical challenges to the interpreter when dealing with a time-migrated section. The accuracy of the depth-migrated data does however rely on the velocity model.

Deep seismic lines can be collected where the energy emitted is sufficiently high to penetrate deep parts of the crust and even the upper mantle. Such lines are useful for exploring the large-scale structure of the lithosphere. While widely spaced deep seismic lines and regional seismic lines are called two-dimensional (2-D) seismic data, more and more commercial (petroleum company) data are collected as a three-dimensional (3-D) cube where line spacing is close enough (c. 25 m) that the data can be processed in three dimensions, and where sections through the cube can be made in any direction. The lines parallel to the direction of collection are sometimes called **inlines**, those orthogonal to inlines are referred to as **crosslines**, while other vertical lines are **random lines**. Horizontal sections are called **time slices**, and can be useful during fault interpretation.

Three-dimensional seismic data provide unique opportunities for 3-D mapping of faults and folds in the subsurface. However, seismic data are restricted by **seismic resolution**, which means that one can only distinguish

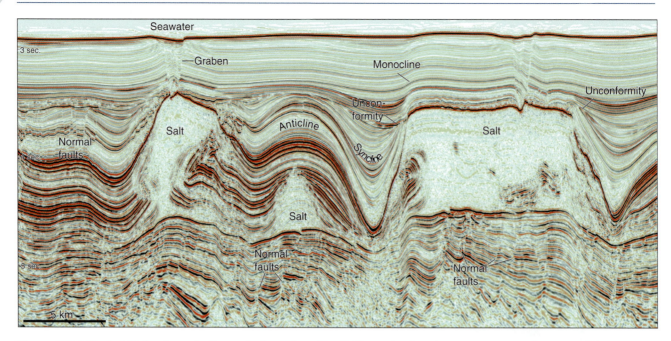

Figure 1.6 Seismic 2-D line from the Santos Basin offshore Brazil, illustrating how important structural aspects of the subsurface geology can be imaged by means of seismic exploration. Note that the vertical scale is in seconds. Some basic structures returned to in later chapters are indicated. Seismic data courtesy of CGGVeritas.

layers that are a certain distance apart (typically around 5–10 m), and only faults with a certain minimum offset can be imaged and interpreted. The quality and resolution of 3-D data are generally better than those of 2-D lines because the reflected energy is restored more precisely through 3-D migration. The seismic resolution of high-quality 3-D data depends on depth, acoustic impedance of the layer interfaces, data collection method and noise, but would typically be at around 15–20 m for identification of fault throw.

Sophisticated methods of data analysis and visualization are now available for 3-D seismic data sets, helpful for identifying faults and other structures that are underground. Petroleum exploration and exploitation usually rely on seismic 3-D data sets interpreted on computers by geophysicists and structural geologists. The interpretation makes it possible to generate structural contour maps and geologic cross-sections that can be analyzed structurally in various ways, e.g. by structural restoration (Chapter 20).

3-D seismic data form the foundation of our structural understanding of hydrocarbon fields.

Other types of seismic data are also of interest to structural geologists, particularly seismic information from earthquakes. This information gives us important information about current fault motions and tectonic regime, which in simple terms means whether an area is undergoing shortening, extension or strike-slip deformation.

1.8 Experimental data

Physical modeling of folding and faulting have been performed since the earliest days of structural geology (Figure 1.7), and since the middle part of the twentieth century such modeling has been carried out in a more systematic way. Buckle folding, shear folding, reverse, normal and strike-slip faulting, fault populations, fault reactivation, porphyroclast rotation, diapirism and boudinage are only some of the processes and structures that have been modeled in the laboratory. The traditional way of modeling geologic structures is by filling a box with clay, sand, plaster, silicone putty, honey and other media and applying extension, contraction, simple shear or some other deformation. A ring shear apparatus is used when large amounts of shear are required. In this setup, the outer part of the disk-shaped volume is rotated relative to the inner part. Many models can be filmed and photographed during the deformation history or scanned using computer tomography. Another tool is the centrifuge, where material is deformed under the influence of the centrifugal force. Here the centrifugal force plays the same role in the models as the force of gravity does in geologic processes.

Ideally we wish to construct a **scale model**, where not only the size and geometry of the natural object or structure that it refers to are shrunk, but where also physical properties are scaled proportionally. Hence we

Figure 1.7 Experimental work in 1887, carried out by means of clay and a simple contractional device. This and similar models were made by H. M. Cadell to illustrate the structures of the northwest Scottish Highlands. With permission of the Geological Survey of Britain.

need a geometrically similar model where its lengths are proportional to the natural example and where equality of angles is preserved. We also need kinematic similarity, with comparability of changes in shape and position and proportionality of time. Dynamic similarity requires proportional values of cohesion or viscosity contrast and similar angles of internal friction.

In practice, it is impossible to scale down every aspect or property of a deformed part of the Earth's crust. Sand has grains that, when scaled up to natural size, may be as large as huge boulders, preventing the replication of small-scale structures. The grain size of clay may be more appropriate, but we may find that the fine grain size of clay makes it too cohesive. Plaster has properties that change during the course of the experiment and are thus difficult to describe accurately. Obviously, physical models have their limitations, but observations of progressive deformation under known boundary conditions still provide important information that can help us to understand natural structures.

For a small physical model to realistically reproduce a natural example, we must proportionally scale down physical proportions and properties as best we can.

Experimental deformation of rocks and soils in a deformation rig under the influence of an applied pressure (stress) is used to explore how materials react to various stress fields and strain rates. The samples can be

Figure 1.8 Section through a sandstone sample deformed in a triaxial deformation rig. The light bands are called deformation bands (see Chapter 7), the sandstone is the Scottish Locharbrigg Sandstone and the diameter of the cylindrical sample is 10 cm. You can read about these experiments in Mair *et al.* (2000). Photo: Karen Mair.

a few tens of cubic centimeters in size (Figure 1.8), and are exposed to uniaxial compression or tension (uniaxial means that a force is applied in only one direction) with a fluid-controlled confining pressure that relates to the crustal depth of interest. Triaxial tests are also performed, and the resulting deformation may be both plastic and brittle. For plastic deformation we run into problems with strain rate. Natural plastic strains accumulate over thousands or millions of years, so we have to apply higher temperatures to our laboratory samples to produce plastic structures at laboratory strain rates. We are thus back to the challenge of scaling, this time in terms of temperature, time and strain rate.

1.9 Numerical modeling

Numerical modeling of geologic processes has become increasingly simple with the development of increasingly faster computers. Simple modeling can be performed using mathematical tools such as spreadsheets or Matlab™. Other modeling requires more sophisticated and expensive software, often building on finite element and finite difference methods. The models may range from microscale, for instance dealing with mineral grain deformation, to the deformation of the entire lithosphere. We can model such things as stress field changes during faulting and fault interaction, fracture formation in rocks, fold formation in various settings and conditions, and microscale diffusion processes during plastic deformation. However, nature is complex, and when the degree of complexity is increased, even the fastest supercomputer at some point reaches its physical limitations. Nor can every aspect of natural deformation be described by today's numerical theory. Hence, we have to consider our simplifications very carefully and use field and experimental data both during the planning of the modeling and during the evaluation of the results. Therefore there is a need for geologists who can combine field experience with a certain insight into numerical methodology, with all of its advantages and limitations.

1.10 Other data sources

There is a long list of other data sources that can be of use in structural analysis. **Gravimetric** and **magnetic data** (Figure 1.9) can be used to map large-scale faults and fault patterns in sedimentary basins, covered crust and subsea oceanic crust. Magnetic anisotropy as measured from oriented hand samples can be related to finite strain. Thin section studies and electron microscope images reveal structural information on the microscale. Earthquake data and focal mechanism solutions give valuable information about intraplate stresses and neotectonism and may be linked with *in situ* stress measurements by means of strain gauges, borehole breakouts,

hydraulic fracturing, overcoring etc. Radiometric data can be used to date tectonic events. Sedimentological data and results of basin analysis are closely related to fault activity in sedimentary basins (Figure 1.1). Dike intrusions and their orientations are related to the stress field and preexisting weaknesses, and geomorphologic features can reveal important structures in the underground. The list can be made longer, illustrating how the different geologic disciplines rely on each other and should be used in concert to solve geologic problems.

1.11 Organizing the data

Once collected, geologic data need to be analyzed. Structural field data represent a special source of data because they directly relate to the product of natural deformation in all its purity and complexity. Because of the vastness of information contained in a field area or outcrop, the field geologist is faced with the challenge of sorting out the information that is relevant to the problem in question. Collecting too much data slows down both collection and analyses of the data. At the same time an incomplete data set prevents the geologist from reaching sound and statistically significant conclusions.

There are several examples where general structural mapping was done and large databases were constructed for future unknown purposes and needs. However, later problems and studies commonly require one or more key parameters that are missing or not ideally recorded in preexisting data sets. Consequently, new and specifically planned fieldwork commonly has to be carried out to obtain the type, quality and consistency of data that are required in each case.

Always have a clear objective during data sampling.

Collecting the wrong type of data is of course not very useful, and the quality of the data must be acceptable for further use. The quality of the analysis is limited by the quality of the data upon which it is based. It is therefore essential to have a clear and well-defined objective during data collection. The same is the case for other data types, such as those gathered by seismic methods or remote sensing.

Once collected, data must be grouped and sorted in a reasonable way for further analysis. In some cases field data are spatially homogeneous, and they can all be represented in a single plot (Figure 1.10a). In other cases data show some type of heterogeneity (Figure 1.10b–e), in

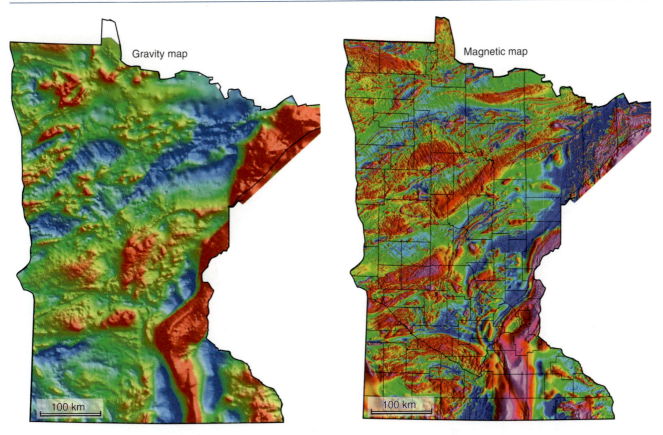

Figure 1.9 Gravity and magnetic maps of the state of Minnesota, where most of the bedrock and its structures are covered by glacial deposits. Modern structural mapping of the bedrock in this state has therefore involved extensive use of gravity and magnetic data. Warm (reddish) colors indicate high density (left) or high magnetic intensity (right). Maps reproduced courtesy of Minnesota Geological Survey.

which case it may be useful to subdivide the data set into subsets or **subpopulations**.

It is sometimes useful to subdivide data into subsets based on geographic occurrence or distribution. A **structural sub-area** is a geographic area within which the structural data set is approximately homogeneous (Figure 1.10a) or where it shows a systematic change (Figure 1.10b–e). Completely non-systematic or chaotic structural data are very unusual; there is usually some fabric or systematic orientation of minerals or fractures resulting from rock deformation.

As an example, Figure 1.11 shows the overall pattern of lineations in a part of the Caledonian orogenic wedge in Scandinavia. Each lineation arrow represents the local average orientation of many field measurements. The lineation pattern is far from homogeneous, so the region should be subdivided into subareas of more homogeneous character, as shown in Figure 1.11c. We can then study each subarea individually, and the variation within each subarea can be displayed by means of stereographic presentation of individual measurements (Figure 1.12). We could also distinguish between

different types of lineations, as discussed in Chapter 13. Figure 1.12 illustrates the distribution of observations as well as the mean orientation of lineations by means of individual poles (points). In addition, rose diagrams (yellow) are presented that reflect the trend of the observations. In this example lineations are thought to reflect the motion of Caledonian thrust nappes, and the plots can be used to say something about movement patterns within the lower parts of an orogenic belt. As usual in structural geology, there are different ways of interpreting the data.

A second example is taken from the petroleum province of the North Sea (Figure 1.13). It shows how fault populations look different at different scales and must therefore be treated at the scale suitable to serve the purpose of the study. The Gullfaks oil field itself (Figure 1.13b) is dominated by N–S oriented faults with 100–300 m offset. This is a bit different from the regional NNE–SSE trend seen from Figure 1.13a. It is also different from the large range in orientations shown by small-scale faults within the Gullfaks oil field. These small faults can be subdivided based on orientation, as shown in Figures 1.13c–g. At this point each

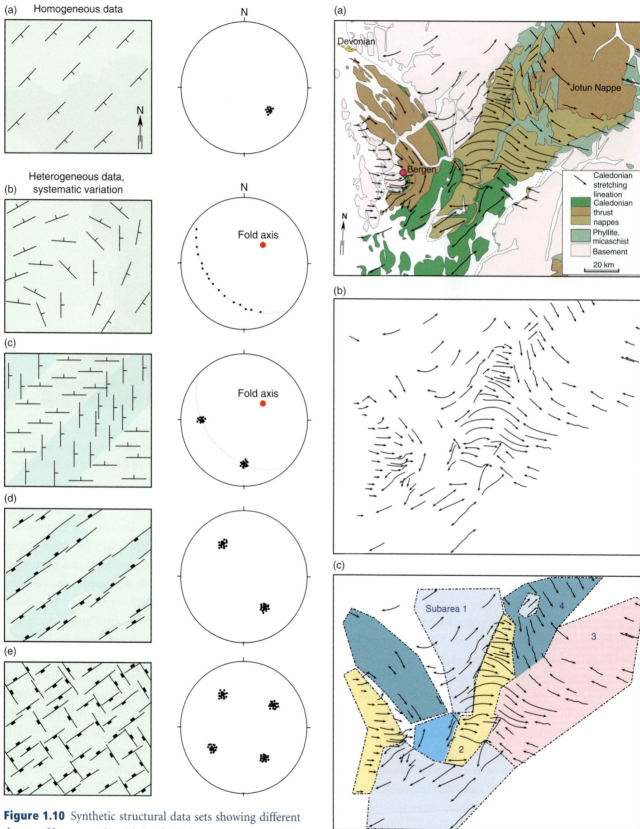

Figure 1.10 Synthetic structural data sets showing different degree of homogeneity. (a) Synthetic homogeneous set of strike and dip measurements. (b) Systematic variation in layer orientation measurements. (c) Homogeneous subareas due to kink or chevron folding. (d, e) Systematic fracture systems. Note how the systematics is reflected in the stereonets.

Figure 1.11 (a, b) Caledonian lineation pattern in the Scandinavian Caledonides east of Bergen, Norway. To analyze this pattern, subareas of approximately uniform orientation are defined (c).

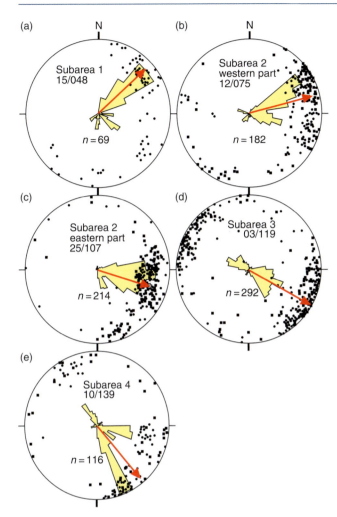

Figure 1.12 Lineation data from subareas defined in the previous figure. The plots show the variations within each subarea, portrayed by means of poles, rose diagrams, and an arrow indicating the average orientation. The number of data within each subarea is indicated by "*n*". From Fossen (1993).

subgroup can be individually analyzed with respect to orientation (stereo plots), displacement, sealing properties, or other factors, depending on the purpose of the study.

1.12 Structural analysis

Many structural processes span thousands to millions of years, and most structural data describe the final product of a long deformation history. The history itself can only be revealed through careful analysis of the data. When looking at a fold, it may not be obvious whether it formed by layer parallel shortening, shearing or passive bending (see Chapter 11). The same thing applies to a fault. What part of the fault formed first? Did it form by linking of individual segments, or did it grow from a single point outward, and if so, was this point in the central part of the present fault surface? It may not always

be easy to answer such questions, but the approach should always be to analyze the field information and compare with experimental and/or numerical models.

Geometric analysis

The analysis of the geometry of structures is referred to as geometric analysis. This includes the shape, geographic orientation, size and geometric relation between the main (first-order) structure and related smaller-scale (second-order) structures. The last point emphasizes the fact that most structures are composite and appear in certain **structural associations** at various scales. Hence, various methods are needed to measure and describe structures and structural associations.

Geometric analysis is the classic descriptive approach to structural geology that most secondary structural geologic analytical methods build on.

Shape is the spatial description of open or closed surfaces such as folded layer interfaces or fault surfaces. The shape of folded layers may give information about the fold-forming process or the mechanical properties of the folded layer (Chapter 11), while fault curvature may have implications for hanging-wall deformation (Figure 20.6) or could give information about the slip direction (Figure 8.3).

Orientations of linear and planar structures are perhaps the most common type of structural data. Shapes and geometric features may be described by mathematical functions, for instance by use of vector functions. In most cases, however, natural surfaces are too irregular to be described accurately by simple vector functions, or it may be impossible to map faults or folded layers to the extent required for mathematical description. Nevertheless, it may be necessary to make geometric interpretations of partly exposed structures. Our data will always be incomplete at some level, and our minds tend to search for geometric models when analyzing geologic information. For example, when the Alps were mapped in great detail early in the twentieth century, their major fold structures were generally considered to be cylindrical, which means that fold axes were considered to be straight lines. This model made it possible to project folds onto cross-sections, and impressive sections or geometric models were created. At a later stage it became clear that the folds were in fact non-cylindrical, with curved hinge lines, requiring modification of earlier models.

In geometric analysis it is very useful to represent orientation data (e.g. Figures 1.10 and 1.12) by means

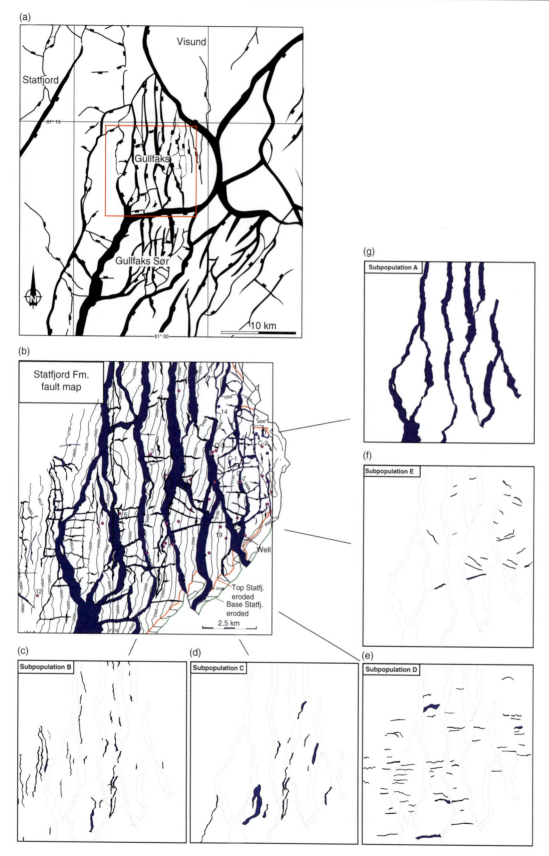

Figure 1.13 This set of figures from the northern North Sea Gullfaks oil field area illustrates how fault patterns may change from one scale to another. Note the contrast between the N–S faults dominating the Gullfaks area and the different orientations of small (<100 m displacement) faults in lower panels. They should be separated as shown for further analysis. See Fossen and Rørnes (1997) for more details.

of **stereographic projection** (see Appendix B). Stereo-graphic projection is used to show or interpret both the orientation and geometry of structures. The method is quick and efficient, and the most widely used tool for presenting and interpreting spatial data. In general, geometry may be presented in the form of maps, profiles, stereographic projections, rose diagrams or three-dimensional models based on observations made in the field, from geophysical data, satellite information or laser scanning equipment. Any serious structural geologist needs to be familiar with the stereographic projection method.

Strain and kinematic analysis

Geometric description and analysis may form the basis for strain quantification or strain analysis. Such quantification is useful in many contexts, e.g. in the restoration of geologic sections through deformed regions. Strain analysis commonly involves **finite strain analysis**, which concerns changes in shape from the initial state to the very end result of the deformation. Structural geologists are also concerned with the deformation history, which can be explored by **incremental strain analysis**. In this case only a portion of the deformation history is considered, and a sequence of increments describes the deformation history.

By definition, strain applies to **ductile deformation**, i.e. deformation where originally continuous structures such as bedding or dikes remain continuous after the deformation. Ductile deformation occurs when rocks **flow** (without fracture) under the influence of stress. The opposite, **brittle deformation**, occurs when rocks break or **fracture**. However, modern geologists do not restrict the use of strain to ductile deformation. In cases where fractures occur in a high number and on a scale that is significantly smaller than the discontinuity each of them causes, the discontinuities are overlooked and the term **brittle strain** is used. It is a simplification that allows us to perform strain analysis on brittle structures such as fault populations.

Geometric description also forms the foundation of **kinematic analysis**, which concerns how rock particles have moved during deformation (the Greek word *kinema* means movement). Striations on fault surfaces (Figure 1.14) and deflection of layering along faults and in shear zones are among the structures that are useful in kinematic analysis.

To illustrate the connection between kinematic analysis and geometric analysis, consider the fault depicted in Figure 1.15a. We cannot correlate the layers from one side to the other, and we do not know whether this is a normal or reverse fault. However, if we find a deflection of the layering along the fault, we can use that geometry to interpret the sense of movement on the fault. Figure 1.15b, c

Figure 1.14 Abrasive marks (slickenlines) on fault slip surfaces give local kinematic information. Seismically active fault in the Gulf of Corinth.

shows the different geometries that we would expect for normal and reverse movements. In other words, a field-based kinematic analysis relies on geometric analysis. More examples of kinematic analysis are given in Chapters 9 and 15, while strain analysis is dealt with in Chapters 3 and 20.

Dynamic analysis

Dynamics is the study of forces that cause motion of particles (kinematics). Forces acting on a body generate **stress**, and if the level of stress becomes high enough, rocks start to move. Hence dynamics in the context of structural geology is about the interplay between stress and kinematics. When some particles start to move relative to other particles we get deformation, and we may be able to see changes in shape and the formation of new structures.

Dynamic analysis explores the stresses or forces that cause structures to form and strain to accumulate.

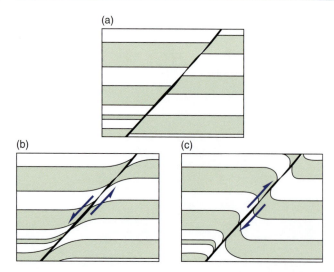

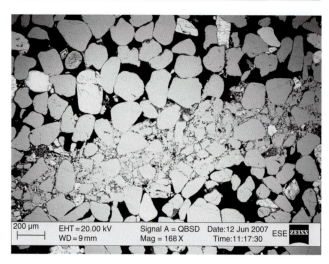

Figure 1.15 An example of how geometric analysis can lead to a kinematic model, in this case of sense of movement on a fault. (a) A fault where stratigraphy cannot be correlated across the fault. (b, c) Relative movement can be determined if layer rotation can be observed close to the fault. The geometry shown in (b) supports a normal fault movement, while (c) illustrates the geometry expected along a reverse fault.

Figure 1.16 Scanning electron microphotograph of a millimeter-thin zone of grain deformation (deformation band) in the Entrada Sandstone near Goblin Valley State Park, Utah. Photo: Anita Torabi.

In most cases dynamic analysis seeks to reconstruct the orientation and magnitude of the stress field by studying a set of structures, typically faults and fractures. Returning to the example shown in Figure 1.15, it may be assumed that a strong force or stress acted in the vertical direction in case (b), and in the horizontal direction in case (c). In practice, the exact orientations of forces and stress axes (see Chapters 4 and 5) are difficult or impossible to estimate from a single fault structure, but can be estimated for populations of faults forming in a uniform stress field. This is dealt with in Chapter 9, where it becomes clear that several assumptions have to be made to relate stress and kinematics.

Applying stress to syrup gives a different result than stressing a cold chocolate bar: the syrup will flow, while the chocolate bar will break. We are still dealing with dynamic analyses, but the part of dynamics related to the flow of rocks is referred to as **rheologic analysis**. Similarly, the study of how rocks (or sugar) break or fracture is the field of **mechanical analysis**. In general, rocks flow when they are warm enough, which usually means when they are buried deep enough. "Deep enough" means little more than surface temperatures for salt, around 300 °C for a quartz-rich rock, perhaps closer to 550 °C for a feldspathic rock, and even more for olivine-rich rocks. Pressure also plays an important role, as does water content and strain rate. It is important to realize that different rocks behave differently under any given conditions, but also that the same rock reacts differently to stress under different physical conditions.

Rheological testing is done in the laboratory in order to understand how different rocks flow in the lithosphere.

Tectonic analysis

Tectonic analysis involves dynamic, kinematic and geometric analysis at the scale of a basin or orogenic belt. This kind of analysis may therefore involve additional elements of sedimentology, paleontology, petrology, geophysics and other subdisciplines of geoscience. Structural geologists involved in tectonic analysis are sometimes referred to as **tectonicists**. On the opposite end of the scale range, some structural geologists analyze the structures and textures that can only be studied through the microscope. This is the study of how deformation occurs between and within individual mineral grains and is referred to as **microstructural analysis** or **microtectonics**. Both the optical microscope and the scanning electron microscope (SEM) (Figure 1.16) are useful tools in microstructural analysis.

1.13 Concluding remarks

Structural geology has changed from being a descriptive discipline to one where analytical methods and physical and numerical modeling are increasingly important. Many new data types and methods have been applied over the last few decades, and more new methods will probably appear in this field in the years to come. Nevertheless, it is hard to overemphasize the importance of field studies even where the most sophisticated numerical algorithms are being used or where the best 3-D seismic data set is available. The connection between field observations and modeling must be tight. It is the lithosphere and the processes acting in it that we seek to understand. It is the rocks themselves that contain the information that can reveal their structural or tectonic history. Numerical and physical modeling help us create simple models that capture the main features of a deformed region or a structural problem. Models can also help us understand what is a likely and what is an unlikely or impossible interpretation. However, they must always comply with the information retrievable from the rocks.

Review questions

1. What is structural geology all about?

2. Name the four principal ways a structural geologist can learn about structural geology and rock deformation. How would you rank them?

3. How can we collect structural data sets? Name important data types that can be used for structural analysis.

4. What are the advantages and disadvantages of seismic reflection data sets?

5. What is a scale model?

6. What is kinematic analysis?

E-MODULE

The e-learning module called *Stereographic projection, Structural geology* and Appendix B are recommended for this chapter.

FURTHER READING

Traditional field methods
Lisle, R. J., 2003, Geological Structures and Maps: A Practical Guide. Amsterdam: Elsevier.
McClay, K., 1987, *The Mapping of Geological Structures.* New York: John Wiley and Sons.

Modern field methods
Jones, R. R., McCaffrey, K. J. W., Clegg, P., Wilson, R. W. and Holliman, N. S., 2009, Integration of regional to outcrop digital data: 3D visualisation of multi-scale geological models. *Computers and Geosciences* **35**: 4–18.
McCaffrey, K., Jones, R. R., Holdsworth, R. E., Wilson, R. W., Clegg, P., Imber, J., Hollinman, N. and Trinks, I., 2005, Unlocking the spatial dimension: digital technologies and the future of geoscience fieldwork. *Journal of the Geological Society* **162**: 927–938.

Remote sensing
Hollenstein, M. D., Müller, A. and Geiger, H. -G. K., 2008, Crustal motion and deformation in Greece from a decade of GPS measurements, 1993–2003. *Tectonophysics* **449**: 17–40.

Kreemer, C., Holt, W. E. and Haines, A. J., 2003, An integrated global model of present-day plate motions and plate boundary deformation. *Geophysical Journal International* **154**: 8–34.
Zhang, P.-Z. et al., 2004, Continuous deformation of the Tibetan Plateau from global positioning system data. *Geology* **32**: 809–812.

Physical and numerical modeling
Hubbert, M. K., 1937, Theory of scale models as applied to the study of geologic structures. *Bulletin of the Geological Society of America* **48**: 1459–1520.
Huismans, R. S. and Beaumont, C., 2003, Symmetric and asymmetric lithospheric extension: Relative effects of frictional-plastic and viscous strain softening. *Journal of Geophysical Research* **108**: doi:10.1029/2002JB 002026.
Maerten, L. and Maerten, F., 2006, Chronologic modeling of faulted and fractured reservoirs using geomechanically based restoration: Technique and industry applications. *American Association of Petroleum Geologists Bulletin* **90**, 1201–1226.

Chapter 2

Deformation

Deformed rocks and their structures and fabrics can be studied and mapped, and we had a glimpse of some methods and techniques in the previous chapter. Each structure reflects a change in shape and perhaps transport within a given reference frame. We generally refer to these changes as deformation, and as we inspect deformed rocks we automatically start to imagine what the rock could have looked like before the deformation started and what it has gone through. If we want to understand the structures we need to understand the fundamentals of deformation, including some useful definitions and mathematical descriptions. That is the topic of this chapter.

2.1 What is deformation?

The term **deformation** is, like several other structural geology terms, used in different ways by different people and under different circumstances. In most cases, particularly in the field, the term refers to the distortion (strain) that is expressed in a (deformed) rock. This is also what the word literally means: a change in form or shape. However, rock masses can be translated or rotated as rigid units during deformation, without any internal change in shape. For instance, fault blocks can move during deformation without accumulating any internal distortion. Many structural geologists want to include such rigid displacements in the term deformation, and we refer to them as **rigid body deformation**, as opposed to **non-rigid body deformation** (strain or distortion).

Deformation is the transformation from an initial to a final geometry by means of rigid body translation, rigid body rotation, strain (distortion) and/or volume change.

It is useful to think of a rock or rock unit in terms of a continuum of particles. Deformation relates the positions of particles before and after the deformation history, and the positions of points before and after deformation can be connected with vectors. These vectors are called **displacement vectors**, and a field of

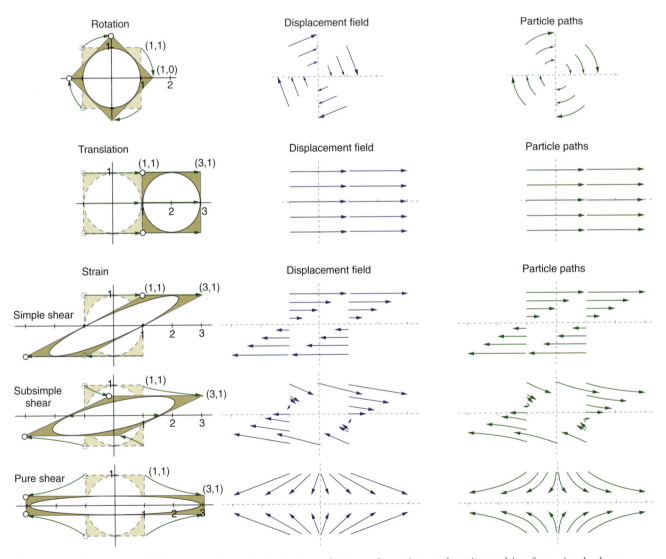

Figure 2.1 Displacement field and particle paths for rigid translation and rotation, and strain resulting from simple shear, subsimple shear and pure shear (explained later in this chapter). Particle paths trace the actual motion of individual particles in the deforming rock, while displacement vectors simply connect the initial and final positions. Hence, displacement vectors can be constructed from particle paths, but not the other way around.

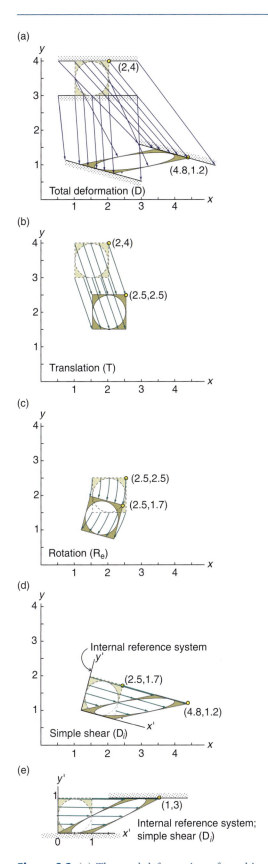

such vectors is referred to as the **displacement field**. Displacement vectors, such as those displayed in the central column of Figure 2.1, do not tell us *how* the particles moved during the deformation history – they merely link the undeformed and deformed states. The actual path that each particle follows during the deformation history is referred to as a **particle path**, and for the deformations shown in Figure 2.1 the paths are shown in the right column (green arrows). When specifically referring to the progressive changes that take place during deformation, terms such as **deformation history** or **progressive deformation** should be used.

2.2 Components of deformation

The displacement field can be decomposed into various components, depending on the purpose of the decomposition. The classic way of decomposing it is by separating rigid body deformation in the form of rigid translation and rotation from change in shape and volume. In Figure 2.2 the translation component is shown in (b), the rotation component in (c) and the rest (the strain) in (d). Let us have a closer look at these expressions.

Translation

Translation moves every particle in the rock in the same direction and the same distance, and its displacement field consists of parallel vectors of equal length. Translations can be considerable, for instance where thrust nappes (detached slices of rocks) have been transported several tens or hundreds of kilometers. The Jotun Nappe (Figure 2.3) is an example from the Scandinavian Caledonides. In this case most of the deformation is rigid translation. We do not know the exact orientation of this nappe prior to the onset of deformation, so we cannot estimate the rigid rotation (see below), but field observations reveal that the change in shape, or strain, is largely confined to the lower parts. The total deformation thus consists of a huge translation component, an unknown but possibly small rigid rotation component and a strain component localized to the base of the nappe.

Figure 2.2 (a) The total deformation of an object (square with an internal circle). Arrows in (a) are displacement vectors connecting initial and final particle positions. Arrows in (b)–(e) are particle paths. (b, c) Translation and rotation components of the deformation shown in (a). (d) The strain component. A new coordinate system (x', y') is introduced (d). This internal system eliminates the translation and rotation (b, c) and makes it easier to reveal the strain component, which is here produced by a simple shear (e).

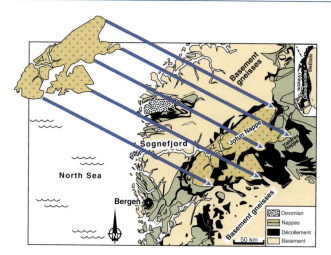

Figure 2.3 The Jotun Nappe in the Scandinavian Caledonides seems to have been transported more than 300 km to the southeast, based on restoration and the orientation of lineations. The displacement vectors are indicated, but the amount of rigid rotation around the vertical axis is unknown. The amount of strain is generally concentrated to the base.

On a smaller scale, rock components (mineral grains, layers or fault blocks) may be translated along slip planes or planar faults without any internal change in shape. One model where there is only translation and rigid rotation is the classic domino fault model, which we will explore in Chapter 17.

Rotation

Rotation is here taken to mean rigid rotation of the entire deformed rock volume that is being studied. It should not be confused with the rotation of the (imaginary) axes of the strain ellipse during progressive deformation, as discussed in Section 2.25. Rigid rotation involves a uniform physical rotation of a rock volume (such as a shear zone) relative to an external coordinate system.

Large-scale rotations of a major thrust nappe or entire tectonic plate typically occur about vertical axes. Fault blocks in extensional settings, on the other hand, may rotate around horizontal axes, and small-scale rotations may occur about any axis.

Strain

Strain or distortion is non-rigid deformation and relatively simple to define:

Any change in shape, with or without change in volume, is referred to as strain, and it implies that particles in a rock have changed positions relative to each other.

A rock volume can be transported (translated) and rotated rigidly in any way and sequence, but we will never be able to tell just from looking at the rock itself. All we can see in the field or in samples is strain, and perhaps the way that strain has accumulated. Consider your lunch bag. You can bring it to school or work, which involves a lot of rotation and translation, but you cannot see this deformation directly. It could be that your lunch bag has been squeezed on your way to school – you can tell by comparing it with what it looked like before you left home. If someone else prepared your lunch and put it in your bag, you would use your knowledge of how a lunch bag should be shaped to estimate the strain (change in shape) involved.

The last point is very relevant, because with very few exceptions, we have not seen the deformed rock in its undeformed state. We then have to use our knowledge of what such rocks typically look like when unstrained. For example, if we find strained ooliths or reduction spots in the rock, we may expect them to have been spherical (circular in cross-section) in the undeformed state.

Volume change

Even if the shape of a rock volume is unchanged, it may have shrunk or expanded. We therefore have to add volume change (area change in two dimensions) for a complete description of deformation. Volume change, also referred to as dilation, is commonly considered to be a special type of strain, called **volumetric strain**. However, it is useful to keep this type of deformation separate if possible.

2.3 System of reference

For studies of deformation, a reference or coordinate system must be chosen. Standing on a dock watching a big ship entering or departing can give the impression that the dock, not the ship, is moving. Unconsciously, the reference system gets fixed to the ship, and the rest of the world moves by translation relative to the ship. While this is fascinating, it is not a very useful choice of reference. Rock deformation must also be considered in the frame of some reference coordinate system, and it must be chosen with care to keep the level of complexity down.

We always need a reference frame when dealing with displacements and kinematics.

It is often useful to orient the coordinate system along important geologic structures. This could be the base of a thrust nappe, a plate boundary or a local shear zone (see Chapter 15). In many cases we want to eliminate translation and rigid rotation. In the case of shear zones we normally place two axes parallel to the shear zone with the third being perpendicular to the zone. If we are interested in the deformation in the shear zone as a whole, the origin could be fixed to the margin of the zone. If we are interested in what is going on around any given particle in the zone we can "glue" the origin to a particle within the zone (still parallel/perpendicular to the shear zone boundaries). In both cases translation and rigid rotation of the shear zone are eliminated, because the coordinate system rotates and translates along with the shear zone. There is nothing wrong with a coordinate system that is oblique to the shear zone boundaries, but visually and mathematically it makes things more complicated.

2.4 Deformation: detached from history

Deformation is the difference between the deformed and undeformed states. It tells us nothing about what actually happened during the deformation history.

A given strain may have accumulated in an infinite number of ways.

Imagine a tired student (or professor for that matter) who falls asleep in a boat while fishing on the sea or a lake. The student knows where he or she was when falling asleep, and soon figures out the new location when waking up, but the exact path that currents and winds have taken the boat is unknown. The student only knows the position of the boat before and after the nap, and can evaluate the strain (change in shape) of the boat (hopefully zero). One can map the deformation, but not the deformation history.

Let us also consider **particle flow**: Students walking from one lecture hall to another may follow infinitely many paths (the different paths may take longer or shorter time, but deformation itself does not involve time). All the lecturer knows, busy between classes, is that the students have moved from one lecture hall to the other. Their history is unknown to the lecturer (although he or she may have some theories based on cups of hot coffee etc.). In a similar way, rock particles may move along a variety of paths from the undeformed to the deformed state. One difference between rock particles and individual students is of course that students are free to move on an individual basis, while rock particles, such as mineral grains in a rock, are "glued" to one another in a solid continuum and cannot operate freely.

2.5 Homogeneous and heterogeneous deformation

Where the deformation applied to a rock volume is identical throughout that volume, the deformation is homogeneous. Rigid rotation and translation by definition are homogenous, so it is always strain and volume or area change that can be heterogeneous. Thus **homogeneous deformation** and **homogeneous strain** are equivalent expressions.

For homogeneous deformation, originally straight and parallel lines will be straight and parallel also after the deformation, as demonstrated in Figure 2.4. Further, the strain and volume/area change will be constant throughout the volume of rock under consideration. If not, then the deformation is **heterogeneous** (inhomogeneous). This means that two objects with identical initial shape and orientation will end up having identical shape and orientation after the deformation. Note, however, that the initial shape and orientation in general will differ from the final shape and orientation. If two objects have identical shapes but different orientations before deformation, then they will generally have different shapes after deformation even if the deformation is homogeneous. An example is the deformed brachiopods in Figure 2.4. The difference reflects the strain imposed on the rock.

Homogeneous deformation: Straight lines remain straight, parallel lines remain parallel, and identically shaped and oriented objects will also be identically shaped and oriented after the deformation.

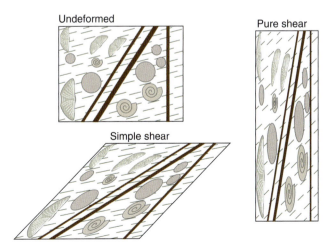

Figure 2.4 Homogeneous deformations of a rock with brachiopods, reduction spots, ammonites and dikes. Two different deformations are shown (pure and simple shear). Note that the brachiopods that are differently oriented before deformation obtain different shapes.

A circle will be converted into an ellipse during homogeneous deformation, where the **ellipticity** (ratio between the long and short axes of the ellipse) will depend on the type and intensity of the deformation. Mathematically, this is identical to saying that homogeneous deformation is a linear transformation. Homogeneous deformation can therefore be described by a set of first-order equations (three in three dimensions) or, more simply, by a transformation matrix referred to as the deformation matrix.

Before looking at the deformation matrix, the point made in Figure 2.5 must be emphasized:

A deformation that is homogeneous on one scale may be considered heterogeneous on a different scale.

A classic example is the increase in strain typically seen from the margin toward the center of a shear zone. The strain is heterogeneous on this scale, but can be subdivided into thinner elements or zones in which strain is approximately homogeneous. Another example is shown in Figure 2.6, where a rock volume is penetrated by faults. On a large scale, the deformation may be considered homogeneous because the discontinuities represented by the faults are relatively small. On a smaller scale, however, those discontinuities become more apparent, and the deformation must be considered heterogeneous.

2.6 Mathematical description of deformation

Deformation is conveniently and accurately described and modeled by means of elementary linear algebra. Let us use a local coordinate system, such as one attached to a shear

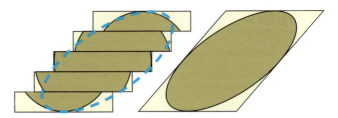

Figure 2.6 Discrete or discontinuous deformation can be approximated as continuous and even homogeneous in some cases. In this sense the concept of strain can also be applied to brittle deformation (brittle strain). The success of doing so depends on the scale of observation.

zone, to look at some fundamental deformation types. We will think in terms of particle positions (or vectors) and see how particles change positions during deformation. If (x, y) is the original position of a particle, then the new position will be denoted (x', y'). For homogeneous deformation in two dimensions (i.e. in a section) we have that

$$x' = D_{11}x + D_{12}y$$
$$y' = D_{21}x + D_{22}y \tag{2.1}$$

These equations can be written in terms of matrices and position vectors as

$$\begin{bmatrix} x' \\ y' \end{bmatrix} = \begin{bmatrix} D_{11} & D_{12} \\ D_{21} & D_{22} \end{bmatrix} \begin{bmatrix} x \\ y \end{bmatrix} \tag{2.2}$$

which can be written

$$\mathbf{x}' = \mathbf{D}\mathbf{x} \tag{2.3}$$

The matrix $\mathbf{D}$ is called the deformation matrix or the position gradient tensor, and the equation describes a linear transformation or a homogeneous deformation.

There is a corresponding or inverse matrix $\mathbf{D}^{-1}$ (where the matrix product $\mathbf{D}\mathbf{D}^{-1} = \mathbf{I}$ and $\mathbf{I}$ is the identity matrix) that represents the **reciprocal** or **inverse deformation**. $\mathbf{D}^{-1}$ reverses the deformation imposed by $\mathbf{D}$:

$$\mathbf{x} = \mathbf{D}^{-1}\mathbf{x}' \tag{2.4}$$

The reciprocal or inverse deformation takes the deformed rock back to its undeformed state.

The deformation matrix $\mathbf{D}$ is very useful if one wants to model deformation using a computer. Once the deformation matrix is defined, any aspect of the deformation itself can be found. Once again, it tells us nothing about the deformation history, nor does it reveal how a given deforming medium responds to such a deformation. For more information about matrix algebra, see Box 2.1.

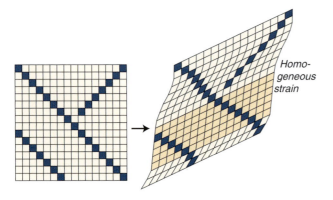

Homogeneous strain

Figure 2.5 A regular grid in undeformed and deformed state. The overall strain is heterogeneous, so that some of the straight lines have become curved. However, in a restricted portion of the grid, the strain is homogeneous. In this case the strain is also homogeneous at the scale of a grid cell.

BOX 2.1 | MATRIX ALGEBRA

Matrices contain coefficients of systems of equations that represent linear transformations. In two dimensions this means that the system of equations shown in Equation 2.1 can be expressed by the matrix of Equation 2.2. A linear transformation implies a homogeneous deformation. The matrix describes the shape and orientation of the strain ellipse or ellipsoid, and the transformation is a change from a unit circle, or a unit sphere in three dimensions.

Matrices are simpler to handle than sets of equations, particularly when applied in computer programs. The most important matrix operations in structural geology are multiplications and finding eigenvectors and eigenvalues:

Matrix multiplied by a vector:

$$\begin{bmatrix} D_{11} & D_{12} \\ D_{21} & D_{22} \end{bmatrix} \begin{bmatrix} x \\ y \end{bmatrix} = \begin{bmatrix} D_{11}x + D_{12}y \\ D_{21}x + D_{22}y \end{bmatrix}$$

Matrix–matrix multiplication:

$$\begin{bmatrix} D_{11} & D_{12} \\ D_{21} & D_{22} \end{bmatrix} \begin{bmatrix} d_{11} & d_{12} \\ d_{21} & d_{22} \end{bmatrix} = \begin{bmatrix} D_{11}d_{11} + D_{12}d_{21} & D_{11}d_{12} + D_{12}d_{22} \\ D_{21}d_{11} + D_{22}d_{21} & D_{21}d_{12} + D_{22}d_{22} \end{bmatrix}$$

Transposition, meaning shifting of columns and rows in a matrix:

$$\begin{bmatrix} D_{11} & D_{12} \\ D_{21} & D_{22} \end{bmatrix}^{T} = \begin{bmatrix} D_{11} & D_{21} \\ D_{12} & D_{22} \end{bmatrix}$$

The inverse of a matrix $\mathbf{D}$ is denoted $\mathbf{D}^{-1}$ and is the matrix that gives the identity matrix $\mathbf{I}$ when multiplied by $\mathbf{D}$:

$$\begin{bmatrix} D_{11} & D_{12} \\ D_{21} & D_{22} \end{bmatrix}^{-1} \begin{bmatrix} D_{11} & D_{12} \\ D_{21} & D_{22} \end{bmatrix} = \begin{bmatrix} 1 & 0 \\ 0 & 1 \end{bmatrix} = \mathbf{I}$$

Matrix multiplication is non-commutative:

$$\mathbf{D}_1\mathbf{D}_2 \neq \mathbf{D}_2\mathbf{D}_1$$

The determinant of a matrix $\mathbf{D}$ is

$$\det \mathbf{D} = \begin{vmatrix} D_{11} & D_{12} \\ D_{21} & D_{22} \end{vmatrix} = \mathbf{D}_{11}\mathbf{D}_{22} - \mathbf{D}_{21}\mathbf{D}_{12}$$

The determinant describes the area or volume change: If det $\mathbf{D} = 1$ then there is no area or volume change involved for the transformation (deformation) represented by $\mathbf{D}$.

Eigenvectors (x) and eigenvalues (λ) of a matrix $\mathbf{A}$ are the vectors and values that fulfill

$$\mathbf{A}x = \lambda x$$

If $\mathbf{A} = \mathbf{D}\mathbf{D}^{T}$, then the deformation matrix has two eigenvectors for two dimensions and three for three dimensions. The eigenvectors describe the orientation of the ellipsoid (ellipse), and the eigenvalues describe its shape (length of its principal axes) (see Appendix A). Eigenvalues and eigenvectors are easily found by means of a spreadsheet or computer program such as MatLab^{TM}.

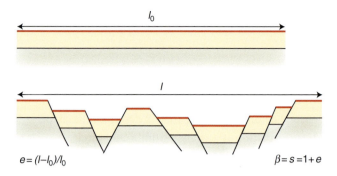

$e = (l - l_0)/l_0$ $\beta = s = 1 + e$

Figure 2.7 Extension of layers by faulting. The red layer has an original (l_0) and a new length, and the extension e is found by comparing the two. The beta-factor (β) is commonly used to quantify extension across extensional basins.

2.7 One-dimensional strain

In one dimension (a single direction), strain is about stretching and shortening (negative stretching) of lines or approximately linear (straight) objects. One might say that one-dimensional strain makes no sense, since an extending straight line does not change shape, just length. On the other hand, a change in shape, such as a circle changing into an ellipse, can be described by the change in length of lines of different orientations. It is therefore convenient to include change of line lengths in the concept of strain.

There are special terms in use, such as elongation, extension, stretching, contraction, shortening, and, as any other strain quantity, they are dimensionless.

Elongation (e or ε) of a line is defined as $e = (l - l_0)/l_0$, where l_0 and l are the lengths of the line before and after deformation, respectively (Figure 2.7). The line may represent a horizontal line or bedding trace in a cross-section, the long axis of a belemnite or some other fossil on which a line can be defined, the vertical direction in a rock mechanics experiment, and many other things. The logarithmic or **natural elongation** $\bar{e} = \ln(e)$ is also in use.

Extension of a line is identical to elongation (e) and is used in the analysis of extensional basins where the elongation of a horizontal line in the extension direction indicates the extension. Negative extension is called **contraction** (the related terms compression and tension are reserved for stress).

Stretching of a line is designated $s = 1 + e$, where s is called the stretch. Hence, $s = l/l_0$. **Stretching factors** are commonly referred to in structural analysis of rifts and extensional basins. These are sometimes called β-factors, but are identical to s.

Quadratic elongation, $\lambda = s^2$, is identical to the eigenvalues of the deformation matrix **D**. Quadratic stretch would be a better name, because we are looking at the square value of the stretch, not of the elongation. Nevertheless, quadratic elongation is the term in common use.

Natural strain, $\bar{e}$, is simply $\ln(s)$ or $\ln(1 + e)$.

2.8 Strain in two dimensions

Observations of strain in planes or sections are described by the following dimensionless quantities:

Angular shear, ψ, which describes the change in angle between two originally perpendicular lines in a deformed medium (Figures 2.8 and 2.9).

Shear strain, $\gamma = \tan \psi$, where ψ is the angular shear (Figure 2.8). The shear strain can be found where objects of known initial angular relations occur. Where a number of such objects occur within a homogeneously strained area, the strain ellipse can be found.

The **strain ellipse** is the ellipse that describes the amount of elongation in any direction in a plane of homogeneous deformation (Box 2.2). It represents the deformed shape of an imaginary circle on the undeformed section. The strain ellipse is conveniently described by a long (X) and a short (Y) axis. The two

(a)

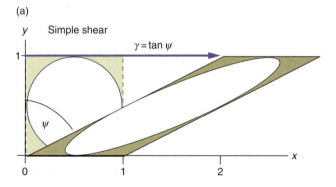

(b)

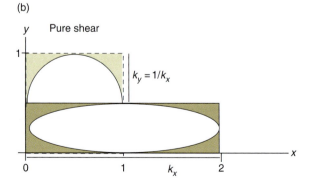

Figure 2.8 Simple and pure shear.

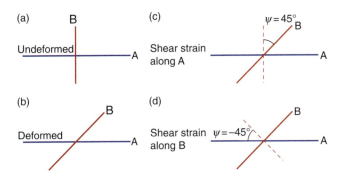

Figure 2.9 Angular shear strain is the change in angle between two initially perpendicular lines, and is positive for clockwise rotations and negative for anti-clockwise rotations. In this example the angular shear strain is 45° along line A and −45° along line B.

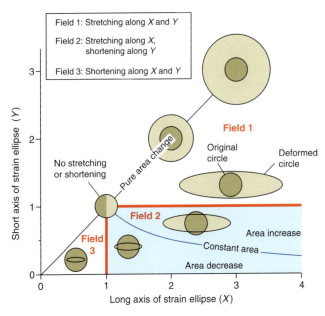

Figure 2.10 Classification of strain ellipses. Only the lower part of the diagram is in use because $X \geq Y$. Note that Field 2 is divided in two by the constant area line. The plot is called an X–Y plot, but we could also call it an X–Z plot if we plot the largest and smallest principal strains. Based on Ramsay and Huber (1983).

BOX 2.2 | THE SECTIONAL STRAIN ELLIPSE

X, Y and Z are the three principal strains or strain axes in three-dimensional strain analysis. However, when considering a section, X and Y are commonly used regardless of the orientation of the section relative to the strain ellipsoid. It would perhaps be better to name them X' and Y' or something similar, and reserve the designations X, Y and Z for the true principal strains in three dimensions. An arbitrary section through a deformed rock contains a strain ellipse that is called a **sectional strain ellipse**. It is important to specify which section we are describing at any time.

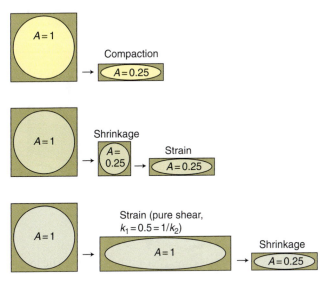

Figure 2.11 Compaction involves strain. The deformation can be considered as a combination of uniform shrinking and strain (middle drawings). The lower drawings illustrate that the order (strain versus dilation) is irrelevant (only true for coaxial deformations): the final strain ellipses are identical for the three cases.

axes have lengths $1 + e_1$ and $1 + e_2$, and the ratio $R = X/Y$ or $(1 + e_1)/(1 + e_2)$ describes the ellipticity or eccentricity of the ellipse and thus the strain that it represents. For a circle (no strain), $R = 1$.

Area change: For area change without any strain, $R = X/Y = 1$. A circle drawn on the initial section remains a circle after a pure area change, albeit with a smaller or larger radius. In a simple diagram where X is plotted against Y, **isochoric** deformations will plot along the main diagonal (Figure 2.10). The same diagram illustrates strain fields characteristic for different combinations of area change and strain. We will later look at the different types of structures formed in these different fields.

It will always be possible to decompose a deformation into some combination of area change and strain, i.e. to isolate the strain and area change components. Figure 2.11 shows how compaction can be decomposed into a strain and an area change.

2.9 Three-dimensional strain

The spectrum of possible states of strain widens significantly if we allow for stretching and contraction in three dimensions. Classic reference situations are known as uniform extension, uniform flattening and plane strain, as illustrated in Figure 2.12. **Uniform extension**, also referred to as axially symmetric extension, is a state of strain where stretching in X is compensated for by equal shortening in the plane orthogonal to X. **Uniform flattening** (axially symmetric flattening) is the opposite, with shortening in a direction Z compensated for by identical stretching in all directions perpendicular to Z. These two reference states are end-members in a continuous spectrum of deformation types. Between uniform flattening and extension lies **plane strain**, where stretching in one direction is perfectly compensated by shortening in a single perpendicular direction. The strain is "plane" or two-dimensional because there is no stretching or shortening in the third principal direction, i.e. along the Y-axis.

Strain is said to be plane (two-dimensional) where there is no length change along the Y-axis, while three-dimensional strain implies a length change along X, Y and Z.

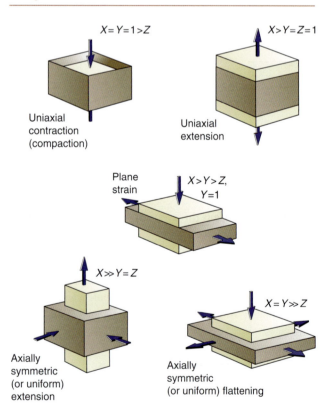

Figure 2.12 Some reference states of strain. The conditions are uniaxial (top), planar (middle) and three-dimensional (bottom).

2.10 The strain ellipsoid

The finite spatial change in shape that is connected with deformation is completely described by the **strain ellipsoid**. The strain ellipsoid is the deformed shape of an imaginary sphere with unit radius that is deformed along with the rock volume under consideration.

The strain ellipsoid has three mutually orthogonal planes of symmetry, the **principal planes of strain**, which intersect along three orthogonal axes that are referred to as the **principal strain axes**. Their lengths (values) are called the **principal stretches**. These axes are commonly designated X, Y and Z, but the designations $\sqrt{\lambda_1}$, $\sqrt{\lambda_2}$ and $\sqrt{\lambda_3}$, S_1, S_2 and S_3 as well as ε_1, ε_2 and ε_3 are also used. We will use X, Y and Z in this book, where X represents the longest, Z the shortest and Y the intermediate axis:

$$X > Y > Z$$

When the ellipsoid is fixed in space, the axes may be considered vectors of given lengths and orientations. Knowledge of these vectors thus means knowledge of both the shape and orientation of the ellipsoid. The vectors are named $\mathbf{e}_1$, $\mathbf{e}_2$ and $\mathbf{e}_3$, where $\mathbf{e}_1$ is the longest and $\mathbf{e}_3$ the shortest, as shown in Figure 2.13.

If we place a coordinate system with axes x, y and z along the principal strain axes X, Y and Z, we can write the equation for the strain ellipse as

$$\frac{x^2}{\lambda_1^2} + \frac{y^2}{\lambda_2^2} + \frac{z^2}{\lambda_3^2} = 1 \tag{2.4}$$

It can be shown that λ_1, λ_2 and λ_3 are the eigenvalues of the matrix product $\mathbf{DD}^\mathrm{T}$, and that $\mathbf{e}_1$, $\mathbf{e}_2$ and $\mathbf{e}_3$ are the corresponding eigenvectors (see Appendix A). So if $\mathbf{D}$ is known, one can easily calculate the orientation and shape of the strain ellipsoid or vice versa. A deformation matrix would look different depending on the choice of coordinate system. However, the eigenvectors and eigenvalues will always be identical for any given state of strain. Another way of saying the same thing is that they are **strain invariants**. Shear strain, volumetric strain and the kinematic vorticity number (W_k) are other examples of strain invariants. Here is another characteristic related to the strain ellipsoid:

Lines that are parallel with the principal strain axes are orthogonal, and were also orthogonal in the undeformed state.

This means that they have experienced no finite shear strain. No other set of lines has this property. Thus,

(a)

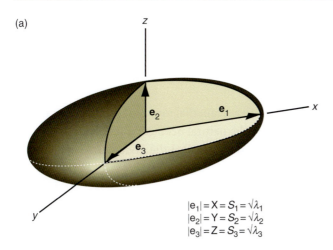

$$|e_1| = X = S_1 = \sqrt{\lambda_1}$$
$$|e_2| = Y = S_2 = \sqrt{\lambda_2}$$
$$|e_3| = Z = S_3 = \sqrt{\lambda_3}$$

(b)

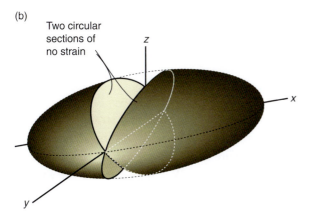

Two circular sections of no strain

Figure 2.13 (a) The strain ellipsoid is an imaginary sphere that has been deformed along with the rock. It depends on homogeneous deformation and is described by three vectors, e_1, e_2 and e_3, defining the principal axes of strain (X, Y and Z) and the orientation of the ellipsoid. The length of the vectors thus describes the shape of the ellipsoid, which is independent of choice of coordinate system. (b) The ellipsoid for plane strain, showing the two sections through the ellipsoid that display no strain.

estimating shear strain from sets of originally orthogonal lines gives information about the orientation of X, Y and Z (see next chapter). This goes for two- as well as three-dimensional strain considerations.

2.11 More about the strain ellipsoid

Any strain ellipsoid contains two **surfaces of no finite strain**. For constant volume deformations, known as **isochoric deformations**, these surfaces are found by connecting points along the lines of intersection between the ellipsoid and the unit sphere it was deformed from. For plane strain, where the intermediate principal strain axis has unit length, these surfaces happen to be planar

(Figure 2.13b). In general, when strain is three-dimensional, the surfaces of no finite strain are non-planar. Lines contained in these surfaces have the same length as in the undeformed state for constant volume deformations, or are stretched an equal amount if a volume change is involved. This means that:

A plane strain deformation produces two planes in which the rock appears unstrained.

It also means that physical lines and particles move through these theoretical planes during progressive deformation.

The shape of the strain ellipsoid can be visualized by plotting the axial ratios X/Y and Y/Z as coordinate axes. As shown in Figure 2.14a, logarithmic axes are commonly used for such diagrams. This widely used diagram is called the **Flinn diagram**, after the British geologist Derek Flinn who first published it in 1962. The diagonal of the diagram describes strains where $X/Y = Y/Z$, i.e. planar strain. It separates **prolate** geometries or cigar shapes of the upper half of the field from **oblate** geometries or pancake shapes of the lower half. The actual shape of the ellipsoid is characterized by the Flinn k-value: $k = (R_{XY} - 1)/(R_{YZ} - 1)$, where $R_{XY} = X/Y$ and $R_{YZ} = Y/Z$.

The horizontal and vertical axes in the Flinn diagram represent axially symmetric flattening and extension, respectively. Any point in the diagram represents a unique combination of strain magnitude and three-dimensional shape or **strain geometry**, i.e. a strain ellipsoid with a unique Flinn k-value. However, different types of deformations may in some cases produce ellipsoids with the same k-value, in which case other criteria are needed for separation. An example is pure shear and simple shear (see below), which both plot along the diagonal of the Flinn diagram ($k = 1$). The orientation of the strain ellipse is different for simple and pure shear, but this is not reflected in the Flinn diagram. Thus, the diagram is useful within its limitations.

In the Flinn diagram, strain magnitude generally increases away from the origin. Direct comparison of strain magnitude in the various parts of the diagram is, however, not trivial. How does one compare pancake-shaped and cigar-shaped ellipsoids? Which one is more strained? We can use the radius (distance from the origin; dashed lines in Figure 2.14), although there is no good mathematical or physical reason why this would be an accurate measure of strain magnitude. An alternative parameter is given by the formula

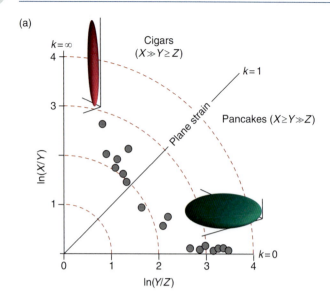

(a)

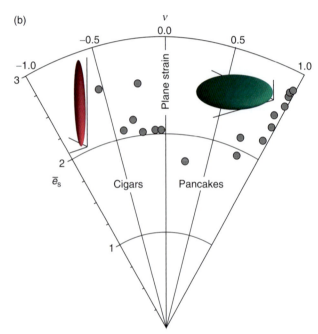

(b)

performed during the deformation history. It does however not take into consideration the rotation of the strain ellipse that occurs for non-coaxial deformations (see Section 2.12) and is therefore best suited for coaxial deformations.

An alternative strain diagram can be defined by means of $\bar{\gamma}_{oct}$, where the natural octahedral unit shear is plotted against a parameter v called the Lode parameter, where

$$v = \frac{2\bar{e}_2 - \bar{e}_1 - \bar{e}_3}{\bar{e}_1 - \bar{e}_3} \tag{2.7}$$

This diagram is shown in Figure 2.14b, and is known as the **Hsü diagram**. The radial lines in this diagram indicate equal amounts of strain, based on the natural octahedral unit shear.

2.12 Volume change

A pure volume change or **volumetric strain** of an object is given by $\Delta = (V - V_0)/V_0$, where V_0 and V are volumes of the object before and after the deformation, respectively. The volume factor Δ is thus negative for volume decrease and positive for volume increase. The deformation matrix that describes general volume change is

$$\begin{bmatrix} D_{11} & 0 & 0 \\ 0 & D_{22} & 0 \\ 0 & 0 & D_{33} \end{bmatrix} = \begin{bmatrix} 1+\Delta_1 & 0 & 0 \\ 0 & 1+\Delta_2 & 0 \\ 0 & 0 & 1+\Delta_3 \end{bmatrix} \tag{2.8}$$

The product $D_{11}D_{22}D_{33}$, which is identical to the determinant of the matrix in Equation 2.8 (see Box 2.3), is always different from 1. This goes for any deformation that involves a change in volume (or area in two dimensions). The closer det **D** is to 1, the smaller the volume (area) change. Volume and area changes do not involve any internal rotation, meaning that lines parallel to the principal strain axes have the same orientations that they had in the undeformed state. Such deformation is called **coaxial**.

A distinction is sometimes drawn between isotropic and anisotropic volume change. **Isotropic volume change** (Figure 2.15) is real volume change where the object is equally shortened or extended in all directions, i.e. the diagonal elements in Equation 2.8 are equal and det **D** $\neq 1$. This means that any marker object has decreased or increased in size, but retained its shape. So, strictly speaking, there is no change in shape involved in isotropic volume change, and the only strain involved is a volumetric strain. In two dimensions, there is

Figure 2.14 Strain data can be represented in (a) the Flinn diagram (linear or logarithmic axes) or (b) the Hsü diagram. The same data are plotted in the two diagrams for comparison. Data from Holst and Fossen (1987).

$$\bar{e}_s = \frac{\sqrt{3}}{2}\bar{\gamma}_{oct} \tag{2.5}$$

The variable $\bar{e}_s$ is called the natural octahedral unit shear, and

$$\bar{\gamma}_{oct} = \frac{2}{3}\sqrt{(\bar{e}_1 - \bar{e}_2)^2 + (\bar{e}_2 - \bar{e}_3)^2 + (\bar{e}_3 - \bar{e}_1)^2} \tag{2.6}$$

where the $\bar{e}$'s are the natural principal strains. This unit shear is directly related to the mechanical work that is

BOX 2.3 | THE DETERMINANT OF THE DEFORMATION MATRIX D

The determinant of a matrix **D** is generally found by the following formula:

$$\det \begin{bmatrix} D_{11} & D_{12} & D_{13} \\ D_{21} & D_{22} & D_{23} \\ D_{31} & D_{32} & D_{33} \end{bmatrix} = D_{11}(D_{22}D_{33} - D_{23}D_{32}) - D_{12}(D_{21}D_{33} - D_{23}D_{31}) + D_{13}(D_{21}D_{32} - D_{22}D_{31})$$

If the matrix is diagonal, meaning that it has non-zero values along the diagonal only, then det **D** is the product of the diagonal entries of **D**. Fortunately, this is also the case for triangular matrices, i.e. matrices that have only zeros below (or above) the diagonal:

$$\det \begin{bmatrix} D_{11} & D_{12} & D_{13} \\ 0 & D_{22} & D_{23} \\ 0 & 0 & D_{33} \end{bmatrix} = D_{11}D_{22}D_{33}$$

The deformation matrices for volume change and pure shear are both examples of diagonal matrices, and those for simple and subsimple shear are triangular matrices. When det **D** = 1, then the deformation represented by the matrix is isochoric, i.e. it involves no change in volume.

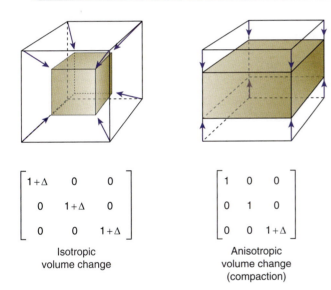

$$\begin{bmatrix} 1+\Delta & 0 & 0 \\ 0 & 1+\Delta & 0 \\ 0 & 0 & 1+\Delta \end{bmatrix}$$

Isotropic
volume change

$$\begin{bmatrix} 1 & 0 & 0 \\ 0 & 1 & 0 \\ 0 & 0 & 1+\Delta \end{bmatrix}$$

Anisotropic
volume change
(compaction)

Figure 2.15 The difference between isotropic volume change, which involves no strain, and anisotropic volume change represented by uniaxial shortening (compaction).

isotropic area change in which an initial circle remains a circle, albeit with a different radius.

Isotropic volume increase: $X = Y = Z > 1$
Isotropic volume decrease: $X = Y = Z < 1$

Anisotropic volume change involves not only a volume (area) change but also a change in shape because its effect on the rock is different in different directions. The most obvious examples are compaction or uniaxial

contraction and uniaxial extension, as shown in Figure 2.11 and discussed in the next section.

One may argue that anisotropic volume change is a redundant term, because any anisotropic strain can be decomposed into a combination of (isotropic) volume change and change in shape. The fact that deformation is not concerned with the deformation history makes any decomposition of the deformation into such components mathematically correct, even though they have nothing to do with the actual process of deformation in question. However, if we think about how compaction of sediments and sedimentary rocks comes about, it makes sense to consider it as an anisotropic volume change rather than a combination of isotropic volume change and a strain. Sediments compact by vertical shortening (Figure 2.11, top), not discretely by shrinking and then straining (Figure 2.11, middle). As geologists, we are concerned with reality and retain the term anisotropic volume change where we find it useful.

Anisotropic volume increase: $XYZ \neq 1$, where two or all of X, Y and Z are different.

2.13 Uniaxial strain (compaction)

Uniaxial strain is contraction or extension along one of the principal strain axes without any change in length along the other two. Such strain requires a reorganization, addition or removal of rock volume. If volume is lost, we have **uniaxial contraction** and volume reduction. This happens through grain reorganization during physical compaction of porous sediments and tuffs near the surface, leading to a denser packing of grains. Only water, oil or gas that filled the pore space leaves the rock volume, not the rock minerals themselves.

In calcareous rocks and deeply buried siliciclastic sedimentary rocks, uniaxial strain can be accommodated by (pressure) solution, also referred to as chemical compaction. In this case, minerals are dissolved and transported out of the rock volume by fluids. Removal of minerals by diffusion can also occur under metamorphic conditions in the middle and lower crust. This can result in cleavage formation or can lead to compaction across shear zones. **Uniaxial extension** implies expansion in one direction. This may occur by the formation of tensile fractures or veins or during metamorphic reactions.

Uniaxial contraction: $X = Y > Z$, $X = 1$
Uniaxial extension: $X > Y = Z$, $Z = 1$

Uniaxial strain may occur in isolation, such as during compaction of sediments, or in concert with other deformation types such as simple shear. It has been found useful to consider many shear zones as zones of simple shear with an additional uniaxial shortening across the zone.

Uniaxial shortening or compaction is such an important and common deformation that it needs some further attention. The deformation matrix for uniaxial strain is

$$\begin{bmatrix} 1 & 0 & 0 \\ 0 & 1 & 0 \\ 0 & 0 & 1+\Delta \end{bmatrix} \tag{2.9}$$

where Δ is the elongation in the vertical direction (negative for compaction) and $1 + \Delta$ is the vertical stretch (Figure 2.16). The fact that only the third diagonal element is different from unity implies that elongation or shortening only occurs in one direction. The matrix gives the strain ellipsoid, which is oblate or pancake-shaped for compaction. It can also be used to calculate how planar features, such as faults and bedding, are affected by compaction (Figure 2.16).

If we can estimate the present and initial porosity Φ_0 of a compacted sediment or sedimentary rock, then we can use the equation

$$\Phi = \Phi_0 e^{-CZ} \tag{2.10}$$

to find $1 + \Delta$, where Z is the burial depth and C is a constant that typically is about 0.29 for sand, 0.38 for silt

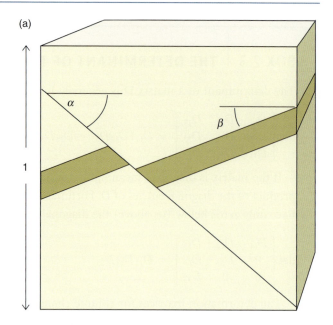

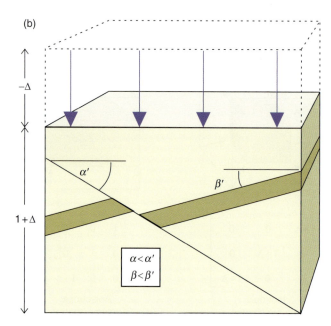

Figure 2.16 Compaction reduces the dips of both layers and faults. The effect depends on the amount of post-faulting compaction and can be estimated using the deformation matrix for compaction (Equation 2.9).

and 0.42 for shale; e is now the exponential function, not the extension factor. Equation 2.10 tells us that the porosity Φ changes with depth Z, and we are looking at a matrix of the form

$$\begin{bmatrix} 1 & 0 & 0 \\ 0 & 1 & 0 \\ 0 & 0 & 1+f(Z) \end{bmatrix} \tag{2.11}$$

It can be shown that $\Delta = (1 - \Phi_0)/(1 - \Phi_0\,e^{-CZ})$, and the deformation matrix then becomes

$$\begin{bmatrix} 1 & 0 & 0 \\ 0 & 1 & 0 \\ 0 & 0 & 1 + (1 - \Phi_0)/(1 - \Phi_0 e^{-CZ}) \end{bmatrix} \qquad (2.12)$$

Matrix (2.12) helps us predict the compaction at any point in a sedimentary basin, and it also predicts how structures such as folds and faults are modified by compaction. A relationship that can be found from matrix (2.12) relates the original dip (α) to the new dip (α') after the compaction:

$$\alpha' = \tan^{-1}[(1 + \Delta)\tan\alpha] \qquad (2.13)$$

In metamorphic rocks, uniaxial shortening or compaction can be estimated by comparing portions of the rock affected by compaction with those believed to be unaffected. If the concentration of an immobile mineral, such as mica or an opaque phase, is C in the compacted part of the rock, and is believed to have been C_0 before compaction, then the compaction factor is given by the relationship

$$1 + \Delta = \frac{C_0}{C}\begin{bmatrix} k_x & 0 \\ 0 & k_y \end{bmatrix} \qquad (2.14)$$

C_0 is found outside of the deformation zone, which could be a millimeter-thick cleavage-related microlithon, as discussed in Chapter 12 on cleavages and foliations. The deformation zone could also be a mesoscopic shear zone, where the wall rock is assumed to be unaffected by both shearing and compaction. As we will see in Chapter 15 on shear zones, ideal shear zones can only accommodate compaction in addition to simple shear.

2.14 Pure shear and coaxial deformations

Pure shear (Figure 2.8b) is a perfect **coaxial deformation**. This means that a marker that is parallel to one of the principal axes has not rotated away from its initial position. Uniaxial strain, where the rock shortens or extends in one direction, is another example of coaxial deformation.

Coaxial deformation implies that lines along the principal strain axes have the same orientation as they had in the undeformed state.

Pure shear is here considered a plane (two-dimensional) strain with no volume change, although some geologists also apply the term to three-dimensional coaxial deformations. Pure shear is identical to balance shortening in one

direction with extension in the other, as expressed by its deformation matrix:

$$\begin{bmatrix} k_x & 0 \\ 0 & k_y \end{bmatrix} \qquad (2.15)$$

where k_x and k_y are the stretch and shortening along the x and y coordinate axes, respectively. Since pure shear preserves area (volume) we have that $k_y = 1/k_x$.

2.15 Simple shear

Simple shear (Figure 2.8) is a special type of constant-volume plane strain deformation. There is no stretching or shortening of lines or movement of particles in the third direction. Unlike pure shear, it is a **non-coaxial deformation**, meaning that lines parallel to the principal strain axes have rotated away from their initial positions. This **internal rotation** component of the strain has caused several geologists to refer to simple shear and other non-coaxial deformations as **rotational deformations**. By internal rotation we here mean the difference between the orientation of a line along the longest finite strain axis and the orientation of this material line prior to deformation. While pure shear has no internal rotation component, the internal rotation component of a simple shear strain depends on the amount of strain.

Another characteristic of non-coaxial deformations relates the orientation of the strain ellipsoid and the amount of strain:

For non-coaxial deformations, the orientations of the principal strain axes are different for different amounts of strain, while for coaxial deformations they always point in the same directions (same orientation, different lengths).

As for any plane strain deformation type, the strain ellipsoid produced by simple shear has two circular sections (Figure 2.13b). One of them is parallel to the shear plane, regardless of the amount of strain involved. By **shear plane** we mean the plane on which shear occurs, as shown in Figure 2.17. The shear plane is similar to the slip plane for faults, and for simple shear it is a plane of no strain.

The consideration of coaxiality and internal rotations is easier to discuss in terms of progressive deformation (below). For now, we will look at the deformation matrix for simple shear:

$$\begin{bmatrix} 1 & \gamma \\ 0 & 1 \end{bmatrix} \qquad (2.16)$$

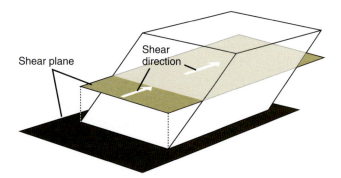

Shear plane

Shear
direction

Figure 2.17 Illustration of the meaning of the terms shear plane and shear direction, by means of a deformed cube. The terms relate to simple shear or the simple shear component of a more general deformation type, such as subsimple shear.

The factor γ is called the shear strain and $\gamma = \tan \psi$, where ψ is the angle of rotation of a line that was perpendicular to the shear plane in the undeformed state (Figure 2.8). Lines and planes that lie within (parallel to) the shear plane do not change orientation or length during simple shear. Lines and planes with any other orientation do. It is noteworthy that deformation matrices describing coaxial deformations are symmetric, while those describing non-coaxial deformations are asymmetric.

2.16 Subsimple shear

Between pure shear and simple shear is a spectrum of planar deformations, commonly referred to as **subsimple shear** (also referred to as general shear, although these deformations are just a subset of planar deformations and thus not very general). Subsimple shear can be considered as a mix of pure and simple shear because the internal rotation involved is less than for simple shear. Mathematically we have to combine the deformation matrices for simple and pure shear, which is not as trivial as it may sound. It turns out that the matrix can be written as

$$\begin{bmatrix} k_x & \Gamma \\ 0 & k_y \end{bmatrix} \qquad (2.17)$$

where $\Gamma = \gamma[(k_x - k_y)] / [\lambda_n(k_x - k_y)]$. If there is no area change in addition to the pure and simple shear components, then $k_y = 1/k_x$ and $\Gamma = \gamma(k_x - 1/k_x)/2\ln(k_x)$. An example of subsimple shear and its displacement field and particle paths is shown in Figure 2.1.

2.17 Progressive deformation and flow parameters

Simple shear, pure shear, volume change and any other deformation type relate the undeformed to the deformed

state only. The history that takes place between the two states is a different matter, and is the focus of the study of **flow** and **progressive deformation of rocks**.

It is useful to consider individual particles in the rock or sediment when discussing progressive deformation. If we keep track of a single particle during the deformation history, we can get a picture of a single **particle path**. If we map the motion of a number of such particles we get an impression of the **flow pattern**.

The flow pattern is the sum of particle paths in a deforming medium.

Particle paths can be recorded directly in experiments where individual particles or (colored) grains can be traced throughout the deformation history. The particle paths shown in Figure 2.18 are not quite complete, because they are based on points of intersection between faults and markers. The flow pattern is also affected by faults, which are displacement discontinuities (see Chapter 8), but the overall pattern can be seen to be close to that of pure shear. Filming an ongoing experiment enables the scientist to reconstruct the flow pattern during the experiment. In the field, things are different and we can only see the final stage of the deformation, i.e. the last picture of the film roll.

Let us imagine that we can photograph a deformation from the start to the end. Then the difference between any two adjacent pictures represents a small interval of the total deformation. Based on the differences between the two pictures we may be able to find the size and orientation of the **incremental strain ellipsoid** for this interval and describe this increment of the deformation history. When such an interval becomes very small, we get the **infinitesimal** or **instantaneous deformation parameters**. Parameters that act instantaneously during the deformation history are called **flow parameters**. As indicated by Figure 2.19, the flow parameters include the infinitesimal or instantaneous stretching axes, the flow apophyses, vorticity and the velocity field, all of which deserve some extra attention.

The Instantaneous Stretching Axes (ISA) are the three perpendicular axes (two for plane deformations) that describe the directions of maximum and minimum stretching at any time during deformation. They are all called instantaneous stretching axes although the minimum stretching axis (ISA$_3$) is the direction of maximum shortening, or negative stretching. Lines along the longest axis (ISA$_1$) are stretching faster than any other line orientation. Similarly, no line is stretching slower

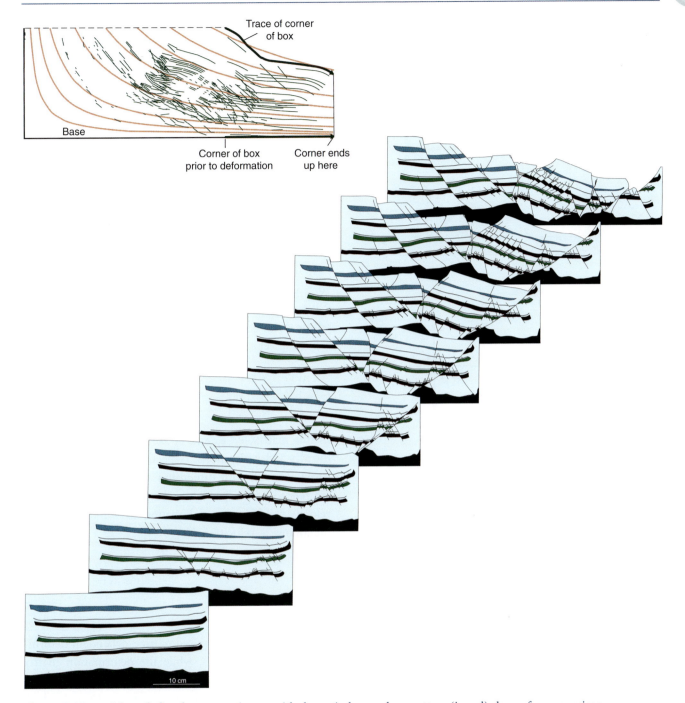

Figure 2.18 Particle path for plaster experiment, with theoretical pure-shear pattern (in red) shown for comparison. Drawings of pictures taken during the experiment are also shown. Particle paths are found by connecting corner points and points where faults intersect bedding. In this case the deformation is heterogeneous and brittle, but can be compared to homogeneous flow (pure shear) because the discontinuities are many and distributed. The experiment is described in Fossen and Gabrielsen (1996).

(or shortening faster) than those along the shortest instantaneous stretching axis (ISA₃).

Flow apophyses separate different domains of particle paths (AP in Figure 2.19). Particles located along the apophyses rest or move along the straight apophyses. Other particle paths are curved. No particle can cross a flow apophysis unless the conditions change during the deformation history.

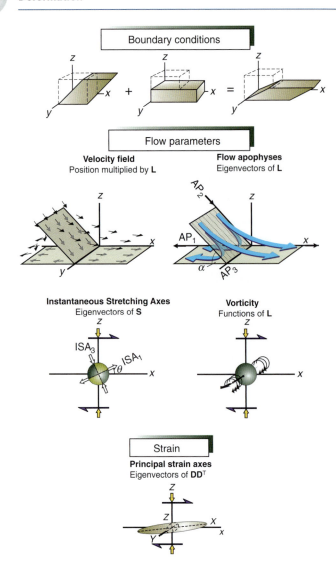

Figure 2.19 The most important deformation parameters. Boundary conditions control the flow parameters, which over time produce strain. Modified from Fossen and Tikoff (1997).

Vorticity describes how fast a particle rotates in a soft medium during the deformation (Section 2.20). A related quantity is the **kinematic vorticity number** (W_k), which is 1 for simple shear and 0 for pure shear and somewhere between the two for subsimple shear.

The **velocity field** describes the velocity of the particles at any instance during the deformation history. Let us have a closer look.

2.18 Velocity field

The velocity (gradient) matrix (or tensor) **L** describes the velocity of the particles at any instant during the deformation. In three dimensions the velocity field is described by the equations

$$v_1 = L_{11}x + L_{12}y + L_{13}z$$
$$v_2 = L_{21}x + L_{22}y + L_{23}z$$
$$v_3 = L_{31}x + L_{32}y + L_{33}z$$

which in matrix notation become

$$\begin{bmatrix} v_1 \\ v_2 \\ v_3 \end{bmatrix} = \begin{bmatrix} L_{11} & L_{12} & L_{13} \\ L_{21} & L_{22} & L_{23} \\ L_{31} & L_{32} & L_{33} \end{bmatrix} \begin{bmatrix} x \\ y \\ z \end{bmatrix} \quad (2.18)$$

or

$$\mathbf{v} = \mathbf{L}\mathbf{x}$$

Here, the vector **v** describes the velocity field and the vector **x** gives the particle positions.

If we consider flow with a 3-D coaxial component, such as axially symmetric flattening or extension (Figure 2.12) in combination with a progressive simple shear whose shear plane is the x–y plane, then we have the following velocity matrix:

$$\mathbf{L} = \begin{bmatrix} \dot{e}_x & \dot{\gamma} & 0 \\ 0 & \dot{e}_y & 0 \\ 0 & 0 & \dot{e}_z \end{bmatrix} \quad (2.19)$$

In this matrix, $\dot{e}_x$, $\dot{e}_y$ and $\dot{e}_z$ are the elongation rates in the x, y and z directions, respectively, and $\dot{\gamma}$ is the shear strain rate (all with dimension s^{-1}). These strain rates are related to the particle velocities and thereby to the velocity field. Inserting Equation 2.19 into 2.18 gives the following velocity field:

$$v_1 = \dot{e}_x x + \dot{\gamma}y$$
$$v_2 = \dot{e}_y y \quad (2.20)$$
$$v_3 = \dot{e}_z z$$

The matrix **L** is composed of time-dependent deformation rate components, while the deformation matrix has spatial components that do not involve time or history. The matrix **L** for progressive subsimple shear now becomes

$$\mathbf{L} = \begin{bmatrix} \dot{e}_x & 0 \\ 0 & \dot{e}_y \end{bmatrix} + \begin{bmatrix} 0 & \dot{\gamma} \\ 0 & 0 \end{bmatrix}$$
$$= \begin{bmatrix} 0 & \dot{\gamma} \\ 0 & 0 \end{bmatrix} + \begin{bmatrix} \dot{e}_x & 0 \\ 0 & \dot{e}_y \end{bmatrix} = \begin{bmatrix} \dot{e}_x & \dot{\gamma} \\ 0 & \dot{e}_y \end{bmatrix} \quad (2.21)$$

This equation illustrates a significant difference between deformation rate matrices and ordinary deformation matrices: while deformation matrices are non-commutative, strain rate matrices can be added in any order without changing the result. The disadvantage is that information about the strain ellipse is more cumbersome to extract from deformation rate matrices, since we then need to integrate with respect to time over the interval in question.

L can be decomposed into a symmetric matrix $\dot{S}$ and a so-called skew-symmetric matrix W:

$$L = \dot{S} + W \tag{2.22}$$

$\dot{S}$ is the stretching matrix (or tensor) and describes the portion of the deformation that over time produces strain. W is known as the vorticity or spin matrix (tensor) and contains information about the internal rotation during the deformation. For progressive subsimple shear the decomposition becomes

$$L = S + W = \begin{bmatrix} \dot{\varepsilon}_x & \frac{1}{2}\dot{\gamma} \\ \frac{1}{2}\dot{\gamma} & \dot{\varepsilon}_y \end{bmatrix} + \begin{bmatrix} 0 & \frac{1}{2}\dot{\gamma} \\ -\frac{1}{2}\dot{\gamma} & 0 \end{bmatrix} \tag{2.23}$$

The eigenvectors and eigenvalues to $\dot{S}$ give the orientations and lengths of the ISA (instantaneous stretching axes). The eigenvectors of L describe the flow apophyses, which are discussed in the next section. Whether one wants to work with strain rates or simple deformation parameters such as k and γ is a matter of personal preference in many cases. Both are in use, and both have their advantages and disadvantages.

2.19 Flow apophyses

Flow apophyses, shown as blue lines along with green particle paths in Figure 2.20, are **theoretical lines** (meaning that they are invisible "ghost lines" that are free to rotate independently of material lines) that separate different fields of the flow. Particles cannot cross an apophysis, but they can move along them or rest on them. For the case of simple shearing (progressive simple shear), the particles will always move straight along the shear direction. This occurs because there is no shortening or extension perpendicular to the shear plane, and it tells us that one of the apophyses is parallel to the shear direction (Figure 2.20, simple shear). As we will see, there is only this one apophysis for simple shear. For pure shear there are two orthogonal apophyses along which particles move straight toward or away from the origin. Subsimple shear has two oblique apophyses; one parallel to the shear direction and one at an angle α to the first one. The angle α between the two apophyses varies from 90° for pure shear to 0° for simple shear.

For simple shear, pure shear and subsimple shear, the apophyses occur in the plane perpendicular to the shear plane and parallel to the shear direction, which means

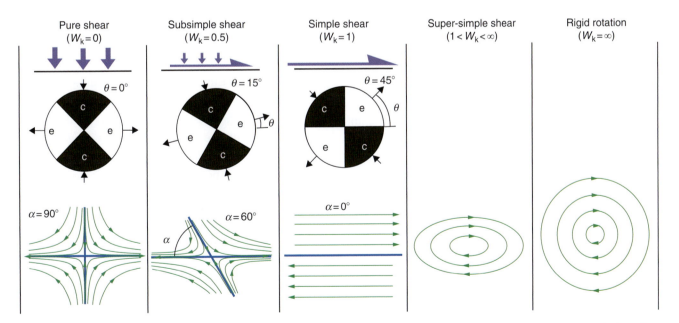

Figure 2.20 Particle paths (green) and flow apophyses (blue) for planar deformations. The two flow apophyses, which describe the flow pattern, are orthogonal for pure shear, oblique for subsimple shear and coincident for simple shear. For deformations with more internal rotation, particles move along elliptical paths. The end-member is rigid rotation, where particles move along perfect circles. Rigid rotation involves perfect rotation without strain, while pure shear is simply strain with no rotation. Note that ISA are generally oblique to the flow apophyses for $W_k > 0$.

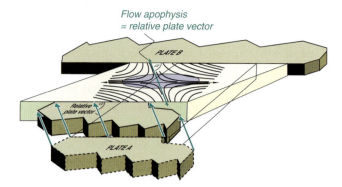

Flow apophysis
= relative plate vector

PLATE B

Relative
plate vector

PLATE A

Figure 2.21 Two rigid plates (A and B) and an intermediate deforming zone (yellow). Standing on plate B, we will observe plate A moving obliquely towards us. If this shortening is compensated by lateral extension, then the oblique flow apophysis is parallel to the plate vector and the particle path is known. W_k can be found from Figure 2.24. Modified from Fossen and Tikoff (1998).

that we can express the vectors in two dimensions if we wish:

$$\begin{bmatrix} 1 \\ 0 \end{bmatrix}, \begin{bmatrix} \frac{-\gamma}{\ln(k_x/k_y)} \\ 1 \end{bmatrix} \tag{2.24}$$

The first of these vectors or apophyses is the one that is parallel to the shear direction (here chosen to be along the x-axis of our coordinate system), while the other one is oblique. The angle α between the apophyses is directly related to how close to simple shear or pure shear the deformation is. α is zero for simple shear and 90° for pure shear, and W_k thus depends on α:

$$W_k = \cos(\alpha) \tag{2.25}$$

To illustrate how flow apophyses can be useful in tectonics, consider the convergent motion of one tectonic plate relative to another. It actually turns out that the oblique apophysis is parallel to the convergence vector, which becomes apparent if we recall that straight particle motion can only occur in the direction of the apophyses. In other words, for oblique plate convergence (transpression, see Chapter 18) α describes the angle of convergence and is 90° for head-on collision and 0° for perfect strike-slip. Head-on collision is thus a pure shear on a large scale, while strike-slip or conservative boundaries deform by overall simple shear. Figure 2.21 illustrates a theoretical example of oblique convergence, and if we know the plate motion vector of one of the plates relative to the other, we can, at least in principle, estimate the average orientations of the oblique flow apophysis,

because the two will be parallel. Once we know the flow apophysis, we also know W_k (from Equation 2.25) and we can use this information to model or evaluate deformation structures along the plate boundary. Or we can analyze old structures along a (former) plate boundary and say something about the paleo-convergence angle. Hence, matrix calculus and field geology meet along plate boundaries, among other places, in very useful ways.

2.20 Vorticity and W_K

We separate **non-coaxial deformation histories**, where material lines (imaginary lines drawn on a section through the deforming rock) that in one instance are parallel to ISA and in the next instance have rotated away from them, and **coaxial deformation histories**, where the material lines along ISA remain along these axes for the entire deformation history. The degree of (internal) rotation or coaxiality is denoted by the kinematic vorticity number W_k. This number is 0 for perfectly coaxial deformation histories, 1 for progressive simple shear, and between 0 and 1 for subsimple shear. Values between 1 and ∞ are deformation histories where the principal strain axes rotate continuously around the clock. $W_k > 1$ deformations are therefore sometimes termed spinning deformations, and the result of such deformations is that the strain ellipsoid records a cyclic history of being successively strained and unstrained.

To get a better understanding of what W_k actually means, we need to explore the concept of vorticity. **Vorticity** is a measure of the internal rotation during the deformation. The term comes from the field of fluid dynamics, and the classic analogy is a paddle wheel moving along with the flow, which we can imagine is the case with the paddle wheel shown in Figure 2.22. If the paddle wheel does not turn, then there is no vorticity. However, if it does turn around, there is a vorticity, and the vector that describes the velocity of rotation, the angular velocity vector $\boldsymbol{\omega}$, is closely associated with the vorticity vector $\mathbf{w}$:

$$\mathbf{w} = 2\omega = \mathrm{curl}\ \mathbf{v} \tag{2.26}$$

where $\mathbf{v}$ is the velocity field.

Another illustration that may help is one where a spherical volume of the fluid freezes, as in Figure 2.23. If the sphere is infinitely small, then the vorticity vector $\mathbf{w}$ will represent the axis of rotation of the sphere, and its length will be proportional to the speed of the rotation.

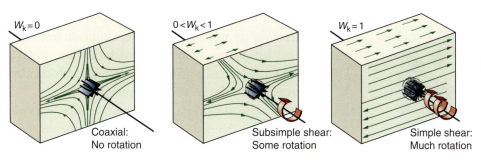

Figure 2.22 The paddle-wheel interpretation of flow. The axis of the paddle wheel is parallel to the vorticity vector and is not rotating for coaxial deformation ($W_k = 0$) and shows an increasing tendency to rotate for increasing W_k.

The vorticity vector can be interpreted as: (1) the average rotation of all lines in the plane perpendicular to **w** relative to the ISA; (2) the speed of rotation of a set of physical lines that are parallel to the ISA; (3) the average speed of rotation of two orthogonal physical lines in the plane perpendicular to **w**; or (4) half the speed of rotation of a rigid spherical inclusion in a ductile matrix where there is no slip along the edge of the sphere and where the viscosity contrast is infinitely high.

As an example, let us put a rigid sphere in a deformation box filled with a softer material that contains strain markers. If we apply a coaxial deformation by squeezing the box in one direction and letting it extend in the other (s), the sphere will not rotate and the vorticity is 0. However, if we add a simple shear to the content of the box, the sphere will rotate, and we will observe that there is a relationship between the strain and the rotation, regardless of how fast we perform the experiment: the more strain, the more rotation. This relation between strain and (internal) rotation is the kinematic vorticity number W_k. For a simultaneous combination of pure shear and simple shear, i.e. subsimple shear, there is less rotation of the sphere for a certain strain accumulation than for simple shear. Therefore W_k is less than 1 by an amount that depends on the relative amount of simple versus pure shear.

W_k **is a measure of the relation between the vorticity (internal rotation) and how fast strain accumulates during the deformation.**

Mathematically, the kinematic vorticity number is defined as

$$W_k = \frac{w}{\sqrt{2(s_x^2 + s_y^2 + s_z^2)}} \qquad (2.27)$$

where s_n are the principal strain rates, i.e. the strain rates along the ISA. This equation can be rewritten in terms of

pure and simple shear components if we assume steady flow (see next section) during the deformation:

$$W_k = \frac{\gamma}{\sqrt{2\left[(\ln k_x)^2 + (\ln k_y)^2\right] + (\gamma)^2}} \qquad (2.28)$$

which for constant area becomes

$$W_k = \cos[\arctan(2\ln k/\gamma)] \qquad (2.29)$$

This expression is equivalent to the simpler equation $W_k = \cos(\alpha)$ (Equation 2.25), where α and α' are the acute and obtuse angles between the two flow apophyses, respectively, and k and γ are the pure shear and simple shear components, as above. The relation between W_k and α is shown graphically in Figure 2.24, where also the corresponding relation between W_k and ISA$_1$ is shown.

2.21 Steady-state deformation

If the flow pattern and the flow parameters remain constant throughout the deformation history, then we have **steady-state flow** or deformation. If, on the other hand, the ISA rotate, W_k changes value or the particle paths

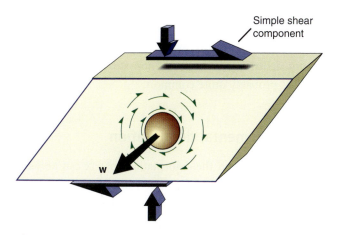

Figure 2.23 The vorticity vector (**w**) in progressive subsimple shear.

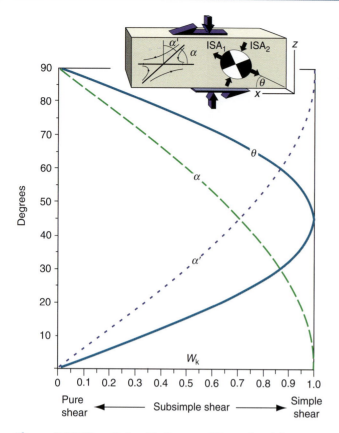

Figure 2.24 The relationship between W_k, α, α' and θ.

a portion of the deformation history. When using an incremental approach, each deformation increment is represented by an **incremental deformation matrix**, and the product of all of the incremental deformation matrices equals the finite deformation matrix.

There is an important thing about matrix multiplication that we should be aware of: The order by which we multiply matrices (apply deformations) is not arbitrary. For example, a pure shear followed by a simple shear does not result in the same deformation as a simple shear followed by a pure shear:

$$\begin{bmatrix} k_x & 0 \\ 0 & k_y \end{bmatrix}\begin{bmatrix} 1 & \gamma \\ 0 & 1 \end{bmatrix} \neq \begin{bmatrix} 1 & \gamma \\ 0 & 1 \end{bmatrix}\begin{bmatrix} k_x & 0 \\ 0 & k_y \end{bmatrix} \qquad (2.30)$$

Interestingly, the matrix representing the first deformation or deformation increment is the last of the matrices to be multiplied. For example, if a deformation is represented by three increments, D_1 representing the first part of the history and D_3 the last part, then the matrix representing the total deformation is the product $D_{tot} = D_3 D_2 D_1$.

Another point to note is that since we are here operating in terms of kinematics only (not time), it does not make any difference whether D_1, D_2 and D_3 represent different deformation phases or increments of the same progressive history.

One can define as many increments as one needs to model progressive deformation. When the incremental matrices represent very small strain increments, the principal strain axes of the matrices approximate the ISA and other flow parameters can be calculated for each increment and compared. Modeling progressive deformation numerically can provide useful information about the deformation history. It may be difficult, however, to extract information from naturally deformed rocks that reveal the actual deformation history. For this reason, steady-state deformations represent useful reference deformations, although natural deformations are not restricted by the limitations of steady-state flow.

change during the course of deformation, then we have a non-steady-state flow.

> During steady-state deformation the ISA and flow apophyses retain their initial orientation throughout the deformation history, and W_k is constant.

An example of non-steady deformation is a subsimple shear that moves from being close to simple shearing towards a more pure shear dominated flow. For practical reasons, and because non-steady-state deformations are difficult to identify in may cases, steady-state flow is assumed. In nature, however, non-steady-state deformation is probably quite common.

2.22 Incremental deformation

The theory around the deformation matrix can be used to model progressive deformation in a discrete way. We make a distinction between **finite deformation** or **finite strain** on one hand, which is the result of the entire deformation history, and **incremental deformation** or **incremental strain** on the other, which concerns only

2.23 Strain compatibility and boundary conditions

If we deform a chunk of soft clay between our hands, there will be free surfaces where the clay can extrude. The situation is quite different for a volume of rock undergoing natural deformation in the crust. Most deformation happens at depth where the rock volume is surrounded by other rock and under considerable pressure. The resulting

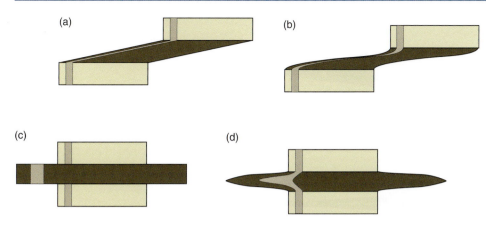

Figure 2.25 Homogeneous (a) and heterogeneous (b) simple shear create no compatibility problems between the deformed and undeformed rocks. Homogeneous pure shear (c) does, because lateral extrusion of material creates discontinuities. The space problem is also apparent for heterogeneous pure shear (d), but the discontinuities can be eliminated.

strain will depend on the anisotropy of the rocks. Anisotropy, such as a weak layer or foliation, may control the deformation just as much as the stress field.

Let us illustrate this by a ductile shear zone with straight and parallel margins and undeformed walls, as shown in Figure 2.25a, b. In principle, simple shearing can continue "forever" in such a zone without creating any problem with the wall rock (challenges may arise around shear zone termination points, but we will neglect those here).

Adding shortening across the shear zone as a result of anisotropic volume change is also easy to accommodate. However, adding a pure shear means that the shear zone will have to extrude laterally while the wall rock remains undeformed (Figure 2.25c, d). This causes a classic **strain compatibility problem** since the continuity across the shear zone boundaries is lost. The strain in the shear zone is no longer compatible with the undeformed walls.

For the state of strain in two adjacent layers to be compatible, the section through their respective strain ellipsoids parallel to their interface must be identical.

For our shear zone example, the rock on one side is undeformed, and the section must be circular, which it is for simple shear with or without shear zone perpendicular compaction (Figure 2.26). However, if a pure shear component is involved, compatibility between the deformed and undeformed volumes is not maintained.

A practical solution to this problem is to introduce a discontinuity (slip surface or fault) between the zone and each of the walls. The shear zone material can then be squeezed sideways in the direction of the shear zone. The problem is that it has nowhere to go, since the neighboring part of a parallel-sided shear zone will try to do exactly the same thing. We can therefore conclude that

a shear zone with parallel boundaries is not compatible with (cannot accommodate) pure shear. Besides, strain compatibility as a concept requires that the deformed volume is coherent and without discontinuities, overlaps or holes. Discontinuities and non-parallel shear zones are, as we know, common in many deformed rocks. The requirement of continuity in strain compatibility is thus negotiable.

2.24 Deformation history from deformed rocks

It is sometimes possible to extract information about the deformation history from naturally deformed rocks. The key is to find structures that developed during a limited part of the total deformation history only. In some cases the deformation has moved from one part of the deformed rock volume to another, leaving behind deformed rock that has recorded the conditions during earlier increments of the deformation. For example, strain may after some time localize to the central part of the shear zone. Thus the margins bear a record of the first increment of the deformation. But shear zones can

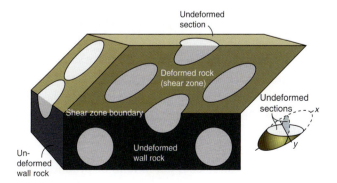

Figure 2.26 The compatibility between undeformed wall rock and a simple shear zone. Any section parallel to the shear zone wall will appear undeformed.

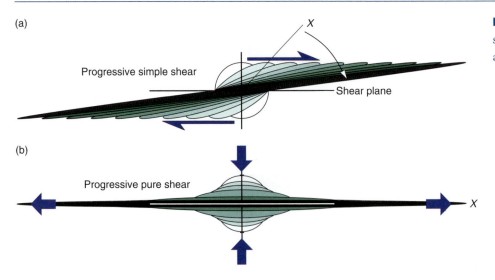

(a)

Progressive simple shear

X

Shear plane

(b)

Progressive pure shear

X

Figure 2.27 The evolution of strain during progressive simple and pure shear.

also initiate as narrow zones and widen over time. In this case the outer portion of the shear zone records the last increments of deformation. The search for deformation history is not necessarily an easy task!

In some cases there is evidence for mineral fiber growth at various stages or veins forming successively during deformation. The orientation of fibers and veins reflects the orientation of ISA and thus gives information about the flow parameters.

If the metamorphic conditions change during the course of deformation, structures that carry information about incremental deformation may be separated by means of metamorphic mineralogy.

2.25 Coaxiality and progressive simple shear

Earlier in this chapter we defined certain reference deformation types such as simple shear, pure shear, sub-simple shear and volume change. The same deformations define progressive deformation if every increment, large or small, represents the same deformation type as the total one. Hence, a deformation that not only ends up as simple shear, but where any interval of the deformation history is also a simple shear is a progressive simple shear or **simple shearing**. Similarly we have progressive pure shear or **pure shearing**, progressive subsimple shear or **subsimple shearing**, and progressive volume change or **dilating**.

A given simple shear does not have to be produced by progressive simple shear, but may for example be a pure shear followed by rigid rotation. The rigid rotation must then be exactly arctan(1/2 tan ψ).

Progressive deformations are separated into coaxial and non-coaxial ones. A **non-coaxial deformation history** implies that the orientation of the progressive strain ellipsoid is different at any two points in time during the deformation, as clearly is the case in Figure 2.27a where X rotates progressively. Another characteristic feature is that lines that are parallel to ISA or the principal strain axes rotate during deformation. These rotational features allow us to call them rotational deformation histories. During a **coaxial deformation history** the orientation of the strain ellipsoid is constant throughout the course of the deformation and lines parallel to ISA do not rotate. Coaxial deformation histories, such as the pure shearing shown in Figure 2.27b, are therefore referred to as non-rotational.

Simple shearing involves no stretching, shortening or rotation of lines or planar structures parallel to the shear plane. Other lines will rotate towards the shear direction as they change length, and planes will rotate toward the shear plane. The long axis of the strain ellipse also rotates toward the shear direction during the shearing history, although it will never reach this direction.

In order to improve our understanding of simple shearing we will study a circle with six physical lines numbered from 1 to 6 (Figure 2.28). We will study what happens to three orthogonal pairs of lines, numbered 1 and 4, 2 and 5, and 3 and 6, at various stages during the shearing. At the moment the shearing starts, the rock (and circle) is stretched fastest in the direction of ISA_1 and slowest along ISA_3 (negative stretching, which actually means shortening along ISA_3). The ISA are constant during the entire history of deformation, since we are considering a steady-state deformation. Two important

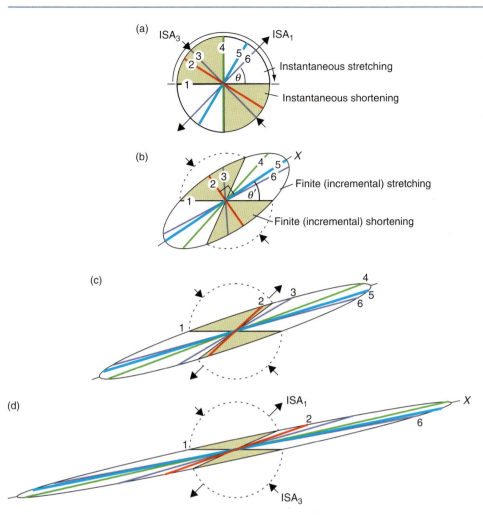

Figure 2.28 Simple shearing of a circle and three sets of orthogonal lines (1–6). The arrow along the circle in (a) indicates the rotation direction for lines during the deformation.

pairs of fields occur in Figure 2.28a. In the two white fields, lines are continuously being stretched, and we can call them the **fields of instantaneous stretching**. Similarly, lines in the two yellow fields continuously experience shortening and we call them the **fields of instantaneous shortening**. The borderlines between these fields are the **lines of no stretching or shortening**. In three dimensions the fields become volumes and the borderlines become surfaces of no stretching or shortening. These fields remain constant during the deformation history while lines may rotate from one field to the next.

Lines 1 and 4 start out parallel to the boundary between active stretching and shortening. Line 1 lies in the shear plane and maintains its original orientation and length, i.e. remains undeformed. Line 4 quickly rotates into the stretching field toward ISA$_1$ (Figure 2.28b). In Figure 2.28c, line 4 has rotated through ISA$_1$ and will also pass the long axis X of the strain ellipse if the deformation keeps going.

Lines 2 and 5 are located in the fields of shortening and stretching, respectively. In Figure 2.28b, line 2 is shortened and line 5 extended, but the two lines are orthogonal at this point. They were orthogonal also before the deformation started, which means that the angular shear along these two directions is now zero. This also means that lines 2 and 5 are parallel to the strain axes at this point. They do, however, rotate faster than the deformation ellipse, and in Figure 2.28c line 5 has passed X and line 2 has passed the shortest axis of the strain ellipse. In fact, line 2 is now parallel with ISA$_1$ and has passed the boundary between the fields of instantaneous shortening and stretching. This means that the line has gone from a history of shortening to one of stretching. The total shortening still exceeds the total stretching, which is why this line is still in the contractional field of the cumulative strain ellipsoid. On the way toward the last stage (Figure 2.28d), line 2 passes ISA$_1$ as well as the borderline between total stretching and shortening.

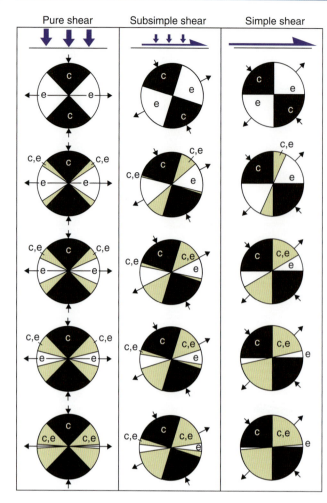

Figure 2.29 The development of sectors where lines experience a qualitatively common history: c, contractional field; e, extension field. c, e indicate that lines in this field were first shortened and then extended. Note the symmetric picture produced by pure shear and the asymmetry created by non-coaxial deformation histories. Field observations of deformed dikes and veins can sometimes be used to construct the sectors and thus the degree of coaxiality.

The last pair of lines (3 and 6) start out parallel with ISA_3 and ISA_1, respectively. Both lines immediately rotate away from the principal stretching directions because of the rotational or non-coaxial nature of simple shearing. Line 3 experiences shortening until stage (b). Soon thereafter it is stretched and its original length is restored as it passes through the borderline between the fields of instantaneous stretching and shortening (between stages b and c). It keeps stretching as it rotates toward the shear direction. Line 6 experiences stretching during the entire deformation history, but less and less so as it rotates toward the shear direction and away from

ISA_1. Clearly, line 6 rotates faster than X, with which it was parallel at the onset of deformation.

The deformation history can also be represented in terms of sectors, in which lines share a common history of contraction, extension or contraction followed by extension (Figure 2.29). If we start out with a randomly oriented set of lines, we will soon see a field containing lines that first shortened, then stretched (yellow fields in Figure 2.29). This field increases in size and will cover more and more of the field of instantaneous stretching. Note that the fields are asymmetric for simple shearing and other non-coaxial deformations, but symmetric for coaxial deformations such as pure shearing (left column in Figure 2.29).

Even though the line rotations discussed here were in a plane perpendicular to the shear plane, simple shearing can cause lines and planes to rotate along many other paths. Figure 2.30 shows how lines (or particles) move along great circles when exposed to simple shearing. This is yet another characteristic feature of simple shearing that may help in separating it from other deformations.

Based on these observations we can list the following characteristics of simple shearing:

Lines along the shear plane do not deform or rotate.
Physical (material) lines rotate faster than the axes of the strain ellipse (generally true).
The sense of line (and plane) rotation is the same for any line orientation.
Lines that are parallel with the ISA rotate (generally true for non-coaxial progressive deformations).
Lines can rotate from the field of instantaneous shortening to that of instantaneous stretching (to produce boudinaged folds) but never the other way (never folded boudins) for steady-state simple shearing.
In the Schmidt net, lines rotate along great circles toward the shearing direction.

2.26 Progressive pure shear

Progressive pure shear or pure shearing is a two-dimensional coaxial deformation, i.e. the strain ellipsoid does not rotate and is fixed with respect to the ISA throughout the deformation (Figures 2.27b and 2.29). Coaxial deformation histories result in coaxial (finite) deformations, which are characterized by symmetric deformation matrices.

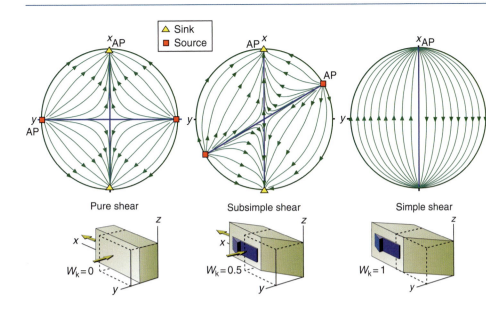

Figure 2.30 Stereographic representation of line rotation during pure shearing, subsimple shearing and simple shearing. AP are flow apophyses, where one is an attractor (sink) and the other is a repeller (source).

For pure shearing the ISA and the fields of instantaneous stretching and shortening are symmetrically arranged with respect to the principal strain axes (axes of the strain ellipse). The following characteristics of pure shearing can be found by studying the different stages shown in Figure 2.31:

The greatest principal strain axis (X) does not rotate and is always parallel with ISA_1.

Lines that are parallel with the ISA do not rotate during the deformation.

Any other line rotates toward X and ISA_1.

The lines rotate both clockwise and anti-clockwise in a symmetric pattern about the ISA.

Lines that are parallel with ISA do not rotate (generally true for coaxial deformation histories).

Lines can rotate from the shortening field and into the stretching field, but not the other way.

2.27 Progressive subsimple shear

Progressive subsimple shear or **subsimple shearing** can be described as a simultaneous combination of simple and pure shearing, and the component of simple shearing gives it a non-coaxial nature.

The same sets of orthogonal lines that were discussed for simple and pure shearing can also be deformed under subsimple shearing. We have chosen a subsimple shearing where $W_k = 0.82$ in Figure 2.32, which means that it has a substantial component of simple shearing. An important difference from simple shearing is that lines in the sector approximately between lines 1 and 2

rotate against the shearing direction. The size of this sector is identical to the angle between the flow apophyses (α) and depends on W_k and θ (see Figure 2.24). The smaller the pure shearing component, the larger the sector of back-rotation. For pure shearing ($W_k = 0$) the two sectors of oppositely rotating lines are of equal size (Figure 2.31a), while for simple shearing one of them has vanished.

Line 1 is parallel to the shear plane and does not rotate, but in contrast to the case of simple shearing it is stretched in the shearing direction. Line 4 rotates clockwise and demonstrates the high degree of non-coaxiality of this version of subsimple shearing. Line 4 rotates from the instantaneous shortening field into that of stretching, where it retains its original length at stage (b). From this point line 4 grows longer while it, together with all the other lines, rotates toward the horizontal apophysis, which is the shearing direction of the simple shearing component.

Line 2 rotates counterclockwise as it shortens, while lines 5 and 6 are stretched on their way toward the horizontal apophysis. At stage (e), line 6 has rotated through the theoretical X-axis, while line 5 has not quite reached that point.

Line 3 is in the field of instantaneous shortening. At stage (c) it parallels ISA_3 and rotates through this axis due to the non-coaxial nature of this deformation. From this point it follows the path already taken by line 4.

These are some of the characteristic features of subsimple shearing:

Lines of any orientation rotate toward a flow apophysis (the shear direction).

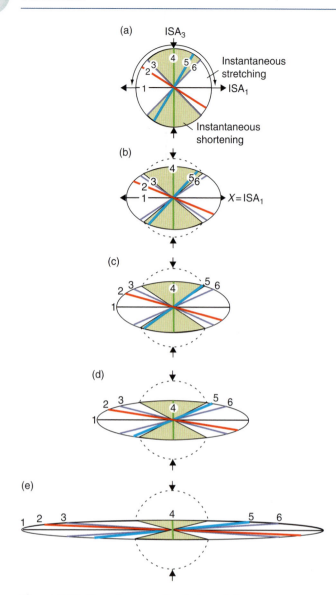

Figure 2.31 Pure shearing of a circle and three sets of orthogonal lines (1–6). The arrows along the circle in (a) indicate the rotation directions for lines during the deformation.

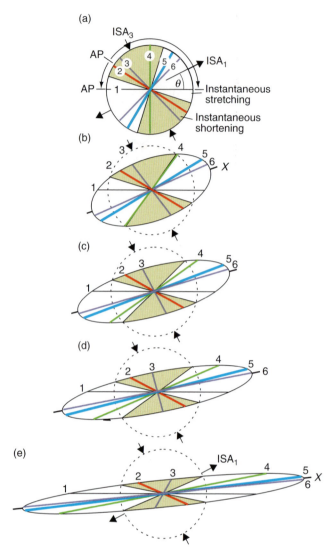

Figure 2.32 Subsimple shearing of a circle and three sets of orthogonal lines (1–6). The arrows along the circle in (a) indicate the rotation directions for lines during the deformation. AP: flow apophysis.

There are two sectors of opposite line rotation.
The size and asymmetry of the fields is controlled by the flow apophyses and indicate W_k and one of the flow apophyses (shear direction).
Lines of any orientation are being stretched or shortened during subsimple shearing.
Lines parallel with ISA rotate. Only those parallel with the shear plane do not rotate.
The long axis X of the strain ellipsoid rotates, but slower than for simple shearing.
Lines rotate from the field of instantaneous shortening into that of instantaneous stretching, but never the other way.

2.28 Simple and pure shear and their scale dependence

Simple and pure shear(ing) have been defined mathematically in this chapter. The practical use of these terms depends on the choice of coordinate system and scale. As an example, consider a simple shear zone in the crust that is 50 m thick and 5 km long. If we want to study this zone it would be natural to place a coordinate system with the x-axis in the shear direction and another axis perpendicular to the zone. We could then study or model the effect of simple shear in this zone. Now, if there should be some tens of such shear zones on a bigger scale dipping in opposite directions, then it would be

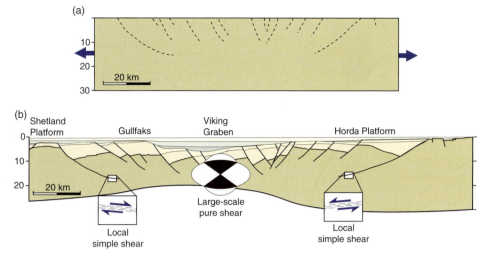

Figure 2.33 Profile across the northern North Sea rift (restored and current). Locally the deformation is simple shear, but is better treated as pure shear on a larger scale.

better to orient the *x*-axis horizontally, which is parallel to the surface (or base) of the crust. In this case the deformation is closer to pure shear, although it contains simple shear zones and perhaps discrete faults on a smaller scale. This can be applied to a profile across a rift, such as the North Sea in Figure 2.33, where the total assemblage of faults and shear zones generates an overall strain that is close to pure shear.

This example illustrates how deformation can be considered as simple shear on one scale and as pure shear on a larger scale. The opposite can also be the case, where there is a partitioning of simple shear on a smaller scale. Some shear zones contain small-scale domains of different deformation types that together constitute an overall simple shear. Hence, pure and simple shear are scale-dependent concepts.

2.29 General three-dimensional deformation

Strain measurements in deformed rocks typically indicate that strain is three-dimensional, i.e. they plot off the diagonal in the Flinn diagram (Figure 2.14). Three-dimensional deformation theory tends to be considerably more complex than that of plane deformation. Pure shear, which is a two-dimensional coaxial deformation, is replaced by a wealth of coaxial deformations, where uniform extension and uniform flattening are end-members (Figure 2.34). Furthermore, there is room for several simple shear components in different directions, all of which can be combined with coaxial strain.

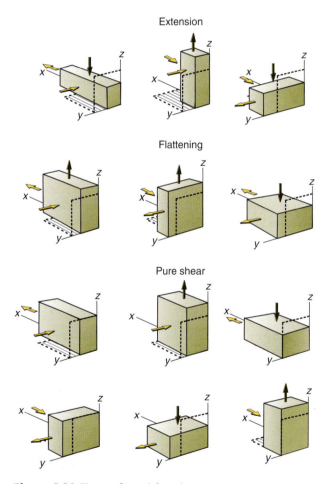

Figure 2.34 Types of coaxial strain.

It is therefore useful to define a few simple three-dimensional deformation types, where two of the principal strain axes of the coaxial strain coincide with the shear plane(s) involved. Figure 2.35 shows a spectrum of three-dimensional deformations that arises

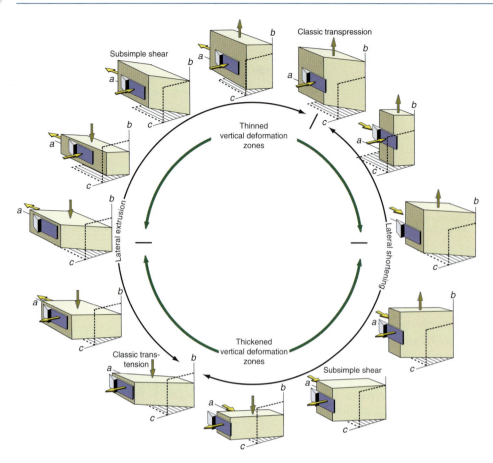

Figure 2.35 Spectrum of deformations based on combinations of a single simple shear (violet arrows) and orthogonal coaxial deformations (yellow). Thinning shear zones occur in the upper half and thickening zones in the lower half of the circle. Based on Tikoff and Fossen (1999).

from a combination of coaxial deformations shown in Figure 2.34 with a simple shear. Transpression and transtension are some of the deformations found in this spectrum and will be discussed later in this book (Chapter 18). Also note that subsimple shear separates different types of three-dimensional deformations in Figure 2.35.

2.30 Stress versus strain

One would perhaps think that the deformation type (pure shear, simple shear, general flattening etc.) is given when the magnitudes and orientations of the three principal stresses are known. This is not so, and strain information combined with observable structures in deformed rocks usually gives more information about the deformation type than does stress information alone.

As an illustration, consider a homogeneous medium that is exposed to linear-viscous (Newtonian) deformation (Chapter 6). In such an idealized medium there will be a simple relationship between stress and strain, and the

ISA will parallel the principal stresses (this is the case for any isotropic medium that deforms according to a so-called power-law stress–strain rate).

Natural rocks are rarely (if ever) homogeneous, and linear-viscous deformation is an idealization of natural flow in rocks. Thus, even if we constrain our considerations to planar deformations, the orientation of the principal stresses does not predict the type of plane strain caused by the stresses in a heterogeneous rock. For a given state of stress, the deformation may be pure shear, simple shear or subsimple shear, depending on boundary conditions or heterogeneities of the deforming material. Figure 2.36 shows how the introduction of a rigid wall rock completely changes the way the rock deforms. In this case the boundary between the rigid rock and the weak, deforming rock is important. A related example is the deformation that can occur along plate boundaries. There will usually be a weak zone between the two plates along which most strain is accommodated, and the orientation of the plate boundaries will be important in addition to their relative motions. Further, the weak, deforming

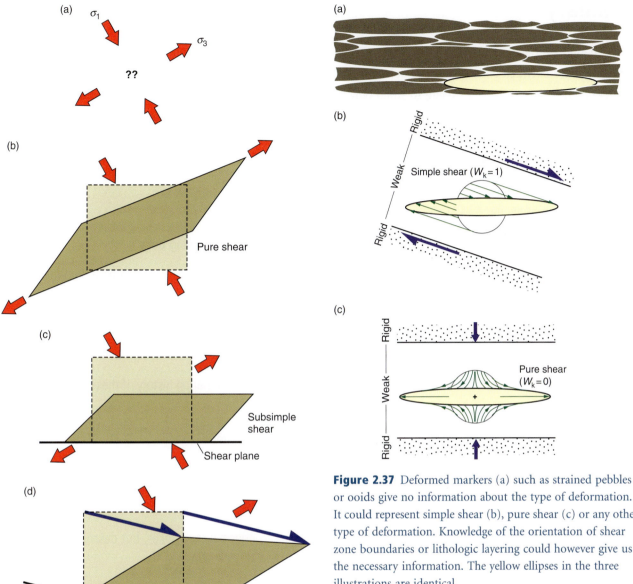

Figure 2.36 Knowledge about the orientations of the principal stresses (a) is not sufficient to predict the resulting deformation. In a perfectly isotropic medium the deformation will be a pure shear as shown in (b). However, if our boundary conditions involve a plane of weakness (potential shear plane), then we could have a subsimple (c) shear. In the special case where the angle between σ_1 and the weak plane is 45° a simple shear could result.

zone may contain faults, soft layers or other heterogeneities that may cause the deformation to partition into domains dominated by coaxial and non-coaxial strains.

Figure 2.37 Deformed markers (a) such as strained pebbles or ooids give no information about the type of deformation. It could represent simple shear (b), pure shear (c) or any other type of deformation. Knowledge of the orientation of shear zone boundaries or lithologic layering could however give us the necessary information. The yellow ellipses in the three illustrations are identical.

If it is difficult or impossible to predict strain from stress alone, can we more easily go from strain to stress? If we know only the shape and orientation of the strain ellipse (ellipsoid in three dimensions), then Figure 2.37 reveals that we have no clue about even the deformation type. However, if we can relate the strain to such things as shear zone boundaries, then we can find the orientation of the ISA (if we assume steady-state deformation). The question then is whether the ISA correspond to the principal stresses. Paleostress analyses rely on the assumption that they are equal, which is not necessarily quite true. We will return to this topic in Chapter 9.

Summary

In this chapter we have covered the basic theory of deformation and strain. Make sure you understand that strain is the difference between undeformed and deformed states, that flow parameters describe any instant during a deformation history and that they relate to strain only when the flow that they describe is considered over a specified time period. The deformation theory of this chapter forms the foundation for most of the later chapters, where structures resulting from strain are treated. Concepts such as pure and simple shear, ISA and coaxiality will reappear throughout the text, and the next chapter gives some ideas about how to retrieve information about strain from deformed rocks. Before that, we review some of the many important points from this chapter:

- Deformation is strictly the sum of strain, rigid rotation and translation.

- Each of these components results in a change in position of particles in a deforming body.

- Mathematically, homogeneous deformation is a linear transformation and can be represented by a deformation (or transformation) matrix that connects the deformed and undeformed states.

- Deformation history deals with the evolution from the undeformed to the deformed state.

- Strain alone tells us nothing about the deformation history (the way that strain accumulated over time).

- Flow parameters describe the situation or flow at any instant during the deformation history. If these parameters stay fixed during deformation we have steady-state deformation.

- Flow apophyses separate different fields of flow, or different domains of particle paths. They are orthogonal for coaxial strain, parallel for simple shear and oblique for subsimple shear.

- The fastest and slowest stretching directions (ISA_1 and ISA_3) are fixed at $45°$ to the shear plane for progressive simple shear. They bisect the fields of instantaneous stretching and shortening.

- The instantaneous stretching axes (ISA) are not necessarily equal to principal stress axes, but describe how strain accumulates at a given instance during the deformation history.

- The deformation history can be described by intervals, each with its own incremental deformation matrix and strain ellipsoid.

- Flow parameters describe deformation at an instant, while strain accumulates over time and is directly related to flow parameters only if the latter remain constant during the deformation history.

Review questions

1. List and explain the flow parameters discussed in this chapter.
2. What is the deformation called if flow parameters are constant throughout the deformation history?
3. Are ISA equal to stress axes?
4. What is the difference between angular shear and shear strain?
5. What is plane strain and where does it plot in the Flinn diagram?
6. Give examples of plane strain.

7. What is meant by particle paths?

8. What happens to the principal strain axes during pure shearing?

9. What is meant by the expression non-coaxial deformation history?

10. What is the kinematic vorticity number?

11. What set of material lines do not rotate or change length during simple shear?

E-MODULE

 The first part of the e-learning module called *Deformation* is recommended for this chapter. Also see Appendix A for more details about the deformation matrix.

FURTHER READING

General deformation theory

Means, W. D., 1976, *Stress and Strain: Basic Concepts of Continuum Mechanics for Geologists.* New York: Springer-Verlag.

Means, W. D., 1990, Kinematics, stress, deformation and material behavior. *Journal of Structural Geology* **12**: 953–971.

Ramsay, J. G., 1980, Shear zone geometry: a review. *Journal of Structural Geology* **2**: 83–99.

Tikoff, B. and Fossen, H., 1999, Three-dimensional reference deformations and strain facies. *Journal of Structural Geology* **21**: 1497–1512.

The deformation matrix

Flinn, D., 1979, The deformation matrix and the deformation ellipsoid. *Journal of Structural Geology* **1**: 299–307.

Fossen, H. and Tikoff, B., 1993, The deformation matrix for simultaneous simple shearing, pure shearing, and volume change, and its application to transpression/transtension tectonics. *Journal of Structural Geology* **15**: 413–422.

Ramberg, H., 1975, Particle paths, displacement and progressive strain applicable to rocks. *Tectonophysics* **28**: 1–37.

Sanderson, D. J., 1976, The superposition of compaction and plane strain. *Tectonophysics* **30**: 35–54.

Sanderson, D. J., 1982, Models of strain variations in nappes and thrust sheets: a review. *Tectonophysics* **88**: 201–233.

The strain ellipse

Flinn, D., 1963, On the symmetry principle and the deformation ellipsoid. *Geological Magazine* **102**: 36–45.

Treagus, S. H. and Lisle, R. J., 1997, Do principal surfaces of stress and strain always exist? *Journal of Structural Geology* **19**: 997–1010.

The Mohr circle for strain

Passchier, C. W., 1988, The use of Mohr circles to describe non-coaxial progressive deformation. *Tectonophysics* **149**: 323–338.

Volumetric strain

Passchier, C. W., 1991, The classification of dilatant flow types. *Journal of Structural Geology* **13**: 101–104.

Xiao, H. B., and Suppe, J., 1989, Role of compaction in listric shape of growth normal faults. *American Association of Petroleum Geologists Bulletin* **73**: 777–786.

Vorticity and deformation history

Elliott, D., 1972, Deformation paths in structural geology. *Geological Society of America Bulletin* **83**: 2621–2638.

Ghosh, S. K., 1987. Measure of non-coaxiality. *Journal of Structural Geology* **9**: 111–113.

Jiang, D., 1994, Vorticity determination, distribution, partitioning and the heterogeneity and non-steadiness of natural deformations. *Journal of Structural Geology* **16**: 121–130.

Passchier, C. W., 1990, Reconstruction of deformation and flow parameters from deformed vein sets. *Tectonophysics* **180**: 185–199.

Talbot, C. J., 1970, The minimum strain ellipsoid using deformed quartz veins. *Tectonophysics* **9**: 47–76.

Tikoff, B. and Fossen, H., 1995, The limitations of three-dimensional kinematic vorticity analysis. *Journal of Structural Geology* **17**: 1771–1784.

Truesdell, C., 1953, Two measures of vorticity. *Journal of Rational Mechanics and Analysis* **2**: 173–217.

Wallis, S. R., 1992, Vorticity analysis in a metachert from the Sanbagawa Belt, SW Japan. *Journal of Structural Geology* **12**: 271–280.

The stress–strain relationship

Tikoff, B. and Wojtal, S. F., 1999, Displacement control of geologic structures. *Journal of Structural Geology* **21**: 959–967.

Chapter 3

Strain in rocks

Strain can be retrieved from rocks through a range of different methods. Much attention has been paid to one-, two- and three-dimensional strain analyses in ductilely deformed rocks, particularly during the last half of the twentieth century, when a large portion of the structural geology community had their focus on ductile deformation. Strain data were collected or calculated in order to understand such things as thrusting in orogenic belts and the mechanisms involved during folding of rock layers. The focus of structural geology has changed and the field has broadened during the last couple of decades. Today strain analysis is at least as common in faulted areas and rift basins as in orogenic belts. While we will return to strain in the brittle regime in Chapter 20, we will here concentrate on classic aspects of how strain is measured and quantified in the ductile regime

3.1 Why perform strain analysis?

It can be important to retrieve information about strain from deformed rocks. First of all, strain analysis gives us an opportunity to explore the state of strain in a rock and to map out strain variations in a sample, an outcrop or a region. Strain data are important in the mapping and understanding of shear zones in orogenic belts. Strain measurements can also be used to estimate the amount of offset across a shear zone. As will be discussed in Chapter 15, it is possible to extract important information from shear zones if strain is known.

In many cases it is useful to know if the strain is planar or three dimensional. If planar, an important criterion for section balancing is fulfilled, be it across orogenic zones or extensional basins. The shape of the strain ellipsoid may also contain information about how the deformation occurred. Oblate (pancake-shaped) strain in an orogenic setting may, for example, indicate flattening strain related to gravity-driven collapse rather than classic push-from-behind thrusting.

The orientation of the strain ellipsoid is also important, particularly in relation to rock structures. In a shear zone setting, it may tell us if the deformation was simple shear or not (Chapter 15). Strain in folded layers helps us to understand fold-forming mechanism(s) (Chapter 11). Studies of deformed reduction spots in slates give good estimates on how much shortening has occurred across the foliation in such rocks (Chapter 12), and strain markers in sedimentary rocks can sometimes allow for reconstruction of original sedimentary thickness. Strain will follow us through most of this book.

3.2 Strain in one dimension

One-dimensional strain analyses are concerned with changes in length and therefore the simplest form of strain analysis we have. If we can reconstruct the original length of an object or linear structure we can also calculate the amount of stretching or shortening in that direction. Objects revealing the state of strain in a deformed rock are known as **strain markers**. Examples of strain markers indicating change in length are boudinaged dikes or layers, and minerals or linear fossils such as belemnites or graptolites that have been elongated, such as the stretched Swiss belemnites shown in Figure 3.1. Or it could be a layer shortened by folding. It could even be a faulted reference horizon on a geologic or seismic profile, as will be discussed in Chapter 20. The horizon

Figure 3.1 Two elongated belemnites in Jurassic limestone in the Swiss Alps. The different ways that the two belemnites have been stretched give us some two-dimensional information about the strain field: the upper belemnite has experienced sinistral shear strain while the lower one has not and must be close to the maximum stretching direction.

may be stretched by normal faults or shortened by reverse faults, and the overall strain is referred to as **brittle strain**. One-dimensional strain is revealed when the horizon, fossil, mineral or dike is restored to its pre-deformational state.

3.3 Strain in two dimensions

In **two-dimensional strain analyses** we look for sections that have objects of known initial shape or contain linear markers with a variety of orientations (Figure 3.1). Strained reduction spots of the type shown in Figure 3.2 are perfect, because they tend to have spherical shapes where they are undeformed. There are also many other types of objects that can be used, such as sections through conglomerates, breccias, corals, reduction spots, oolites, vesicles, pillow lavas (Figure 3.3), columnar basalt, plutons and so on. Two-dimensional strain can also be calculated from one-dimensional data that represent different directions in the same section. A typical example would be dikes with different orientations that show different amounts of extension.

Strain extracted from sections is the most common type of strain data, and sectional data can be combine to estimate the three-dimensional strain ellipsoid.

Changes in angles

Strain can be found if we know the original angle between sets of lines. The original angular relations between structures such as dikes, foliations and bedding

Figure 3.2 Reduction spots in Welsh slate. The light spots formed as spherical volumes of bleached (chemically reduced) rock. Their new shapes are elliptical in cross-section and oblate (pancake-shaped) in three dimensions, reflecting the tectonic strain in these slates.

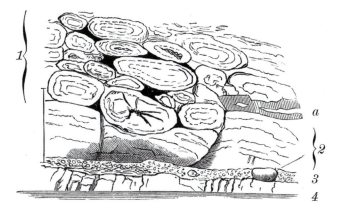

Figure 3.3 Section through a deformed Ordovician pahoe-hoe lava. The elliptical shapes were originally more circular, and Hans Reusch, who made the sketch in the 1880s, understood that they had been flattened during deformation. The R_f/ϕ, center-to-center, and Fry methods would all be worth trying in this case.

are sometimes found in both undeformed and deformed states, i.e. outside and inside a deformation zone. We can then see how the strain has affected the angular relationships and use this information to estimate strain. In other cases orthogonal lines of symmetry found in undeformed fossils such as trilobites, brachiopods and worm burrows (angle with layering) can be used to determine the angular shear in some deformed sedimentary rocks. In general, all we need to know is the change in angle between sets of lines and that there is no strain partitioning due to contrasting mechanical properties of the objects with respect to the enclosing rock.

If the angle was 90° in the undeformed state, the change in angle is the local angular shear ψ (Section 2.8). If, as we recall from the previous chapter, the two originally orthogonal lines remain orthogonal after the deformation, then they must represent the principal strains and thus the orientation of the strain ellipsoid. Observations of variously oriented line sets thus give information about the strain ellipse or ellipsoid. All we need is a useful method. Two of the most common methods used to find strain from initially orthogonal lines are known as the Wellman and Breddin methods, and are presented in the following sections.

The Wellman method

This method dates back to 1962 and is a geometric construction for finding strain in two dimensions (in a section). It is typically demonstrated on fossils with orthogonal lines of symmetry in the undeformed state. In Figure 3.4a we use the hinge and symmetry lines of brachiopods. A line of reference must be drawn (with arbitrary orientation) and pairs of lines that were orthogonal in the unstrained state are identified. The reference line must have two defined endpoints, named A and B in Figure 3.4b. A pair of lines is then drawn parallel to both the hinge line and symmetry line for each fossil, so that they intersect at the endpoints of the reference line. The other points of intersection are marked (numbered 1–6 in Figure 3.4b, c). If the rock is unstrained, the lines will define rectangles. If there is a strain involved, they will define parallelograms. To find the strain ellipse, simply fit an ellipse to the numbered corners of the parallelograms (Figure 3.4c). If no ellipse can be fitted to the corner points of the rectangles the strain is heterogeneous or, alternatively, the measurement or assumption of initial orthogonality is false. The challenge with this method is, of course, to find enough fossils or other features with initially orthogonal lines – typically 6–10 are needed.

The Breddin graph

We have already stated that the angular shear depends on the orientation of the principal strains: the closer the deformed orthogonal lines are to the principal strains, the lower the angular shear. This fact is utilized in a method first published by Hans Breddin in 1956 in German (with some errors). It is based on the graph shown in Figure 3.5, where the angular shear changes with orientation and strain magnitude R. Input are the

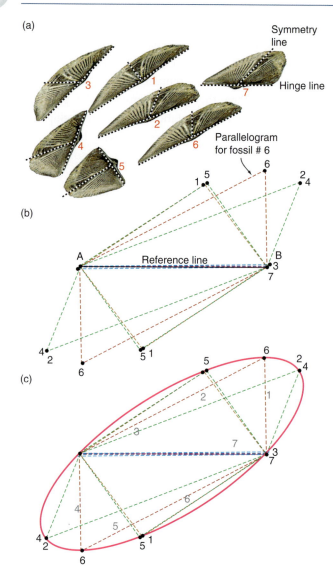

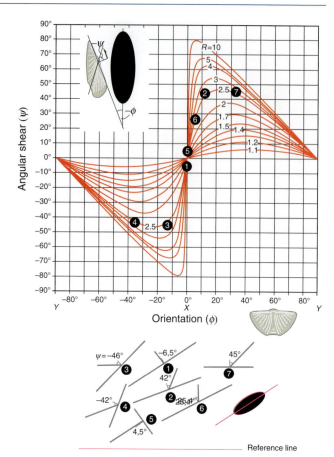

Figure 3.4 Wellman's method involves construction of the strain ellipse by drawing parallelograms based on the orientation of originally orthogonal pairs of lines. The deformation was produced on a computer and is a homogeneous simple shear of $\gamma = 1$. However, the strain ellipse itself tells us nothing about the degree of coaxiality: the same result could have been attained by pure shear.

Figure 3.5 The data from the previous figure plotted in a Breddin graph. The data points are close to the curve for $R = 2.5$.

angular shears and the orientations of the sheared line pairs with respect to the principal strains. These data are plotted in the so-called Breddin graph and the R-value (ellipticity of the strain ellipse) is found by inspection (Figure 3.5). This method may work even for only one or two observations.

In many cases the orientation of the principal axes is unknown. In such cases the data are plotted with respect to an arbitrarily drawn reference line. The data are then moved horizontally on the graph until they fit one of the curves, and the orientations of the strain axes are then found at the intersections with the horizontal axis (Figure 3.5). In this case a larger number of data are needed for good results.

Elliptical objects and the R_f/ϕ-method

Objects with initial circular (in sections) or spherical (in three dimensions) geometry are relatively uncommon, but do occur. Reduction spots and ooliths perhaps form the most perfect spherical shapes in sedimentary rocks. When deformed homogeneously, they are transformed into ellipses and ellipsoids that reflect the local finite strain. Conglomerates are perhaps more common and contain clasts that reflect the finite strain. In contrast to oolites and reduction spots, few pebbles or cobbles in a conglomerate are spherical in the undeformed state. This will of course influence their shape in the deformed state and causes a challenge in strain analyses. However, the clasts tend to have their long axes in a spectrum of orientations in the undeformed state, in which case

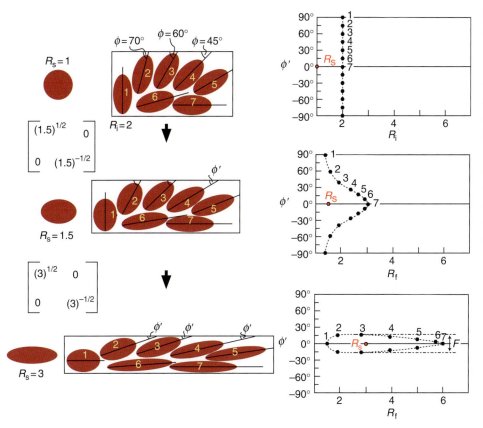

Figure 3.6 The R_f/ϕ method illustrated. The ellipses have the same ellipticity (R_i) before the deformation starts. The R_f–ϕ diagram to the right indicates that $R_i = 2$. A pure shear is then added with $R_s = 1.5$ followed by a pure shear strain of $R_s = 3$. The deformation matrices for these two deformations are shown. Note the change in the distribution of points in the diagrams to the right. R_s in the diagrams is the actual strain that is added. Modified from Ramsay and Huber (1983).

methods such as the R_f/ϕ-method may be able to take the initial shape factor into account.

The R_f/ϕ-method was first introduced by John Ramsay in his well-known 1967 textbook and was later improved. The method is illustrated in Figure 3.6. The markers are assumed to have approximately elliptical shapes in the deformed (and undeformed) state, and they must show a significant variation in orientations for the method to work.

The R_f/ϕ-method handles initially non-spherical markers, but the method requires a significant variation in the orientations of their long axes.

The ellipticity (X/Y) in the undeformed (initial) state is called R_i. In our example (Figure 3.6) $R_i = 2$. After a strain R_s the markers exhibit new shapes. The new shapes are different and depend on the initial orientation of the elliptical markers. The new (final) ellipticity for each deformation marker is called R_f and the spectrum of R_f-values is plotted against their orientations, or more specifically against the angle ϕ' between the long axis of the ellipse and a reference line (horizontal in Figure 3.6). In our example we have applied two

increments of pure shear to a series of ellipses with different orientations. All the ellipses have the same initial shape $R_i = 2$, and they plot along a vertical line in the upper right diagram in Figure 3.6. Ellipse 1 is oriented with its long axis along the minimum principal strain axis, and it is converted into an ellipse that shows less strain (lower R_f-value) than the true strain ellipse (R_s). Ellipse 7, on the other hand, is oriented with its long axis parallel to the long axis of the strain ellipse, and the two ellipticities are added. This leads to an ellipticity that is higher than R_s. When $R_s = 3$, the true strain R_s is located somewhere between the shape represented by ellipses 1 and 7, as seen in Figure 3.6 (lower right diagram).

For $R_s = 1.5$ we still have ellipses with the full spectrum of orientations ($-90°$ to $90°$; see middle diagram in Figure 3.6), while for $R_s = 3$ there is a much more limited spectrum of orientations (lower graph in Figure 3.6). The scatter in orientation is called the fluctuation F. An important change happens when ellipse 1, which has its long axis along the Z-axis of the strain ellipsoid, passes the shape of a circle ($R_s = R_i$,) and starts to develop an ellipse whose long axis is parallel to X. This happens when $R_s = 2$, and for larger strains the data points define a circular shape. Inside this shape is the strain R_s that we are interested in. But where exactly is R_s? A simple average of the maximum and minimum R_f-values would depend on the original distribution of orientations. Even if the initial distribution is random, the average R-value would be too high, as high values tend to be overrepresented (Figure 3.6, lower graph).

To find R_s we have to treat the cases where $R_s > R_i$ and $R_s < R_i$ separately. In the latter case, which is represented by the middle graph in Figure 3.6, we have the following expressions for the maximum and minimum value for R_f:

$$R_{fmax} = R_s R_i$$

$$R_{fmin} = R_i / R_s$$

Solving for R_i and R_s gives

$$R_s = (R_{fmax} / R_{fmin})^{1/2}$$

$$R_i = (R_{fmax} R_{fmin})^{1/2}$$

which represent expressions for both the strain related to the deformation and the initial ellipticity.

For higher-strain cases, where $R_s < R_i$, we obtain

$$R_{fmax} = R_s R_i$$
$$R_{fmin} = R_s / R_i$$

Solving for R_s gives

$$R_s = (R_{fmax} R_{fmin})^{1/2}$$
$$R_i = (R_{fmax} / R_{fmin})^{1/2}$$

In both cases the orientation of the long (X) axis of the strain ellipse is given by the location of the maximum R_f-values. Strain could also be found by fitting the data to pre-calculated curves for various values for R_i and R_s. In practice, such operations are most efficiently done by means of computer programs.

The example shown in Figure 3.6 and discussed above is idealized in the sense that all the undeformed elliptical markers have identical ellipticity. What if this were not the case, i.e. some markers were more elliptical than others? Then the data would not have defined a nice curve but a cloud of points in the R_f/ϕ-diagram. Maximum and minimum R_f-values could still be found and strain could be calculated using the equations above. The only change in the equation is that R_i now

represents the maximum ellipticity present in the undeformed state.

Another complication that may arise is that the initial markers may have had a restricted range of orientations. Ideally, the R_f/ϕ-method requires the elliptical objects to be more or less randomly oriented prior to deformation. Conglomerates, to which this method commonly is applied, tend to have clasts with a preferred orientation. This may result in an R_f–ϕ plot in which only a part of the curve or cloud is represented. In this case the maximum and minimum R_f-values may not be representative, and the formulas above may not give the correct answer and must be replaced by a computer-based iterative retrodeformation method where X is input. However, many conglomerates have a few clasts with initially anomalous orientations that allow the use of R_f/ϕ analysis.

Center-to-center method

This method, here demonstrated in Figure 3.7, is based on the assumption that circular objects have a more or less statistically uniform distribution in our section(s). This means that the distances between neighboring particle centers were fairly constant before deformation. The particles could represent sand grains in well-sorted sandstone, pebbles, ooids, mud crack centers, pillow-lava or pahoe-hoe lava centers, pluton centers or other objects that are of similar size and where the centers are easily definable. If you are uncertain about how closely your section complies with this criterion, try anyway. If the method yields a reasonably well-defined ellipse, then the method works.

The method itself is simple and is illustrated in Figure 3.7: Measure the distance and direction from the center of an ellipse to those of its neighbors. Repeat this for all ellipses and graph the distance d' between the centers and the angles α' between the center tie lines and a reference line. A straight line occurs if the section is unstrained, while a deformed section yields a curve with maximum (d'_{max}) and minimum values (d'_{min}). The ellipticity of the strain ellipse is then given by the ratio: $R_s = (d'_{max})/(d'_{min})$.

The Fry method

A quicker and visually more attractive method for finding two-dimensional strain was developed by Norman Fry at the end of the 1970s. This method, illustrated in Figure 3.8, is based on the center-to-center method and

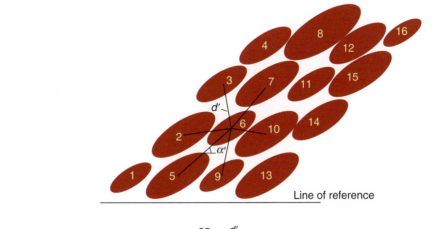

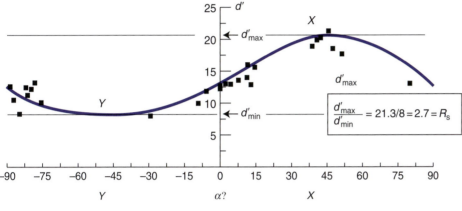

is most easily dealt with using one of several available computer programs. It can be done manually by placing a tracing overlay with a coordinate origin and pair of reference axes on top of a sketch or picture of the section. The origin is placed on a particle center and the centers of all other particles (not just the neighbors) are marked on the tracing paper. The tracing paper is then moved, without rotating the paper with respect to the section, so that the origin covers a second particle center, and the centers of all other particles are again marked on the tracing paper. This procedure is repeated until the area of interest has been covered. For objects with a more or less uniform distribution the result will be a visual representation of the strain ellipse. The ellipse is the void area in the middle, defined by the point cloud around it (Figure 3.8c).

The Fry method, as well as the other methods presented in this section, outputs two-dimensional strain. Three-dimensional strain is found by combining strain estimates from two or more sections through the deformed rock volume. If sections can be found that each contain two of the principal strain axes, then two

sections are sufficient. In other cases three or more sections are needed, and the three-dimensional strain must be calculated by use of a computer.

3.4 Strain in three dimensions

A complete strain analysis is **three-dimensional**. Three-dimensional strain data are presented in the Flinn diagram or similar diagrams that describe the shape of the strain ellipsoid, also known as the **strain geometry**. In addition, the orientation of the principal strains can be presented by means of stereographic nets. Direct field observations of three-dimensional strain are rare. In almost all cases, analysis is based on two-dimensional strain observations from two or more sections at the same locality (Figure 3.9). A well-known example of three-dimensional strain analysis from deformed conglomerates is presented in Box 3.1.

In order to quantify ductile strain, be it in two or three dimensions, the following conditions need to be met:

(a)   (b)   (c)

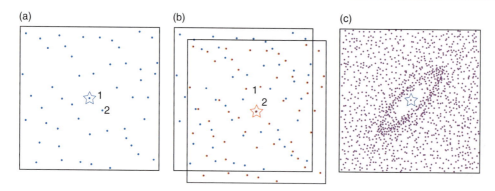

Figure 3.8 The Fry method performed manually. (a) The centerpoints for the deformed objects are transferred to a transparent overlay. A central point (1 on the figure) is defined. (b) The transparent paper is then moved to another of the points (point 2) and the centerpoints are again transferred onto the paper (the overlay must not be rotated). The procedure is repeated for all of the points, and the result (c) is an image of the strain ellipsoid (shape and orientation). Based on Ramsay and Huber (1983).

The strain must be homogeneous at the scale of observation, the mechanical properties of the objects must have been similar to those of their host rock during the deformation, and we must have a reasonably good knowledge about the original shape of strain markers.

The first point is obvious. If the strain is heterogeneous we have to look at another scale, either a larger one where the heterogeneities vanish, or subareas where strain can be considered to be approximately homogeneous. The latter was done in the example in Box 3.2, where a strain pattern was mapped out that could be related to a larger structure. This example shows how mapping of the state of strain can help us to understand the formation of larger-scale structures, in this case the history of folding.

The second point is an important one. For ductile rocks it means that the object and its surroundings must have had the same competence or viscosity (see Chapter 5). Otherwise the strain recorded by the object would be different from that of its surroundings. This effect is one of several types of **strain partitioning**, where the overall strain is distributed unevenly in terms of intensity and/or geometry in a rock volume. As an example, we mark a perfect circle on a piece of clay before flattening it between two walls. The circle transforms passively into an ellipse that reveals the two-dimensional strain if the deformation is homogeneous. If we embed a colored sphere of the same clay, then it would again deform along with the rest of the clay, revealing the three-dimensional strain involved. However, if we put a stiff marble in the clay the result is

quite different. The marble remains unstrained while the clay around it becomes more intensely and heterogeneously strained than in the previous case. In fact, it causes a more heterogeneous strain pattern to appear. Strain markers with the same mechanical properties as the surroundings are called **passive strain markers** because they deform passively along with their surroundings. Those that have anomalous mechanical properties respond differently than the surrounding medium to the overall deformation, and such markers are called **active strain markers**.

An example of data from active strain markers is shown in Figure 3.10. These data were collected from a deformed polymictic conglomerate where three-dimensional strain has been estimated from different clast types in the same rock and at the same locality. Clearly, the different clast types

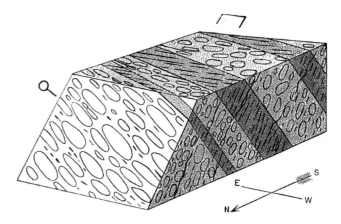

Figure 3.9 Three-dimensional strain expressed as ellipses on different sections through a conglomerate. The foliation (*XY*-plane) and the lineation (*X*-axis) are annotated. This illustration was published in 1888, but what are now routine strain methods were not developed until the 1960s.

BOX 3.1 | DEFORMED QUARTZITE CONGLOMERATES

Quartz or quartzite conglomerates with a quartzite matrix are commonly used for strain analyses. The more similar the mineralogy and grain size of the matrix and the pebbles, the less deformation partitioning and the better the strain estimates. A classic study of deformed quartzite conglomerates is Jake Hossack's study of the Norwegian Bygdin conglomerate, published in 1968. Hossack was fortunate – he found natural sections along the principal planes of the strain ellipsoid at each locality. Putting the sectional data together gave the three-dimensional state of strain (strain ellipsoid) for each locality. Hossack found that strain geometry and intensity varies within his field area. He related the strain pattern to static flattening under the weight of the overlying Caledonian Jotun Nappe. Although details of his interpretation may be challenged, his work demonstrates how conglomerates can reveal a complicated strain pattern that otherwise would have been impossible to map.

Hossack noted the following sources of error:

- Inaccuracy connected with data collection (sections not being perfectly parallel to the principal planes of strain and measuring errors).

- Variations in pebble composition.

- The pre-deformational shape and orientation of the pebbles.

- Viscosity contrasts between clasts and matrix.

- Volume changes related to the deformation (pressure solution).

- The possibility of multiple deformation events.

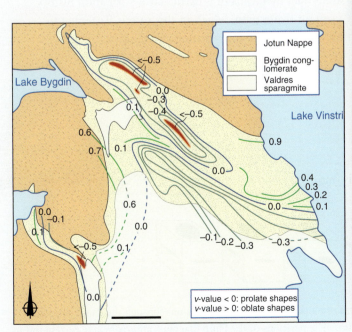

Hossack's strain map from the Bygdin area, Norway.

The Bygdin conglomerate.

BOX 3.2 | STRAIN AROUND A FOLD

Deformed conglomerates are an important source of strain data in deformed rocks because conglomerates are relatively common and contain large numbers of objects (clasts). An example is shown where strain was evaluated at several stations around a folded conglomerate layer, deformed under greenschist facies conditions. It was found that the long limbs were totally dominated by flattening strain (oblate strain geometry) while there was a change toward constrictional strain in the hinge and short limb area. This information would have been difficult to achieve without mesoscopic strain markers, because the rock is recrystallized so that the original sand grain boundaries are obliterated.

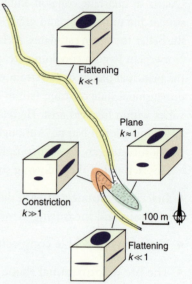

Map of the conglomerate layer. Note the thickened short fold limb.

The strain distribution then had to be explained, and was found to fit a model where an already flattened conglomerate layer is rotated into the field of shortening during shearing. A dextral shear rotates the foliation and the oblate clasts into the shortening field, which makes the Y-axis shrink. This takes the strain ellipsoid across the plane strain diagonal of the Flinn diagram and into the constrictional field ($k > 1$). At this point we are on the inverted limb or at the lower fold hinge. The process continues, and the strain ellipse again becomes flattened. This model explains strain data by means of a particular deformation history, defining a certain strain path, which in this case is flattening to constriction and then back to flattening strain again.

The conglomerate in a constrictional state of strain.

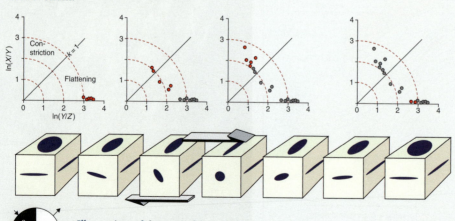

Illustration of the strain history in terms of block diagrams showing sections through the strain ellipsoid and strain data plotted in Flinn diagrams. The positions of the Flinn diagrams approximately correspond to those of the block diagrams below. Also shown is the direction of the instantaneous stretching axes and fields of instantaneous contraction (black) and extension for dextral simple shear.

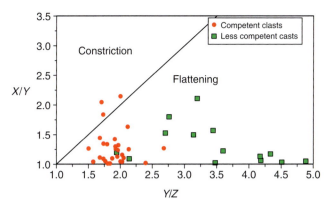

Figure 3.10 Strain obtained from deformed conglomerates, plotted in the Flinn diagram. Different pebble types show different shapes and finite strains. Polymict conglomerate of the Utslettefjell Formation, Stord, southwest Norway. Data collected by D. Kirschner, R. Malt and the author.

have recorded different amounts of strain. Competent (stiff) granitic clasts are less strained than less competent greenstone clasts. This is seen using the fact that strain intensity generally increases with increasing distance from the origin in Flinn space. But there is another interesting thing to note from this figure: It seems that competent clasts plot higher in the Flinn diagram (Figure. 3.10) than incompetent ("soft") clasts, meaning that competent clasts take on a more prolate shape. Hence, not only strain intensity but also strain geometry may vary according to the mechanical properties of strain markers.

The way that the different markers behave depends on factors such as their mineralogy, preexisting fabric, grain size, water content and temperature–pressure conditions

at the time of deformation. In the case of Figure 3.10, the temperature–pressure regime is that of lower to middle greenschist facies. At higher temperatures, quartz-rich rocks are more likely to behave as "soft" objects, and the relative positions of clast types in Flinn space are expected to change.

The last point above also requires attention: the initial shape of a deformed object clearly influences its postdeformational shape. If we consider two-dimensional objects such as sections through oolitic rocks, sandstones or conglomerates, the R_f/ϕ method discussed above can handle this type of uncertainty. It is better to measure up two or more sections through a deformed rock using this method than dig out an object and measure its three-dimensional shape. The single object could have an unexpected initial shape (conglomerate clasts are seldom perfectly spherical or elliptical), but by combining numerous measurements in several sections we get a statistical variation that can solve or reduce this problem.

***

Three-dimensional strain is usually found by combining two-dimensional data from several differently oriented sections.

***

There are now computer programs that can be used to extract three-dimensional strain from sectional data. If the sections each contain two of the principal strain axes everything becomes easy, and only two are strictly needed (although three would still be good). Otherwise, strain data from at least three sections are required.

Summary

Strain markers in deformed rocks reveal how much the rock has been strained, and information about the nature of the deformation (e.g. flattening versus constriction, and the direction of strain axes) can be obtained. This is very useful when trying to understand what has happened to a deformed region, and when searching for a model for the deformation. The way that rocks accumulate strain depends on the stress field, boundary conditions, physical properties of the rock as well as external factors such as temperature, pressure and state of stress. In the next few chapters we will explore some of these relationships. Some key points to review and questions to address:

- Strain is only revealed by means of new deformation structures (cleavage, shear zones, fractures) or by means of preexisting markers that have changed shape during deformation.

- Strain analysis requires knowledge about the shape or geometry of strain markers before the deformation initiated.

- Objects that were circular or spherical before deformation are ideal, but non-spherical objects with a spread in initial orientations can also be used.

- Several techniques and computer codes exist that can help us extract strain from deformed rocks.

- Always look for objects, layers or linear features that can reveal strain in deformed rocks.

Review questions

1. What is meant by the term "strain marker"? Give examples.

2. What information can we obtain from linear or planar strain markers?

3. What is the effect of a viscosity (competence) difference between strain markers and the matrix?

4. How can we deal with pre-deformational fabrics, for example in conglomerate pebbles?

5. What is needed to find shear strain in a rock?

6. Give some serious concerns (pitfalls) regarding strain analysis.

7. How can we find three-dimensional strain from a deformed conglomerate?

8. Shear zones are expression of heterogeneous strain. How can we perform strain analyses in shear zones?

9. What are meant by passive and active strain markers?

10. What is meant by strain partitioning in this context?

E-MODULE

The e-learning module called *Strain in rocks* is recommended for this chapter.

FURTHER READING

Strain techniques in more detail

Lisle, R., 1985, *Geological Strain Analysis*. Amsterdam: Elsevier.

Ramsay, J. G. and Huber, M. I., 1983, *The Techniques of Modern Structural Geology. Vol. 1: Strain Analysis*. London: Academic Press.

Strain ellipsoid from sectional data

De Paor, D. G., 1990, Determination of the strain ellipsoid from sectional data. *Journal of Structural Geology* **12**: 131–137.

Erslev, E. A., 1988. Normalized center-to-center strain analysis of packed aggregates. *Journal of Structural Geology* **10**: 201–209.

Three-dimensional strain

Bhattacharyya, P. and Hudleston, P., 2001, Strain in ductile shear zones in the Caledonides of northern Sweden: a three-dimensional puzzle. *Journal of Structural Geology* **23**: 1549–1565.

Holst, T. B. and Fossen, H., 1987, Strain distribution in a fold in the West Norwegian Caledonides. *Journal of Structural Geology* **9**: 915–924.

Hossack, J., 1968, Pebble deformation and thrusting in the Bygdin area (Southern Norway). *Tectonophysics* **5**: 315–339.

Strine, M. and Wojtal, S. F. 2004, Evidence for non-plane strain flattening along the Moine thrust, Loch Srath nan Aisinnin, North-West Scotland. *Journal of Structural Geology* **26**, 1755–1772.

Strain associated with cleavage

Goldstein, A., Knight, J. and Kimball, K., 1999, Deformed graptolites, finite strain and volume loss during cleavage formation in rocks of the taconic slate belt, New York and Vermont, U.S.A. *Journal of Structural Geology* **20**: 1769–1782.

Chapter 4

Stress

In the previous chapter we looked at how strain can be observed and measured in deformed rocks. The closely related concept of stress is a much more abstract concept, as it can never be observed directly. We have to use observations of strain (preferentially very small strains) to say something about stress. In other words, the deformation structures that we can observe tell us something about the stress field that the rock experienced. The relation is not straightforward, and not even the most precise knowledge of the state of stress can predict the resulting deformation structures unless additional information, such as mechanical or physical properties of the rock, temperature, pressure and physical boundary conditions, is added. The most basic concepts of stress are presented here, before looking at stress in the lithosphere and the relations between stress, strain and physical properties in the following two chapters.

4.1 Definitions, magnitudes and units

The terms pressure and stress are often used interchangeably, but as structural geologists we need to use these terms more carefully. In geology, use of the term **pressure** (*p*) is generally limited to media with no or very low shear resistance (fluids), while **stress** (*σ*) is used when dealing with media with a minimum of shear resistance (rocks). To check if a medium has a **shear resistance**, put some of it between your hands and move them in parallel but opposite directions. The resistance you feel reflects the shear resistance. Water will have no shear resistance (repeat the above exercise in a swimming pool, with just water between your hands), while clay and loose sand will resist shearing.

In a buried porous sandstone layer we can talk about both pressure and stress: it has a certain **pore pressure** and it is in a certain state of **stress**. They are both related to the external forces that affect the rock volume.

There are two different types of forces. One type affects the entire volume of a rock, the outside as well as the inside, and is known as **body forces**. Body forces define three-dimensional fields. The most important type of body force in structural geology is gravity. Another example is magnetic forces.

The other type of force acts on surfaces only and is referred to as **surface forces**. Surface forces originate when one body pushes or pulls another body. The force that acts across the contact area between the two bodies is a surface force. Surface forces are of great importance during deformation of rocks. In a similar way we can talk about stress on a surface and state of stress at a point. Stress on a plane is a vector quantity, while stress at a point is a second-order tensor.

Stress on a surface is a vector (first-order tensor), while state of stress at a point is a second-order tensor.

Engineers and rock mechanics-oriented geologists may refer to stress on a surface as **traction** and reserve the term stress to mean the state of stress at a point in a body. Also, as geologists we need to be aware of these two different uses and avoid confusion between those two meanings of stress.

4.2 Stress on a surface

The **stress on a surface** such as a fracture or a grain–grain contact is a vector (**σ**) that can be defined as the ratio between a force (*F*) and the area (*A*) across which the force acts. The stress that acts on a point on the surface can be formulated as

$$\vec{\sigma} = \lim_{\Delta A \to 0} (\Delta F / \Delta A) \tag{4.1}$$

This formulation indicates that the stress value may change from place to place on a surface. The SI unit for force (*F*) is the **newton** (N) = m kg/s^2: 1 N is the force at the surface of the Earth that is created by a mass of 102 g. Some geologists use the unit dyne, where 1 dyne (g cm/s^2) = 10^{-5} N. Differential stress or pressure is given in **megapascals** (MPa), where

1 Pa = 1N/m^2 = 1 kg/(m s^2)

1 MPa = 10 bar = 10.197 kp/cm^2 = 145 lb/in^2

100 MPa = 1 kbar

Compressive stresses are normally considered positive in the geologic literature, while tension is regarded as negative (see Box 4.1). In material science the definition is reversed, so that tension becomes positive. This is related to the fact that the tensile strength of a material is lower than its compressional strength. Tensile strength thus becomes more important during the evaluation of constructions such as bridges and buildings. The crust, on the other hand, is dominated by compression. Remember the difference between stress and strain: Both extension and contraction can result from a stress field where all stress axes are compressional.

Stresses in the lithosphere are almost everywhere compressional, even in rifts and other areas undergoing extension.

Normal stress and shear stress

A stress vector oriented perpendicular to a surface is called the **normal stress** on that surface, while a stress vector that acts parallel to a surface is referred to as the **shear stress**. In general, stress vectors act obliquely on planes. The stress vector can then be resolved into normal and shear stress components. It is emphasized that the concept of normal and shear stress has a meaning only when related to a specific surface.

While the decomposition of forces is quite simple, the decomposition of stress vectors is slightly more complicated. The complications relate to the fact that stress depends on the area across which it acts while forces do not. Therefore, simple vector addition does not work for stress vectors. As shown in Figure 4.1 we have the relationships

BOX 4.1 | SIGN CONVENTIONS

Compressive normal forces are always positive in geology, while tensile ones are negative (in engineering geology the sign convention is opposite). Geologists like this convention because stresses tend to be compressive in the crust. Note, however, that shear stresses are subjected to at least two conventions. For the Mohr circle construction, we have the following convention: Shear stresses consistent with clockwise rotation are negative. For tensor notation the sign convention is different for the shear stresses (absolute values are identical): If the shear components on the negative (hidden) sides of the cube shown in the figures act in the positive directions of the coordinate axes, then the sign is positive and vice versa. This may be confusing, so always check if the sign of an output of a calculation makes sense.

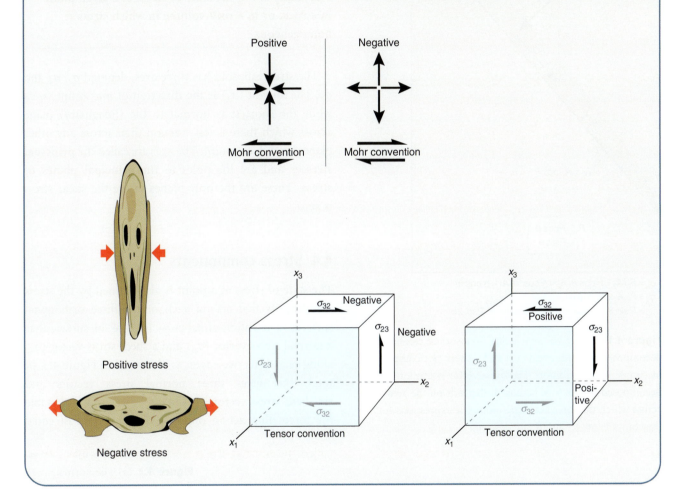

$$\sigma_n = \sigma \cos^2 \theta, \quad \sigma_s = (\sigma \sin 2\theta)/2 \qquad (4.2)$$

where θ is the angle between the stress vector and the surface in question, or the dip of the surface if the stress vector is vertical. For comparison, decomposition of the force vector into normal and shear force vectors (Figure 4.1) gives

$$F_n = F \cos \theta, \quad F_s = F \sin \theta \qquad (4.3)$$

These four functions are illustrated graphically in Figure 4.2.

4.3 Stress at a point

We leave the concept of stress on a single plane to consider the state of stress at a given point in a rock, for example a point within a mineral grain. We may imagine that there are planes in an infinite number of orientations through this point. Perpendicularly across each of the planes there are two oppositely directed and equally long traction or stress vectors. Different pairs of

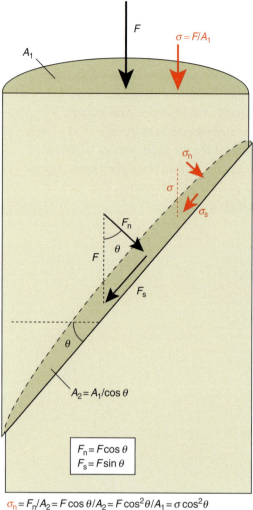

$$F_n = F\cos\theta$$
$$F_s = F\sin\theta$$

$\sigma_n = F_n/A_2 = F\cos\theta/A_2 = F\cos^2\theta/A_1 = \sigma\cos^2\theta$

$\sigma_s = F_s/A_2 = F\sin\theta/(A_1/\cos\theta)$

$\quad = F\sin\theta\cos\theta/A_1 = \sigma\sin\theta\cos\theta = \sigma 1/2\sin 2\theta$

Figure 4.1 A force vector **F** acting on a surface can be decomposed into a normal (F_n) and a shear (F_s) component by simple vector addition. The stress vector σ cannot be decomposed in this way, because it depends on the area across which the force acts. Trigonometric expressions for the components σ_n and σ_s are derived.

stress vectors may be of different lengths, and when a representative family of such vectors is drawn about the point an ellipse emerges in two dimensions (Figure 4.3), and an ellipsoid is defined in three dimensions (Figure 4.4). An obvious requirement for expressing stress as an ellipse (ellipsoid) is that there is not a combination of positive and negative tractions. The ellipse is called the **stress ellipse**, and the ellipsoid is the **stress ellipsoid**.

The stress ellipsoid and its orientation tell us everything about the state of stress at a given point in a rock, or in a rock volume in which stress is homogeneous.

The stress ellipsoid has three axes, denoted σ_1, σ_2 and σ_3. The longest (σ_1) is the direction of maximum stress while the shortest is normal to the (imaginary) plane across which there is less traction than across any other plane through the point. The axes are called the **principal stresses** and are the poles to the **principal planes of stress**. These are the only planes where the shear stress is zero.

4.4 Stress components

The state of stress at a point is also defined by the stress components that act on each of the three orthogonal surfaces in an infinitesimal cube. Each of the surfaces has a normal stress vector (σ_n) and a shear stress vector (σ_s) along each of its two edges, as illustrated in Figure 4.5. In total, this gives three normal stress vectors and six shear stress vectors. If the cube is at rest and stable the forces that act in opposite directions are of equal

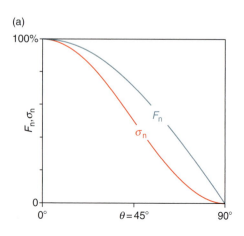

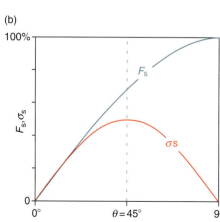

Figure 4.2 (a) The normal components of the force (F_n) and stress (σ_n) vectors acting on a surface, plotted as a function of the orientation of the vectors relative to the surface (θ, see Figure 4.1). Note the difference between the two. (b) The same for the shear components. Note that the shear stress is at its maximum at 45° to the surface while maximum shear force is obtained parallel to the surface.

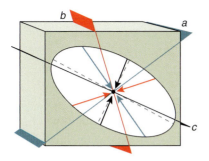

Figure 4.3 Two-dimensional illustration of stress at a point. Three planes (a, b and c) are oriented perpendicular to the section in question, and their normal stresses are represented in the form of vectors (corresponding colors). The stress vectors define an ellipse, whose ellipticity depends on the state of stress.

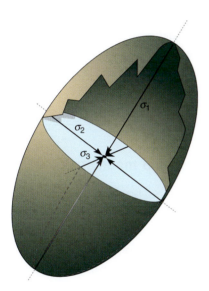

Figure 4.4 The stress ellipsoid.

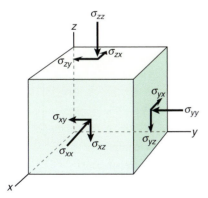

Figure 4.5 The stress components acting on the faces of a small cube. Positive stress components are shown – corresponding stress components exist on the negative and hidden faces of the cube. σ_{xx}, σ_{yy} and σ_{zz} are normal stresses, the others are shear stresses and are parallel to the edges of the cube.

magnitude and hence cancel each other out. This implies that

$$\sigma_{xy} = -\sigma_{yx}, \sigma_{yz} = -\sigma_{zy} \quad \text{and} \quad \sigma_{xz} = -\sigma_{zx} \tag{4.4}$$

and we are left with six independent stress components.

The cube can be oriented so that all of the shear stresses are zero, in which case the only non-zero components are the three normal stress vectors. In this situation these vectors represent the **principal stress** directions and are the **principal stresses** or **principal axes of the stress ellipsoid**. The three surfaces that define the cube are the **principal planes of stress** that divide the stress ellipsoid into three.

4.5 The stress tensor (matrix)

It is useful to put the nine components of stress into a matrix (second-order tensor) known as the **stress tensor** or **stress matrix** (Box 4.2):

$$\begin{bmatrix} \sigma_{11} & \sigma_{12} & \sigma_{13} \\ \sigma_{21} & \sigma_{22} & \sigma_{23} \\ \sigma_{31} & \sigma_{32} & \sigma_{33} \end{bmatrix} \tag{4.5}$$

The normal stresses σ_{11}, σ_{22} and σ_{33} occupy the diagonal while the off-diagonal terms represent the shear stresses. We have that $|\sigma_{11}| = |\sigma_{xx}|$, $|\sigma_{12}| = |\sigma_{xy}|$, $|\sigma_{13}| = |\sigma_{xz}|$ etc., but they may have different signs because of different conventions used for tensor components. In the stable situation where forces are balanced we have $\sigma_{12} = \sigma_{21}$, $\sigma_{31} = \sigma_{13}$ and $\sigma_{23} = \sigma_{32}$, and the stress tensor can be written as

$$\begin{bmatrix} \sigma_{11} & \sigma_{12} & \sigma_{13} \\ \sigma_{12} & \sigma_{22} & \sigma_{23} \\ \sigma_{13} & \sigma_{23} & \sigma_{33} \end{bmatrix} \tag{4.6}$$

This is now a symmetric matrix (a matrix where changing columns to rows does not change anything), but the values will change with choice of coordinate system or how we orient our little cube from Figure 4.5. If we are lucky or orient our little cube (coordinate system) with care, we have the principal stresses along the edges of the box, in which case the matrix becomes

$$\begin{bmatrix} \sigma_{11} & 0 & 0 \\ 0 & \sigma_{22} & 0 \\ 0 & 0 & \sigma_{33} \end{bmatrix} = \begin{bmatrix} \sigma_1 & 0 & 0 \\ 0 & \sigma_2 & 0 \\ 0 & 0 & \sigma_3 \end{bmatrix} \tag{4.7}$$

BOX 4.2 | VECTORS, MATRICES AND TENSORS

A **scalar** is a real number, reflecting temperature, mass, density, speed or any other physical magnitude that has no direction. A **vector** has both magnitude (length) and direction, such as force, traction (stress vector) or velocity. A **matrix** is a two-dimensional array of numbers (3×3 or 2×2 in most geologic applications, meaning that they have 9 or 4 components). Matrices can represent the state of stress or strain in a medium.

The term **tensor** is, in rock mechanics, applied to vectors and particularly matrices. We can consider scalars as tensors of order zero, vectors as first-order tensors and matrices as second-order tensors. Hence, for our purposes, the terms matrix and second-order tensor are identical. However, there are other cases where numbers are arranged in matrices that are not tensors, such as in the field of economics.

An important tensor property is that they are independent of any chosen frame of reference, meaning that the "quantity" represented by the tensor (such as the state of stress or strain at any point in a volume) remains the same regardless of the choice of coordinate system. Hence a vector will be of the same length and of the same magnitude in two different coordinate systems, even though it is represented by different numbers.

A tensor may be defined at a single point or collection of isolated points, or it may be defined over a continuum of points and thus form a field (scalar field, vector field etc.). In the latter case, the elements of the tensor are functions of position and the tensor forms what is called a **tensor field**. This simply means that the tensor is defined at every point within a region of space (or space-time), rather than just at a point or collection of isolated points.

Being the only non-zero entries, the principal stresses can now readily be extracted from the matrix. The three principal stress vectors are the three columns (σ_{11}, 0, 0), (0, σ_{22}, 0) and (0, 0, σ_{33}). In other words:

The stress tensor is composed of the three principal stress vectors.

In other cases we have to find the eigenvectors and eigenvalues of the matrix, which are the principal stress vectors and principal stresses, respectively. This is easily done by means of readily available computer programs. It is important to know that even if the elements of the stress tensor vary for different choices of coordinate system, the eigenvalues and eigenvectors of the tensor remain the same – they are **invariant**.

Stress tensors represent the same state of stress (same shape and orientation of the stress ellipsoid) regardless of our choice of coordinate system.

Since the state of stress will vary from point to point in the lithosphere, so will the stress ellipsoid and the stress tensor. This leads us into the concept of tensor fields. Hence, a complete description of the state of stress in a volume of rock is given by a tensor field.

4.6 Deviatoric stress and mean stress

Any stress tensor can be split into two symmetric matrices, where the first part represents the mean stress and the second is called the deviatoric stress. This is not just another boring mathematical exercise, but a very useful decomposition that allows us to distinguish two very important components of stress, which we also can denote as the isotropic and anisotropic components. The decomposition is

$$
\begin{bmatrix} \sigma_{11} & \sigma_{12} & \sigma_{13} \\ \sigma_{12} & \sigma_{22} & \sigma_{23} \\ \sigma_{13} & \sigma_{23} & \sigma_{33} \end{bmatrix} =
$$
total stress tensor

$$
\begin{bmatrix} \sigma_{m} & 0 & 0 \\ 0 & \sigma_{m} & 0 \\ 0 & 0 & \sigma_{m} \end{bmatrix} + \begin{bmatrix} \sigma_{11} - \sigma_{m} & \sigma_{12} & \sigma_{13} \\ \sigma_{12} & \sigma_{22} - \sigma_{m} & \sigma_{23} \\ \sigma_{13} & \sigma_{23} & \sigma_{33} - \sigma_{m} \end{bmatrix} \tag{4.8}
$$

isotropic component + anisotropic component

(mean stress tensor)   (deviatoric stress tensor)

The σ_{m} in this decomposition is called the **mean stress** and is simply the arithmetic mean of the three principal stresses. Thus, $\sigma_{m} = (\sigma_1 + \sigma_2 + \sigma_3)/3$ gives an average measure of stress.

If there is no deviatoric stress so that the anisotropic component is zero (in which case the deviatoric stress tensor becomes the identity matrix), then the stress or traction is identical on any plane through the point, regardless of the orientation of the plane. Furthermore, the stress ellipsoid is a perfect sphere, $\sigma_1 = \sigma_2 = \sigma_3$, there is no shear stress "anywhere" and there is no off-diagonal stress in the total stress tensor. Such a condition is commonly referred to as **hydrostatic stress** or **hydrostatic pressure**, and represents an isotropic state of stress. In the lithosphere, the mean stress is closely related to **lithostatic pressure**, which is controlled by burial depth and the density of the overlying rock column. We will return to this discussion in Chapter 5.

Deviatoric stress is the difference between the mean stress and the total stress: $\sigma_{dev} = \sigma_{tot} - \sigma_m$, or $\sigma_{tot} = \sigma_m + \sigma_{dev}$. The deviatoric stress tensor represents the **anisotropic** component of the total stress and the deviatoric stress is generally considerably smaller than the isotropic mean stress, but of greater significance when it comes to the formation of geologic structures in most settings. While isotropic stress results in dilation (inflation or deflation), only the anisotropic component results in strain. The relationship between its principal stresses influences what type of structures are formed.

4.7 Mohr circle and diagram

Before looking at stress states in the crust in the next chapter we will consider a practical graphical way of presenting and dealing with stress that is based on a diagram referred to as the **Mohr diagram**. In the nineteenth century, the German engineer Otto Mohr found a particularly useful way of dealing with stress. He constructed the diagram shown in Figure 4.6, now known as the Mohr diagram, where the horizontal and vertical axes represent the normal (σ_n) and shear (σ_s) stresses that act on a plane through a point. The value of the maximum and minimum principal stresses (σ_1 and σ_3, also denoted σ_1 and σ_2 for two dimensional cases) are plotted on the horizontal axis, and the distance between σ_1 and σ_3 defines the diameter of a circle centered at $((\sigma_1 + \sigma_3)/2, 0)$. This circle is called the **Mohr circle**,

---

The Mohr circle describes the normal and shear stress acting on planes of all possible orientations through a point in the rock.

More specifically, for any given point on the circle a normal stress and a shear stress value can be read off the axes of the diagram. These are the normal and shear stresses acting on the plane represented by that point.

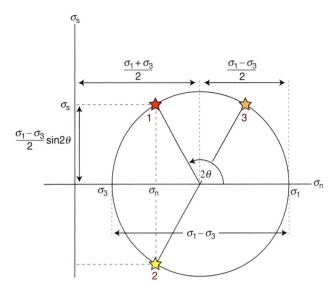

Figure 4.6 The Mohr circle. θ is the angle between the largest stress and a given plane. Note the use of double angles.

How do we know the orientation of the plane represented by a given point on the Mohr circle? In two dimensions and with σ_1 and σ_3 plotted on the horizontal axis, the planes represented on the circle contain σ_2. If θ is the angle between the normal to the plane and σ_1, as shown in Figure 4.1, then the angle between the radius to the point on the circle and the horizontal axis is 2θ.

The difference between the maximum and minimum principal stresses ($\sigma_1 - \sigma_3$) is the diameter of the circle. This difference is called **differential stress** and is important in fracture mechanics. In general, great differential stress promotes rock fracturing.

Angle θ and other angles are measured in the same sense on the Mohr diagram as in physical space, but the angles are doubled in Mohr space. Two points representing perpendicular planes are thus separated by 180° in the Mohr diagram. This is why the two principal stresses both plot on the horizontal axis. Another reason is that principal planes have no shear stress, which is fulfilled only along the horizontal axis. This illustrates how the Mohr space is different from physical space, and it is important to understand the connection between the two. Let us explore some more. The doubling of angles in Mohr space means that any plane, such as that indicated by point 1 in Figure 4.6, has a complementary plane (point 3 in the same figure) with identical shear stress and different normal stress. Point 1 in Figure 4.6 also has another complementary plane (point 2) of identical normal stress and a shear stress that differs by sign only. Maximum shear stress occurs for planes where $2\theta = \pm 90°$, or where

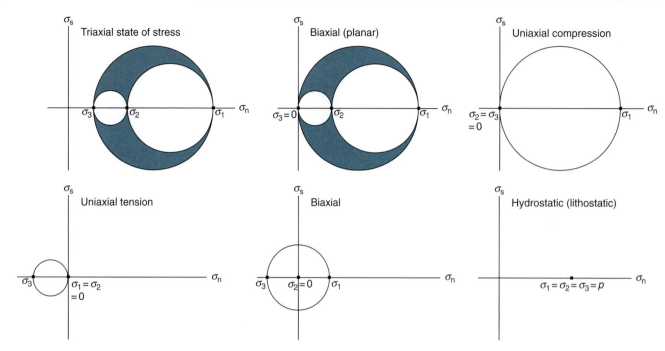

Figure 4.7 Characteristic states of stress are illustrated in the Mohr diagram for 3-D stress. The 3-D state of stress is illustrated by means of three circles connecting the three principal stresses. The largest circle contains σ_1 and σ_3. The three circles reduce to two or one for special states of stress.

the angle to σ_1 is 45° (Figure 4.2b). The Mohr diagram used in geologic applications is generally constructed so that compression is positive and tension is negative, while the opposite convention is common in the engineering literature. In most cases all principal stresses are positive in the lithosphere, but not always. For tensile stress, the Mohr circle is moved to the left of the origin into the tensile field. If all principal stresses are tensile (a most unusual case in geology), then the entire circle is located to the left of the origin.

The Mohr diagram can also be used in three dimensions, where all three principal stresses are plotted along the horizontal axis. In this way we can represent three Mohr circles in a single Mohr diagram. A number of important states of stress can be represented in the three-dimensional Mohr diagram, as shown in Figure 4.7.

Summary

In this chapter we have looked at the fundamentals of stress. It is important to understand the difference between forces, stress on a plane (both vector quantities) and stress at a point (second-order tensor). Stress at a point and the stress ellipsoid are in many ways similar to strain and the strain ellipsoid. The principal stresses correspond in a similar way to the principal strains. But despite the similarities it is important to understand that stress is something that may or may not lead to strain, and if it does, it cannot be expected to produce a strain ellipsoid of similar shape and orientation. At this point we should be able to understand and answer the following statements and questions:

- The term stress can be used for stress on a plane or stress at a point (local state of stress).

- Stress acting on a plane (surface) is a vector determined by the applied force and the area that it acts on. An oblique vector decomposes into a normal and a shear stress component.

- Stress at a point (state of stress) describes the total state of stress at that point and is a second-order tensor (3×3 matrix in three dimensions).

- A complete description of the state of stress in a body is given by the stress tensor field, which describes how the three-dimensional state of stress varies in the body (rock).

- Stress cannot be decomposed in the same way as force, because stress also depends on area.

Review questions

1. When is it appropriate to use the term pressure in geology?

2. How can we graphically visualize the state of stress in two and three dimensions?

3. Where could we expect to find tensile stress in the crust?

4. How will the shape and orientation of the stress ellipsoid change if we define a different coordinate system?

5. Will the stress tensor (matrix) look different if we choose a different coordinate system?

6. A diagonal tensor has numbers on the diagonal running from the upper left to the lower right corner, with all other entries being zero. What does a diagonal stress tensor imply?

7. If the diagonal entries in a diagonal stress matrix are equal, what does the stress ellipsoid look like, and what do we call this state of stress?

8. If we apply a stress vector at various angles to a given surface, at what angle is the shear stress at its maximum? How does that compare to applying a force (also a vector) to the same surface, i.e. at what orientation would the shear component of the force be maximized?

E-MODULE

 The first part of the e-learning module called *Stress* is recommended for this chapter.

FURTHER READING

Means, W. D., 1976, *Stress and Strain: Basic Concepts of Continuum Mechanics for Geologists*. New York: Springer-Verlag.

Oertel, G. F., 1996, *Stress and Deformation: A Handbook on Tensors in Geology*. Oxford: Oxford University Press.

Price, N. J. and Cosgrove, J. W., 1990, *Analysis of Geological Structures*. Cambridge: Cambridge University Press.

Turcotte, D. L. and Schubert, G., 2002, *Geodynamics*. Cambridge: Cambridge University Press.

Twiss, R. J. and Moores, E. M., 2007, *Structural Geology*, 2nd edition. New York: H. W. Freeman and Company.

Chapter 5

Stress in the lithosphere

With a basic understanding of the nature of stress, we will now look at how we get information about stress in the crust and how to understand it. A large number of stress measurements have been performed globally over the last few decades. These measurements indicate that the stress conditions in the crust are complex, partly because of geologic heterogeneities (faults, fracture zones and compositional contrasts), and partly because many areas have been exposed to multiple phases of deformation, each associated with different stress fields. The latter is of importance because the crust has the ability to "freeze in" a state of stress and preserve remnants of it over geologic time. Knowledge of the local and regional stress fields has a number of practical applications, including evaluation of tunneling operations, drilling and stimulation of petroleum and water wells. Besides, knowledge of the present and past states of stress provides important information about tectonic processes, then and now.

5.1 Importance of stress measurements

Knowledge of stress is important for many purposes, for instance during drilling and blasting in highly stressed rock during underground construction, tunneling, quarrying and mining operations. In these cases, high stresses may cause pieces of rock to literally shoot off the walls or the roof, obviously a serious safety issue. As we will see, there is always a stress concentration associated with underground openings. Such concentrations may result in roof closure, sidewall movement and ground subsidence. Unlined pressure tunnels and shafts used in hydroelectric and water supply systems may leak (by hydrofracture) if the internal water pressure exceeds the minimum *in situ* principal stress in the surrounding rock mass. High rock stresses counteract the water pressure and help keep fractures closed. Hence, in this case, high stresses are advantageous.

At greater depths, in oil fields for example, the *in situ* stress field helps steer the drill head in the desired direction during well drilling, prevents sand production and maintains borehole stability. Stresses around wells must also be monitored during production because pore pressure reduction may reduce horizontal stresses by a factor large enough to cause formations to collapse and the seafloor to subside. Hydrofracturing of reservoirs in order to increase permeability around producing wells also requires information about the stress field.

At any level in the crust, stresses are related to the formation and orientation of geologic structures, i.e. the accumulation of strain. Any deformation can be related to some stress field that deviates from the "normal" stress situation. In the deeper portions of the crust, stress cannot be measured or estimated, except for the information obtained from focal mechanisms. There are, however, ways to estimate paleostress in rocks that have been exhumed and exposed at the surface.

5.2 Stress measurements

A challenging aspect of stress is that it cannot be observed directly. Only the effect of stress in the form of elastic or permanent strain, if any, can be observed. Obviously, different media (rock types) react differently to stress (Chapter 6), and if the medium is anisotropic the relations tend to be complex. However, because the strains typically involved in measurements of current stress fields are very small, the connection between the two is close, and useful estimates of stress can be obtained.

A series of different methods are applied, depending on where stress data are to be collected. Some are applied in boreholes (borehole breakouts and hydraulic fracturing), some are more commonly used at the surface or in tunnels (overcoring), and one is related to the first motion generated by stress release during the rupture of faults (focal mechanisms). In addition, neotectonic or recent geologic structures affecting the surface can give useful information about the present state of stress in an area.

Borehole breakouts are zones of failure of the wall of a well that give the borehole an irregular and typically elongated shape, as illustrated in Figure 5.1a. It is assumed that the spalling of fragments from the wellbore occurs preferentially parallel to the minimum horizontal stress (σ_h) and orthogonal to the maximum horizontal stress (σ_H).

The ellipticity of the hole indicates the local orientation of the horizontal stress axes in the wellbore.

Information about the shape of the hole is obtained by **dipmeter tools** or **well imaging tools**. These are tools with arms that are pressed against the borehole wall as the tool is moved along the wellbore. The tool thus records the geometry of the hole, and the orientation of the tool is also recorded. Hence, in addition to the measurements of the orientation of planar structures intersecting the wellbore (the primary purpose of a dipmeter tool), a record of the shape of the borehole is produced, yielding information about the horizontal stresses.

Borehole breakout data are primarily collected in wells drilled for petroleum exploration and production. The shapes of holes drilled for road and tunnel blasting operations have also been used for stress analyses, although this method is not regarded as very reliable. Even in tubular tunnels, preferred spalling of fragments sometimes indicates the orientation of the stress field (Figure 5.2). The principle is the same for all cases: the borehole takes on an "elliptical" shape where the elongated direction is assumed to parallel σ_h.

Overcoring (Figure 5.1b) is a strain relaxation method where, in principle, a sample (core or block) is extracted from a rock unit, measured, and then released so that it can freely expand. The change in shape that occurs reflects the compressive stresses that have been released, but also depends on the rock's elasticity. In general, maximum expansion occurs in the direction of σ_h.

Overcoring is done to map the state of stress at or near the surface. Less commonly, it is done by bringing a sample from a deep bore hole in a confined condition to the surface, and measuring the three-dimensional expansion as the sample is unconfined. Surface or

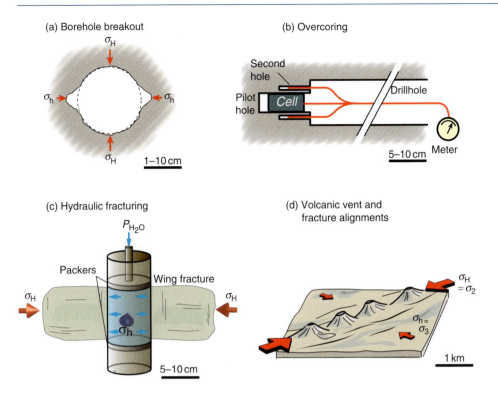

Figure 5.1 Examples of stress determination. (a) Borehole breakouts as illustrated by a horizontal section through a vertical wellbore. Maximum and minimum horizontal stresses σ_H and σ_h are generally equal to or close to two of the principal stresses. (b) The overcoring method. A pilot hole is drilled at the end of the main hole into which stress meters or a strain cell is placed. A wider core is cut outside the cell, and strain is calculated by comparing measurements before and after the drilling. Strain is related to stress by means of elastic theory, and the state of stress is found. (c) Hydraulic fracturing. (d) Recent surface structures related to the current stress field.

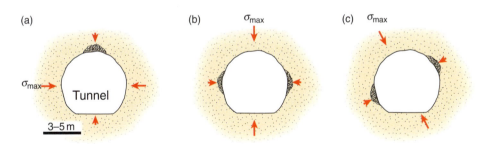

Figure 5.2 Spalling of rock fragments in certain parts of a tunnel gives information about the orientation of the principal stresses and the differential stress.

near-surface overcoring is done by drilling a hole (typically 76 mm in diameter) into the rock and adding a small (36 mm) pilot hole at the end of the main hole. Stressmeters or a strain cell are placed in the pilot hole before it is overcored using a larger coring bit, which relieves the stress in the overcored hollow core. This stress release causes elastic deformation that is recorded by the stressmeters or strain cell. The fact that the unit of microstrain ($\mu e = 10^{-6} e$) is used indicates that the strains involved are minute. At least six stressmeters are required in the hole to completely record the three-dimensional stress field. Calculation of the stress tensor is based on elasticity theory. Elasticity is about how a rock responds to stress

below the limit where strain becomes permanent, and elasticity theory is applied to calculate the orientations and magnitudes of the principal stresses. To do this, the elastic properties known as Young's modulus (E) and Poisson's ratio (v) are measured in the laboratory.

In short, Young's modulus describes the relationship between stress and strain ($E = \sigma/e$) while Poisson's ratio characterizes how much an object that is shortening extends perpendicular to the direction of shortening (or how much it shortens perpendicular to the direction of extension if we extend the object). We will deal with these elastic properties and elasticity more closely in the next chapter.

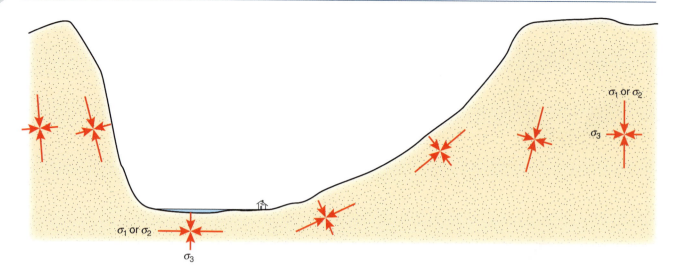

Figure 5.3 State of stress around a valley or fjord. One of the principal stresses will always be perpendicular to the free surface of the Earth, because the shear stress is zero along any free surface. Thus, a non-planar surface causes the orientation of the stresses to rotate as shown on the figure. Note that these deviations occur near the surface only, but must be considered when stress is measured at or near the surface or other free surfaces (tunnel walls, etc.).

At the surface, topography creates local stresses that must be accounted for. Mountains and valleys create stress effects near the surface that influence regional stress patterns, as shown in Figure 5.3. In tunnels and rock chambers, the effect of free space in the rock on the stress field must be taken into consideration, or holes must be drilled far enough away from the tunnel that this effect is negligible. Even the perturbation of the stress field by the drill hole itself must be corrected for, although such a correction is easily done using modern equipment. Weak faults and fracture zones, weathering, and contacts between rocks of contrasting physical properties are examples of geologic structures that are likely to distort the stress field locally. The effect is shown schematically in Figure 5.4: σ_3 will try to orient itself parallel to the weak structure.

Hydraulic fracturing (hydrofracturing, "hydrofracking"; Figure 5.1c) means increasing the fluid pressure until the rock fractures. The technique is frequently applied to petroleum reservoirs to increase the near-well permeability. In this case the interval of the wellbore that is to be fractured is sealed off and pressure is pumped up until tensile fractures form. The pressure that is just enough to keep the fracture(s) open equals σ_h in the formation. Knowing the tensile strength of the rock, it is possible to calculate σ_H. Furthermore, the vertical stress is assumed to be a principal stress and equal to $\rho g z$. Petroleum engineers use knowledge of the stress field to plan hydrofracturing of reservoir units to take advantage of the predicted direction of fracture propagation.

Figure 5.4 Deflection of the stress field near a fault or fracture zone. The structure is weaker than the surrounding rock and can support lower shear stresses than its surroundings. The situation is similar to that where an open surface exists, e.g. the free surface of the Earth (see Figure 5.3).

Earthquake focal mechanisms (Box 9.1) give information about the Earth's immediate response to stress release along new or preexisting fractures. They provide information about the stress regime (Section 5.6) as well as the relative magnitude of the principal stresses. The main problem with this method is that the P- and T-axes do not necessarily parallel principal stress axes. Combining focal mechanisms of faults of different orientation helps reduce this problem.

Geologic structures formed by active tectonic processes also give reliable indications of certain aspects of the present day stress field. The orientation and pattern of recent fault scarps, fold traces, tensile fractures (Figure 5.5) and volcanic vent alignments (Figure 5.1d) all indicate the orientation of the principal stresses.

Figure 5.5 Active vertical fractures on the surface of Holocene lava flows in southeast Iceland indicate the orientation of σ_h. Because the fractures occur at the surface, $\sigma_h = \sigma_3$, and σ_1 must be vertical. A historic basalt flow in the background is less influenced by the fractures.

We have now seen how stress can be measured *in situ* from rocks in the upper crust, i.e. without taking samples away to a laboratory for stress determinations. The deepest reliable stress measurement reported so far was made at a depth of about 9 km in the German Continental Deep Drilling Project (hydrofracturing). Information about the stress conditions deeper in the crust can only be inferred from focal mechanisms, theoretical considerations, and through the use of paleostress methods, which will be treated in Chapter 9.

Our information about the current stress field below a few (4–5) kilometers depth is indirect, inaccurate and incomplete.

5.3 Reference states of stress

Various theoretical models exist that describe how the state of stress changes through the crust, and three of them are presented below. Such models are referred to as reference models or **reference states of stress**. They assume a planet with only one lithospheric shell, without the complications of plate tectonics. In fact, no tectonic forces are included in the reference states of stress. Thus, to find tectonic stress we need to look at deviations from the reference models.

Reference states of stress define idealized states of stress in the crust as if the crust were a static planet with no tectonic processes.

Lithostatic/hydrostatic reference state

The **lithostatic reference state** is the simplest general stress model for the interior of the Earth. It is based on an idealized situation where the rock has no shear strength ($\sigma_s = 0$). A rock volume with this condition cannot support differential stress over geologic time ($\sigma_1 - \sigma_3 = 0$), which means that its state of stress is described as a point on the horizontal axis of the Mohr diagram (Figure 4.7, hydrostatic/lithostatic). This means that stress is independent of direction:

$$\sigma_1 = \sigma_2 = \sigma_3 = \rho g z \tag{5.1}$$

The lithostatic reference state is an isotropic state of stress, where the vertical and horizontal stresses are equal.

The stress is, according to this model, completely controlled by the height and density of the overlying rock column. For continental rocks, which have an average density of ~2.7 g/cm^3, this means a vertical stress gradient of 26.5 MPa/km, which fits the data shown in Figure 5.6 (blue symbols) quite well. Porous rocks have lower density (2.1–2.5 g/cm^3) depending on porosity and mineralogy, and the gradient becomes somewhat lower in sedimentary basins.

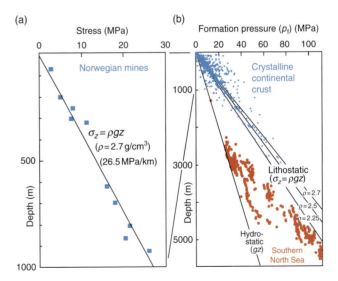

Figure 5.6 (a) Vertical stress measurements compared to the theoretical curve for lithostatic stress ($\rho g z$) in Norwegian mines down to 1 km depth (crystalline rocks). (b) Pressure data from crystalline rocks worldwide and North Sea sedimentary rocks. The North Sea data plot between the gradients for hydrostatic and lithostatic pressure. Individual linear trends are seen, indicating overpressured formations and multiple pressure regimes. Note that these pressure data are for formation pressures, meaning fluid pressures. Data in (a) from Myrvang (2001), in (b) from Darby *et al.* (1996) (North Sea) and many other sources.

BOX 5.1 | PRESSURE AND METAMORPHISM

Metamorphic petrologists tend to talk about pressure rather than stress (commonly in terms of kilobars, where 1 kbar = 100 MPa), while structural geologists reserve the term pressure for fluids. Any rock in the lithosphere has a shear strength even over geologic time: rocks can sustain anisotropic stress as long as the melting point is not reached. Metamorphic petrologists use the term pressure to discuss phase transitions and stability of metamorphic minerals, such as the Al_2SiO_5 system or the transformation of graphite into diamond. This use closely matches our lithostatic reference state. They can do this because they operate at relatively large depths, where anisotropic (tectonic) stresses do not make a big difference when, for example, andalusite changes into kyanite or graphite becomes diamond. Anisotropic stresses do however determine the tectonic regime and the structures formed.

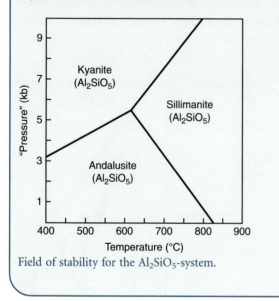

Field of stability for the Al_2SiO_5-system.

Some similar numbers: Common stress gradient, 27 MPa/km; common temperature gradient, 27 °C/km; common rock density, 2.7 g/cm³.

No real, solid rock experiences a perfectly lithostatic reference state. Only magma and other fluids do, in which case the term **hydrostatic pressure** is more appropriate (also see Box 5.1). This is also relevant in sedimentary basins where the formation fluid is generally water. The density contrast between water and rock forces us to operate with two different stress situations. One is the hydrostatic pressure, which is $P_{H_2O} = \rho g z = g z$ (using a water density of 1 g/cm³). The other is the lithostatic stress, which is larger by a factor of around 2.7 (using 2.7 g/cm³ for rock density). If the rock contains oil and/or gas, the densities of hydrocarbons must also be considered.

In a rock column where the rock is porous, the lithostatic stress is distributed over the grain contact area, and this stress is called the **effective stress** $\bar{\sigma}$. In addition we have the pore pressure of the water p_f (or perhaps hydrocarbons) in the pore volume. Hence, we have to operate with two different stress systems in porous media, and the sum of the two is the vertical stress at any given depth:

$$\sigma_v = \bar{\sigma} + p_f \tag{5.2}$$

Pore fluid pressure reduces the effective stress, which is the stress at grain contacts in porous rocks.

If the fluid pressure p_f, often referred to as the formation pressure, equals the hydrostatic pressure $\rho g z$, then the fluid pressure is normal or hydrostatic. In this case pores are interconnected all the way to the surface, and the pore fluid forms a continuous column. This is not always the situation, and deviations from hydrostatic pressure are common. The pore fluid pressure p_f is routinely measured in oil fields and exploration wells, and many reservoir formations are found to be overpressured.

Overpressure forms when formation fluid in porous formations is trapped between non-permeable layers. Sandstones sandwiched between shale layers typically become overpressured during burial because the pore fluid is trapped at the same time as the sandstone carries an increasingly heavy load. The deeper the burial, the larger the deviation between the actual pore pressure and the (theoretical) hydrostatic pressure. This explains the deviations from hydrostatic pressure shown by the red data points in Figure 5.6b (North Sea data). This figure shows evidence of several pressure regimes in southern North Sea reservoirs, each with their own trends between the hydrostatic and lithostatic gradients.

Deviation from hydrostatic pore pressure is important. A high deviation (overpressure) may indicate that the sandstone is poorly compacted, which could mean sand production (sand flowing into the well together with oil during production) and unstable well conditions. This may occur because, during overpressure, p_f in Equation 5.2 increases, and the effective stress across the grain contact points decreases. Where the pore pressure approaches lithostatic pressure, very loose sand (stone) may be expected even at several kilometers depth.

Anomalously high pore pressure can have consequences for deformation. Overpressured layers are weak and may act as detachments during deformation. Thrust faults in foreland settings or accretionary prisms preferentially form in overpressured formations, and extensional detachments are known to develop along such zones. On a smaller scale, grain reorganization rather than cataclasis is promoted by overpressure during deformation of sandstones and other porous rocks.

Artificial overpressure in a formation can be created by increasing the mud weight (hydrostatic pressure) in a chosen interval of a wellbore. The rock responds by fracturing at some critical pressure level, and the operation is known as hydraulic fracturing (Figure 5.1c).

Uniaxial-strain reference state

The lithostatic state of stress is easy and convenient, but not necessarily realistic. A somewhat related model is called the **uniaxial-strain reference state**. This reference state is based on the boundary condition that no elongation (positive or negative) occurs in the horizontal directions (Figure 5.7). Strain only occurs in the vertical

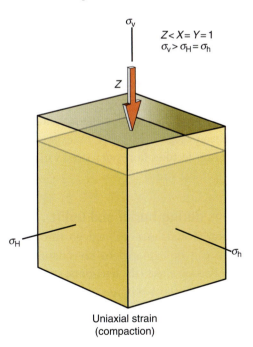

$$Z < X = Y = 1$$
$$\sigma_v > \sigma_H = \sigma_h$$

Uniaxial strain
(compaction)

Figure 5.7 Uniaxial-strain reference state of lithostatic stress. Note the difference between the principal stresses (σ_v, σ_H and σ_h) and principal strains (X, Y and Z). The strain is uniaxial (one component different from zero), while stress is not. In this model the vertical stress comes from overburden while the horizontal one is influenced by the uniaxial-strain boundary condition. The model fits well the effect of compaction in sedimentary basins.

direction (strain is uniaxial), and the stress has to comply with this condition. Make sure you do not confuse uniaxial strain with uniaxial stress, which was defined in Figure 4.7. The uniaxial strain model discussed here results in a triaxial stress. It is interesting to note that stress in this case is prescribed by strain (boundary conditions), whereas often we are inclined to think of strain being the product of stress. Is this really realistic?

The answer is, as you may suspect, yes. This is related to the fact that there is a free surface at the top of any rock column, i.e. the surface of the Earth. In the upper part of the crust this is important, because it is possible to lift or lower this free surface. Lowering is perhaps easier than lifting, and compaction is a deformation that conforms to this uniaxial strain field. Hence, a rock or rock column can shorten (compact) in the vertical direction, but not in the horizontal plane.

Uniaxial strain is characteristic of compaction of sediments where tectonic stresses are absent or negligible. During burial, the horizontal stresses are equal ($\sigma_H = \sigma_h$) and will increase as a function of increasing burial depth or σ_v, but, as shown in Section 6.3, the vertical stress will increase faster than the horizontal stress if the crust is modeled as a linearly elastic medium.

The vertical stress is identical to that predicted by the lithostatic reference state, i.e. $\sigma_v = \rho g z = \sigma_1$, and the horizontal stress $\sigma_H = \sigma_h = \sigma_2 = \sigma_3$ is given by the expression

$$\sigma_H = \frac{\nu}{1-\nu}\sigma_v = \frac{\nu}{1-\nu}\rho g z \tag{5.3}$$

where ν is Poisson's ratio (see Section 6.3 for derivation). In contrast to the lithostatic model, the horizontal stress depends on the physical properties of the rock.

Let us explore Equation 5.3 in more detail. Rocks typically have ν-values in the range 0.25–0.33. For $\nu = 0.25$ the equation gives $\sigma_H = (1/3)\sigma_v$ and, for $\nu = 0.33$, $\sigma_H = (1/2)\sigma_v$. In other words, the horizontal stress is predicted to be between half and one-third of the vertical stress, i.e. considerably less than what is predicted by the lithostatic reference state. The two models are identical ($\sigma_H = \sigma_h = \sigma_v$) only if the lithosphere is completely incompressible, i.e. if $\nu = 0.5$. As stated in Chapter 6, rocks do not even come close to incompressible, but for a sediment that progressively becomes more and more cemented and lithified, its elastic property changes (ν increases) and the uniaxial-strain model approaches the lithostatic one.

The uniaxial-strain reference state predicts that the vertical stress is considerably larger than the horizontal stress.

This fact predicts a state of stress characteristic of extensional regimes ($\sigma_v > \sigma_H > \sigma_h$). However, stress regimes where $\sigma_H > \sigma_h > \sigma_v$ are very common even in the upper crust, so the uniaxial-strain reference state alone does not sufficiently explain the state of stress in many cases. Besides, it predicts somewhat unrealistically large changes in σ_H during thermal changes and uplift events in the lithosphere. A third model has thus been proposed, called the constant-horizontal-stress reference state.

Constant-horizontal-stress reference state

The **constant-horizontal-stress reference state** is based on the assumption that the average stress in the lithosphere is everywhere the same to the depth of isostatic compensation under the thickest lithosphere (z_1 in Figure 5.8). Below z_1, the Earth is assumed to behave like a fluid ($\sigma_H = \sigma_h = \sigma_v = \sigma_m$ in Figure 5.8), where the lithostatic stress σ_m is generated by the overburden. This is a plane strain model with strain in the vertical and one horizontal direction only. In general, this model is probably the most realistic one for a lithosphere unaffected by tectonic forces.

The constant horizontal stress requirement is maintained by isostatic equilibrium. After erosion, but before isostatic reequilibration, the average horizontal stress (σ_h^*) in the thinner portion of the lithosphere must be higher than that in the thicker portion (σ_h) if the horizontal forces are balanced. The depth of isostatic compensation is z_1, and it is assumed that the mantle below has no shear strength over geologic time. Hence, the state of stress is lithostatic below z_1. The average horizontal stress σ_h^* can then be expressed by the equation:

$$\sigma_h^* = \sigma_h[z_1/(z_1 - z)] \\ - \rho_1 g z(\rho_1/\rho_m)[(z_1 - z/2)/(z_1 - z)] \tag{5.4}$$

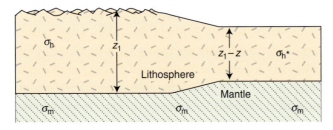

Figure 5.8 Schematic illustration of the relationship between erosion, isostasy and stress for a constant-horizontal-stress reference state, as indicated in Equation 5.4. Erosion of the right-hand side caused upward movement of the base of the lithosphere until isostatic equilibrium was reached. The mantle is considered as a fluid where $\sigma_H = \sigma_h = \sigma_v = \sigma_m$. Based on Engelder (1993).

where $\sigma_h z_1/(z_1 - z)$ describes the horizontal stress increase resulting from lithospheric thinning (horizontal forces must balance, and stress increases as the force acts across a smaller area), and $\rho_1 g z(\rho_1/\rho_m)[(z_1 - z/2)/(z_1 - z)]$ expresses the stress reduction caused by isostasy. Lithospheric thickening can be considered by making $z_1 > z$.

Calculations indicate that the constant-horizontal-stress model predicts lower stress changes during uplift related to lithospheric thinning than the uniaxial-strain model.

5.4 The thermal effect on horizontal stress

Temperature changes occur as rocks are buried, uplifted or exposed to local heat sources (intrusions and lavas) and must be added to the three reference states of stress discussed above. The effect of temperature changes on horizontal stress can be significant, and can be calculated using the following equation for uniaxial strain behavior:

$$\Delta\sigma_h^T = \frac{E\alpha_T(\Delta T)}{1 - \nu} \tag{5.5}$$

where E is Young's modulus, α_T is the linear thermal expansion coefficient, ΔT is the temperature change and ν is Poisson's ratio. As an example, a temperature change of 100 °C on a rock with $\nu = 0.25$, $E = 100$ and $\alpha_T = 7 \times 10^{-6}\,°C^{-1}$ results in a reduction of the horizontal stress of 93 MPa. Cooling during uplift thus has the potential to cause extension fractures in rocks and may in part explain why many uplifted rocks tend to be extensively jointed. Joints are particularly common in competent layers in uplifted sedimentary sequences, such as the sandstones of the Colorado Plateau (Figure 5.9). Let us explore this feature in terms of non-tectonic horizontal stress variations before turning to tectonic stress.

Stress variations during burial and uplift

Rocks that are buried and later uplifted go through a stress history that can be explored by considering the thermal effect, the Poisson effect (the horizontal stress generated due to a change in the overburden, see Section 6.3) and the effect of overburden. Modifying the horizontal stress from Equation 5.3 and adding the thermal effect from Equation 5.5 gives the following equation for a change in depth (change in vertical stress and temperature):

$$\sigma_H = \sigma_h = \left(\frac{\nu}{1 - \nu}\right)\Delta\sigma_v + \left(\frac{E}{1 - \nu}\right)\alpha\Delta T \tag{5.6}$$

where $\Delta\sigma_v$ is the change in vertical stress.

Figure 5.9 Densely jointed Permian sandstones of the Colorado Plateau, exposed by the Colorado River. Such joints would not occur in a reservoir sandstone unless it was uplifted and cooled substantially.

Equation 5.6 can be used to estimate changes in horizontal stress as a rock moves from one crustal depth to another. The result depends on the mechanical properties of the rock (E and v), which means that adjacent sandstone and shale layers will develop different stress histories during burial and uplift. To simplify the calculations, we use one set of mechanical properties during burial and another during uplift, assuming that lithification occurs at the deepest point of the burial curve. In other words, sand and clay layers go down, and sandstone and shale layers come back up. The result is illustrated in Figure 5.10, which indicates that the clay/shale is always in the compressional regime while the sandstone is predicted to enter the tensional regime during uplift. The exact path depends on the elastic properties, and the effect of pore pressure is not considered. Furthermore, tensile stresses are rarely recorded in the crust, but in spite of the simplistic aspect of this model it illustrates how adjacent layers may develop different states of stress and therefore different fracture patterns during uplift.

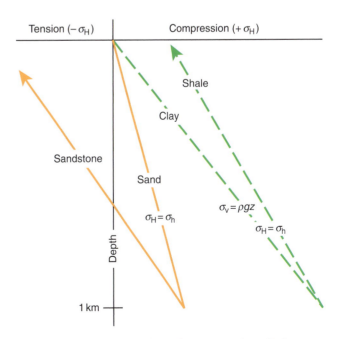

Figure 5.10 Simple modeling of stress variations during burial and uplift for sand(stone) and clay (shale). Lithification is assumed to occur instantaneously at maximum burial depth. Based on Engelder (1985).

Tensile fractures or joints are more likely to develop in rock layers with the highest Young's modulus and the lowest Poisson's ratio, which in simple terms means that stiff and competent layers (e.g. sandstones and limestones) build up more differential stress than surrounding layers.

Joints are more likely to initiate in sandstones than in shale during uplift of clastic sedimentary rocks.

This can be important to geologists exploring petroleum reservoirs in uplifted areas, where vertical tensile fractures that may cause leakage of oil traps are more likely. It also means that a smaller overpressure is needed to produce fractures in sandstone than in claystone (Figure 5.11), which is why hydrofractures tend to be confined to sandstone layers rather than adjacent claystones or shales.

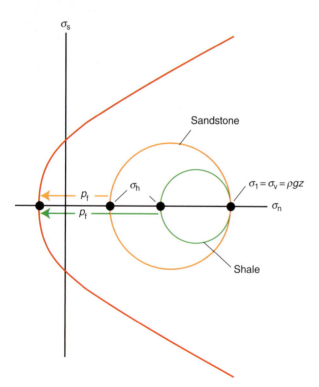

Figure 5.11 Stresses in alternating shale–sandstone layers. Sandstone is stronger and can sustain a higher differential stress than shale. The critical pore pressure needed to generate tensile fractures in the sandstone is less than for shale, since the vertical stress is the same for the layers. The red curve defines the fracture criterion that describes the conditions at which the rocks fracture. Fracturing happens when the circle for sandstone or shale touches the red curve. This can happen due to increased pore fluid pressure (see next chapter) or due to uplift. In either case, the sandstone first touches the red line, and it happens at the tensile side of the normal stress axis, meaning that tensile fractures form. Fracture criteria are treated in Chapter 7 (Section 7.3).

5.5 Residual stress

Stress can be locked in and preserved after the external force or stress field has been changed or removed, and is then referred to as **residual stress**. In principle, any kind of stress can be locked into a rock if, for some reason, elastic strain remains after the external stress field is removed. The causes for the external stress may be overburden, tectonic stress or thermal effects.

Let us look at how residual stress can form in sandstone during compaction, cementation and uplift. During burial and physical loading, stress builds up across grain contact areas. Assume that cementation occurs prior to removal of the external stress field or the overburden. If uplift and erosion later exposes the sandstone at the surface and thereby causes a stress decrease, then the elastic deformation of the grains caused by the now removed overburden will start to relax. However, relaxation is partly prevented by the cement. Hence, some of the stress is transferred to the cement while the rest remains in the sand grains as locked-in stress. In this way, stress that was imposed on the sand during burial thus remains locked in as residual stress.

Residual stress may also be caused by metamorphic transformations that involve volumetric changes, by intrusions where magma cooling sets up stresses that are locked into the crust, by changes in temperature and/or pressure, or by past tectonic episodes. There is therefore a close connection between residual stress, thermal stress and tectonic stress, which we will look at next.

5.6 Tectonic stress

The reference states of stress discussed above relate to natural factors such as rock density, boundary conditions (uniaxial versus plane strain), thermal effects and the physical properties of rock. Natural deviations from a reference state are generally caused by **tectonic stress**. On a large scale, tectonic stress in many cases means stress related to plate movements and plate tectonics. Locally, however, tectonic stresses may be influenced by such things as bending of layers, e.g. ahead of a propagating fault, fault interference and other local effects. Hence, local tectonic stress may be quite variable with respect to orientation, while regional tectonic stress patterns are often found to be consistent over large areas.

Tectonic stresses are those parts of the local stress state that deviate from the reference state of stress as a consequence of tectonic processes.

Somewhat simplified, tectonic stress is the deviation from any chosen reference state of stress. There are also other components of stress, including thermal and residual stress, that we may want to distinguish from present tectonic stress.

Although individual components may be difficult to separate, the total state of stress in any given point in the lithosphere can be separated into a reference state of stress, residual stress, thermal stress, tectonic stress and terrestrial stress (stress related to seasonal and daily temperature changes, earth tide etc.):

Current tectonic stress = Total stress − (reference state of stress + non-tectonic residual stress + thermal stress + terrestrial stress).

If we are close to the surface, we should also eliminate the effect of topography shown in Figure 5.3.

It is seldom easy to separate the tectonic component of stress from the other contributions. Again we turn to ideal models, and we will start by looking at Anderson's classic classification of tectonic stress before looking at actual data.

Anderson's classification of tectonic stress

The traditional classification of tectonic stress regimes into normal, thrust and strike-slip regimes was coined in Anderson's famous 1951 publication. Anderson made the assumption that, since there is no shear stress at the Earth's surface (shear stress cannot occur in fluids), one of the principal stresses has to be vertical, implying that the other two are horizontal. Depending on which of the three principal stresses is the vertical one, Anderson defined three regimes, as illustrated in Figure 5.12:

$\sigma_v = \sigma_1$; normal-fault regime

$\sigma_v = \sigma_2$; strike-slip fault regime

$\sigma_v = \sigma_3$; thrust-fault regime

Anderson's classification is strictly valid only in coaxial deformational regimes, where lines parallel to ISA and principal strain axes do not rotate. Furthermore, the deforming rock must be isotropic. The vertical stress can be related to the weight and density of the overlying rock column:

$$\sigma_v = \rho g z. \tag{5.7}$$

One of the reference states of stress must be chosen to calculate the horizontal stresses. Let us use the thrust-fault regime as an example. A horizontal tectonic stress acts in this regime, which we will call σ_t^*. This stress adds to that specified by the reference state of stress (Equation 5.7). For a lithostatic state of stress, σ_H thus becomes

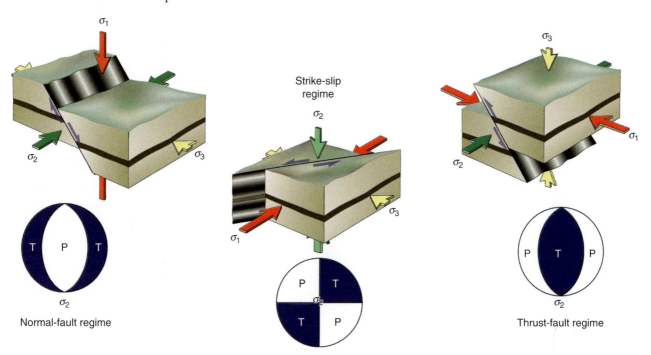

Figure 5.12 Relationships between the orientation of the principal stresses (stress regimes) and tectonic regimes according to Anderson (1951). Stereonets show fields of compression (P) and tension (T).

$$\sigma_H = \rho g z + \sigma_t^* \tag{5.8}$$

If instead we consider the uniaxial-strain reference state of stress, then we implicitly assume that the stress condition also depends on the physical properties of the rock. In this case (excluding any thermal effect) we use the designation σ_t for horizontal tectonic stress, and by adding this tectonic stress to the uniaxial-strain reference state (Equation 5.3) we have

$$\sigma_H = [\nu/(1-\nu)]\rho g z + \sigma_t \tag{5.9}$$

It follows by comparing the two expressions that, since $[\nu/(1-\nu)] < 0$, $\sigma_t > \sigma_t^*$. This means that the magnitude of the tectonic stress depends on our choice of reference state of stress. Note that when ν approaches 0.5, Equation 5.8 approaches Equation 5.9, i.e. σ_t approaches σ_t^*. These considerations also hold for the normal-fault regime, except that σ_t^* is tensional and thus becomes negative (σ_t is only negative or tensional in areas of very active normal faulting).

The examples illustrate that the definition of tectonic stress is dependent on the choice of reference state of stress. Hence, the absolute value of tectonic stress is not always easy to estimate.

5.7 Global stress patterns

Stress is estimated around the world in mines, during construction and tunneling work, during onshore and offshore drilling operations, and in relation to earthquake monitoring. Together, these data are evaluated, compiled in **The World Stress Map Project**, and are available on the World Wide Web (Figure 5.13).

In order to compile a worldwide stress map, several different sources of information are used. They are grouped into (1) earthquake focal mechanisms, (2) borehole breakouts and drilling-induced fractures, (3) *in situ* stress measurements (overcoring, hydraulic fracturing) and (4) neotectonic geologic structural data (from fault-slip analysis and volcanic vent alignments). The data are ranked according to reliability, and it is assumed that one principal stress is vertical and the other two horizontal. Focal mechanism data completely dominate the data set, particularly the deeper (4–20 km) portion of the data, and are most frequent where earthquakes are common, i.e. along plate boundaries. At shallower levels there is a dominance of data from breakouts, hydrofractures and overcoring.

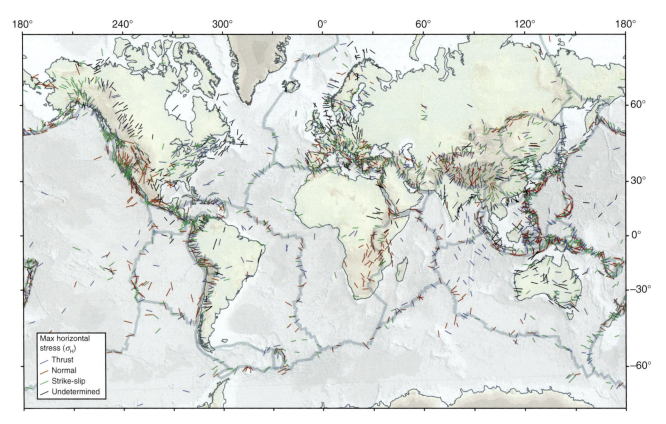

Figure 5.13 World Stress Map, based on stress measurements from around the world. Lines indicate the orientation of σ_H, and line colors indicate tectonic regime (normal-fault, strike-slip, thrust-fault regimes shown in Figure 5.12). Based on data from www.world-stress-map.org.

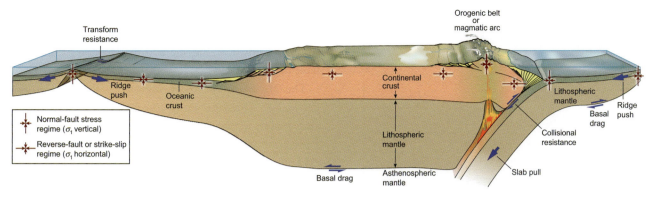

Figure 5.14 Forces related to plate tectonics (blue arrows) and stress regimes expected from these forces. The maximum stress axis in continental plates is expected to be horizontal except for the upper part of rift zones (continental rift not shown), passive margins and elevated parts of orogenic belts.

Figure 5.13 shows that there are large areas of little or no stress information, onshore as well as offshore. The correlation between the orientation of σ_H and plate motion is also obvious many places, but with many deviations that tell us that the current stress field is influenced by many different mechanisms and sources of stress. Regardless, tectonic processes at plate margins are thought to have a significant influence on the regional stress pattern, and the main sources are thought to be slab pull, ridge push, collisional resistance and basal drag (frictional drag along the base of the lithosphere) (Figure 5.14). **Slab pull** is the gravitational pull exerted by the sinking slab on the rest of the plate. It is largest for old and cold, dense oceanic lithosphere, and negative if light and buoyant continental crust is subducted. **Ridge push** is simply the push from the topographically high oceanic ridge that marks divergent plate boundaries. This ridge rises several kilometers from the ocean floor and thus produces a significant lateral force. The related stress regime promoted by ridge push is one of normal faulting in the elevated ridge area, and reverse faulting farther away. These are probably the two most important sources of plate tectonic forces and thus the most important factors in the shaping of the global-scale stress pattern. The effect of **basal drag**, which is the frictional resistance or shear force acting at the base of the lithosphere, is uncertain. This force could drive or resist plate motions, depending on plate kinematics relative to the behavior of local mantle convection cells. A similar force acts on top of the subducting slab. The effect of the **collisional resistance** depends on the strength of the coupling between the two colliding plates, but is greatest for continent–continent collision zones.

Plate tectonic processes are responsible for a global stress pattern that is locally modified by gravity-controlled second-order sources of stress.

So-called second-order sources of stress are continental margins influenced by sediment loading, areas of glacial rebound, areas of thin crust and upwelling hot mantle material, ocean–continent transitions, orogenic belts and large, weak faults such as the San Andreas Fault that can deflect the stress field (Figure 5.4). The many sources of stress may be difficult to identify from the present world stress database, but a combination of new stress data and modeling results may improve our understanding of stress in the lithosphere in the years to come.

A characteristic feature of the stress data shown in Figure 5.13 is that the orientation of σ_H is fairly consistent within broad regions. Consistent patterns can exist within continents, far from plate boundaries. One such area is the North American continent east of the Rockies, which is dominated by a NE–SW oriented σ_H. Another is Northern Europe and Scandinavia, where σ_H is oriented NW–SE. The orientation of the latter can perhaps be explained by ridge push from the North Atlantic mid-ocean ridge, while the more oblique (with respect to the ridge axis) orientation of the North American stress field may be influenced by other effects, perhaps basal drag. Even if we cannot fully explain intraplate stress orientation patterns yet, the consistent orientation of σ_H over large areas suggests that plate tectonic forces play an important role.

If we go closer to the plate boundaries, we see at least some consistency between orientations of the boundaries and that of σ_H, for example along the convergent plate boundary along the western coast of South America,

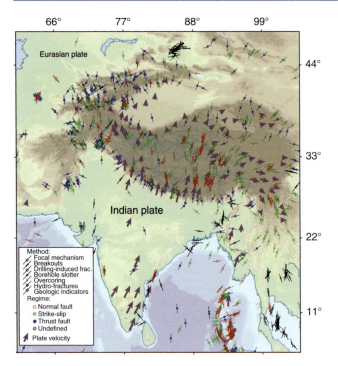

Figure 5.15 Stress data from the Himalaya area – the collision zone between the Eurasian and Indian plates, together with plate velocity vectors determined from GPS measurements. Note the close connection between kinematics and σ_H, and the occurrence of all three tectonic regimes in the collision zone. Data from www.world-stress-map.org and http://jules.unavco.org/Voyager/ILP_GSRM.

where σ_H makes a high angle to the plate boundary (Figure 5.13). In contrast, σ_H is much more oblique in the dextral strike-slip dominated San Andreas Fault area in California, an obliquity that fits with dextral sense of shear. Also in the Himalaya, where the Indian plate moves northward into the Eurasian plate, there is a good correlation between plate motions and σ_H (Figure 5.15).

When we look at the global distribution of the three different fault regimes (normal, strike-slip and thrust regimes) the pattern may seem quite complex, with two or even three of the regimes occurring in the same region. However, there are some general features that explain important aspects of this pattern. An important observation is that the maximum principal stress is horizontal ($\sigma_H = \sigma_1$), implying a thrust or strike-slip regime within large parts of continents such as eastern North America, South America, and Scandinavia/Northern Europe. We could say that, in general, $\sigma_H = \sigma_1$ in large parts of the continents, if not the entire brittle upper crust. As mentioned above, we could credit or blame plate tectonics and the forces that come with it for this situation. Within this pattern, we find the **strike-slip stress regime** represented in several places, but primarily in areas of significant strike-slip faulting, such as the San Andreas Fault in California and the Dead Sea transform fault in the Middle East.

The **thrust-fault stress regime** is expected to be particularly common along convergent plate boundaries and major active orogenic zones. Examples are the Himalayan orogenic belt and the Andes, where geologic evidence of thrust tectonics is widespread. These zones also have significant elements of strike-slip and even normal-fault regimes, and there are a couple of basic explanations for this. One is the concept called strain (or stress) partitioning, which in simple terms means that the perpendicular component goes into thrusting and other contractional deformation and the boundary-parallel component is taken out by strike-slip motion and related states of stress (see Chapter 18). Another is the fact that elevated areas of the crust, such as the high Andes or the Tibetan Plateau, develop a vertical σ_1 and even tend to expand laterally due to gravity. Consequently, the high portions of an orogenic belt can extend while the deeper part is shortening, as discussed in Chapter 16.

As an example, the stress situation in and around the Himalayan orogen shows the presence of all three regimes (Figure 5.15). The thrust-fault regime dominates the lower parts along the plate boundary (southern foothills), while the normal-fault regime characterizes the high Tibetan Plateau. The strike-slip regime is particularly common in the eastern part, where crust is being squeezed sideways as India intrudes into Eurasia (see Chapter 18). Hence, relative plate motion (India moving into Eurasia), gravitational potential of overthickened crust (Tibetan Plateau) and boundary conditions (stiff Indian plate forced into heated and thus softened material immediately north of the Indian and Eurasian plate boundary) are three factors that help explain the occurrence and distribution of the different stress regimes in the Himalayan collision zone.

Stress fields consistent with Anderson's **normal-fault regime stresses** are found along divergent plate boundaries, but are more pronounced in areas of active continental rifting and extension. The East African rift zone, the Aegean area and the Basin and Range province of the western USA are obvious examples, but also elevated areas in convergent settings, such as the Tibetan Plateau, the high Andes, and the western US Cordillera, tend to be under extension. One of the most popular models for the extension in these areas is related to gravitational collapse of topographically high areas, which we will return to in Chapter 17.

5.8 Differential stress, deviatoric stress and some implications

The amount of stress increases downwards from the surface into the lithosphere. How much stress can a rock withstand before deformation occurs? The reference states predict an increase in stress all the way to the center of the Earth. Rock-forming minerals undergo phase changes and metamorphic reactions in response to this increase, and it takes a deviation from the reference state for rocks to deform by fracturing or shearing. Anderson gave us an idea of how the relative orientation of the tectonic stresses influences the style of faulting (close to the surface). We also want to know how faulting occurs in the lithosphere. Here it is not the absolute level of stress, but rather the difference between the maximum and minimum principal stresses that causes the rock to fracture or flow. This difference is called the **differential stress**:

$$\sigma_{\text{diff}} = \sigma_1 - \sigma_3 \qquad (5.10)$$

For **lithostatic stress** (Figure 5.7) the principal stresses are all equal, and we have

$$\sigma_{\text{diff}} = 0 \qquad (5.11)$$

Hence, the lithostatic model itself provides no differential stress to the lithosphere, regardless of depth of burial.

For a **uniaxial-strain reference state** of stress the situation is

$$\sigma_{\text{H}} = \sigma_{\text{h}} < \sigma_{\text{v}} \qquad (5.12)$$

and the differential stress becomes

$$\sigma_{\text{diff}} = \sigma_1 - \sigma_3 = \sigma_{\text{v}} - \sigma_{\text{h}} = \sigma_{\text{v}}[(1 - 2\nu)/(1 - \nu)] \qquad (5.13)$$

For a rock with $\nu = 0.3$, $\sigma_{\text{diff}} = 0.57\sigma_{\text{v}}$ and σ_{diff} increases with depth at a rate of $\sim$13 MPa/km for continental crust. While the uniaxial-strain reference state of stress may be reasonably realistic in sedimentary basins, it may not be very realistic deep in the lithosphere. Again we see that our choice of reference model has significant implications.

Regardless of the choice of reference state of stress, tectonic stress adds to the total differential stress in a rock. The amount of differential stress that can exist in the lithosphere is, however, limited by the strength of the rock itself. When a rock deforms by brittle fracturing, its strength changes and σ_{diff} is reduced. Hence, while the vertical stress in the lithosphere is governed by the weight of the overburden, the horizontal stresses are limited by the local rock strength.

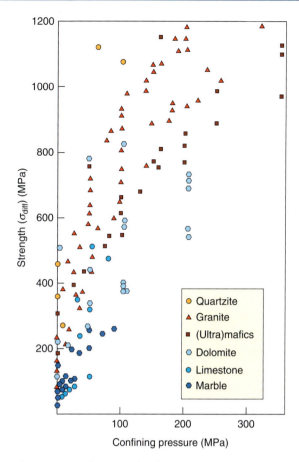

Figure 5.16 The strength of various rock types, plotted against confining pressure (burial depth). The data indicate that the strength of the brittle crust increases with depth, and that the absolute strength depends on lithology (mineralogy). Data compiled from a range of sources.

Differential stress at any given point in the Earth is limited by the strength of the rock itself. Any attempt to increase the differential stress above the ultimate rock strength will lead to deformation.

This does not mean that differential stress is independent of overburden. In fact, there is a very important positive relationship in the upper part of the crust between the amount of overburden and the differential stress that any given rock can support, as reflected by the experimental data shown in Figure 5.16. Thus, the strength increases from the surface and down toward the depth at which the rock starts to flow plastically. For granitic rocks this generally means mid-crustal depths (10–15 km). This transition is controlled by temperature rather than σ_{v}, and is related to the brittle–plastic transition in the crust (Section 6.9).

The strength of rocks in the crust is in practice controlled by anisotropic features, particularly weak fractures and shear zones. We will return to crustal strength

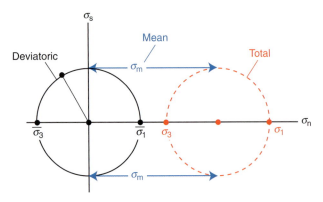

Figure 5.17 The total state of stress can be considered as consisting of an isotropic component, the mean normal stress, and an anisotropic component, the deviatoric stress. The center of the Mohr circle is moved to the origin when the mean stress is subtracted.

toward the end of the next chapter. We will conclude the present chapter by looking at deviatoric stress.

Deviatoric stress (σ_{dev}) was defined in Chapter 4 as the difference between the total stress tensor and the mean stress tensor:

$$\sigma_{diff} = \sigma_{tot} - \sigma_m \tag{5.14}$$

where

$$\sigma_m = (\sigma_1 + \sigma_2 + \sigma_3)/3 \tag{5.15}$$

In three dimensions deviatoric stress is thus defined as

$$
\begin{bmatrix} \sigma_{11dev} & \sigma_{12dev} & \sigma_{13dev} \\ \sigma_{21dev} & \sigma_{22dev} & \sigma_{23dev} \\ \sigma_{31dev} & \sigma_{32dev} & \sigma_{33dev} \end{bmatrix} = \begin{bmatrix} \sigma_{11} & \sigma_{12} & \sigma_{13} \\ \sigma_{21} & \sigma_{22} & \sigma_{23} \\ \sigma_{31} & \sigma_{32} & \sigma_{33} \end{bmatrix} - \begin{bmatrix} \sigma_m & 0 & 0 \\ 0 & \sigma_m & 0 \\ 0 & 0 & \sigma_m \end{bmatrix}
$$

$$
= \begin{bmatrix} \sigma_{11} - \sigma_m & \sigma_{12} & \sigma_{13} \\ \sigma_{21} & \sigma_{22} - \sigma_m & \sigma_{23} \\ \sigma_{31} & \sigma_{32} & \sigma_{33} - \sigma_m \end{bmatrix}
$$

$$\tag{5.16}$$

Or, if the principal stresses are oriented along the coordinate axes of our reference system:

$$
\begin{bmatrix} \sigma_{1dev} & 0 & 0 \\ 0 & \sigma_{2dev} & 0 \\ 0 & 0 & \sigma_{3dev} \end{bmatrix} = \begin{bmatrix} \sigma_1 & 0 & 0 \\ 0 & \sigma_2 & 0 \\ 0 & 0 & \sigma_3 \end{bmatrix} - \begin{bmatrix} \sigma_m & 0 & 0 \\ 0 & \sigma_m & 0 \\ 0 & 0 & \sigma_m \end{bmatrix}
$$

$$
= \begin{bmatrix} \sigma_1 - \sigma_m & 0 & 0 \\ 0 & \sigma_2 - \sigma_m & 0 \\ 0 & 0 & \sigma_3 - \sigma_m \end{bmatrix}
$$

$$\tag{5.17}$$

This implies that the mean stress by definition is the *isotropic* component of the total stress, while the deviatoric stress is the *anisotropic* component at the same point. This becomes clearer if we rearrange the above equation (also see Equation 4.8):

$$
\begin{bmatrix} \sigma_1 & 0 & 0 \\ 0 & \sigma_2 & 0 \\ 0 & 0 & \sigma_3 \end{bmatrix} = \begin{bmatrix} \sigma_m & 0 & 0 \\ 0 & \sigma_m & 0 \\ 0 & 0 & \sigma_m \end{bmatrix}
$$

$$
+ \begin{bmatrix} \sigma_{1dev} & 0 & 0 \\ 0 & \sigma_{2dev} & 0 \\ 0 & 0 & \sigma_{3dev} \end{bmatrix} \tag{5.18}
$$

Total stress = Isotropic + Anisotropic part

Deviatoric stress in two dimensions is visualized in Mohr space in Figure 5.17. The distance from the center of the Mohr circle to the origin is the mean stress. The two deviatoric stresses are positive ($\sigma_1 - \sigma_m$) and negative ($\sigma_3 - \sigma_m$), respectively, and their directions indicate the tectonic regime (normal, thrust or strike-slip). Note that even though one of the anisotropic stresses is negative or tensional ($\sigma_3 - \sigma_m < 0$), the isotropic component is generally large enough in the lithosphere that all principal stresses of the total state of stress become positive (compressive).

Summary

It is both challenging and interesting to try to measure and understand the present stress pattern in the crust. It is clear that we need more measurements and a better understanding of how stresses originate and accumulate in rocks. Nevertheless, the models and concepts presented should form a useful fundament for dealing with stress. Here are some important points to remember:

- Stress is not directly observable, but is revealed by strain in one way or another.

- Reference states of stress in the crust are very generalized models that can be used to detect anomalies.

- Anomalies could be due to local conditions, such as overpressured units (in sedimentary basins), thermal effects, stress refraction near mechanically weak zones (e.g. fracture zones), residual stress or the effect of topography near the surface.

- When these factors are accounted for, the remaining deviation from the general reference state of stress is likely to be tectonic stress.

- Tectonic stress is ideally classified by three end-member states or regimes: the normal (σ_1 vertical), strike-slip (σ_2 vertical) and thrust (σ_3 vertical) regimes.

- For deformation to occur in any of these regimes requires the differential stress to exceed the strength of the rock.

- The amount of differential stress a rock can support increases downward through the brittle upper crust.

Review questions

1. How can we get information about the stress field near the surface? Some kilometers down? Even deeper down?

2. Which of the three reference states of stress are, by nature, isotropic?

3. Is the uniaxial-strain reference state a stress or strain state? How are stress and strain related in this model?

4. What physical factors control the state of stress in a rock that is being uplifted through the upper crust?

5. Why does sandstone fracture more easily than shale when uplifted?

6. How can we define tectonic stress?

7. What conditions must apply for Anderson's classification of tectonic stress to be strictly valid?

8. What is the differential stress at 5 km depth for continental crust if we have a perfect lithostatic state of stress?

9. What forces related to plate tectonics can cause tectonic stress?

10. Why do we find evidence of strike-slip and normal-fault stress regimes in addition to the thrust-fault regime in active (contractional) orogens such as the Himalaya and Andes?

11. What stress regime(s) would we expect along strike-slip faults such as the San Andreas Fault?

12. Why does the differential stress increase downwards in the brittle crust?

13. If we increase the fluid pressure in a sandstone unit, will the effective stress increase or decrease?

E-MODULE

The second part of the e-learning module called *Stress* is recommended for this chapter.

FURTHER READING

Amadei, B and Stephansson, O., 1997, *Rock Stress and its Measurement*. London: Chapman & Hall.

Engelder, J. T., 1993, *Stress Regimes in the Lithosphere*. Princeton: Princeton University Press.

Fjær, E., Holt, R. M., Horsrud, P., Raaen, A. M. and Risnes, R., 1992, *Petroleum Related Rock Mechanics*. Amsterdam: Elsevier.

Turcotte, D. L. and Schubert, G., 2002, *Geodynamics*. Cambridge: Cambridge University Press.

Chapter 6

Rheology

Stress and strain are related, but the relationship depends on the properties of the deforming rock, which themselves depend on physical conditions such as state of stress, temperature and strain rate. A rock that fractures at low temperatures may flow like syrup at higher temperatures, and a rock that fractures when hit by a hammer may flow nicely at low strain rates. When discussing rock behavior it is useful to look to material science, where ideal behaviors or materials (elastic, Newtonian and perfectly plastic) are defined. These reference materials are commonly used when modeling natural deformation. This is what we will do in this chapter, and we will focus on a very useful arena for exploring related rock deformation, which is the rock deformation laboratory. Experimenting with different media has greatly increased our knowledge about rock deformation and rheology.

6.1 Rheology and continuum mechanics

Rheology is the study of the mechanical properties of solid materials as well as fluids and gases. The name derives from the Greek word "rheo", which means "to flow". But what have flow and fluids got to do with solid rocks? In answering this question, it is interesting to consider the Greek philosopher Heraclitus' aphorism "Panta Rhei", meaning "everything flows". He argued that everything is in constant change, which is easier to accept if geologic time is involved.

It is not only water that flows, but also oil, syrup, asphalt, ice, glass and rocks. The flow of oil and syrup can be studied over time spans of minutes, while it takes days, months or years to study the flow of ice (Figure 6.1) and salt glaciers, which again flow considerably faster than glass. While glass flows too slowly to produce observable changes in shape even over a few centuries, glass clearly flows fairly quickly when heated by a glassmaker. Temperature influences most solids, including rock.

When hot rocks flow, they accumulate strain gradually, like a very slow-moving glacier or cake of syrup, without the formation of fractures or other discontinuities.

The effect of temperature is the main reason why flow mostly occurs in the middle and lower crust rather than in the cool upper crust. The upper crust tends to fracture, a behavior that strictly speaking falls outside of the field of rheology but still within the realm of **rock mechanics**. In addition to external factors, such as stress, temperature, pressure and the presence of fluids, the properties of the rock itself are of course important. There are many different types of rocks and minerals, and the upper crust tends to fracture only because that is the way common upper-crustal minerals such as quartz and feldspar react to stress under upper-crustal conditions. However, at any depth where we have thick salt layers, these layers will flow rather than fracture, as we will see in Chapter 20. Even clay or sand layers may flow, particularly when the pore pressure becomes high, so flow is not completely restricted to the lower part of the crust. However, deep earthquakes and field evidence may indicate that dry lower-crustal rocks can also fracture under certain conditions, particularly in "dry" rocks more or less devoid of fluids.

If we consider rock as a continuous medium, neglecting heterogeneities such as microfractures, mineral grain boundaries and pore space, and consider physical properties to be constant or evenly changing through the rock volume, simple mathematics and physics can be used to describe and analyze rock deformation in the framework of **continuum mechanics**. Equations that mathematically describe the relationship between stress and strain or strain rate will be important in this chapter. Such equations are called **constitutive laws** or **constitutive equations**. The term *constitutive* emphasizes the importance of the constitution or composition of the material.

Rheology and continuum mechanics deal with the flow of rocks, while rock mechanics primarily deals with the way rocks respond to stress by brittle faulting and fracturing.

Figure 6.1 Ice in glaciers flows similarly to a viscous fluid, but the many fractures at the surface of glaciers tell us that this may not be a perfect model for its uppermost part. Southeast Greenland.

6.2 Idealized conditions

In a simple and idealized continuum mechanics context, materials can be said to react to stress in three fundamentally different ways: by elastic, plastic and viscous deformation. In addition there is brittle deformation and cataclastic flow, but these are beyond the field of continuum mechanics. As the physical conditions change during the deformation history, a given material can deform according to each of these types of flow, and eventually enter the field of brittle deformation.

Deformations are commonly analyzed by plotting a stress–strain or stress–strain rate curve, where strain or strain rate is plotted along the horizontal axis and stress along the vertical axis, as shown in Figure 6.2. Time-dependent deformations are described by means of stress–time and strain–time graphs, where time is plotted along the horizontal axis. Several curves can be plotted for different external conditions or for different materials. Each curve can also be subdivided into stages, where each stage has its own slope. We will start by looking at the elastic response to stress and then move to permanent non-brittle deformation or flow.

It is always useful to start out simple, so let us consider a perfectly **isotropic medium** (rock). By isotropic we mean a medium that has the same mechanical properties in all directions, so that it reacts identically to stress regardless of its orientation. Many of the strains we will consider in this chapter are small, less than a few percent for elastic deformation. This contrasts to strains that we often face

when studying rocks in the field. The advantage is that a simple relationship occurs between stress and strain for small strains in such an ideal medium. In particular, the instantaneous stretching axes (ISA) will coincide with the principal stresses, as will be assumed throughout this chapter.

6.3 Elastic materials

An **elastic material** resists a change in shape, but strains as more stress is applied. Ideally, it returns to its original shape once the applied stress (force) is removed.

Elastic strain is recoverable because it involves stretching rather than breaking of atomic bonds.

Most rubber bands fit this definition very well: more stretching requires more force, and the band recovers its original shape once the force is removed. Rubber is, however, not a linear elastic material.

Linear elasticity and Hooke's law

A **linear elastic material** shows a linear relationship between stress (or force) and strain. This means that if it flattens twice as much under two tons weight as under one, it will flatten four times as much under four tons weight. An analogy is often made with a simple spring (Figure 6.2a): If the weight on the spring is doubled then the change in length also doubles and so on. In other words, the elongation of the spring is proportional to the force applied, and the spring will return to its original

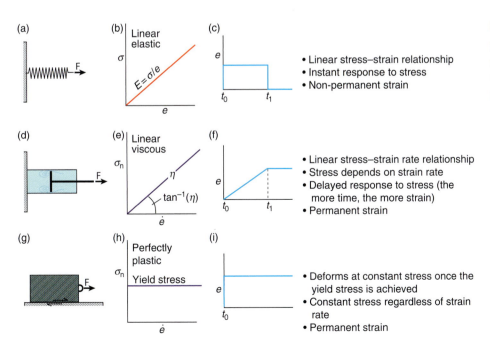

(a)

(b) Linear elastic
σ
$E = \sigma/e$
e

(c)
e
t_0 t_1
- Linear stress–strain relationship
- Instant response to stress
- Non-permanent strain

(d)

(e) Linear viscous
σ_n
η
$\tan^{-1}(\eta)$
$\dot{e}$

(f)
e
t_0 t_1
- Linear stress–strain rate relationship
- Stress depends on strain rate
- Delayed response to stress (the more time, the more strain)
- Permanent strain

(g)

(h) Perfectly plastic
σ_n
Yield stress
$\dot{e}$

(i)
e
t_0
- Deforms at constant stress once the yield stress is achieved
- Constant stress regardless of strain rate
- Permanent strain

Figure 6.2 Elastic, viscous and plastic deformation illustrated by mechanical analogs, stress–strain (rate) curves (center) and strain history curves (right).

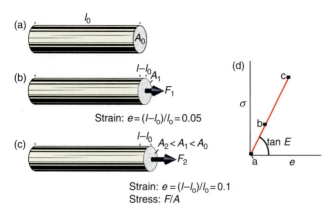

Figure 6.3 Elastic deformation illustrated (a–c) by uniaxial extension of a rod. The stronger the force F that acts on the end area A, the longer the rod (length l). If the material is linear elastic, then the relationship between the extension e and σ ($= F/A$) is linear and forms a line in e–σ-space (d). The gradient of the line is E (Young's modulus). When the force is relaxed, the material returns to its original length (the origin).

Table 6.1 Representative values of Young's modulus (E) and Poisson's ratio (v) for some rocks, minerals and familiar media

Medium	E (GPa)	v (Poisson's ratio)
Iron	196	0.29
Rubber	0.01–0.1	almost 0.5
Quartz	72	0.16
Salt	40	~0.38
Diamond	1050–1200	0.2
Limestone	80	0.15–0.3
Sandstone	10–20	0.21–0.38
Shale	5–70	0.03–0.4
Gabbro	50–100	0.2–0.4
Granite	~50	0.1–0.25
Amphibolite	50–110	0.1–0.33
Marble	50–70	0.06–0.25

length once the force is removed. A similar example is shown in Figure 6.3, where a rod of some elastic material is pulled. Such a linear relationship between stress and strain is expressed by **Hooke's law**:

$$\sigma = Ee \tag{6.1}$$

where σ = stress, e = extension (i.e. one-dimensional strain), and E = **Young's modulus** or the **elastic modulus** (also denoted Y) or, less formally, the **stiffness** of a material. Hooke's law is a constitutive equation for elastic materials.

Young's modulus can also be viewed as the stress/strain ratio:

$$E = \sigma/e \tag{6.2}$$

and is closely related to the **shear modulus** μ (also denoted G and called the rigidity modulus, not to be confused with the friction coefficient introduced in Chapter 2). For uniaxial strain the relationship is simple:

$$E = 2\mu \tag{6.3}$$

The shear modulus μ is related to the shear strain γ, and Hooke's law can be written:

$$\sigma_s = \mu\gamma \tag{6.4}$$

or

$$\sigma = 2\mu e \tag{6.5}$$

Young's modulus E expresses the ratio between the normal stress and the related elastic extension or shortening in the same direction, and describes how hard it is to deform a certain elastic material or rock. Similarly, μ quantifies how hard it is to deform a rock elastically under simple shear (for very small finite strains).

A rock with a low E-value is mechanically weak, as its resistance to deformation is small. Since strain is dimensionless, Young's modulus has the same dimension as stress, and is typically given in GPa (10^9 Pa). Young's modulus for diamond is more than 1000 GPa (very hard to strain), and for iron it is 196 GPa (under axial tension). It takes less force to squeeze aluminum (E = 69 GPa), and rubber is really easy to stretch elastically with typical values of E in the range 0.01–0.1 GPa. Table 6.1 gives some examples of experimentally determined strengths and their characteristic values of E.

Non-linear elasticity

Several minerals are linearly elastic, and we can see from Figure 6.4 that quartz and dolomite are two of them. Even some granites and dolomites obey Hookean elasticity for small strains, but most elastic materials do not, meaning that the line in σ–e-space is not straight. This implies that there is no constant stress–strain relationship, no single Young's modulus. The curves defined during straining (loading) and unstraining (unloading) may still be identical, in which case the material is **perfect elastic** (Figure 6.5b). In this case, the word *perfect* relates to the fact that the material perfectly recovers to its original shape. In many cases of experimental rock deformation, the stress–strain curves during elastic loading and unloading differ, and the material is said to be **elastic with hysteresis** (Figure 6.5c).

Elastic deformation and Poisson's ratio

Before looking at permanent deformation, let us return to our example of the elastically extended rod in Figure 6.3. Here, the axial stretching is accompanied by thinning of the rod. Therefore the area A_0 in Figure 6.3 shrinks as the rod extends. The same effect can be seen when pulling a rubber band: the more it is stretched, the thinner it gets. This effect is known as the **Poisson effect**.

If we consider an isotropic material, the shortening will be the same in any direction perpendicular to the elongation direction (long axis of the rod). If we put our rod in a coordinate system with the long axis along z and assume that volume is preserved, then the elongation along z is balanced by the elongations in the directions represented by the x- and y-axes (remember that negative elongations imply shortening):

$$e_z = -(e_x + e_y) \tag{6.6}$$

where e_z is the elongation parallel to the long axis of the rod and e_x is the perpendicular elongation. Since $e_x = e_y$ (assuming isotropy) we can write the equation as

$$e_z = -2e_x \tag{6.7}$$

or

$$0.5\,e_z = -e_x \tag{6.8}$$

In our example e_x becomes negative, because the rod gets thinner during stretching. We could of course shorten the rod in Figure 6.3. In this case the shortening causes the sample to expand in the perpendicular direction.

Equation 6.8 tells us that shortening in one direction is perfectly balanced by elongation in the plane perpendicular to the shortening direction. This holds true only for perfectly **incompressible materials**, i.e. materials that do not change volume during deformation. Rubber is a familiar material that is almost incompressible. In low-strain rock deformation there is always some volume change involved, and the 0.5 in Equation 6.8 must be replaced by a constant v that relates the axial and perpendicular extensions. This constant is known as

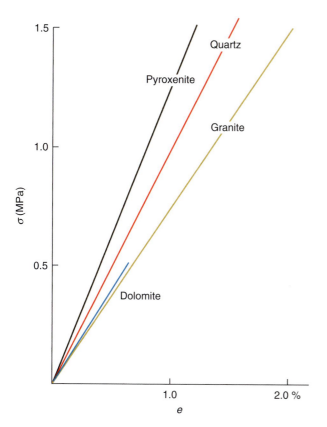

Figure 6.4 Some minerals and rocks show linear elasticity, which means that they follow the same linear path in stress–strain space during stress build-up as during unloading. Data from Griggs and Handin (1960) and Hobbs *et al.* (1972).

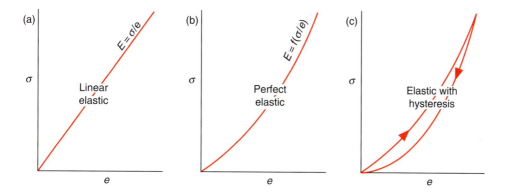

Figure 6.5 The three types of elasticity. (a) Linear elasticity where the loading (straining) and unloading (unstraining) paths are both linear and identical and where the gradient is described by Young's modulus. (b) Perfect elastic deformation follows the same non-linear path during loading and unloading. (c) Elasticity with hysteresis is where the path is non-linear and different during loading and unloading.

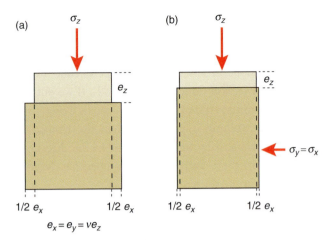

Figure 6.6 (a) A vertical stress (σ_z) applied to an unconstrained rod (unconstrained uniaxial compression). The dashed rectangle indicates the shape of the material prior to the uniaxial deformation. The horizontal elongation e_x is directly related to the vertical shortening through Poisson's ratio. (b) Adding a confining pressure gives a more realistic situation where horizontal stresses arise that counteract the effect of the vertical stress.

Poisson's ratio, which gives the ratio between the extensions normal and parallel to the stress vector σ_z in Figure 6.6:

$$\nu = -e_x/e_z \tag{6.9}$$

The minus is commonly omitted when referring to Poisson's ratio for rocks. The closer the Poisson's ratio gets to 0.5, the less compressible the material. Even steel changes volume during elastic deformation, and most steels have ν-values around 0.3, meaning that a contraction in one direction is not fully compensated for by perpendicular elongation. Most rocks have ν-values between 0.2 and 0.33. For comparison, ν for cork is close to 0, meaning that it hardly expands or shortens perpendicular to an applied stress. Curiously enough, some materials, such as polymer foams, have a negative Poisson's ratio; if these materials are stretched in one direction, they become thicker in perpendicular directions, i.e. a quite unusual scenario for a rock. Some ν-values for rocks and familiar media are given in Table 6.1.

In the crust, rocks are confined, and this puts restrictions on how much a rock volume can extend or shorten, particularly in the horizontal direction. For instance, a volume of sediment or sedimentary rock under burial would experience a vertical shortening that only to a small extent could be compensated for by horizontal extension. We could simulate this situation in the laboratory by confining our sample, and we would see that horizontal stresses arise that counteract the axial shortening e_z.

We now have the following expression for the component of vertical strain resulting from the vertical stress:

$$e_{z'} = \sigma_z/E \tag{6.10}$$

The horizontal stresses will give rise to vertical strains that counteract the effect of Equation 6.10:

$$e_{z''} = \nu\sigma_x/E \tag{6.11}$$

and

$$e_{z'''} = \nu\sigma_y/E \tag{6.12}$$

Thus, the total axial strain is

$$e_z = e_{z'} - e_{z''} - e_{z'''} \tag{6.13}$$

By substituting Equations 6.10, 6.11 and 6.12 into Equation 6.13 we obtain the following expression for the axial strain:

$$\begin{aligned} e_z &= (\sigma_z/E) - (\nu\sigma_x/E) - (\nu\sigma_y/E) \\ &= \frac{1}{E}\left[\sigma_z - \nu(\sigma_x + \sigma_y)\right] \end{aligned} \tag{6.14}$$

Similar expressions can be found for the horizontal stresses:

$$e_y = \frac{1}{E}\left[\sigma_y - \nu(\sigma_z + \sigma_x)\right] \tag{6.15}$$

and

$$e_x = \frac{1}{E}\left[\sigma_x - \nu(\sigma_z + \sigma_y)\right] \tag{6.16}$$

The generation of stresses perpendicular to the loading direction is an effect that is very relevant to rocks in the crust, and contributes to the state of stress in buried rocks. A consequence of confinement is that there will be no or very little horizontal strain ($e_x = e_y \approx 0$), which reduces the vertical shortening e_z. In fact, e_z will depend on both ν and σ_z. To find the vertical stress e_z we apply the boundary condition $e_x = 0$ to Equation 6.16 and obtain:

$$\frac{1}{E}\left[\sigma_x - \nu(\sigma_z + \sigma_y)\right] = 0 \tag{6.17}$$

Multiplying by E on each side gives:

$$\sigma_x - \nu(\sigma_z + \sigma_y) = 0 \tag{6.18}$$

Rearranging and using $\sigma_x = \sigma_y$ gives

$$\sigma_x = \sigma_y = \frac{\nu}{1-\nu}\sigma_z \tag{6.19}$$

Note that we have already seen this equation in the previous chapter during our discussion of the uniaxial-strain reference state (Equation 5.3). Also note that it is possible

BOX 6.1 | POISSON'S RATIO AND SOUND WAVES

Poisson's ratio, named after the French mathematician Simeon Poisson (1781–1840), is a measure of a medium's compressibility perpendicular to an applied stress and can be expressed in terms of velocities of P-waves (V_P) and S-waves (V_S). P-waves or compressional waves are waves of elastic deformation or energy in which particles oscillate in the direction of wave propagation. Conventional seismic sections are based on P-waves. S-waves or shear waves are elastic body waves where particles oscillate perpendicular to the propagation direction. These are different ways of elastic deformation and their relation to Poisson's ratio is:

$$\nu = (V_P^2 - 2V_S^2)/2(V_P^2 - V_S^2)$$

Hence, if V_P and V_S can be measured, Poisson's ratio can be calculated. The ratio is useful in estimating rock and fluid properties in a petroleum reservoir. For example, if $V_S = 0$, then $\nu = 0.5$, indicating either a fluid (shear waves do not pass through fluids) or an incompressible material (which as already stated is not found in the crust). V_S approaching zero is characteristic of a gas reservoir.

to estimate Poisson's ratio from sound waves, as discussed in Box 6.1.

If pressure changes cause elastic deformation rather than directed force, then the **bulk modulus** K relates the pressure change Δp to volume change (volumetric strain):

$$K = \frac{\Delta p}{\Delta V / V_0} \tag{6.20}$$

The bulk modulus is the inverse of the compressibility of a medium, which is a measure of the relative volume change (volumetric strain) of a fluid or solid as a response to a pressure or mean stress change. The higher the value of bulk modulus, the more pressure is needed for the material to compress. Equation 6.20 is a specific form of Hooke's law and K is related to Young's modulus (E) and to the shear modulus (μ) by

$$K = \frac{E}{3(1 - 2\nu)} = \frac{2(1 + \nu)}{3(1 - 2\nu)} \mu \tag{6.21}$$

Even though many low-strain or incipient deformation structures actually represent permanent or inelastic strain that forms beyond the stage of elasticity (see next section), elastic theory is commonly applied when, for example, fracture initiation and growth are modeled. This is to some extent justified when strain is small (a few percent or less), in which case the deviations from elasticity generally are small enough that elastic theory can successfully be used. In particular, **linear elastic fracture mechanics** is used as a simple way to explore and model the state of stress around fractures. It describes stress orientations and stress concentrations based on the geometry of the object (fracture) and the overall (remote) state of stress. However, we will not go deeper into this field here, but continue by looking at deformation beyond elasticity.

6.4 Plasticity and flow: permanent deformation

While elastic theory may work well for very small strains in the upper crust, heated rocks tend to flow and accumulate permanent deformation, and sometimes very large permanent strains. In this context it is useful to consider how fluids respond to stress. Rocks can never quite become fluids unless they melt, but at high temperatures and over geologic time they may get fairly close in terms of rheology.

Viscous materials (fluids)

The ease with which fluids flow is described in terms of their **viscosity** η. Viscosity and laminar flowing fluids were first explored quantitatively by Sir Isaac Newton. He found that the shear stress and shear strain rate are closely related:

$$\sigma_s = \eta \dot{\gamma} \tag{6.22}$$

where η is the viscosity constant, and $\dot{\gamma}$ is the shear strain rate, as indicated in Figure 6.7. A material that deforms according to this equation is a **Newtonian fluid** or a **linear** or **perfectly viscous material**. The constitutive equation for viscous materials can also be expressed in terms of normal stress and elongation rate:

$$\sigma_n = \eta \dot{e} \tag{6.23}$$

Viscous deformation implies dependence of stress on strain rate: higher stress means faster flow or more rapid strain accumulation.

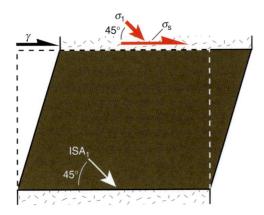

Figure 6.7 Shearing of a medium (fluid) implies that the maximum principal stress is acting at 45° to the surface. For small strains this is equal to the orientation of ISA₁. Increasing the stress results in faster shearing if the material is viscous. The relation between the two is determined by the viscosity of the material.

These two equations state that there is a simple, linear relationship between stress and strain rate (not strain): the higher the stress, the faster the flow. So, while stress was proportional to strain for elastic deformation, it is proportional to strain rate for viscous media. Viscous deformation can therefore be said to be **time-dependent deformation**; strain is not instant but accumulates over time, at rates discussed in Box 6.2.

A perfectly viscous material flows like a fluid when influenced by an external force. This means that there is no elastic deformation involved. Hence, when the force is removed, a viscous material does not recover to its original shape. Viscous deformation is therefore said to be **irreversible** and creates **permanent strain**.

A physical analogy to a perfectly viscous material is an oil-filled cylinder with a perforated piston (Figure 6.2d). When the piston is pulled, it moves through the oil at a constant speed that is proportional to the stress (Figure 6.2e). When the force is removed, the piston stops and remains where it is. If the oil is replaced by a

---

BOX 6.2 | HOW QUICKLY DO ROCKS DEFORM?

Strain rate is a measure of how fast a rock object changes length or shape. Since strain is dimensionless, strain rate has the somewhat peculiar dimension s^{-1} (per second). There are generally two different types of strain rate that we must consider. The simplest is the **rate of elongation** and is denoted $\dot{\varepsilon}$ or $\dot{e}$. This is elongation per unit time (second):

$$\dot{e} = \frac{e}{t} = \left(\frac{l - l_0}{t l_0} \right)$$

We can also call this the extension rate or contraction rate. In experiments this is closely related to the speed at which we squeeze a sample. For an axial compression test that lasts for one hour, where the sample is shortened at 10%, the elongation rate becomes

$$\dot{e} = \frac{-0.1}{3600 \text{ s}} = -2.778 \cdot 10^{-5} \text{ s}^{-1}$$

In some geologic settings the **shear strain rate** may be more appropriate. Here the change in shear strain over time is considered, and this is denoted $\dot{\gamma}$. Its dimension (s^{-1}) is the same as for elongation rate, and they are clearly related, since shear deformation also results in elongation. The two are, however, not linearly related, and it is important to clearly distinguish between the two.

Typical natural geologic strain rates are 10^{-14}–10^{-15} s^{-1} and thus much slower than the ones we observe in the rock deformation laboratory (10^{-7} s^{-1} or faster). Clearly, this is a challenge when applying experimental results to naturally deformed rocks. In many cases temperature is increased in the laboratory to "speed up" plastic deformation mechanisms and thus increase strain rates. Experimental strain rates must then be scaled down together with temperature before they can be compared to natural rock deformation. Alternatively, the processes must be studied at smaller length scales.

more viscous fluid, such as syrup or warm asphalt, then the force must be increased for strain rate to be maintained. Otherwise, the piston will move at a lower speed. If the oil is heated, viscosity goes down and the force must be decreased to keep the strain rate constant. Thus, temperature is an important variable when viscosity is considered.

In layered rocks the **relative viscosity** is also of great interest, as the most viscous (stiff) layers tend to boudinage/fracture or buckle under layer-parallel extension or shortening. Relative viscosity is related to **competency**, where a competent layer is stiffer or more viscous than its surroundings.

Competency is resistance of layers or objects to flow. The term is qualitative and relative to that of its neighboring layers or matrix.

Only fluids are truly viscous, so that in geology only magma, salt and perhaps overpressured (fluidized) mud can be modeled as truly viscous media. However, viscosity is a useful reference when dealing with certain aspects of plastic deformation. We will therefore return to viscosity in a discussion of folding and boudinage in later chapters. Note that **non-linear viscous behavior** has been recorded experimentally for deforming hot rocks and is perhaps more applicable to rocks than linear viscosity. Non-linear behavior in this context simply means that the viscosity changes with strain rate, as illustrated in Figure 6.8. For numerical modeling of folds, both linear

and non-linear viscosity is assumed, while theoretical modeling of boudins require non-linear viscosity.

Viscosity is stress divided by strain rate, and is thus measured in the unit of stress multiplied by time, represented by Pa s or $kg\,m^{-1}\,s^{-1}$ in the SI system. The unit poise was used in the past, where 1 poise = 0.1 Pa s. While the viscosity of water is around 10^{-3} Pa s, that of a glacier (ice) is about 10^{11}–10^{12} Pa s. Glass flows very slowly at room temperature, with a viscosity of 10^{14} Pa s. Viscosity estimates of rocks are greatly variable, but are generally many orders of magnitude larger. Even salt, with a viscosity of $\sim 10^{17}$ Pa s (grain-size dependent) as it flows to form salt diapirs and other salt structures discussed in Chapter 19, is 1000 times more viscous than glass. The mantle underneath the lithosphere is considered as a fluid in many calculations, and it can be estimated to have a viscosity of $\sim 10^{21}$ Pa s on average, increasing downward.

Plastic deformation (flow of solid rock)

Ideally, viscous materials (fluids) react to stress according to Equations 6.22 and 6.23 no matter how small the stress may be. Under most realistic conditions a certain amount of stress is required for permanent strain to accumulate. In fact, the most important difference between fluids and solids is that solids can sustain shear stresses while fluids cannot. For rocks and other solids, elastic deformation occurs for strains up to a few percent.

Beyond the **elastic limit** or the **yield stress**, permanent strain is added to the elastic strain (Figure 6.9). If permanent strain keeps accumulating under a constant stress condition, then we have perfect **plastic deformation** (Figure 6.2g–i). When the stress is removed after a history of elastic–plastic deformation only the plastic strain will remain (the elastic component is by definition nonpermanent). Another requirement for a deformation (strain) to be called plastic is that of continuity or coherency, i.e. the material must not fracture at the scale of observation.

Plastic strain is the permanent change in shape or size of a body without fracture, accumulated over time by a sustained stress beyond the elastic limit (yield point) of the material.

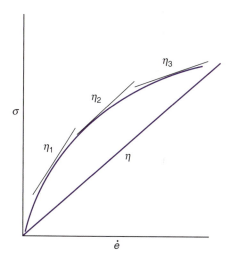

Figure 6.8 Linear (straight line) and non-linear viscous rheology in stress–strain rate space. The slope of the straight line is the viscosity (stress over strain). The non-linear curve has a gradually changing gradient, which is called the effective viscosity. The steepest gradient implies the highest viscosity, which means that it deforms relatively slowly for any given stress condition.

Plastic strain is associated with microscale deformation mechanisms such as dislocation movements, diffusion or twinning (Chapter 10). Because of the many mechanisms involved at the atomic level, plastic flow

(a)

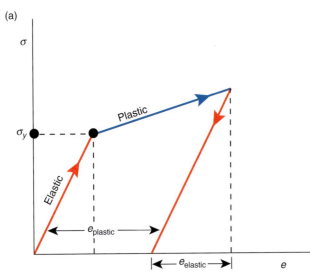

(b)

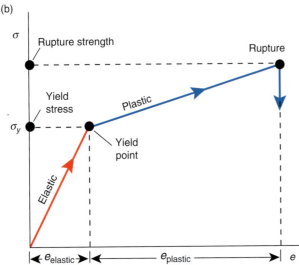

Figure 6.9 Stress–strain curve for elastic–plastic deformation. (a) Elastic strain is replaced by plastic strain as the yield stress (σ_y) is reached. When stress is removed the elastic strain is released, and the plastic or permanent strain remains. (b) In this case the stress is increased to the point where brittle rupture occurs.

does not lend itself to simple physical parameters the way elastic and viscous deformations do. Instead, there are different equations or **flow laws** for different plastic flow mechanisms. A general example is the power-law equation on the form

$$\dot{e} = A\sigma^n \exp(-Q/RT) \qquad (6.24)$$

where A is a constant, R is the gas constant, and T and Q are the absolute temperature and activation energy, respectively. For $n = 1$ the material flows as a perfectly viscous fluid, and the flow is linear. For rocks this is approximated at high temperatures. We will return to flow laws and their underlying deformation mechanisms in Chapter 10.

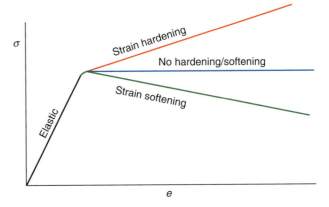

Figure 6.10 Stress–strain curve for elastic–plastic material with hardening, softening, and no hardening/softening properties.

Perfectly plastic materials

A **perfectly plastic material**, or **Saint Venant material**, is one where the stress cannot rise above the yield stress and strain can continue to accumulate without any change in the stress level (Figure 6.2h). The strength is not strain rate sensitive: it does not matter how fast you force the material to flow – the stress–strain curve will not change. A perfectly plastic material is also incompressible. Where there is an additional component of elastic deformation, then the material is called **elastic perfect plastic** (such as the blue middle curve in Figure 6.10). A mechanical analog to perfect plastic deformation is a rigid object resting on a friction surface. Force is increased without any deformation until the frictional resistance between the object and the surface (corresponding to the yield strength) is exceeded (Figure 6.2g). From this point on, the force cannot exceed the frictional resistance except during acceleration, regardless of the velocity.

Strain hardening and softening

Rocks do not generally behave as perfectly plastic materials during plastic deformation. Strain rate is likely to have an effect, and the stress level is likely to change during the deformation history. If we have to increase the applied stress for additional strain to accumulate (Figure 6.10, red curve), then we are dealing with a material science phenomenon called **work hardening** or **strain hardening**.

Strain hardening means that the stress necessary to deform the rock must be increased for strain to accumulate, because the rock becomes stronger and harder to deform.

Strain hardening is particularly apparent in metals, which can be made harder or more resistant through plastic deformation. Just bend a metal wire, and then try to bend it exactly back to its original shape. It will be difficult, because the bent part of the metal has hardened: it takes more stress to retrodeform the bend than it takes to deform an adjacent area. If you are stubborn and keep bending the wire back and forth, the hardening is going to be more and more pronounced until the wire eventually breaks: strain hardening can result in a transition from plastic to brittle deformation if the level of stress is increased. In geology, strain will try to relocalize to an adjacent zone, which may explain why many shear zones get wider as strain accumulates (see Chapter 15).

Strain hardening (during plastic deformation) is related to deformation at the atomic scale. During deformation, atomic-scale defects known as **dislocations** (see Chapter 10) form and move. These dislocations intertangle, which makes it harder to accumulate strain. Hence, more stress is needed to drive deformation, and we have strain hardening. Elevated temperature eases the motion of dislocations and thus reduces the effect of strain hardening. In other words, heating the bent wire makes rebending easier.

If there is no strain hardening and the material keeps deforming without any increase in the applied force or stress, then the process is called **creep**. If, in addition, the strain rate

$$\dot{e} = \frac{de}{dt} \tag{6.25}$$

is constant, then we have **steady-state flow**. Steady-state flow may imply that dislocation movements are quick enough that strain can accumulate at a constant rate for any given stress level. The speed at which dislocations can move around in a crystal depends on differential stress, temperature, and the activation energy that it takes to break atomic bonds. The relationship between dislocation-related strain and these variables can therefore be expressed in a flow law of the type portrayed in Equation 6.24 (see Chapter 10).

Work softening or **strain softening** is the case when less stress is required to keep the deformation going. A geologic example is the effect of grain size reduction during plastic deformation (mylonitization). Grain size reduction makes deformation mechanisms such as grain boundary sliding more effective because of the increase in grain surface area. Other factors that can lead to softening are the recrystallization into new and weaker

minerals, the introduction of fluid(s) and, as already mentioned, an increase in temperature.

The terms strain hardening/softening are also used in connection with the growth of brittle deformation structures, such as deformation bands, although originally defined within the context of plastic deformation. Strain hardening can, for instance, occur during deformation of unconsolidated sand or soil, where interlocking of grains may lead to strain hardening.

6.5 Combined models

Rocks and other natural materials are rheologically complex and generally do not behave as perfect elastic, viscous or plastic materials. It may therefore be useful to combine these three types of deformations in order to describe natural rock deformation. Such combinations are commonly illustrated by means of a spring (elastic), a dashpot (viscous) and a rigid block that will not slide until some critical stress is exceeded (plastic) (Figure 6.2a, d, g).

We have already briefly touched upon one combined model, which is the combination of elastic and plastic deformation. This is the situation where stress and elastic strain increase until the yield point is reached, beyond which the deformation is plastic. A material that responds in this way is **elastic–plastic** or a **Prandtl material**. The typical mechanical analog is shown in Figure 6.11a. The elastic–plastic model is commonly applied to large-scale deformation of the entire crust and the mantle.

A **viscoplastic** or **Bingham material** is one that flows as a perfectly viscous material, but only above a certain yield stress (a characteristic of plastic behavior). Below this yield stress there is no deformation at all. Both rheological experiments on lava and the actual morphology of real lava flows suggest that, over a significant range of temperatures, (liquid) silicic lava behaves like a viscoplastic fluid; that is, due to its crystal content, lava has a yield stress. Paint is a more common fluid that shows viscoplastic behavior: It takes a certain yield stress for it to flow, which prevents thin layers from running off a newly painted wall. The classic mechanical analog is a serial combination of the dashpot (a perforated piston inside a fluid-filled cylinder) and a rigid object resting on a friction surface (Figure 6.11d).

Viscoelastic models combine viscous and elastic behavior. **Kelvin viscoelastic behavior** is where the deformation process is reversible but both the accumulation and recovery of strain are delayed. Viscoelastic materials can be viewed as intermediate states between

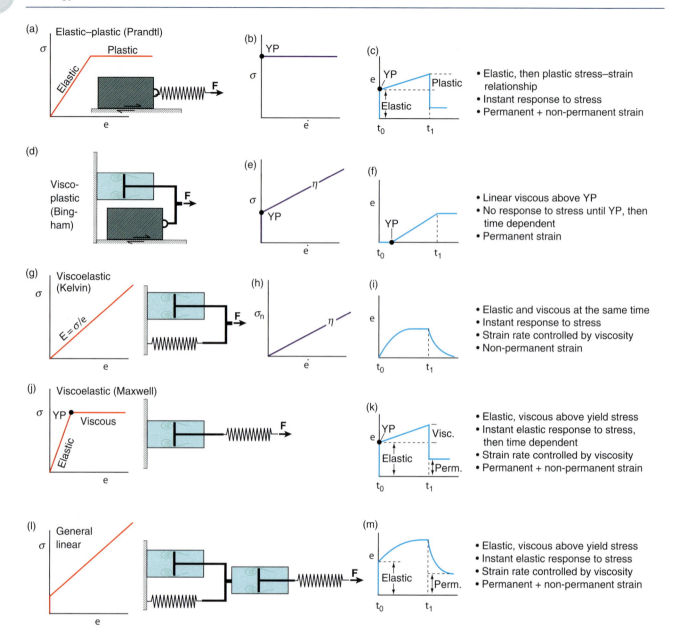

Figure 6.11 Combinations of elastic, viscous and plastic deformation illustrated by mechanical analogs (left), stress–strain (rate) curves and strain history curves (right). Perfect elastic deformation is represented by a spring, while a box with basal friction represents perfect plastic deformation. Perfect viscous deformation is represented by a dashpot. YP, yield point; t_1, time of stress removal.

fluids and solids where both the flow of a fluid and the elastic response of a solid are present. A physical model of a Kelvin viscoelastic material would be a parallel arrangement of a spring and a dashpot (Figure 6.11g). Both systems move simultaneously under the influence of stress, but the dashpot retards the extension of the spring. When the stress is released the spring will return to its original position, but again the dashpot will retard the movement. Such deformation is therefore referred to as **time dependent**.

The constitutive equation for the Kelvin viscoelastic behavior reflects the combination of viscosity and elasticity:

$$\sigma = Ee + \eta\dot{e} \qquad (6.26)$$

A related viscoelastic model is the **Maxwell model**. A Maxwell viscoelastic material accumulates strain from the moment a stress is applied, first elastically and thereafter in a gradually more viscous manner. In other words, its short-term reaction to stress is elastic while its long-term response is viscous, i.e. the strain becomes permanent. This model fits the mantle quite well: It deforms elastically during seismic wave propagation and viscously during mantle convection or flow related to lithospheric loading (e.g. glacial loading). The mechanical analog now consists of a serial arrangement of a dashpot and a spring (Figure 6.11j). A familiar example is the stirring of bread dough. Just a little push creates elastic deformation, while more serious stirring creates permanent deformation. When the stirring is stopped, the dough gradually comes to rest after rotating slightly in the opposite direction due to the release of the elastic component. The constitutive equation for Maxwell viscoelastic deformation is

$$\dot{e} = \sigma/E + \sigma/\eta \qquad (6.27)$$

Viscoelastic models are useful in large-scale modeling of the crust, where the elastic deformation describes its short-term response to stress and the viscous part takes care of the long-term flow.

General linear behavior is a model that more closely approximates the response of natural rocks to stress. Its mechanical analog is shown in Figure 6.11l, where the two viscoelastic models are placed in series. The first application of stress accumulates in the elastic part of the Maxwell model. Continued stress is accommodated within the rest of the model. With the removal of stress the elastic strain is recovered first, followed by the viscoelastic component. However, some strain (from the Maxwell model) is permanent.

While most of these combined idealized models predict a linear stress–strain (rate) relationship, there is no reason to assume that plastically deforming rocks follow such simple relationships during natural deformations. In fact, experimental results indicate that they do not, and a power-law (i.e. non-linear) relationship between stress and strain rate of the form indicated in Equation 6.24 with $n > 1$ exists (curved line in Figure 6.8). This relationship characterizes **non-linear material behavior**. Nevertheless, the idealized models are useful reference models, just as deformations such as simple shear and pure shear are useful reference deformations in strain analysis.

6.6 Experiments

Experiments form the basis for much of our understanding of flow in rock. In the laboratory we can choose the medium and control physical variables such as

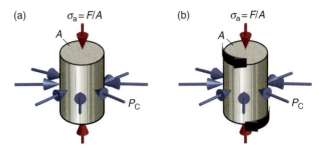

Figure 6.12 (a) The standard loading configuration in triaxial rigs. The axial load σ_a and the confining pressure (P_c) are controlled independently. (b) A configuration where a torsion is added to the axial compression and the confining pressure. This configuration allows for large shear strains to accumulate.

temperature, pressure, stress conditions and strain rate. An obvious disadvantage is that we do not have enough time to apply geologic strain rates, which makes it challenging to compare laboratory results to naturally deformed rocks.

There are many different experimental setups, depending on the property one wants to explore and the physical conditions one wants to impose. The most common one is the triaxial deformation rig where cylindrical samples are exposed to a confining pressure and a principal axial stress (P_c and σ_a in Figure 6.12a). All stresses are compressive, and the sample is shortened when the confining pressure is smaller than the axial compressive stress. If the confining pressure is larger, then the sample extends axially.

There are thus two aspects of stress in this setup. One is the axial or directed stress (anisotropic component), which is the applied force divided by the cross-sectional area of the cylindrical sample. The other is confining pressure (isotropic component), which is created by pumping up the pressure in a confining fluid or soft material. Confining pressures up to 1 GPa are typical, while temperatures may be up to 1400 °C and strain rates 10^{-3} to 10^{-8} s^{-1}. The stress referred to is then typically differential stress (Section 5.8). A furnace around the sample chamber is used to control the temperature of the sample during the deformation.

Besides uniaxial or pure shear deformation, some deformation rigs can impose a rotary shear motion (shear strain) on the sample (Figure 6.12b). Most samples that have been deformed experimentally are monomineralic, such as quartzite or calcite. It is frequently assumed that the properties of single minerals such as quartz (upper crust), feldspar (crust) or olivine (mantle) control the rheological properties of various parts of the lithosphere. More data from deformation of polycrystalline samples are therefore needed.

Constant stress (creep) experiments

Experiments can be sorted into those where strain rate is held constant and those where a constant stress field is maintained throughout the course of the experiment. The latter are referred to as creep experiments and involve the phenomenon called **creep**. Creep is a fairly general term used for low-strain-rate deformations. Hence, it is used in geology for anything from slow down-slope movement of soil, via slow accumulation of displacement along faults (brittle creep), to the slow yielding of solids under the influence of stress (ductile or plastic creep). Plastic creep is what we are interested in here, and in the current context it is defined as follows:

Creep is the plastic deformation of a material that is subjected to a persistent and constant stress when that material is at a high homologous temperature.

Homologous temperature T_H is the ratio of a material's temperature T to its melting temperature T_m using the Kelvin scale:

$$T_H = \frac{T}{T_m} \tag{6.28}$$

For water, with $T_m = 273$ K, the homologous temperature at 0 K is $0/273 = 0$, $273/273 = 1$ at 273 K (0 °C), and $137/273 = 0.5$ at 137 K (-100 °C). The homologous temperatures involved in creep processes are greater than 0.5, and creep processes become more active as T_H approaches 1. This is why glaciers can flow: ice deforms by creep at the high homologous temperatures of natural glaciers. The use of homologous temperature makes it possible to compare solids with different melting points. For instance, it turns out that ice and olivine behave fairly similarly at a homologous temperature of 0.95, which corresponds to -14 °C for ice and 1744 °C for olivine.

Figure 6.13 shows a generalized strain–time diagram for a creep experiment. Stress is rapidly increased to a fixed level and, after accumulation of elastic strain, creep occurs at a decreasing strain rate. This first stage of creep is called **primary** or **transient creep**. After some time, strain accumulates more steadily and the region of **secondary** or **steady-state creep is reached**. Then the **tertiary creep** stage is entered where microfracturing or recrystallization

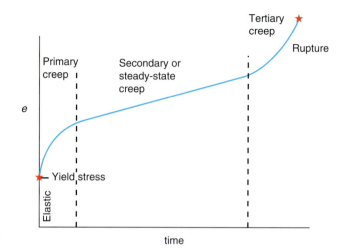

Figure 6.13 Strain–time curve for creep experiment. After initial elastic deformation, three types of creep can be defined. See text for discussion.

causes an increase in strain rate. This stage is terminated when a macroscopic fracture develops.

Steady-state creep is perhaps the most interesting one to structural geologists, because it appears that rocks can deform more or less steadily for extended periods of time. The constitutive equation during steady-state creep is the power law shown in Equation 6.24.

Constant strain rate experiments

During experiments where the strain rate is fixed, the sample first deforms elastically before accumulating permanent strain, i.e. the general behavior of rocks below the level of fracturing. An increase in stress is required at low temperatures in order to maintain a constant strain rate, consistent with the definition of strain hardening. For higher temperatures or low strain rates, strain does not harden and the deformation is close to steady state. A constitutive law in the form of Equation 6.24 is then in effect.

6.7 The role of temperature, water etc.

An increase in **temperature** lowers the yield stress or weakens the rock. We can see this from Figure 6.14 (a and b), where warmer-colored curves shows that marble can sustain less differential stress if temperature goes up. Think of a spring of the type shown in Figure 6.2a. We may be able to pull it quite hard at room temperature, but if we heat it up, we cannot pull it as hard before plastic (permanent deformation) occurs. In both examples, the temperature increase activates microscale crystal-plastic processes such as dislocation movements and diffusion (Chapter 10).

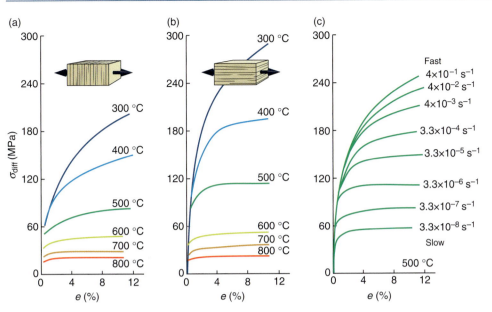

Figure 6.14 Stress–strain curves for Yule marble extended (a) normal and (b) parallel to the foliation. Data from Heard and Raleigh (1972). (c) Stress–strain curves for Yule marble at 500 °C for a variety of strain rates. From Heard (1960).

It also lowers the ultimate rock strength, which is the (differential) stress at which the rock fractures.

Increasing the **strain rate** means increasing the flow stress level, and this is clearly seen from the curves in Figure 6.14c. Since laboratory strain rates must be considerably lower than natural strain rates in most cases, this means that we must apply higher stresses to deform a rock in the laboratory, even if the temperature is the same in the laboratory experiment as in the natural setting we want to explore. As we learned from Figure 6.14a, b, increasing temperature weakens the rock and therefore counteracts this effect. An increased strain rate may also mean that less plastic strain accumulates because the rock may fracture at an earlier stage. Just think about glaciers that consist of slowly flowing ice that can be shattered by a swift hammer stroke. Parts of the Earth behave in a viscous manner because of low strain rates. Increasing the strain rate also makes the rock stronger. Conversely, rocks are weaker at lower strain rates because crystal-plastic processes can then more easily keep up with the applied stress.

Increased **presence of fluids** tends to weaken rocks, lower the yield stress and enhance crystal-plastic deformation. Fluid composition may however also influence rock rheological properties.

Increasing the **confining pressure** allows for larger finite strain to accumulate before failure and thus favors crystal-plastic deformation mechanisms. In simple terms, this is related to the difficulties involved in opening fractures at high confining pressure. The effect of increased confining pressure is counteracted by any increase in pore pressure (for porous rocks), which reduces the effective stress.

The presence of **fluids** (water) in the crystal lattice(s) may lower the strength or yield point significantly. Because of the increasing solubility of water with increasing pressure for many silicates, the effect is pressure dependent.

Non-isotropic features (Box 6.3) such as a preexisting foliation must always be considered. Figure 6.14a, b illustrates how a weak foliation in marble makes foliation-parallel extension more difficult (it takes a higher differential stress to obtain the same amount of strain). Note that the effect decreases with increasing temperature.

Even for monomineralic rocks, **grain size** and **crystallographic fabric** (preferred crystallographic orientation of minerals) may cause the rock to react differently, depending on the orientation of the applied forces. The anisotropy of olivine crystals is illustrated in Figure 6.15. The reaction to stress depends on the orientation of the applied forces relative to the dominant slip systems in olivine. Penetrative crystallographic fabrics may exist in the mantle, which may give the mantle a significant mechanical anisotropy that can influence location of rifting, strike-slip zones and orogeny. The effect of grain size itself depends on the microscale deformation mechanism during deformation, but in the plastic regime where dislocation creep (see Chapter 10) occurs, grain size reduction tends to imply strain weakening. In contrast, grain size reduction in the frictional (brittle) regime almost always implies strain hardening because of interlocking of grains.

Increasing the temperature, increasing the amount of fluid, lowering the strain rate and, in plastically deforming rocks, reducing the grain size all tend to cause strain weakening.

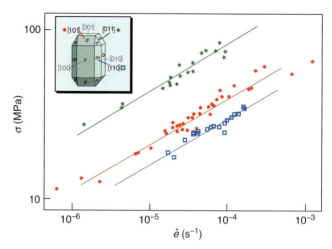

Figure 6.15 Stress–strain rate curves for dry olivine single crystals compressed in three different crystallographic directions. At any strain rate, deformation is easier for crystals shortened in the [110] direction, due to the lower strength of the (010)[100] slip system (see Chapter 10). Data from Durham and Goetze (1977).

BOX 6.3 | ISOTROPIC OR HOMOGENEOUS?

These are two related terms, but with a significant difference. **Homogeneous** means being similar or uniform, while **isotropic** means having properties that do not vary with direction. They are used in various aspects of structural geology, for instance about strain. **Homogeneous strain** means that the state of strain is identical in any one piece of the area or volume in question. It tells us nothing about the relative magnitudes of the principal strains. An **isotropic strain** means that the volume has been shortened or extended by the same amount in every direction. It involves no change in shape, only a change in volume. This is the **isotropic volume change** or volumetric strain from Chapter 3. Recall that volume can change by an equal change of length in all directions (isotropic) or preferentially in one direction (anisotropic).

Isotropic stress is a state where all three principal stresses are of equal magnitude. If they are not, stress can still be homogeneous if the state of stress is the same in every part of the rock.

A **fabric** (penetrative foliation and/or lineation) can be uniform throughout a sample, in which case the rock is homogeneous. However, a fabric represents an anisotropy because it causes the physical properties to be different in different directions. One could envisage that sliding preferentially takes place along the foliation, or that the stress–strain relationship is different in a sample when loaded parallel and perpendicular to the foliation (Figure 6.14a, b). Even a single, perfect crystal, which represents a homogeneous volume, can be anisotropic. This is the case with olivine, which has different mechanical properties along different crystallographic axes.

6.8 Definition of plastic, ductile and brittle deformation

Ductile and brittle are two of the most commonly used terms in structural geology, both within and outside of the fields of rheology and rock mechanics. Again we have the challenge that these terms are given different meanings by different geologists in different contexts.

In the field of rheology and rock mechanics, a **ductile material** is one that accumulates permanent strain (flows) without macroscopically visible fracturing, at least until a certain point where its ultimate strength is exceeded. On the contrary, a **brittle material** is one that deforms by fracturing when subjected to stress beyond the yield point. To a rock-mechanics-oriented geologist, ductile materials show classic stress–strain curves such as the ones shown in Figure 6.9.

Ductile structures are well represented in metamorphic rocks, i.e. rocks that have been deformed in the middle and lower part of the crust. Ductile deformation also occurs in soils and unconsolidated to poorly consolidated sediments where distributed deformation rather than discrete fracturing occurs, even though the **deformation mechanisms** responsible for the ductile deformation in these cases are quite different. Hence, ductile deformation is, as illustrated in Figure 6.16, a scale-dependent structural style and not related to microscale deformation processes:

Ductile deformation preserves continuity of originally continuous structures and layers, and describes a scale-dependent deformation style that can form by a range of deformation mechanisms.

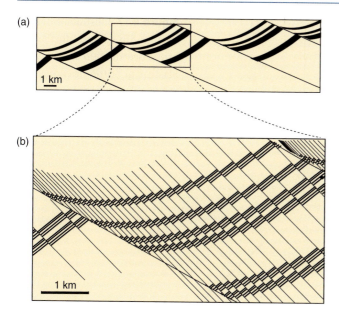

Figure 6.16 The scale-dependent nature of the ductile deformation style illustrated by a regional profile (top), where layers look continuous (ductile deformation style), and close-up (bottom), where it becomes apparent that the deformation is by multiple small faults. This example is directly relevant to seismic versus subseismic deformation.

Hence, a deformed area or volume can be ductile at the seismic or mesoscopic scale and brittle at the subseismic or microscopic scale (Figure 6.16).

In the search for a term that can be used specifically about the ductile deformation that occurs in the middle and lower crust, we turn to the term **plasticity**. The overall physical use of this term is as follows:

Plastic deformation is generally defined as the permanent change in shape or size of a body without fracture, produced by a sustained stress beyond the elastic limit of the material due to dislocation movement.

The discovery that plasticity can be explained in terms of the theory of dislocations (Chapter 10) was made in the 1930s and the theory implies that plasticity initiates where the mineralogy starts to deform by means of dislocation motion, which in general means 10–15 km depth. When needed, the term **crystal-plasticity** or **crystal-plastic deformation** can be used to distinguish this type of plasticity from that used in soil mechanics about water-rich soil. In this text we will restrict the term plasticity to intracrystalline deformation mechanisms other than brittle fracturing, rolling and frictional sliding of grains.

The latter microscale deformation mechanisms are referred to as brittle deformation mechanisms, indicating that the term brittle can be used about both deformation style and microscale deformation mechanisms. We can therefore talk about brittle deformation mechanisms, implying frictional deformation at the microscale, and about the **brittle regime**, where such mechanisms dominate. If we want to use an expression that is not also used about deformation style we can apply the term **frictional deformation** or **frictional regime**.

Plastic or crystal-plastic mechanisms occur at the atomic scale without breaking of atomic bonds by means of creep processes such as dislocation migration. The term plastic is here used in a broad sense. In a strict sense it should be distinguished from diffusion and dissolution, which also are non-brittle (non-frictional) deformation mechanisms. So, if we want to be very specific, there are brittle or frictional deformation mechanisms on one side and plastic, diffusion and dissolution mechanisms on the other.

Figure 6.17 summarizes how the term ductile refers to deformation style while the term brittle can refer to both style and microscale mechanism, and how plasticity and frictional deformation are directly related to microscale mechanism.

6.9 Rheology of the lithosphere

Rocks and minerals react differently to stress and depend on crystallographic anisotropy, temperature, presence of fluids, strain rate and pressure. Three minerals are particularly common in the lithosphere and therefore of particular interest. These are quartz and feldspar, which dominate the crust, and olivine, which controls the rheology of the upper mantle.

Quartz deforms by brittle mechanisms up to about 300–350 °C, which for typical continental temperature gradients corresponds to crustal depths of around 10–12 km. At greater depths, crystal-plastic mechanisms (creep mechanisms) and diffusion dominate. Feldspar, with its well-developed cleavages, is different because of the ease of cleavage-parallel cracking and the difficulty of dislocation glide and climb (crystal-plastic deformation mechanisms). It thus deforms in a brittle manner up to ~500 °C and depths of 20–30 km. However, olivine is brittle down to ~50 km depth.

A simple model subdivides the crust into an upper part dominated by brittle deformation mechanisms and a lower part where plastic flow dominates. The transition

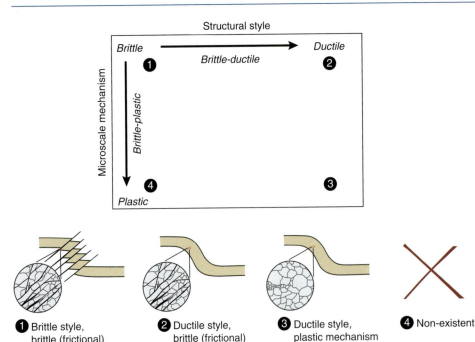

Figure 6.17 Illustration of the relationship between ductile and brittle deformation styles and plastic and brittle (frictional) microscale mechanisms. Example 4 is impossible because we cannot form a brittle deformation style by means of 100% plastic mechanisms.

is referred to as **the brittle–plastic transition** (brittle–ductile transition in some texts). For a theoretical uniform monolithologic crust with a gradual downward increase in temperature, we would expect a single brittle–plastic transitional zone at the depth where temperature-activated flow becomes important. The strength of the lower crust will then follow the constitutive equation shown in Equation 6.24, while the brittle (frictional) upper crust will follow another law known as Byerlee's law. Byerlee's law is based on frictional sliding experiments in the laboratory, since the brittle or frictional strength of the crust is governed by preexisting faults and fractures. Two different strength curves (brittle and plastic laws) can therefore be drawn, based on experimental data. We will return to this subject in Chapter 7.

In Figure 6.18 the plastic curve is based on laboratory experiments on quartz, while the brittle strength turns out to be more or less independent of mineralogy (with the exception of clay minerals). However, the stress required for thrusting is significantly larger than that required for normal faulting, so different curves are indicated in Figure 6.18 for Anderson's three different stress regimes. These curves also depend on the pore fluid pressure, because pore fluid pressure reduces friction.

The crust is not monomineralic, but it is generally assumed that there is enough quartz in the continental crust to control its rheology. This means that even if feldspar is still brittle at 15–20 km, quartz is sufficiently represented and distributed that the crust is

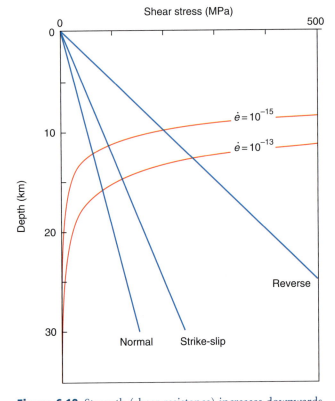

Figure 6.18 Strength (shear resistance) increases downwards through the brittle crust, until the temperature is high enough to activate plastic flow. Brittle and plastic deformation have two different strength profiles, and the intersection between the two defines the brittle–plastic transition. The plastic strength obeys a flow law as shown in Equation 6.24, which depends on strain rate. The plastic flow laws are derived from experimental deformation of quartzite (Gleason and Tullis 1995). Shear resistance for the three different stress regimes is shown.

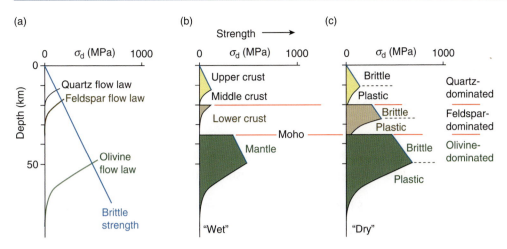

Figure 6.19 Rheologic stratification of continental lithosphere based on a combination of the brittle friction law and the plastic flow law derived experimentally for quartz (quartzite), feldspar (diabase) and olivine (dunite). Brittle–plastic transitions occur where the brittle (frictional) and plastic flow laws intersect. The strength profile depends on mineralogy and lithologic stratification. By choosing a quartz–feldspar–olivine stratification we get three brittle–plastic transitions. Note that dry rocks (c) are considerably stronger (can sustain higher differential stress) than wet rocks (b).

predominantly deforming plastically even at 10–12 km. However, if the stronger mineral feldspar becomes rheologically dominant at some depth, the flow law for feldspar becomes important (Figure 6.19a). In the olivine-dominated mantle, olivine is even stronger and defines a new intercept with the brittle strength curve. Hence, varying mineralogical composition through the lithosphere can lead to several layers of alternating brittle and plastic rheologies, known as **rheologic stratification** (Figure 6.19b, c).

The brittle–plastic transition is generally gradual or recurring over a wide zone in the continental crust.

The transition is also influenced by the presence of fluids, and the fact that "dry" rocks are more resistant to deformation is illustrated in Figure 6.19. Furthermore, strain rate shifts the plastic flow laws (and thus the brittle–plastic transition) vertically, as seen in Figure 6.18.

Because the crust is layered it is important to obtain as much information as possible about composition in order to predict the strength profile or rheological stratigraphy of the crust. Such information goes into the modeling of crustal-scale deformation, such as rifting and orogeny.

Summary

Rheology and its implications for how rocks deform are important to keep in mind when we are studying deformation structures in naturally deformed rocks. A fundamental distinction must be made between elasticity and related laws and moduli on one side, and permanent deformation and plasticity on the other. It is quite useful and fun to explore the concepts presented in this chapter during everyday activities, using rubber, plastics, modeling putty, clay, springs, plaster and many other things. The kitchen, where different types of food (syrup, honey, chocolate, dough, pudding, jelly and more) make a great selection of materials at different temperatures, can also be of good use. Before exploring and eating, here are some important points and review questions that should be familiar by now:

• Elastic theory is used for relatively small strains, from the millimeter scale to lithospheric scale. An example of the latter is the elastic subsidence of the lithosphere caused by ice sheets up to several kilometers thick during regional glaciations. The fact that the lithosphere rebounds when the ice melts tells us that it can be modeled as an elastic plate, and the rate at which it rebounds tells us something about the mantle viscosity and elastic properties of the lithosphere.

- Stress and elastic strain are related through Young's modulus: A low Young's modulus means little resistance to deformation.

- Poisson's ratio describes how much a material that is shortened in one direction expands in the two other directions, or how much a material that is stretched in one direction contracts in the plane perpendicular to the stretching direction.

- Elastic deformation of rocks reaches a critical stress or strain level (yield point) where permanent deformation starts to accumulate.

- Mechanically, plastic deformation occurs when permanent strain keeps accumulating under a constant stress level. More generally, plastic deformation is the deformation of rock by intracrystalline (non-cataclastic) flow.

- For plastic deformation, strain rate is related to stress through a non-linear (power-law) relationship called a flow law.

- Strain hardening and softening mean that the properties of the deforming rock change during deformation.

- The simple model of a predominantly plastically flowing lower crust overlain by a strong, brittle upper crust and underlain by a stronger upper mantle is a simple but useful first approximation to the large-scale rheological stratification of the crust.

- The idealized conditions in models and the laboratory are seldom met in nature, so models such as linear elasticity, viscosity etc. must be used with care.

Review questions

1. What is the difference between rheology and rock mechanics?

2. What is a constitutive law or equation?

3. What does isotropic mean?

4. What is an elastic material?

5. What is an incompressible medium, and what is its Poisson's ratio?

6. Some media are easier to elastically bend, stretch or shorten than others – we could say that there is a difference in stiffness. What constants describe the stiffness of an elastic material or its resistance to elastic deformation?

7. What is the yield stress and what happens if it is exceeded?

8. What is the difference between linear elastic and linear viscous?

9. What types of materials are truly viscous? What parts of the Earth can be modeled as being viscous?

10. What does it mean that a rock layer is more competent than its neighboring layers?

11. What could cause strain softening and strain hardening in a deforming rock?

12. What is the difference between plastic deformation and creep?

13. What controls the locations of brittle–plastic transitions in the lithosphere?

E-MODULE

 The e-learning module called *Rheology* is recommended for this chapter.

FURTHER READING

General

Jaeger, C., 2009, *Rock Mechanics and Engineering.* Cambridge: Cambridge University Press.

Jaeger, J. C. and Cook, N. G. W., 1976, *Fundamentals of Rock Mechanics.* London: Chapman & Hall.

Karato, S. and Toriumi, M. (Eds.), 1989, *Rheology of Solids and of the Earth.* New York: Oxford University Press.

Ranalli, G., 1987, *Rheology of the Earth.* Boston: Allen & Unwin.

Turcotte, D. L. and Schubert, G., 2002, *Geodynamics.* Cambridge: Cambridge University Press.

Strain rates

Pfiffner, O. A. and Ramsay, J. G., 1982, Constraints on geological strain rates: arguments from finite strain states of naturally deformed rocks. *Journal of Geophysical Research* **87**: 311–321.

Chapter 7

Fracture and brittle deformation

Brittle structures such as joints and faults are found almost everywhere at the surface of the solid Earth. In fact, brittle deformation is the trademark of deformation in the upper crust, forming in areas where stress builds up to levels that exceed the local rupture strength of the crust. Brittle structures can form rather gently in rocks undergoing exhumation and cooling, or more violently during earthquakes. In either case, brittle deformation by means of fracturing implies instantaneous breakage of crystal lattices at the atomic scale, and this type of deformation tends to be not only faster, but also more localized than its plastic counterpart. Brittle structures are relatively easily explored in the laboratory, and the coupling of experiments with field and thin-section observations forms the basis of our current understanding of brittle deformation. In this chapter we will look at the formation of various small-scale brittle structures and the conditions under which they form.

7.1 Brittle deformation mechanisms

Once the differential stress in an unfractured rock exceeds a certain limit, the rock may accumulate permanent strain by plastic flow, as discussed in Chapter 6. In the **frictional regime** or **brittle regime**, however, the rock will deform by fracturing once its rupture strength is reached. During brittle fracturing, grains are crushed and reorganized and strain (displacement) becomes more localized.

The brittle regime is where the physical conditions promote brittle deformation mechanisms such as frictional sliding along grain contacts, grain rotation and grain fracture.

In some cases it is important to characterize the amount of fracture in a deformed rock, and a distinction is made between brittle deformation that does and does not involve fracture. Frictional deformation without the generation of fractures typically occurs in relatively poorly consolidated porous rocks and sediments (soils). In such rocks and sediments, frictional slip occurs along existing grain boundaries, and pore space enables grains to move relative to their neighboring grains, as shown in Figure 7.1a. Thus, the grains translate and rotate to accommodate frictional grain boundary slip, and the whole process is called **particulate flow** or **granular flow**. As always in the brittle regime, grain boundary sliding is influenced by friction, and the mechanism is therefore called **frictional sliding**. This means that a certain friction-controlled resistance against sliding must be overcome for frictional sliding to occur. Do not confuse this with the non-frictional grain boundary sliding that takes place in the plastic regime (Chapter 10).

The angle of repose in loose sand is controlled by the friction between individual sand grains. The higher the friction between the grains, the higher the angle of repose. In this case gravity exerts a vertical force on grain contact areas, and the shear stress will depend on the orientation of the surfaces, as discussed in Chapter 4.

Frictional sliding of grains may be widely distributed throughout a rock volume, but can also be localized to millimeter- to decimeter-wide zones or bands. Granular flow results in a ductile shear zone where lamination can be traced continuously from one side of the zone to the other. This is a type of ductile shear zone that is governed by brittle deformation mechanisms.

In other cases new fractures form during deformation. This always happens during brittle permanent deformation of non-porous rocks, but can also occur in porous rocks if the stress on grain contact areas becomes high enough. In the case of porous rocks we often identify **intragranular fractures**, which are fractures restricted to single grains (Figure 7.2a). **Intergranular fractures** are fractures that extend across a number of grains (Figure 7.2b), and characterize brittlely deformed low-porosity or non-porous rocks. The fracture and crushing of grains, coupled with frictional sliding along grain contacts and grain rotation, is called **cataclasis**. Intense cataclasis occurs in thin zones along slip or fault surfaces where extreme grain size reduction goes on. More moderate cataclastic deformation can occur in somewhat wider brittle or cataclastic shear zones. In this case the fragments resulting from grain crushing flow during shearing. This process is referred to as **cataclastic flow** (Figure 7.1b).

Particulate flow involves grain rotation and frictional sliding between grains, while cataclastic flow also involves grain fracturing or cataclasis. Both can give rise to structures that appear ductile at the mesoscopic scale.

Strong grain crushing without evidence of shear offset has also been observed and is called **pulverization**. The process of pulverization is not well understood, but

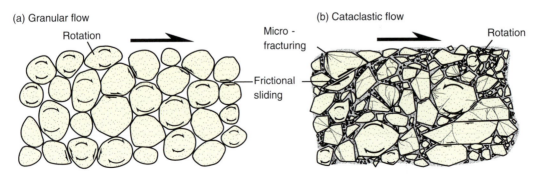

(a) Granular flow
Rotation
Micro-fracturing
Frictional sliding

(b) Cataclastic flow
Rotation

Figure 7.1 Brittle deformation mechanisms. Granular flow is common during shallow deformation of porous rocks and sediments, while cataclastic flow occurs during deformation of well-consolidated sedimentary rocks and non-porous rocks.

(a)   (b)

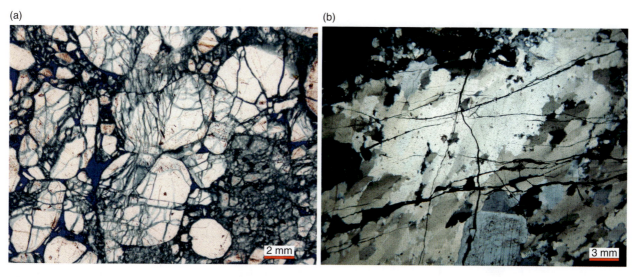

Figure 7.2 (a) Intragranular fractures in cataclastically deformed porous sandstone (Mesa Verde Group, Salina, Utah). Dark blue color is epoxy-filled pore space. (b) Intergranular fractures in metamorphic rock.

seems to be related to very high strain rates (>100 s^{-1}) and may be related to large earthquake events producing very high rupture rates.

7.2 Types of fractures

What is a fracture?

Strictly speaking, a fracture is any planar or subplanar discontinuity that is very narrow in one dimension compared to the other two and forms as a result of external (e.g. tectonic) or internal (thermal or residual) stress. Fractures are discontinuities in displacement and mechanical properties where rocks or minerals are broken, and reduction or loss of cohesion characterizes most fractures. They are often described as surfaces, but at some scale there is always a thickness involved. Fractures can be separated into shear fractures (slip surfaces) and opening or extension fractures (joints, fissures and veins), as illustrated in Figures 7.3 and 7.4. In addition, closing or contraction fractures can be defined.

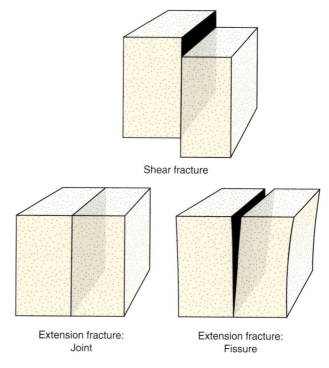

Figure 7.3 Three types of fracture.

Fractures are very narrow zones, often thought of as surfaces, associated with discontinuities in displacement and mechanical properties (strength or stiffness).

A **shear fracture** or **slip surface** is a fracture along which the relative movement is parallel to the fracture. The term shear fracture is used for fractures with small (mm- to dm-scale) displacements, while the term

fault is more commonly restricted to discontinuities with larger offset. The term slip surface is used for fractures with fracture-parallel movements regardless of the amount of displacement and is consistent with the traditional use of the term fault. Fractures are commonly referred to as cracks in material science and rock mechanics oriented literature.

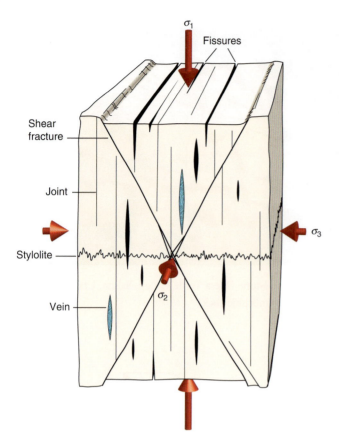

Figure 7.4 The orientation of various fracture types with respect to the principal stresses.

Extension fractures are fractures that show extension perpendicular to the walls. **Joints** have little or no macroscopically detectable displacement, but close examination reveals that most joints have a minute extensional displacement across the joint surfaces, and therefore they are classified as true extension fractures. Extension fractures are filled with gas, fluids, magma or minerals. When filled with air or fluid we use the term **fissure**. Mineral-filled extension fractures are called **veins**, while magma-filled fractures are classified as **dikes**. Joints, veins and fissures are all referred to as extension fractures.

Contractional planar features (anticracks) have contractional displacements across them and are filled with immobile residue from the host rock. Stylolites are compactional structures characterized by very irregular, rather than planar, surfaces. Some geologists now regard stylolites as **contraction fractures** or **closing fractures**, as they nicely define one of three end-members in a complete kinematic fracture framework together with shear and extension fractures. Such structures are known as **anticracks** in the engineering-oriented literature.

Rock mechanics experiments carried out at various differential stresses and confining pressures set a convenient stage for studying aspects of fracture formation (Figure 7.5), and we will refer to experimental rock deformation on several occasions in this chapter (also

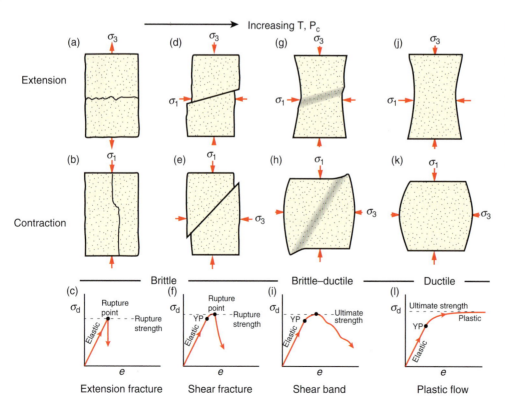

Figure 7.5 Experimental deformation structures that develop under extension and contraction. Initial elastic deformation is seen for all cases, while ductility increases with temperature (T) and confining pressure (P_c). YP, yield point.

BOX 7.1 | DEFORMING ROCKS IN THE LABORATORY

The mechanical properties of rocks are explored in rock mechanics laboratories, where samples are exposed to various stress fields that relate to different depths and stress regimes in the crust. **Uniaxial** rigs can be used to test the uniaxial compressive or tensile strength of rocks. **Triaxial** tests, where $\sigma_1 > \sigma_2 = \sigma_3$, are more common, where rock cylinders are loaded in the axial direction while the sample is confined in fluid that can be pumped up to a certain confining pressure. A typical triaxial rig can build up an axial stress of 2–300 MPa and a confining pressure of up to 50–100 MPa or more. Sample and fluid are commonly separated by a membrane to avoid the fluid entering the sample and changing its mechanical properties. For porous rocks or sediments it may be possible to control the pore pressure (e.g. up to 50 MPa). The distance between the pistons is monitored together with axial loading and confining pressure. The **ringshear** apparatus is used to explore the effect of large shear strain under vertical compression of up to about 25 MPa.

Uniaxial deformation rig, used to find the uniaxial strength of rocks. Experiments show that, in general, fine-grained rocks are stronger than coarse-grained ones, and the presence of phyllosilicates lowers the strength.

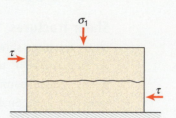

Shearbox experiment where the resistance against shear is explored. The higher the normal stress, the higher the shear stress necessary to activate the fracture. The roughness of the fracture is also important.

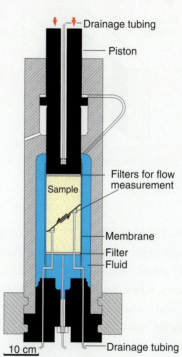

A triaxial rig where fluid (oil or water) can be pumped in and influence the behavior of an existing fracture.

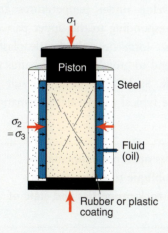

Triaxial rig. Oil pressure is pumped up in a chamber around the sample to increase the confining pressure.

Ringshear apparatus, where the amount of shear strain that can be imposed on the sample is unlimited. Loose sediment is added and processes such as clay smear and cataclasis can be studied.

see Box 7.1). Similarly, numerical modeling has added greatly to our understanding of fracture growth, particularly the field called **linear elastic fracture mechanics**. In the field of fracture mechanics it is common to classify the displacement field of fractures or cracks into three different modes (Figure 7.6). **Mode I** is the opening (extension) mode where displacement is perpendicular to the walls of the crack. **Mode II** (sliding mode) represents slip (shear) perpendicular to the edge and **Mode III** (tearing mode) involves slip parallel to the edge of

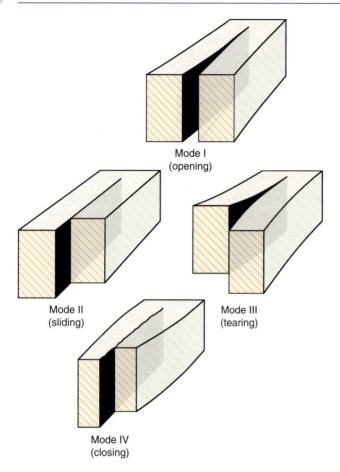

Mode I
(opening)

Mode II
(sliding)

Mode III
(tearing)

Mode IV
(closing)

Figure 7.6 Mode I, II, III and IV fractures.

the crack. Modes II and III occur along different parts of the same shear fracture and it may therefore be confusing to talk about Mode II and Mode III cracks as individual fractures. Combinations of shear (Mode II or III) fractures and tension (Mode I) fractures are called **hybrid cracks** or fractures. Furthermore, the term **Mode IV** (closing mode) is sometimes used for contractional features such as stylolites. The mode of displacement on fractures is an important parameter, for instance when fluid flow through rocks is an issue.

Extension fractures and tensile fractures

Extension fractures ideally develop perpendicular to σ_3 and thus contain the intermediate and maximum principal stresses ($2\theta = 0°$, see Figure 4.6). In terms of strain, they develop perpendicular to the stretching direction under tensile conditions, as shown in Figure 7.5a, and parallel to the compression axis during compression tests (Figure 7.5b). Because of the small strains associated with most extension fractures, stress and strain axes more or less coincide.

Joints are the most common type of extension fracture at or near the surface of the Earth and involve very small strains. Fissures are extension fractures that are more open than joints, and are characteristic of the uppermost few hundred meters of the solid crust, where they may be up to several kilometers long (Figure 7.7).

Extension fractures are typical for deformation under low or no confining pressure, and form at low differential stress. If extension fractures form under conditions where at least one of the stress axes is tensile, then such fractures are true **tensile fractures**. Such conditions are generally found near the surface where negative values of σ_3 are more likely. They can also occur deeper in the lithosphere, where high fluid pressure reduces the effective stress (Section 7.6). Many other joints are probably related to unloading and cooling of rocks, as indicated in Section 5.4.

Shear fractures

Shear fractures show fracture-parallel slip and typically develop at 20–30° to σ_1, as seen from numerous experiments under confined compression (Figure 7.5d, e) (see Box 7.2). Such experiments also show that they commonly develop in conjugate pairs, bisected by σ_1. Shear fractures develop under temperatures and confining pressures corresponding to the upper part of the crust. They can also form near the brittle–plastic transition, where they tend to grow into wider bands or zones of cataclastic flow. Such shear factures result in strain patterns otherwise typical for plastic deformation (Figure 7.5g, h).

While extension fractures open perpendicular to σ_3, shear fractures are oblique to σ_3 by an angle that depends mostly on rock properties and state of stress.

As mentioned in the previous chapter and shown in Figure 7.8, brittle and plastic deformation show different stress–strain curves (blue versus red curves in Figure 7.8): the more ductile the deformation, the greater the amount of plastic deformation prior to fracturing. It is also interesting to note the relationship between confining pressure (depth) and strain regime (contractional or extensional) shown in Figure 7.9. The experimental data indicate that the brittle–plastic transition occurs at higher confining pressure under extension than under contraction. Temperature (Figure 7.9) and

Figure 7.7 Fissures formed in Thingvellir, Iceland, along the rift axis between the Eurasian and Laurentian plates. The fissures are open extension fractures in basalt, but the vertical displacement (right-hand side down) indicates a connection with underlying faults.

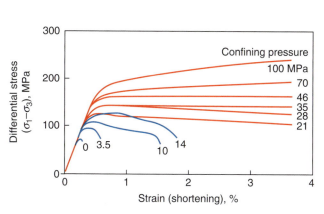

Figure 7.8 Stress–strain curves for triaxial compression of marble for a range of confining pressures. Increasing the confining pressure increases the differential stress that the rock can sustain before failure (blue curves). Above a critical confining pressure the rock retains its strength as it deforms plastically (red curves). From Paterson (1958).

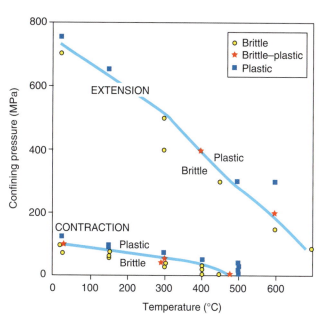

Figure 7.9 Variation of the brittle–plastic transition as a function of confining pressure and temperature for the Solenhofen limestone. From Heard (1960).

BOX 7.2 | WHY SHEAR FRACTURES DO NOT FORM AT 45° TO THE LARGEST PRINCIPAL STRESS

Navier and Coulomb both showed that shear fractures do not simply form along the theoretical surfaces of maximum shear stress. Maximum resolved shear stress on a plane is obtained when the plane is oriented 45° to the maximum principal stress ($\theta = 45°$). This fact is easily extracted from the Mohr diagram, where the value for shear stress is at its maximum when $2\theta = 90°$. However, in this situation the normal stress σ_n across the plane is fairly large. Both σ_s and σ_n decrease as θ increases, but σ_n decreases faster than σ_s. The optimal balance between σ_n and σ_s depends on the angle of internal friction ϕ, and is predicted by the Coulomb criterion to be around 60° for many rock types. At this angle ($\theta = 60°$) σ_s is still large, while σ_n is considerably less. The angle depends also on the confining pressure (depth of deformation), temperature and pore fluid, and experimental data indicate that there is a wide scatter even for the same rock type and conditions.

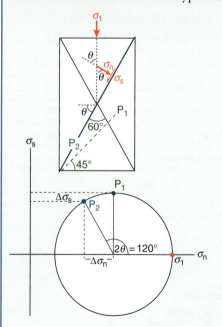

P_1 is the plane of maximum resolved shear stress ($2\theta = 90°$) and forms at 45° to σ_1. The plane P_2 oriented at 30° to σ_1 has a slightly lower shear stress (the difference is $\Delta\sigma_s$), but a much lower normal stress (by $\Delta\sigma_n$). It is therefore easier for a shear fracture to form along P_2 than along P_1.

strain rate are other important factors, as discussed in previous chapters.

7.3 Failure and fracture criteria

We learned in Chapter 6 that a rock's response to stress depends on the level of stress or amount of accumulated strain, and on factors such as anisotropy, temperature, strain rate, pore fluid and confining pressure. In the brittle regime a deforming rock accumulates elastic strain before it ruptures (fractures) at a certain critical stress level. In the brittle–plastic transition there tends to be an intermittent phase of plastic deformation prior to brittle failure, and the failure does not necessarily create an instantaneous through-going fracture, but rather a shear zone or shear band dominated by cataclastic flow. This contrasts with the plastic regime (Figure 7.5j–l) where strain is more broadly distributed and dominated by plastic deformation mechanisms.

While the main focus of Chapter 6 was on elastic–plastic deformation, we now focus on the brittle part. Key questions are *when* and *how* does a rock fracture. Let us look at the first question first. For a given rock under constant temperature and constant positive confining pressure, fracture depends on the differential stress ($\sigma_1 - \sigma_3$) as well as the mean stress (($\sigma_1 + \sigma_3)/2$). If there is no differential stress, then the state of stress is lithostatic and there is no force pulling or pushing our rock volume in any particular direction. The only exception is the potential collapse of the porosity structure in highly porous rocks, but in order to make distinct fractures, differential stress is generally needed.

Fracture initiation requires a differential stress that exceeds the strength of the rock.

The strength of a rock depends on the confining pressure or depth of burial. In the brittle, upper part of the crust, the strength is lowest near the surface and increases downwards. This is easily explored in experiments such as the ones shown in Figure 7.8, where both confining pressure and directed axial stress are varied. We see from this figure that:

Increasing the confining pressure makes it necessary to increase the differential stress in order to fracture a rock.

In the next section we will see how this connection between confining pressure (burial depth) and differential stress can be described by means of a simple relationship between the critical normal and shear stresses. The relationship is known as the Coulomb fracture criterion for confined compression.

The Coulomb fracture criterion

At the end of the seventeenth century, the French physicist Charles Augustin de Coulomb found a criterion that could predict the state of stress at which a given rock under compression is at the verge of failure, commonly described as being **critically stressed**. The criterion considers the critical shear stress (σ_s, or τ) and normal stress (σ_n) acting on a potential fracture at the moment of failure, and the two are related by a constant $\tan\phi$, where ϕ is called the angle of internal friction:

$$\sigma_s = \sigma_n \tan\phi \qquad (7.1)$$

The Coulomb fracture criterion indicates that the shear stress required to initiate a shear fracture also depends on the normal stress across the potential shear plane: the higher the normal stress, the higher the shear stress needed to generate a shear fracture. Tan ϕ is commonly called the **coefficient of internal friction** μ. For loose sand it relates to the friction between sand grains and the critical slope angles of the sand (the angle of repose, $\sim 30°$), but for solid rocks it is merely a constant that varies from 0.47 to 0.7. A value of 0.6 is often chosen for μ for general calculations.

A fracture criterion describes the critical condition at which a rock fractures.

Three centuries after Coulomb, the German engineer Otto Mohr introduced his famous circle (Section 4.7) in σ_s–σ_n-space (Mohr space), and the Coulomb criterion could conveniently be interpreted as a straight line in Mohr space, with μ representing the slope and ϕ the slope angle (note that there is also a second line representing the conjugate shear fracture on which shear stress magnitude is identical but of opposite sign).

Coulomb realized that a fracture only forms if the internal strength or cohesion C of the rock is exceeded. The complete Coulomb fracture criterion (also called the Navier–Coulomb, Mohr–Coulomb or Coulomb–Mohr fracture criterion) is therefore:

$$\sigma_s = C + \sigma_n \tan\phi = C + \sigma_n\mu \qquad (7.2)$$

The constant C represents the critical shear stress along a surface across which $\sigma_n = 0$ (Figure 7.10). C is also called

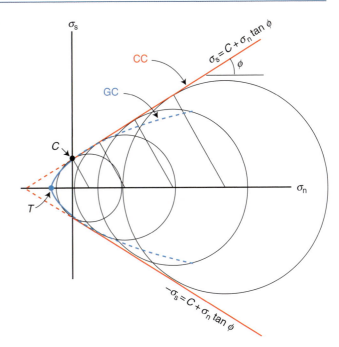

Figure 7.10 The Coulomb fracture criterion occurs as two straight lines (red) in the Mohr diagram. The circles represent examples of critical states of stress. The blue line represents the Griffith criterion for comparison. The combination of the two is sometimes used (GC in the tensile regime and CC in the compressional regime). CC, Coulomb criterion; GC, Griffith criterion; C, the cohesive strength of the rock; T, the tensile strength of the rock.

the **cohesive strength**, and has its counterpart in the **critical tensile strength** T of the rock.

The Mohr diagram provides a convenient way of interpreting the meaning of these constants (Figure 7.10). Because Equation 7.2 is the general formula for a straight line, C represents the intersection with the vertical (σ_s) axis and T denotes the intersection with the horizontal (σ_n) axis. At point C it is clear that $\sigma_n = 0$, while at T, $\sigma_s = 0$. As an example, loose sand has no compressive or tensile strength, which means that $T = C = 0$ and the Coulomb fracture criterion reduces to Equation 7.1. The more the sand lithifies, the higher the C-value. However, lithification will change not only the C-value, but also ϕ and T. In general, C, T and ϕ vary from one rock type to another. For sand(stone), they all increase with increasing degree of lithification.

To see if a rock obeys the Coulomb fracture criterion and, if so, to determine C, T and μ for a given rock or sediment, laboratory tests are performed. A deformation rig is generally used, where the confining pressure as well as the axial load can be adjusted. The state of stress at failure is recorded and plotted in the Mohr diagram. This can be done for many different critical states of stress (Figure 7.10 and 7.11), and for so-called Coulomb materials the line that tangents the Mohr circles represents the Coulomb fracture criterion (Equation 7.2). This line is called the **Coulomb failure envelope** for the rock.

Ideally, the point at which a Mohr circle touches the failure envelope represents the orientation of the plane of failure (remember the 2θ angle in the Mohr diagram), as well as the shear stress and normal stress on the plane at the moment of failure. Any Mohr circle that does not touch the envelope represents a stable state of stress (no fracturing possible; Figure 7.12a). The Coulomb failure

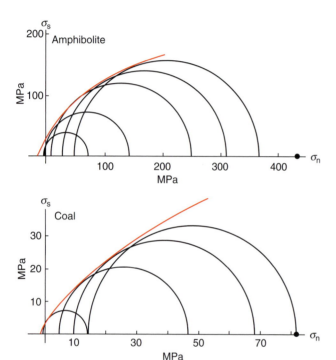

Figure 7.11 The Mohr envelope for amphibolite and coal based on triaxial tests. When the confining pressure is increased, the strength of the rock increases, and a new circle can be drawn in the figure. Note that the envelope diverges from the linear trend defined by the Coulomb criterion. From Myrvang (2001).

envelope is always positive for brittle fracturing. This means that the higher the mean stress (or confining pressure), the higher the differential stress required for failure. In other words:

The deeper into the brittle part of the crust, the stronger the rock, and the larger the differential stress required to fracture it.

Note also that the effect of the intermediate principal stress (σ_2) is ignored and that the fracture plane always contains σ_2, in agreement with Anderson's theory of faulting (Figure 5.12).

The orientation of the fracture can be expressed in terms of the angle of internal friction (φ) and the orientation of the fracture (θ):

$$\phi = 45° - \frac{\theta}{2} \tag{7.4}$$

or

$$\theta = 90° - 2\phi \tag{7.5}$$

Most rocks have $\varphi \approx 30°$ ($\mu \approx 0.6$), which means that the angle between σ_1 and the fracture, which is $90° - \theta$ (see Figure 4.1), is around $30°$. Thus, Andersonian normal and reverse faults dip at $\sim 60°$ and $30°$, respectively (Figure 7.13).

The Mohr failure envelope

The Mohr failure envelope is the envelope or curve in the Mohr diagram that describes the critical states of stress over a range of differential stress, regardless of whether it obeys the Coulomb criterion or not. The envelope, shown as a red line in Figure 7.14, separates the stable field, where the rock does not fracture, and the unstable field, which is in principle unachievable because fracture prevents such states of stress occurring.

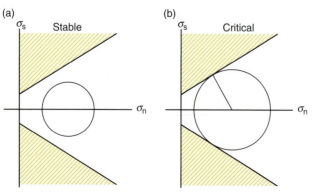

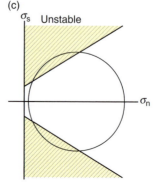

Figure 7.12 (a) Stable state of stress. (b) Critical situation, where the circle touches the envelope. This is when the rock is at the verge of failure, also called critically stressed. (c) Unstable situation where the state of stress is higher than that required for failure.

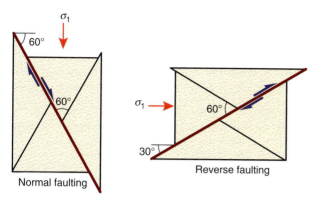

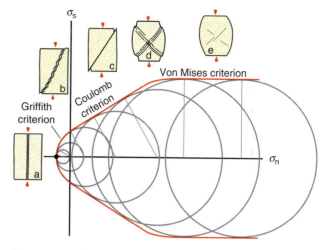

Figure 7.13 The angle between the maximum principal stress and the shear plane is commonly found to lie close to 30°. This implies that normal faults dip steeper (60°) than reverse faults (30°).

Figure 7.15 Three different fracture criteria combined in Mohr space. Different styles of fracturing are related to confining pressure: (a) Tensile fracture, (b) hybrid or mixed-mode fracture, (c) shear fracture, (d) semi-ductile shear bands, (e) plastic deformation.

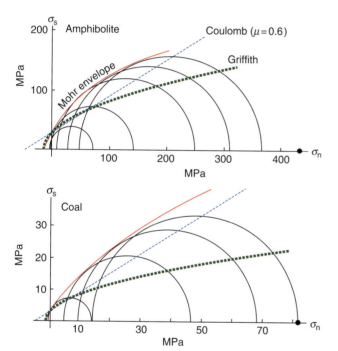

Figure 7.14 The Griffith and Coulomb fracture criteria superimposed on the experimental data presented in Figure 7.11. The criteria are placed so that they intersect the vertical axis together with the Mohr envelope. Neither of the criteria fit the data very well. The Griffith criterion works well for tensile stress (left of origin), but shows a too low slope in the entire compressional regime. The Coulomb criterion approaches the envelope for high confining pressure (right side of the diagram).

Each type of rock has its own failure envelope, and it is found experimentally by fracturing samples of the rock under different confining and differential stress. In some cases the Coulomb fracture criterion is a

reasonably good approximation for a certain stress interval, and in other cases the envelope is clearly non-linear (Figure 7.14). It is common for the envelope to flatten as the ductile regime is approached. In fact, the ductile regime can be approximated by a constant shear stress criterion (horizontal envelope), known as the **von Mises criterion** (σ = constant; Figure 7.15). A consequence of the non-linear shape of the envelope is that the angle θ between σ_1 and the failure plane decreases with increasing value of σ_3.

The tensile regime

The Coulomb criterion predicts the critical state of stress needed to create a shear fracture. Experimental data show that it does not successfully predict tensile fractures. Also, the fact that it relies on the angle of internal friction, which is physically meaningless for tensile normal stress, indicates that the Coulomb criterion is inappropriate in the tensional regime (left of origin in the Mohr diagram). Experiments suggest that the Mohr envelope in the tensile regime is shaped like a parabola. Thus, to cover the full range of stress states in the crust, it is necessary to combine different fracture criteria, such as the parabolic failure criterion for the tensile field, the Coulomb criterion for brittle fracturing in the compressive regime, and the von Mises criterion in the plastic regime (Figure 7.15).

The point where the Mohr envelope intersects the horizontal axis of the Mohr diagram represents the critical stress at which tensile fractures start to grow and represents the **critical tensile stress** T. T is found experimentally to be lower than the cohesive strength C, and varies from rock to rock. Why does T vary so much? Griffith suggested that it is related to the shapes, sizes and distribution of microscopic flaws in the deforming sample.

Griffith's theory of fracture

Around 1920, the British aeronautical engineer Alan Arnold Griffith extended his studies of fracture to the atomic level. He noted a large difference in theoretical strength between perfectly isotropic material and the actual strength of natural rocks measured in the laboratory. Griffith based the theoretical brittle tensile strength on the energy required to break atomic bonds. The uniaxial tensile strength of flawless rock is calculated to be around 1/10 of Young's modulus. For a strong rock E could reach ~100 GPa (Table 6.1), which means a tensile strength of about 10 GPa (10 000 MPa). Experiments indicate that the tensile strength is closer to 10 MPa. Why this enormous discrepancy between theory and practice?

Griffith's answer was that natural rocks and crystals are far from perfect. Rocks contain abundant microscopic flaws, and microcracks, voids, pore space and grain boundaries are all considered as microscopic fractures in this context. For simplicity, Griffith modeled such flaws as strongly elliptical microfractures, now known as **Griffith (micro)cracks**. He considered the stress concentrations associated with these microfractures and the energy that it takes for them to grow and connect. He then obtained much more realistic (although not perfect) estimates of tensile strength.

Microscopic cracks, pores and other flaws weaken rocks.

In contrast to Coulomb, Griffith found a non-linear relationship between the principal stresses for a critically stressed rock (a rock on the verge of fracture). This relationship, which is called the **Griffith fracture criterion**, is given by the equation:

$$\sigma_s^2 + 4T\sigma_n - 4T^2 = 0 \tag{7.6}$$

This equation can also be represented in the Mohr diagram, where it defines a parabola where the tensile strength T is the intersection with the horizontal axis. The intersection between the Griffith parabola and the

vertical axis is found by setting $\sigma_n = 0$ in the equation above, which gives us $\sigma_s = 2T$. This value corresponds to C in the Mohr criterion. In other words, the **cohesive strength** of a rock is twice its tensile strength ($C = 2T$), which is in close agreement with experimental data. We can take advantage of this new information and reformulate the Coulomb fracture criterion as:

$$\sigma_s = 2T + \sigma_n\mu \tag{7.7}$$

With this formulation it is easy to combine the Coulomb criterion for the compressional stress regime with the Griffith criterion for the tensile regime.

Griffith's important contribution is that the brittle strength of rock is controlled by randomly oriented and distributed intragranular microfractures in the rock. Microfractures with orientations close to that of maximum shear stress are then expected to grow faster than others, and will link and eventually form through-going fracture(s) in the rock.

For non-porous rocks the Griffith fracture criterion is a reasonably realistic approximation for the compressional regime as well. However, the Griffith criterion predicts that the uniaxial compressive strength should be eight times the uniaxial tensile strength (Figure 7.16), while experiments indicate that the uniaxial compressive strength of rocks is 10–50 times the uniaxial tensile strength. This discrepancy has resulted in several alternative fracture criteria. However, for porous media, such as sand and sandstone, the Coulomb criterion is quite realistic and can successfully be combined with the Griffith criterion (Figure 7.17).

7.4 Microdefects and failure

Griffith assumed that tensile fractures develop from planar microdefects or microfractures. In Griffith's model a fracture develops by a process where microfractures that are favorably oriented with respect to the external stress field grow and connect to form a through-going macrofracture. As illustrated in Figure 7.18, both tensile fractures (extension fractures) and shear fractures (faults) can form in this manner.

Observations indicate that microfractures occur at anomalously high frequencies near macroscopic fractures. This information suggests that microfractures form in a **process zone** ahead of a propagating macrofracture. In this zone microdefects expand and connect

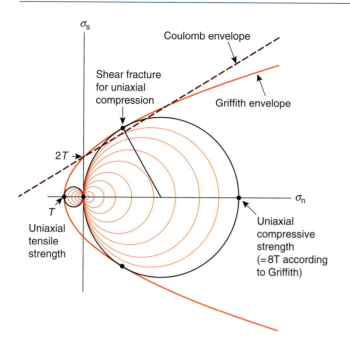

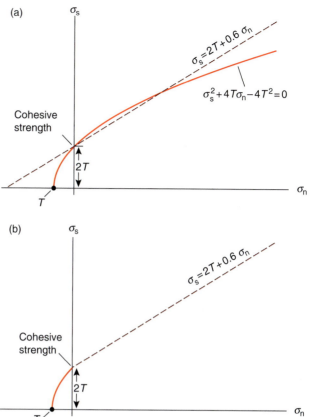

Figure 7.16 Illustration of the meaning of the terms uniaxial tensile and compressive strength in the Mohr diagram. Uniaxial means that only $\sigma_1 \neq 0$, which is obtained in a uniaxial deformation rig where the confining pressure is zero. By gradually compressing the rock sample, the uniaxial compressive strength is reached when a shear fracture first forms. By pulling the sample until a tensile fracture forms, the uniaxial tensile strength is found. Note that the uniaxial compressive strength is much larger than the tensile strength for the same rock and conditions.

Figure 7.17 (a) Comparison of the Griffith and Coulomb fracture criteria (the coefficient of internal friction is chosen to be 0.6). (b) The combined Griffith–Coulomb criterion.

so that the macrofracture can grow. The process zone is in some ways similar to the frontal part of the damage zone that encloses a macroscopic fault, as discussed in Section 8.5. Many interesting things are going on in the process zone, such as the effect of increasing rock volume due to the growth of microfractures, which may lower the local pore pressure and temporarily strengthen the rock. But the most important aspect of microfractures is the stress concentration that occurs at their tips. This explains why they can grow into macroscopic fractures.

The Griffith criterion of fracture is based on the fact that stress is concentrated at the edges of open microfractures in an otherwise non-porous medium. This makes intuitive sense, since the stress that should have been transferred across the open fracture must "find its way" around the edges. If the microfracture is a circular pore space, then the stress concentration at the edge (around the circle) will be three times the remote stress, as illustrated in Figures 7.19 and 7.20.

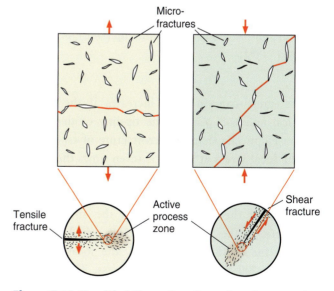

Figure 7.18 Simplified illustration of growth and propagation of extension (left) and shear fractures (right) by propagation and linkage of tensile microfractures (flaws). Propagation occurs in a process zone in front of the fracture tip. Circled figures are centimeter-scale views, while rectangular views illustrate the microscale structure.

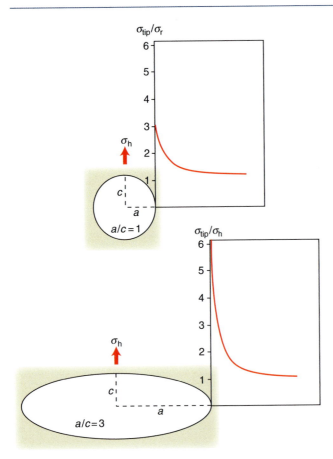

Figure 7.19 Stress concentration around a pore space or microfracture with circular and elliptic geometry in an elastic medium. Increasing the ellipticity a/c increases the stress concentration, as described in Equation 7.8. The far-field stress σ_h is tensile (negative). σ_{tip} is the stress at the circumference of the circle and at the point of maximum curvature on the ellipse (the fracture tip point). Based on Engelder (1993).

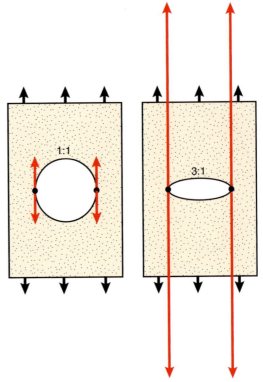

Figure 7.20 Illustration of local stress concentrations in a material with a circular and an elliptical hole. If the material is a sheet of paper it means that the paper with the elliptical hole will be easier to pull apart. Black arrows indicate the remote stress.

The stress concentration increases if the pore is elliptical and will peak at the tip-line of the ellipse. For an elliptically shaped microfracture of aspect ratio 1:3 the **local stress** at the tip is seven times the remote stress. By **remote stress**, also called **far-field stress**, we mean the stress that exists away from the local anomaly, or the state of stress if the anomaly was not there. If the ellipticity is 1:100, which is more realistic for Griffith microcracks, the local stress at the tip is 200 times the remote stress. This concentration may be sufficient to break the local atomic bonds and cause growth of the microcracks. It also implies that once the microcrack starts growing, it increases its length–width ratio, which further increases

the stress concentration at its tips, and continued crack propagation is promoted.

Stress is concentrated at the tips of open microfractures in a rock, and the concentration increases with decreasing thickness/length ratio of the microfracture.

If we model the microcrack as an elliptical pore space, then the stress $-\sigma_{tip}$ at the tip of the pore can be expressed mathematically by the relationship:

$$- \sigma_{tip} = -\sigma_r(1 + (2a/c)) \qquad (7.8)$$

where σ_r is the remote stress and a/c is the ellipticity (aspect ratio of the ellipse). For a circular pore space $a/c = 1$ and $\sigma_r > 0$. The stress σ_t at the tip then becomes $\sigma_{tip} = 3\sigma_r$. The elliptical model of microfractures is of course an approximation. Fractures tend to have a sharply pointed tip zone, which promotes stress concentration even further and increases the likelihood of microfracture growth (Figure 7.21).

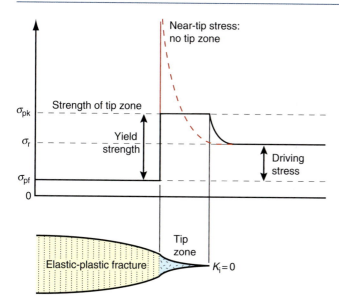

Figure 7.21 Distribution of stress (resolved on the fracture plane) near the tip of an elastic–plastic fracture. Amplified stress due to fracture wall displacements (dashed red curve) decays with distance from fracture tip to σ_r in the surrounding unfractured rock. The length s of the tip zone is defined by a constant value of yield (peak) stress σ_{pk}. The driving stress is the difference between resolved remote stress σ_r and the pore fluid pressure (or residual frictional strength σ_{pf}). The yield strength is the difference between σ_{pk} and internal boundary value σ_i. From Schultz and Fossen (2002).

These considerations suggest that it is likely for fractures to initiate from microdefects in the rock. It also explains why construction, ship and aerospace engineers are so concerned about microdefects and their shapes. Everything depends on whether or not the stress concentration at the tips of the microfractures is high enough to cause them to propagate.

Rock mechanics-oriented geologists sometimes talk about the driving force or **driving stress**. For tensile fractures modeled by means of linear elastic theory, the driving stress is the difference between remote stress resolved on the fracture plane and the internal pore-fluid pressure. Thus, for a tensile fracture to propagate, the driving stress must be large enough to exceed the resolved remote stress. Similarly, for closed shear fractures the driving stress (shear stress) must exceed the resisting forces, such as frictional resistance, for displacements to occur. The **stress intensity factor** K_i considers both the remote stress and the shape and length of the microfracture, and its critical value K_{ic} is called the **fracture toughness**. The fracture toughness can thus be considered as a material's resistance against continued growth of an existing fracture. Naturally, sedimentary rocks have lower values of K_{ic} than igneous rocks. We will not go into the details of linear elastic fracture theory in this book, but rather have a look at how temperature, fabric and sample size can affect strength.

Effects of fabric, temperature, stress geometry and sample size on strength

We have now seen how the presence of microscale heterogeneities in the form of microfractures reduces the strength of rocks. In theory, microfractures can be distributed so that the rock is macroscopically isotropic, i.e. it has the same strength in all directions. Most rocks have an anisotropy stemming from sedimentary or tectonic fabrics such as lamination, bedding, tectonic foliation, lineation and crystallographic fabric (preexisting fractures are treated separately below) and the difference in critical differential stress may vary by several hundred percent, depending on orientation. A rock with a planar anisotropy, such as a slaty cleavage, will fail either along or across the weak cleavage, depending on the orientation of the cleavage with respect to the principal stresses. In Mohr space, there will therefore be two failure envelopes, as shown in Figure 7.22. Which one is applicable depends on the orientation of the foliation.

If the foliation is oriented perpendicular or parallel to σ_1, then there is no resolved shear stress on the foliation, and the upper (blue) envelope in Figure 7.22 applies (Figure 7.22a). A shear fracture forms across the foliation at the characteristic angle of around 30° with σ_1 (i.e. $\theta \approx$ 60°). Foliation-parallel extension fractures along the cleavage plane (longitudinal splitting) can develop even at low confining pressure. When the foliation is oriented closer to that typical of shear fractures in isotropic rocks, shear fractures develop along the foliation at gradually lower differential stress (Figure 7.22b, c). The orientation of the shear fracture and the strength is then controlled by the orientation of the foliation. Minimum strength is obtained where the orientation of the foliation is represented by the point where the Mohr circle touches the "along-foliation failure" envelope on Figure 7.22d. The exact angle will depend on the weakness of the foliation, which determines the slope of the lower failure curve in Figure 7.22.

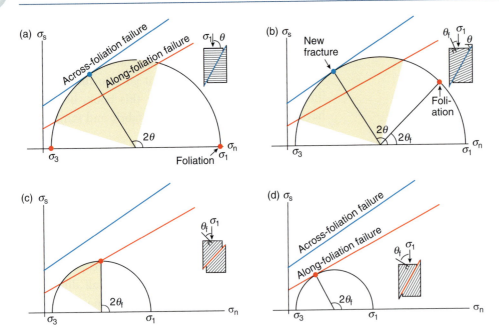

Figure 7.22 Illustration of the role of a preexisting foliation, for constant σ_3. (a) σ_1 acting perpendicular to the foliation, in which case differential stress builds up until the Mohr circle touches the upper envelope and across-foliation failure occurs. Colored sector indicates the range of orientations for along-foliation failure. (b) σ_1 at a high angle to the foliation, still too high for foliation-parallel failure (foliation still outside of the colored sector). (c) σ_1 at 45° to the foliation, causing foliation-parallel failure. Sector indicates the range of foliation orientations where along-foliation failure would occur for this particular state of stress. (d) The angle between σ_1 and the foliation that gives failure at the lowest possible differential stress. This is the weakest direction of a foliated rock.

Whether a rock fails along a weak preexisting fabric or fracture depends on the orientation of the fracture relative to the stress field.

The Mohr diagram and the failure envelopes discussed above only consider confining and differential stress and do not take into consideration σ_2. Experiments show that the influence of σ_2 is small and most pronounced when two of the stress axes are equal in size. For a vertical σ_1, the dip of the shear fracture is lowest when $\sigma_2 = \sigma_1$ and highest when $\sigma_2 = \sigma_3$. For foliated rocks where the foliation does not contain the intermediate principal stress axes, the influence of σ_2 is greater. In this case the resolved normal and shear stress on the foliation depends on all three principal stresses.

Temperature has a major influence on rheology in the plastic regime, but its influence within the brittle regime is relatively small for most common minerals. It does however control the range of the brittle regime as increasing temperature lowers the von Mises yield stress (lowers the yield point or the stress at which rock flows plastically).

An interesting laboratory observation is related to sample size: as the size of the sample increases, its strength is reduced. The reason for this somewhat surprising finding is simply that large samples contain more microfractures than small samples. Because microfractures differ in length and shape, a large sample is likely to contain some microfractures that have a shape that causes larger stress concentrations than any of those in a smaller sample of the same rock.

During a rock experiment, a large sample is likely to fracture before a smaller one.

The dependence on scale is even more pronounced at larger scales. Think of all the joints, faults and other weak structures in the crust that will be activated before the strength of the rock itself is reached. Such weak structures control the strength of the brittle crust, which means that the upper crust is not by far as strong as suggested by experimental testing of unfractured samples in the laboratory. This brings us to another important topic; the reactivation of brittle fractures by frictional sliding.

Growth and morphology of fractures

Shear fractures cannot propagate in their own plane, but rather spawn new tensile cracks (wing cracks) according

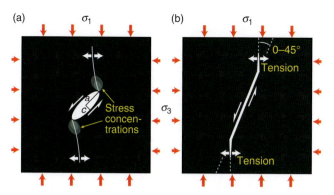

Figure 7.23 (a) Griffith crack modeled as an elliptical void. Tensile stress concentrates near the crack tips (compare with Figure 7. 20). (b) A critically stressed Griffith crack at the tip of a shear fracture. The Griffith crack is oriented between 0 and 45° with respect to σ_1, depending on the ratio σ_1/σ_3. Note that tensile stress develops near the crack tips in spite of the overall compressional stress, that crack growth is accommodated by sliding along the main crack, and that the crack grows toward parallelism with σ_1.

to Griffith's theory (Figure 7.23 and Box 7.3) or develop by activation of already existing extension fractures. In contrast, extension fractures may propagate into long structures. Ideally, an extension fracture will grow radially from a nucleation point so that at any point the propagation front (tipline) has the shape of an ellipse (Figure 7.24a, b). The rate of propagation increases after initiation, and the joint surface gets rougher until it propagates so fast that the stress readjustments or stress oscillations at the crack tip cause it to bend. In detail, the tip bifurcates and off-plane microcracks form because of high stress and/or local heterogeneities in the tip zone. The result is long, narrow planes slightly oblique to the main fracture surface named **hackles** (Figure 7.24c), and the hackles form **plumose** (featherlike) **structures**. Plumose structures reflect the propagation direction along the plume axis, as shown in Figure 7.25.

Locally the main fracture may enter an area with a different stress orientation. This would typically be a bedding interface or some other boundary between two rock types of different mechanical properties, in which case a series of twisted joints or **twist hackles** form in what is called a **fringe zone**. Twist hackles tend to be oriented **en echelon** because of the shear component on the main fracture locally imposed by the new orientation of σ_3. The twist hackles try to orient perpendicular to σ_3, hence the twisting (Figure 7.26).

Extension fractures tend to grow in pulses. Each propagation pulse tends to end with an out-of-plane propagation with a slowing down or complete arrest until

BOX 7.3 | FRACTURE GROWTH AND WING CRACKS

One of the peculiarities of rock mechanics is the fact that, even though a deforming sample develops through-going shear fracture(s) that make an acute (~30°) angle to σ_1, shear fractures cannot grow in their own plane. Instead, a Mode I fracture forms parallel with σ_1. Such fractures are known as wing cracks or edge cracks. In three dimensions wing cracks (Mode I) will form along both the Mode II and Mode III edges of the main fracture.

This development is in agreement with theoretical stress considerations. But how does the shear fracture propagate from this stage? The general answer is that the Mode I wing cracks are broken by a new shear fracture – a process that keeps repeating as the main fracture grows. The result is a zone of minor fractures along and around the main shear fracture, a sort of damage zone akin to that defined for faults in Chapter 8.

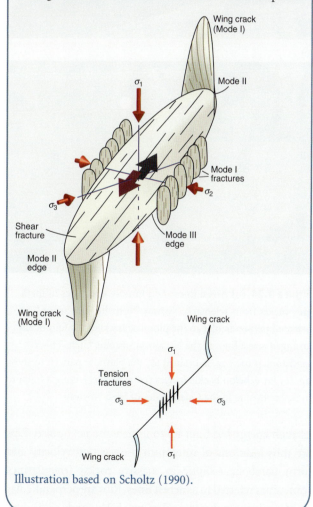

Illustration based on Scholtz (1990).

(a)

(b)

(c)

Figure 7.24 (a) Arrest lines and plumose structures in meta-greywackes from Telemark, Norway. Note the faint arrest lines oriented perpendicular to the plumose hackles. (b) Elliptically arranged arrest lines in the Navajo Sandstone, Utah. This sandstone is too coarse-grained for the plumose pattern to show up. (c) En-echelon hackle fringes (twist hackles) along a fracture in meta-rhyolite in the Caledonides of West Norway.

enough energy has built up to initiate the next pulse. Ribs are thus locations of minimum propagation velocity and form parabolic (elliptic in massive rocks) irregularities sometimes referred to as **arrest lines**. Ribs are perpendicular to the plumose hackles (Figures 7.24a and 7.25) and

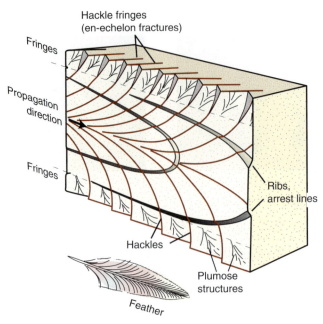

Figure 7.25 Schematic illustration of structures characteristic of joint surfaces. Based on Hodgson (1961).

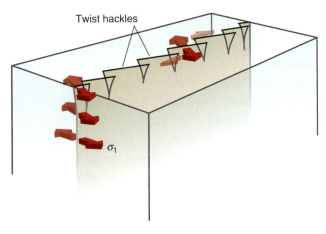

Figure 7.26 The twisting of extension fractures as they reach an interface with a mechanically different rock layer. Note the parallel twisting of σ_1 and the fractures (hackles). Compare with the hackles illustrated in Figure 7.24c.

together these structures provide unique information about the growth history of extension fractures. Plumose structures are characteristic for joints in fine-grained rocks such as siltstones, while arrest lines are also commonly seen in coarser-grained lithologies such as sandstones and granites.

7.5 Fracture termination and interaction

Studies of shear fracture terminations reveal that they sometimes split into two or more fractures with new orientations and, as indicated in Figure 7.27, a spectrum of different fracture geometries can been found. We have

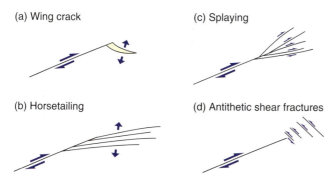

(a) Wing crack (c) Splaying

(b) Horsetailing (d) Antithetic shear fractures

Figure 7.27 Minor fractures at the termination of shear fractures.

Figure 7.28 Horsetailing at the end of a shear fracture in gneiss.

already looked at **wing cracks**, which are tensile fractures at the end of shear fractures (Figure 7.27a). Wing cracks are represented by one or a few tensile fractures at each end of the main fracture and are associated with rapid decrease in displacement toward the tip. In other cases a whole population of minor, typically tensile fractures occur in the tip zone. These are asymmetrically arranged with respect to the main fracture and referred to as **horsetail fractures** (Figures 7.27b and 7.28). If the

secondary fractures in the tip zone represent a fan-shaped splaying of the main fracture, then the commonly used term is **splay faults** (Figure 7.27c). While splay faults are synthetic with respect to the main fault, antithetic fractures may also occur in the tip zone of fractures, as shown in Figure 7.27d.

Most of these tip-zone fractures imply that the energy of the main fracture is distributed onto a number of fracture surfaces. This means that the energy on each fracture is reduced, which hampers continued fracture growth. The evolution of pronounced horsetail fractures or splay faults may thus "arrest" the main fracture and stop or at least pause further propagation of the fracture tip.

The stress field is perturbed around fractures in general and in the tip zone in particular. Thus, when the elastic strain fields around two fractures overlap, the local stress fields around each fracture will interfere and special geometries may develop. If a fracture grows toward an already existing one, the new fracture will curve as it "feels" the effect of the stress perturbation set up by the other fracture. Figure 7.29 indicates that the consequence for the resulting fracture geometry depends on the state of stress along the old fracture. If both fractures approach each other simultaneously, they will mutually affect each other, and the degree of curvature will depend on the general state of stress (Figure 7.30).

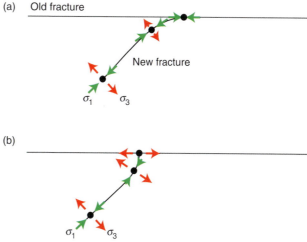

(a) Old fracture

New fracture

σ_1 σ_3

(b)

σ_1 σ_3

Figure 7.29 Local reorientation of fracture propagation direction in the vicinity of an existing fracture. The new fracture grows toward the preexisting one, seeking to maintain a 90° angle to σ_3. The geometry in (a) suggests that σ_1 is compressive with contraction along the preexisting fracture. If the new fracture curves against the preexisting one (b), then σ_1 and σ_3 are likely to be of similar magnitude with tension occurring along the preexisting fracture. Modified from Dyer (1988).

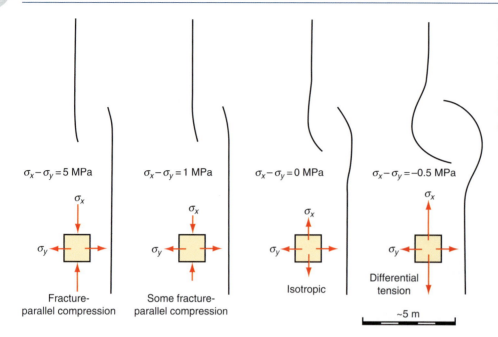

Figure 7.30 Schematic illustration of how fracture tip interaction depends on the differential stress $(\sigma_x - \sigma_y)$ of the remote stress field. Based on Olsen and Pollard (1989) and Cruikshank *et al.* (1991).

$\sigma_x - \sigma_y = 5$ MPa $\sigma_x - \sigma_y = 1$ MPa $\sigma_x - \sigma_y = 0$ MPa $\sigma_x - \sigma_y = -0.5$ MPa

Fracture-parallel compression   Some fracture-parallel compression   Isotropic   Differential tension

~5 m

7.6 Reactivation and frictional sliding

The Coulomb and Griffith fracture criteria, as formulated above, apply until the rock fails. One of the implications of this fact is that Anderson's theory of faulting, which again builds on Coulomb's theory, is only valid for infinitesimal fracture displacements. Once a fracture is formed it represents a plane of weakness. Renewed stress build-up is likely to reactivate existing fractures at a lower level of stress instead of creating a new fracture through the energy-demanding process of growth and linkage of minor flaws in the rock. Reactivation of fractures is a pre-requisite for major faults to develop. Had reactivation not happened, the crust would have been packed with short fractures with small displacements.

The orientation of a preexisting fracture and its friction are the most important parameters, in addition to the stress field itself (Figure 7.31). The orientation determines the resolved shear and normal stresses on the surface. When σ_n is oriented perpendicular to the fracture there is no shear stress on the surface, and the fracture is stable. In the general case there is a (resolved) shear stress on the fracture, and the friction constrains the reactivation potential. That local friction on the fracture is commonly referred to as the **coefficient of sliding friction** (μ_f).

The coefficient of sliding friction is simply the shear stress required to activate slip on the fracture divided by the normal stress acting across the fracture:

$$\mu_f = \frac{\sigma_s}{\sigma_n} \tag{7.9}$$

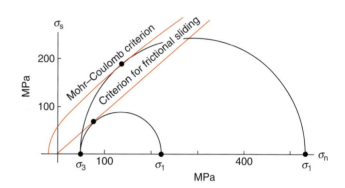

Figure 7.31 The effect of a preexisting fracture (plane of weakness), illustrated in the Mohr diagram. The criterion for reactivation (frictional sliding) is different from that of an unfractured rock of the same kind, and the differential stress required to reactivate the fracture is considerably smaller than that required to generate a new fracture in the rock. This example is based on experiments on crystalline rocks at a confining pressure of 50 MPa (*c.* 2 km depth).

In the Mohr diagram, this is a straight line (Figure 7.31), and it goes through the origin if we assume that the existing fracture has no cohesion. If the fracture has a cohesive strength (C_f) the expression becomes

$$\mu_f = \frac{\sigma_s - C_f}{\sigma_n} \tag{7.11}$$

The magnitude of C_f is usually low, and μ_f is similar for most rocks at moderate to high confining pressures. For low confining pressures the **surface roughness** of the fracture becomes important. Fault asperities resist fault slippage, and may lead to stick-slip deformation (Figure 8.34) at shallow burial depths. At deeper depths

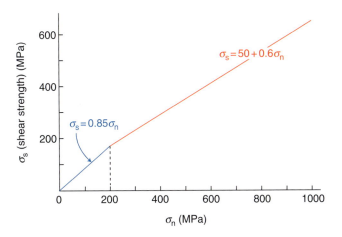

Figure 7.32 Byerlee's law is an empirical law that relates critical shear stress to normal stress. The horizontal scale is related to crustal depth (increasing to the right).

asperities play a much smaller role as far as friction is concerned. After numerous experiments, Byerlee was able to empirically define the critical shear stress at low confining pressure as

$$\sigma_s = 0.85\sigma_n \qquad (\sigma_s < 200 \text{ MPa}) \tag{7.12}$$

while the equation for higher confining pressure was found to be

$$\sigma_s = 0.5 + 0.6\sigma_n \qquad (\sigma_s > 200 \text{ MPa}) \tag{7.13}$$

These equations, shown graphically in Figure 7.32, are known as **Byerlee's law** and hold for most rocks except those that contain abundant H_2O-rich clay minerals.

> Byerlee's law describes the vertical increase in critical shear stress (stress required for faulting) through the frictional upper crust.

7.7 Fluid pressure, effective stress and poroelasticity

One of the great challenges in the field of structural geology in the twentieth century was to explain how gigantic thrust nappes could be transported for hundreds of kilometers without being crushed (Box 16.2). An important part of the explanation has to do with over-pressured thrust zones, i.e. the thrust fault contains fluids with anomalously high pore pressure.

This is one of several examples where fluid pressure plays an important role. We have already looked at overpressured formations in sedimentary sequences (petroleum reservoirs) in Chapter 5, where overpressure can occur if pore

water in a porous and permeable formation is confined between impermeable layers. Overpressure builds up as the weight of the overburden acts on the pore pressure. An additional effect comes from the fact that water expands more quickly than rock minerals during heating. If the water cannot escape, its thermal expansion will further increase the fluid pressure in the permeable unit.

Deeper down, metamorphic reactions release water and carbon dioxide. This may also lead to overpressure if the fluid is unable to escape along fracture networks in the generally impermeable metamorphic rocks. Mineral-filled extension fractures (veins) that occur in many low-grade metamorphic rocks are probably related to the increase of fluid pressure through metamorphic fluid release. Injection of magma under pressure is also a situation where the vertical stress is balanced by fluid (magma) pressure. Finally, an increase in the fluid pressure can cause reactivation of faults and fractures.

> The fluid pressure counteracts the normal stress resolved on the fracture, so that the resolved shear stress may be sufficient for reactivation.

Whether a new fracture forms or an existing one is reactivated is controlled both by the fracture orientation relative to the principal stresses and by the effective stress. **Effective stress** ($\bar{\sigma}$) is the difference between the applied or remote stress and the fluid pressure:

$$\bar{\sigma} = \sigma - p_f \tag{7.14}$$

In three dimensions the effective stress can be expressed as

$$
\begin{bmatrix}
\bar{\sigma}_{11} & \bar{\sigma}_{12} & \bar{\sigma}_{13} \\
\bar{\sigma}_{21} & \bar{\sigma}_{22} & \bar{\sigma}_{23} \\
\bar{\sigma}_{31} & \bar{\sigma}_{32} & \bar{\sigma}_{33}
\end{bmatrix}
=
\begin{bmatrix}
\sigma_{11} & \sigma_{12} & \sigma_{13} \\
\sigma_{21} & \sigma_{22} & \sigma_{23} \\
\sigma_{31} & \sigma_{32} & \sigma_{33}
\end{bmatrix}
-
\begin{bmatrix}
p_f & 0 & 0 \\
0 & p_f & 0 \\
0 & 0 & p_f
\end{bmatrix}
$$

$$
=
\begin{bmatrix}
\sigma_{11} - p_f & \sigma_{12} & \sigma_{13} \\
\sigma_{21} & \sigma_{22} - p_f & \sigma_{23} \\
\sigma_{31} & \sigma_{32} & \sigma_{33} - p_f
\end{bmatrix} \tag{7.15}
$$

or, if the principal stresses coincide with our coordinate axes,

$$
\begin{bmatrix}
\bar{\sigma}_1 & 0 & 0 \\
0 & \bar{\sigma}_2 & 0 \\
0 & 0 & \bar{\sigma}_3
\end{bmatrix}
=
\begin{bmatrix}
\sigma_1 - p_f & 0 & 0 \\
0 & \sigma_2 - p_f & 0 \\
0 & 0 & \sigma_3 - p_f
\end{bmatrix} \tag{7.16}
$$

Fluid pressure will weaken the rock so that deformation can occur at a differential stress that would otherwise be insufficient for failure. For porous sandstone, the pore

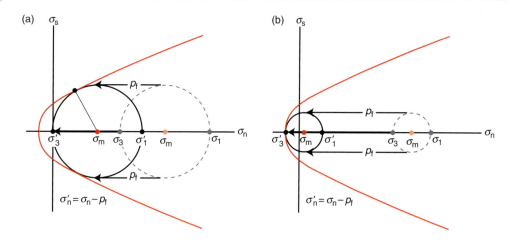

(a) σ_s

(b) σ_s

$\sigma'_n = \sigma_n - p_f$

$\sigma'_n = \sigma_n - p_f$

Figure 7.33 The effect of pumping up the pore fluid pressure p_f in a rock. The Mohr circle is "pushed" to the left (the mean stress is reduced) and a shear fracture will form if the fracture envelope is touched while σ_3 is still positive. A tensile fracture forms if the envelope is reached in the tensile field, as shown in (b) (low differential stress).

(fluid) pressure has the following effect on the Coulomb fracture criterion:

$$\sigma_s = C - \mu(\sigma_n - p_f) \tag{7.17}$$

An increase in pore pressure decreases the mean stress from σ_m to $\sigma_m - p_f$ while the differential stress $(\sigma_1 - \sigma_3)$ is constant (Figure 7.33).

If the effective stress is tensile (negative σ_3), i.e. if

$$\bar{\sigma}_3 = \sigma_3 - p_f < 0 \tag{7.18}$$

tensile fractures can form. In dry or hydrostatically pressured rocks tensile fractures can only be expected to form at very shallow depths (less than a few hundred meters), but fluid overpressure makes it possible to have local tensile stress even at many kilometers depth.

We can rewrite Equation 7.14 so that it becomes clearer that the total stress is the sum of the effective stress and the pore pressure:

$$\sigma = \bar{\sigma} + p_f \tag{7.19}$$

or

$$\begin{bmatrix} \sigma_{11} & \sigma_{12} & \sigma_{13} \\ \sigma_{21} & \sigma_{22} & \sigma_{23} \\ \sigma_{31} & \sigma_{32} & \sigma_{33} \end{bmatrix} = \begin{bmatrix} \bar{\sigma}_{11} & \bar{\sigma}_{12} & \bar{\sigma}_{13} \\ \bar{\sigma}_{21} & \bar{\sigma}_{22} & \bar{\sigma}_{23} \\ \bar{\sigma}_{31} & \bar{\sigma}_{32} & \bar{\sigma}_{33} \end{bmatrix} + \begin{bmatrix} p_f & 0 & 0 \\ 0 & p_f & 0 \\ 0 & 0 & p_f \end{bmatrix} \tag{7.20}$$

To illustrate this relationship, imagine a porous and permeable sandstone (Figure 7.34) that is exposed to a uniaxial-strain reference state of stress inside a container. Let us first assume that the rock is dry and that the grains exert an average stress σ_n^w against the walls of the container. This stress is not evenly distributed along the walls but concentrated at the grain–wall contact points (Figure 7.35). The grain–wall contact stress σ_n^g depends on the area across which it is distributed and can be expressed in terms of the porosity ϕ:

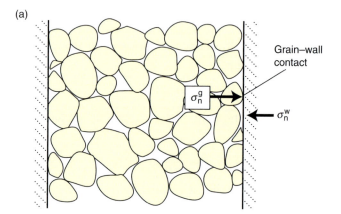

(a)

Grain–wall contact

σ_n^g

σ_n^w

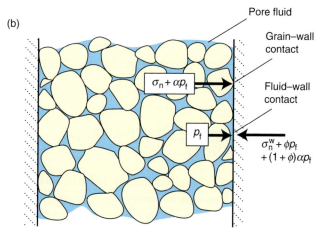

(b)

Pore fluid

Grain–wall contact

$\sigma_n + \alpha p_f$

Fluid–wall contact

p_f

$\sigma_n^w + \phi p_f$
$+ (1 + \phi)\alpha p_f$

Figure 7.34 The effect of increasing the pore pressure p_f on the total stress situation in a porous rock (closed uniaxial-strain stress model). In a dry rock (a), stresses are transmitted across grain–grain or grain–wall contacts only. If pore fluid is added with a low p_f (b), then the increase in normal stress at grain–wall contacts is smaller than the increase in pore pressure because of the absorption of stress by elastic deformation in the grains. This is the poroelastic effect. Modified from Engelder (1993).

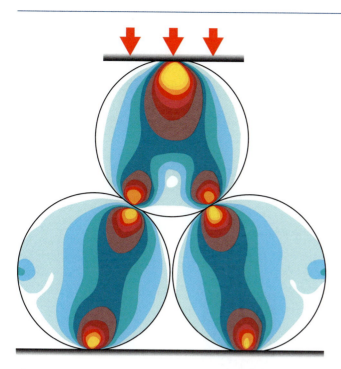

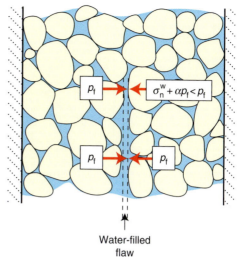

Water-filled
flaw

Figure 7.36 The stress situation in a flaw in a permeable porous rock. The poroelastic effect causes the stress across the grain–flaw part of the flaw walls to be less than the pore fluid pressure. Tensile stress occurs if the pore pressure is high enough.

Figure 7.35 Illustration of stress concentrations (stress bridges) at grain–grain contact areas in a porous rock or sediment. Warm colors indicate high stress. Based on Gallagher *et al.* (1974).

$$\sigma_n^g = \left(\frac{1}{1-\phi}\right)\sigma_n^w \qquad (7.21)$$

We now fill the pores with fluid at some moderate pressure p_f, which causes the pressure against the walls to increase. The parts of the walls that are in contact with the fluid will "feel" the fluid pressure directly. At the grain–wall contact points the normal stress on the wall increases by a fraction of p_f, i.e. by a factor αp_f where $\alpha < 1$. The increase in stress against the wall, which we denote $\Delta\sigma_n$, will not be equal to p_f, but will be less by an amount that depends on the porosity of the sandstone:

$$\Delta\sigma_n = \phi p_f + (1-\phi)\alpha p_f \qquad (7.22)$$

The factor α is known as the Biot poroelastic parameter and characterizes the poroelastic effect. But why is the stress increase less than p_f? Because the cemented grain contacts are elastic. Thus, some of the pore fluid pressure p_f is taken up by elastic deformation.

The poroelastic effect is important when considering the state of stress in sedimentary basins. It may also contribute to the formation and propagation of fractures in porous rocks (Figure 7.36). In keeping with Griffith's theory, a flaw of some kind represents a possible nucleation point for a tensile fracture. Increasing the pore pressure by an amount Δp_f gives a new pore pressure p_f that will be identical within and outside of the flaw,

because the rock is permeable. At the walls between the flaw and the rock, however, the poroelastic effect (Equation 7.22) comes into play. It tells us that the general increase in pore pressure causes the average normal stress at the rock sides of the flaw to increase at a lower rate than in the flaw. As the pore pressure increases, at some point the fluid pressure p_f within the flaw will exceed the average normal stress exerted by the grains on the walls of the flaw. The walls are then under tension and may further separate as the flaw grows into a larger extension fracture. The tensile stress is concentrated at the tips, and its magnitude depends on the shape of the flaw or crack according to Equation 7.8. Once the extension fracture grows, the volume of the fracture increases, the pore pressure drops, and the propagation stops or pauses until the pore pressure is restored. This kind of fracture propagation history is recorded by the formation of arrest lines (Figure 7.25).

7.8 Deformation bands and fractures in porous rocks

Rocks respond to stress in the brittle regime by forming extension fractures and shear fractures (slip surfaces). Such fractures are sharp and mechanically weak discontinuities, and thus prone to reactivation during renewed stress build-up. At least this is how non-porous and low-porosity rocks respond. In highly porous rocks and sediments, brittle deformation is expressed by related, although different, deformation structures referred to as **deformation bands**.

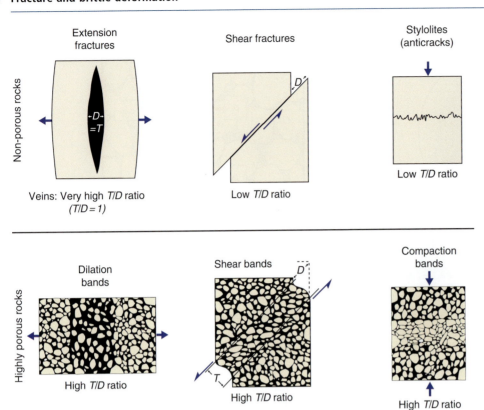

Figure 7.38 Cataclastic deformation band in porous Navajo Sandstone. The thickness of the band seems to vary with grain size, and the shear offset is less than 1 cm (the coin is 1.8 cm in diameter).

Deformation bands are mm-thick zones of localized compaction, shear and/or dilation in deformed porous rocks. Figure 7.37 shows how deformation bands kinematically relate to fractures in non-porous and low-porosity rocks, but there are good reasons why deformation bands should be distinguished from ordinary fractures. One is that they are thicker and at the same time exhibit smaller shear displacements than regular slip surfaces of comparable length (Figure 7.38). This has led to the term **tabular** discontinuities, as opposed to **sharp discontinuities** for fractures. Another is that, while cohesion is lost or reduced across regular fractures, most deformation bands maintain or even show increased cohesion. Furthermore, there is a strong tendency for deformation bands to represent low-permeability tabular objects in otherwise highly permeable rocks. This permeability reduction is related to collapse of pore space, as seen in the band from Sinai portrayed in Figure 7.39. In contrast, most regular fractures increase permeability, particularly in low-permeability and impermeable rocks. This distinction is particularly important to petroleum geologists and hydrogeologists concerned with fluid flow in reservoir rocks. The strain hardening that occurs during the formation of many deformation bands also makes them different from fractures, which are associated with softening.

The difference between brittle fracturing of non-porous and porous rocks lies in the fact that porous rocks have a pore volume that can be utilized during grain reorganization. The pore space allows for effective rolling and sliding of grains. Even if grains are crushed, grain fragments can be organized into nearby pore space.

The kinematic freedom associated with pore space allows the special class of structures called deformation bands to form.

(a)   (b)

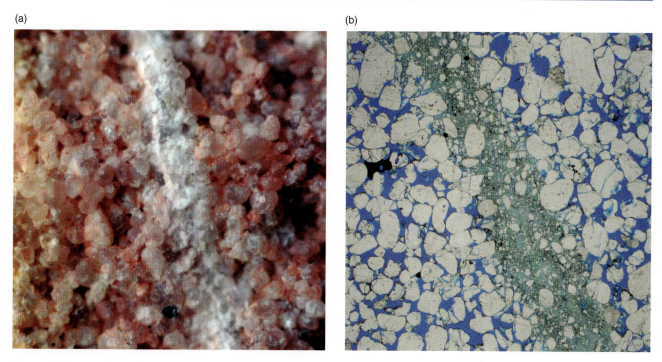

Figure 7.39 Cataclastic deformation band in outcrop (left) and thin section (right) in the Nubian Sandstone, Sinai. Note the extensive crushing of grains and reduction of porosity (pore space is blue in the thin section). Width of bands ∼1 mm.

What is a deformation band?

How do deformation bands differ from regular fractures in non-porous rocks? Here are some characteristics of deformation bands:

- Deformation bands are restricted to highly porous granular media, notably porous sandstones.

- A shear deformation band is a wider zone of deformation than regular shear fractures of comparable displacement.

- Deformation bands do not develop large offsets. Even 100 m long deformation bands seldom have offsets in excess of a few centimeters, while shear fractures of the same length tend to show meter-scale displacement.

- Deformation bands occur as single structures, as clusters, or in zones associated with slip surfaces (faulted deformation bands). This is related to the way that faults form in porous rocks by faulting of deformation band zones (see Chapter 8).

Types of deformation bands

Similar to fractures, deformation bands can be classified in a kinematic framework, where **shear (deformation) bands**, **dilation bands** and **compaction bands** form the end-

members (Figure 7.37). It is also of interest to identify the *mechanisms* operative during the formation of deformation bands. Deformation mechanisms depend on internal and external conditions such as mineralogy, grain size, grain shape, grain sorting, cementation, porosity, state of stress etc., and different mechanisms produce bands with different petrophysical properties. Thus, a classification of deformation bands based on deformation processes is particularly useful if permeability and fluid flow is an issue. The most important mechanisms are:

- Granular flow (grain boundary sliding and grain rotation)

- Cataclasis (grain fracturing)

- Phyllosilicate smearing

- Dissolution and cementation

Deformation bands are named after their characteristic deformation mechanism, as shown in Figure 7.40.

Disaggregation bands develop by shear-related disaggregation of grains by means of grain rolling, grain boundary sliding and breaking of grain bonding cements; the process that we called particulate or granular flow at the beginning of this chapter (Figure 7.1a). Disaggregation bands are commonly found in sand and poorly consolidated sandstones and form the "faults" produced in most

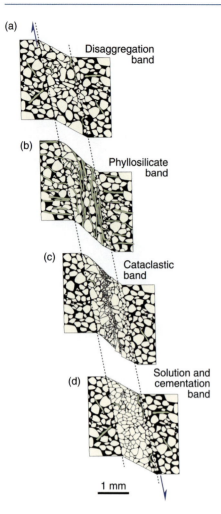

Figure 7.40 The different types of deformation bands, distinguished by dominant deformation mechanism. Modified from Fossen *et al.* (2007).

sandbox experiments. Disaggregation bands can be almost invisible in clean sandstones, but may be detected where they cross and offset laminae (Figure 7.41). Their true offsets are typically a few centimeters and their thickness varies with grain size. Fine-grained sand(stones) develop ~1 mm thick bands, whereas coarser-grained sand(stones) host single bands that may be at least 5 mm thick.

Macroscopically, disaggregation bands are ductile shear zones where sand laminae can be traced continuously through the band. Most pure and well-sorted quartz-sand deposits are already compacted to the extent that the initial stage of shearing involves some dilation (dilation bands), although continued shear-related grain reorganization may reduce the porosity at a later point.

Phyllosilicate bands (also called framework phyllosilicate bands) form in sand(stone) where the content of platy minerals exceeds about 10–15%. They can be considered as a special type of disaggregation band where platy minerals promote grain sliding. Clay minerals tend to mix with other mineral grains in the band while coarser phyllosilicate grains align to form a local fabric within the bands due to shear-induced rotation. Phyllosilicate bands are easy to detect, as the aligned phyllosilicates give the band a distinct color or fabric that may be reminiscent of phyllosilicate-rich laminae in the host rock.

If the phyllosilicate content of the rock changes across bedding or lamina interfaces, a deformation band may change from an almost invisible disaggregation band to a phyllosilicate band. Where clay is the dominant platy mineral, the band is a fine-grained, low-porosity zone that can accumulate offsets that exceed the few centimeters exhibited by other types of deformation bands. This is related to the smearing effect of the platy minerals along phyllosilicate bands that apparently counteracts any strain hardening resulting from interlocking of grains.

If the clay content of the host rock is high enough (more than ~40%), the deformation band turns into a **clay smear**. Clay smears typically show striations and classify as slip surfaces rather than deformation bands. Examples of deformation bands turning into clay smears as they leave sandstone layers are common.

Cataclastic bands form where mechanical grain breaking is significant (Figure 7.39). These are the classic deformation bands first described by Atilla Aydin from the Colorado Plateau in the western USA. He noted that many cataclastic bands consist of a central cataclastic core contained within a mantle of (usually) compacted or gently fractured grains. The core is most obvious and is characterized by grain size reduction, angular grains and significant pore space collapse (Figure 7.39). The crushing of grains results in extensive grain interlocking, which promotes strain hardening. Strain hardening may explain the small shear displacements observed on cataclastic deformation bands (≤3–4 cm). Some cataclastic bands are pure compaction bands (Figure 7.41), while most are shear bands with some compaction across them.

Cataclastic bands occur most frequently in sandstones that have been deformed at depths of about 1.5–3 km, although evidence of cataclasis is also reported from deformation bands deformed at shallower depths. Comparison suggests that shallowly formed cataclastic deformation bands show less intensive cataclasis than those formed at 1.5–3 km depth.

Cementation and dissolution of quartz and other minerals may occur preferentially in deformation bands

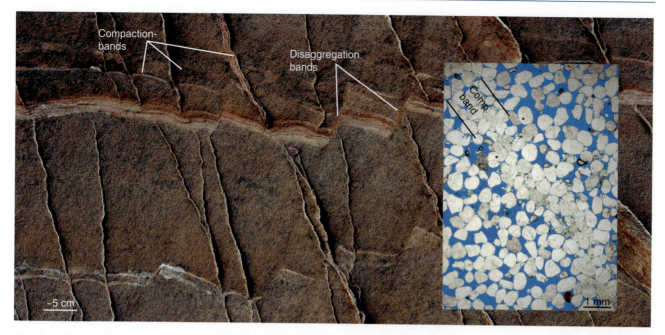

Figure 7.41 Right-dipping compaction bands overprinting left-dipping soft-sedimentary disaggregation bands (almost invisible). The sandstone is very porous except for thin layers, where compaction bands are absent. Hence, the compaction bands only formed in very high porosity sandstone. Thin section photo shows that the compaction is assisted by dissolution and some grain fracture. Navajo Sandstone, southern Utah.

where diagenetic minerals grow on the fresh surfaces formed during grain crushing and/or grain boundary sliding. Such preferential growth of quartz is generally seen in deformation bands in sandstones buried to more than 2–3 km depth (>90 °C) and can occur long after the formation of the bands.

Influence on fluid flow

Deformation bands form a common constituent of porous oil, gas and water reservoirs, where they occur as single bands, cluster zones or in fault damage zones (see next chapter). Although they are unlikely to form seals that can hold significant hydrocarbon columns over geologic time, they can influence fluid flow in some cases. Their ability to do so depends on their internal permeability structure and thickness or frequency. Clearly, the zone of cataclastic deformation bands shown in Figure 7.42 will have a far greater influence on fluid flow than the single cataclastic band shown in Figure 7.38 or 7.39.

Cataclastic deformation bands show the most significant permeability reductions.

Deformation band permeability is governed by the deformation mechanisms operative during their formation, which again depends on a number of lithological and physical factors. In general, disaggregation bands show little porosity and permeability reduction, while phyllosilicate and, particularly, cataclastic bands show permeability reductions up to several orders of magnitude. Deformation bands are thin, so the number of deformation bands (their cumulative thickness) is important when their role in a petroleum reservoir is to be evaluated.

Also important are their continuity, variation in porosity/permeability and orientation. Many show significant variations in permeability along strike and dip due to variations in amount of cataclasis, compaction or phyllosilicate smearing. Deformation bands tend to define sets with preferred orientation (Figure 7.43), for instance in damage zones, and this anisotropy can influence the fluid flow in a petroleum reservoir, for example during water injection. All of these factors make it difficult to evaluate the effect of deformation bands in reservoirs, and each reservoir must be evaluated individually according to local parameters such as time and depth of

Figure 7.42 Very dense cluster of cataclastic deformation bands in the Entrada Sandstone, Utah.

Figure 7.43 Conjugate (simultaneous and oppositely dipping) sets of cataclastic deformation bands in sandstone. Note the positive relief of the deformation bands due to grain crushing and cementation. The bands fade away downward into the more fine-grained and less-sorted unit. Entrada Sandstone, Utah.

deformation, burial and cementation history, mineralogy, sedimentary facies and more.

The influence of deformation bands on petroleum or groundwater production depends on the permeability contrast, cumulative thickness, orientations, continuity and connectivity.

What type of structure forms, where and when?

Given the various types of deformation bands and their different effects on fluid flow, it is important to understand the underlying conditions that control when and where they form. A number of factors are influential, including burial depth, tectonic environment (state of stress) and host rock properties, such as degree of lithification, mineralogy, grain size, sorting and grain shape. Some of these factors, particularly mineralogy, grain size, rounding, grain shape and sorting, are more or less constant for a given sedimentary rock layer. They may, however, vary from layer to layer, which is why rapid changes in deformation band development may be seen from one layer to the next.

Other factors, such as porosity, permeability, confining pressure, stress state and cementation, are likely to change with time. The result may be that early deformation bands are different from those formed at later stages in the same porous rock layer, for example at deeper burial depths. Hence, the sequence of deformation structures in a given rock layer reflects the physical changes that the sediment has experienced throughout its history of burial, lithification and uplift.

To illustrate a typical structural development of sedimentary rocks that go through burial and then uplift, we use the diagram from Figure 5.10 and add characteristic structures (Figure 7.44). The earliest forming deformation bands in sandstones are typically disaggregation bands or phyllosilicate bands. Such structures form at low confining pressures (shallow burial) when forces across grain contact surfaces are low and grain bindings are weak, and are therefore indicated at shallow levels in Figures 7.44 and 7.45. Many early disaggregation bands are related to local, gravity-controlled deformation such as local shale diapirism, underlying salt movement, gravitational sliding and glaciotectonics.

Cataclastic deformation bands can occur in poorly lithified layers of pure sand at shallow burial depths,

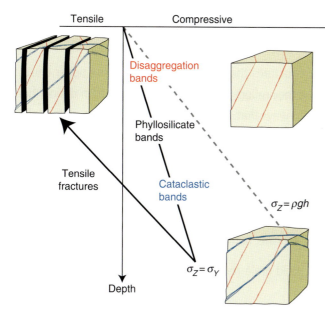

Figure 7.44 Different types of deformation bands form at different stages during burial. Extension fractures (Mode I fractures) are most likely to form during uplift. Also see Figure 5.10. From Fossen *et al.* (2007).

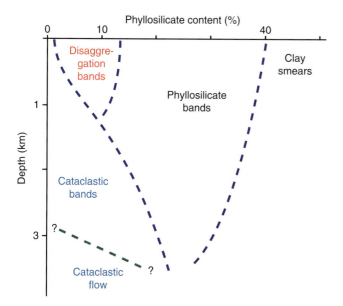

Figure 7.45 Tentative illustration of how different deformation band types relate to phyllosilicate content and depth. Many other factors influence the boundaries outlined in this diagram, and the boundaries should be considered as uncertain. From Fossen *et al.* (2007).

but are much more common in sandstones deformed at 1–3 km depth. Factors promoting shallow-burial cataclasis include small grain contact areas, i.e. good sorting and well-rounded grains, the presence of feldspar or

other non-platy minerals with cleavage and lower hardness than quartz, and weak lithic fragments. Quartz, for instance, seldom develops transgranular fractures under low confining pressure, but may fracture by flaking or spalling. At deeper depths, extensive cataclasis is promoted by high grain contact stresses. Abundant examples of cataclastic deformation bands are found in the Jurassic sandstones of the Colorado Plateau, where the age relation between early disaggregation bands and later cataclastic bands is very consistent (Figure 7.44).

When a sandstone becomes cohesive and loses porosity during lithification (left side of Figure 7.44), deformation occurs by crack propagation instead of pore space collapse, and slip surfaces, joints and mineral-filled fractures form directly without any precursory formation of deformation bands. This is why late, overprinting structures are almost invariably slip surfaces, joints and mineral-filled fractures. Slip surfaces can also form by faulting of low-porosity deformation band zones at any burial depth, according to the model described in the next chapter (Section 8.5).

Joints and veins typically postdate both disaggregation bands and cataclastic bands in sandstones. The transition from deformation banding to jointing may occur as porosity is reduced, notably through quartz dissolution and precipitation. Since the effect of such diagenetically controlled strengthening may vary locally, deformation bands and joints may develop simultaneously in different parts of a sandstone layer, but the general pattern is deformation bands first, then faulted deformation bands (slip surface formation) and finally joints (tensile fractures in Figure 7.44) and perhaps faulted joints (Figure 1.4).

The latest fractures in uplifted sandstones tend to form extensive and regionally mappable joint sets generated or at least influenced by removal of overburden and cooling during regional uplift. Such joints are pronounced where sandstones have been uplifted and exposed, such as on the Colorado Plateau (Figure 1.4), but are unlikely to be developed in subsurface petroleum reservoirs unexposed to significant uplift. It therefore appears that knowing the burial/uplift history of a basin in relation to the timing of deformation events is very useful when considering the type of structures present in, say, a sandstone reservoir. Conversely, examination of the type of deformation structure present also gives information about deformation depth and other conditions at the time of deformation.

Summary

Brittle deformation tends to be extremely localized and result in structures that significantly weaken the upper crust. Separating different types of brittle structures is important because they reflect the state of stress and strain during their formation, and the different types of fractures alter rocks in different ways that affect mechanical properties, potential of reactivation and permeability structure. This has implications for engineering geoscientists, seismologists, hydrogeologists and petroleum geologists alike. The formation of fractures and deformation bands is also essential when it comes to fault formation and fault growth, which is the subject of the next chapter. There are many important points to be made from this chapter, and here are a few:

- Fractures form primarily in the brittle regime where brittle mechanisms dominate.

- Brittle deformation mechanisms are cataclasis (grain fracture), rigid grain rotation and grain translation through frictional grain boundary sliding (grain reorganization).

- Extension fractures such as joints can expand to become extensive structures, while shear fractures cannot expand as such unless small extension structures form ahead of shear fracture tips and weaken the rock. Shear fractures can then expand by coalescence of extension microfractures.

- Stress concentrates at the tip of both small and large fractures and helps them grow.

- High fluid pressure in cracks and pores also promotes fracture and fracture propagation.

- Extension fractures form perpendicular to σ_3.

- Shear fractures typically form at a 20–30° angle to σ_1.

- A fracture criterion relates the shear and normal stress that is required for a rock to fracture, i.e. the critical normal and shear stresses. The Coulomb criterion is linear, which means that there is a constant ratio between the critical shear and normal stress and is therefore represented by a straight line in Mohr space.

- The strength of undeformed rocks as measured experimentally is not representative of the brittle crust because of its many weak faults and fractures.

- A fracture's reactivation potential depends on its frictional resistance to reactivation, the fluid pressure inside the fracture, and its orientation relative to the principal stresses. The latter also determines the mode of reactivation (extension or shear).

- Both fractures and deformation bands are important in terms of permeability in deformed rocks, but generally have opposite functions: fractures increase permeability while deformation bands involve permeability reduction.

Review questions

1. What is the difference between cataclastic and granular flow?

2. What is frictional sliding?

3. What is the process zone that is located at the tip of shear fractures?

4. What is the difference between fractures and deformation bands?

5. Why do shear fractures not form at 45° to σ_1, where the resolved shear stress is at its maximum?

6. What is a wing crack and how do they form?

7. What structures can be found on joints that can reveal their growth history?

8. What does it mean that a rock is critically stressed?

9. What is a failure envelope and how is it established for a rock?

10. What is meant by the term Griffith cracks, and how do they affect rock strength and fracture propagation?

11. Why are large rock samples weaker than small samples of the same rock?

12. What is the coefficient of sliding friction and what is a representative value for this coefficient for the brittle crust?

E-MODULE

 The e-learning module called *Brittle deformation* is recommended for this chapter.

FURTHER READING

Fractures and fracturing

Reches, Z. and Lockner, D. A., 1994, Nucleation and growth of faults in brittle rocks. *Journal of Geophysical Research* **99**: 18159–18172.

Scholz, C. H., 2002, *The Mechanics of Earthquakes and Faulting*. Cambridge: Cambridge University Press.

Schultz, R. A., 1996, Relative scale and the strength and deformability of rock masses. *Journal of Structural Geology* **18**: 1139–1149

Segall, P. and Pollard, D. D., 1983, Nucleation and growth of strike slip faults in granite. *Journal of Geophysical Research* **88**: 555–568.

Joints

Narr, W. and Suppe, J., 1991, Joint spacing in sedimentary rocks. *Journal of Structural Geology* **13**: 1037–1048.

Pollard, D. D. and Aydin, A., 1988, Progress in understanding jointing over the past century. *Geological Society of America Bulletin* **100**, 1181–1204.

Closing fractures

Fletcher, R. C. and Pollard, D. D., 1981, Anticrack model for pressure solution surfaces. *Geology* **9**: 419–424.

Mollema, P. N. and Antonellini, M. A., 1996, Compaction bands: a structural analog for anti-mode I cracks in aeolian sandstone. *Tectonophysics* **267**: 209–228.

Role of fluids

Hubbert, M. K. and Rubey, W. W., 1959, Role of pore fluid pressure in the mechanics of overthrust faulting. I: Mechanics of fluid-filled porous solids and its application to overthrust faulting. *Geological Society of America Bulletin* **70**: 115–205.

Deformation bands

Antonellini, M. and Aydin, A., 1994, Effect of faulting on fluid flow in porous sandstones: petrophysical properties. *American Association of Petroleum Geologists* **78**: 355–377.

Aydin, A. and Johnson, A. M., 1978, Development of faults as zones of deformation bands and as slip surfaces in sandstones. *Pure and Applied Geophysics* **116**: 931–942.

Davis, G. H., 1999, Structural geology of the Colorado Plateau Region of southern Utah. *Geological Society of America Special Paper* **342**: 1–157.

Fossen, H., Schultz, R., Shipton, Z. and Mair, K., 2007, Deformation bands in sandstone: a review. *Journal of the Geological Society, London* **164**, 755–769.

Jamison, W. R., 1989, Fault-fracture strain in Wingate Sandstone. *Journal of Structural Geology* **11**: 959–974.

Rawling, G. C. and Goodwin, L. B., 2003, Cataclasis and particulate flow in faulted, poorly lithified sediments. *Journal of Structural Geology* **25**: 317–331.

Underhill, J. R. and Woodcock, N. H., 1987, Faulting mechanisms in high-porosity sandstones: New Red Sandstone, Arran, Scotland. In *Deformation of Sediments and Sedimentary Rocks*. Special Publication **29**, London: Geological Society, pp. 91–105.

Chapter 8

Faults

Faults disturb layered rock sequences, introducing "faults" or "defects" to the primary stratigraphic framework. While representing challenges to stratigraphers and geologists mapping rocks in the field or interpreting seismic data, faults are extremely intriguing structures that have fascinated structural geologists as much as they have frustrated stratigraphers and miners. We know much more about faults today than just a few decades ago, largely because of their importance to the petroleum industry. Faults also represent challenges associated with waste repositories and tunnel operations, and active faults are closely associated with earthquakes and seismic hazards. In this chapter we will focus on fault geometry, fault anatomy and the evolution of faults and fault populations, with examples and applications relevant to the petroleum industry.

8.1 Fault terminology

While the fractures and related discontinuities covered in the previous chapter are fairly simple structures, faults are much more complex and compound features that can accommodate large amounts of strain in the upper crust. The term **fault** is used in different ways, depending on geologist and context. A simple and traditional definition states:

A fault is any surface or narrow zone with visible shear displacement along the zone.

This definition is almost identical to that of a shear fracture, and some geologist use the two terms synonymously. Sometimes geologists even refer to shear fractures with millimeter- to centimeter-scale offsets as microfaults. However, most geologists would restrict the term shear fracture to small-scale structures and reserve the term fault for more composite structures with offsets in the order of a meter or more.

The thickness of a fault is another issue. Faults are often expressed as planes and surfaces in both oral and written communication and sketches, but close examination of faults reveals that they consist of fault rock material and subsidiary brittle structures and therefore have a definable thickness. However, the thickness is usually much smaller than the offset and several orders of magnitude less than the fault length. Whether a fault should be considered as a surface or a zone largely depends on the scale of observation, objectives and need for precision.

Faults tend to be complex zones of deformation, consisting of multiple slip surfaces, subsidiary fractures and perhaps also deformation bands. This is particularly apparent when considering large faults with kilometer-scale offsets. Such faults can be considered as single faults on a map or a seismic line, but can be seen to consist of several small faults when examined in the field. In other words, the scale dependency, which haunts the descriptive structural geologist, is important. This has led most geologists to consider a fault as a volume of brittlely deformed rock that is relatively thin in one dimension:

A fault is a tabular volume of rock consisting of a central slip surface or core, formed by intense shearing, and a surrounding volume of rock that has been affected by more gentle brittle deformation spatially and genetically related to the fault.

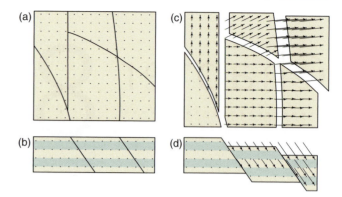

Figure 8.1 Faults appear as discontinuities on velocity or displacement field maps and profiles. The left blocks in the undeformed map (a) and profile (b) are fixed during the deformation. The result is abrupt changes in the displacement field (arrows) across faults.

The term fault may also be connected to deformation mechanisms (brittle or plastic). In a very informal sense, the term fault covers both brittle discontinuities and ductile shear zones dominated by plastic deformation. This is sometimes implied when discussing large faults on seismic or geologic sections that penetrate much or all of the crust. The term **brittle fault** (as opposed to ductile shear zone) can be used if it is important to be specific with regard to deformation mechanism. In most cases geologists implicitly restrict the term fault to slip or shear discontinuities dominated by brittle deformation mechanisms, rendering the term brittle fault redundant:

A fault is a discontinuity with wall-parallel displacement dominated by brittle deformation mechanisms.

By discontinuity we are here primarily referring to layers, i.e. faults cut off rock layers and make them discontinuous. However, faults also represent mechanical and displacement discontinuities. Figure 8.1 illustrates how the displacement field rapidly changes across faults in both map view and cross-section. A kinematic definition, particularly useful for experimental work and GPS-monitoring of active faults can therefore be added:

A fault is a discontinuity in the velocity or displacement field associated with deformation.

As mentioned in the previous chapter, faults differ from shear fractures because a simple shear fracture cannot expand in its own plane into a larger structure. In contrast, faults can grow by the creation of a complex process zone with numerous small fractures, some of

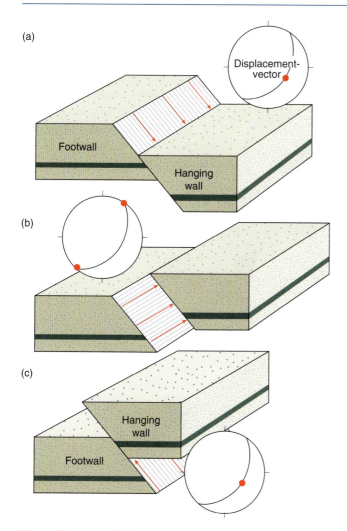

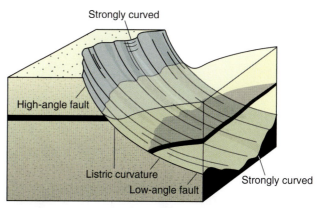

Figure 8.3 Listric normal fault showing very irregular curvature in the sections perpendicular to the slip direction. These irregularities can be thought of as large grooves or corrugations along which the hanging wall can slide.

Figure 8.2 Normal (a), strike-slip (sinistral) (b) and reverse (c) faults. These are end-members of a continuous spectrum of oblique faults. The stereonets show the fault plane (great circle) and the displacement vector (red point).

which link to form the fault slip surface while the rest are abandoned.

Geometry of faults

Non-vertical faults separate the **hanging wall** from the underlying **footwall** (Figure 8.2). Where the hanging wall is lowered or downthrown relative to the footwall, the fault is a **normal fault**. The opposite case, where the hanging wall is upthrown relative to the footwall, is a **reverse fault**. If the movement is lateral, i.e. in the horizontal plane, then the fault is a **strike-slip fault**. Strike-slip faults can be sinistral (left-lateral) or dextral (right-lateral) (from the Latin words *sinister* and *dexter*, meaning left and right, respectively).

Although some fault dip ranges are more common than others, with strike-slip faults typically occurring as steep faults and reverse faults commonly having lower dips than normal faults, the full range from vertical to horizontal faults is found in naturally deformed rocks. If the dip angle is less than 30° the fault is called a **low-angle fault**, while **steep faults** dip steeper than 60°. Low-angle reverse faults are called **thrust faults**, particularly if the movement on the fault is tens or hundreds of kilometers.

A fault that flattens downward is called a **listric fault** (Figure 8.3), while downward-steepening faults are sometimes called antilistric. The terms **ramps** and **flats**, originally from thrust fault terminology, are used for alternating steep and subhorizontal portions of any fault surface. For example, a fault that varies from steep to flat and back to steep again has a **ramp-flat-ramp geometry**.

Irregularities are particularly common in the section perpendicular to the fault slip direction. For normal and reverse faults this means curved fault traces in map view, as can be seen from the faults of the extensional oil field in Figure 8.4. Irregularities in this section cause no conflict during fault slippage as long as the axes of the irregularities coincide with the slip vector. Where irregularities also occur in the slip direction, the hanging wall and/or footwall must deform. For example, a listric normal fault typically creates a hanging-wall rollover (see Chapter 20).

A fault can have any shape perpendicular to the slip direction, but non-linearity in the slip direction generates space problems leading to hanging- or footwall strain.

Figure 8.4 The main faults in the North Sea Gullfaks oil field show high degree of curvature in map view and straight traces in the vertical sections (main slip direction). Red lines represent some of the well paths in this field. From Fossen and Hesthammer (2000).

The term **fault zone** traditionally means a series of subparallel faults or slip surfaces close enough to each other to define a zone. The width of the zone depends on the scale of observation – it ranges from centimeters or meters in the field to the order of a kilometer or more when studying large-scale faults such as the San Andreas Fault. The term fault zone is now also used inconsistently about the central part of the fault where most or all of the original structures of the rock are obliterated, or about the core and the surrounding deformation zone associated with the fault. This somewhat confusing use is widespread in the current petroleum-related literature, so any use of the term fault zone requires clarification.

Two separate normal faults dipping toward each other create a downthrown block known as a **graben** (Figure 8.5). Normal faults dipping away from each other create an upthrown block called a **horst**. The largest faults in a faulted area, called **master faults**, are associated with minor faults that may be antithetic or synthetic. An **antithetic fault** dips toward the master fault, while a **synthetic fault** dips in the same direction as the master fault (Figure 8.5). These expressions are relative and only make sense when minor faults are related to specific larger-scale faults.

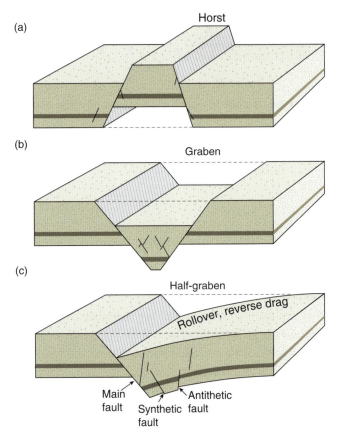

Figure 8.5 A horst (a), symmetric graben (b) and asymmetric graben (c), also known as a half-graben. Antithetic and synthetic faults are shown.

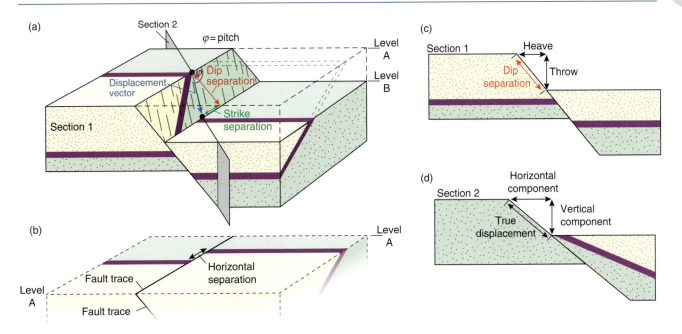

Figure 8.6 Illustration of a normal fault affecting a tilted layer. The fault is a normal fault with a dextral strike-slip component (a), but appears as a sinistral fault in map view (b, which is the horizontal section at level A). (c) and (d) show profiles perpendicular to fault strike (c) and in the (true) displacement direction (d).

Displacement, slip and separation

The vector connecting two points that were connected prior to faulting indicates the local **displacement vector** or **net slip direction** (Figure 8.6). Ideally, a strike-slip fault has a horizontal slip direction while normal and reverse faults have displacement vectors in the dip direction. In general, the total slip that we observe on most faults is the sum of several increments (earthquakes), each with its own individual displacement or slip vector. The individual slip events may have had different slip directions. We are now back to the difference between deformation *sensu stricto*, which only relates the undeformed and deformed states, and deformation *history*. In the field we could look for traces of the slip history by searching for such things as multiple striations, as discussed in the next chapter.

A series of displacement vectors over the slip surface gives us the **displacement field** or **slip field** on the surface. Striations, kinematic indicators (Chapter 9) and offset of layers provide the field geologist with information about direction, sense and amount of slip. Many faults show some deviation from true dip-slip and strike-slip displacement in the sense that the net slip vector is oblique. Such faults are called **oblique-slip faults** (Figure 8.7). The degree of obliquity is given by the **pitch** (also called **rake**), which is the angle between the strike of the slip surface and the slip vector (striation).

Unless we know the true displacement vector we may be fooled by the offset portrayed on an arbitrary section through the faulted volume, be it a seismic section or an outcrop (Figure 8.6b). The apparent displacement that is observed on a section or plane is called the (apparent) separation. **Horizontal separation** is the separation of layers observed on a horizontal exposure or map (Figure 8.6b), while the **dip separation** is that observed in a vertical section (Figure 8.6c). In a vertical section the dip separation can be decomposed into the horizontal and vertical separation. Note that this horizontal separation is different from that shown in Figure 8.6b. These two separations recorded in a vertical section are more commonly referred to as **heave** (horizontal component) and **throw** (vertical component) (Figure 8.6c). Only a section that contains the true displacement vector shows the true displacement or total slip on the fault (Figure 8.6d).

A fault that affects a layered sequence will, in three dimensions, separate each surface (stratigraphic interface) so that two **fault cutoff lines** appear (Figure 8.8). If the fault is non-vertical and the displacement vector is not parallel to the layering, then a map of the faulted surface will show an open space between the two cutoff lines. The width of the open space, which will not have any contours, is related to both the fault dip and the dip separation on the fault. Further, the opening reflects the heave (horizontal separation) seen on vertical sections across the fault (Figure 8.8).

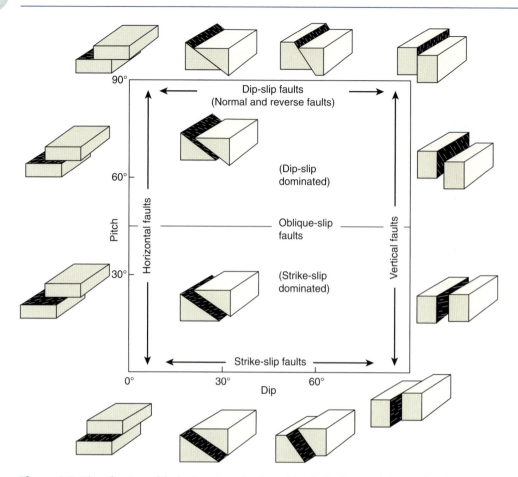

Figure 8.7 Classification of faults based on the dip of the fault plane and the pitch, which is the angle between the slip direction (displacement vector) and the strike. Based on Angelier (1994).

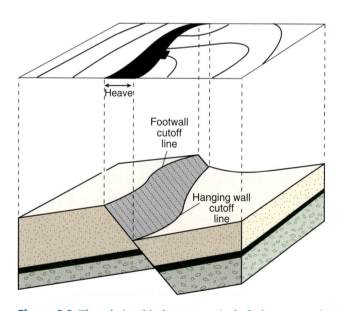

Figure 8.8 The relationship between a single fault, a mapped surface and its two fault cutoff lines. Such structure contour maps are used extensively in the oil industry where they are mainly based on seismic reflection data.

Stratigraphic separation

Drilling through a fault results in either a **repeated section** or a **missing section** at the **fault cut** (the point where the wellbore intersects the fault). For vertical wells it is simple: normal faults omit stratigraphy (Figure 8.9a), while reverse faults cause repeated stratigraphy in the well. For deviated wells where the plunge of the well bore is less than the dip of the fault, such as the well G in Figure 8.9b, stratigraphic repetition is seen across normal faults. The general term for the stratigraphic section missing or repeated in wells drilled through a fault is **stratigraphic separation**. Stratigraphic separation, which is a measure of fault displacement obtainable from wells in subsurface oil fields, is equal to the fault throw if the strata are horizontal. Most faulted strata are not horizontal, and the throw must be calculated or constructed.

8.2 Fault anatomy

Faults drawn on seismic or geologic sections are usually portrayed as single lines of even thickness. In detail,

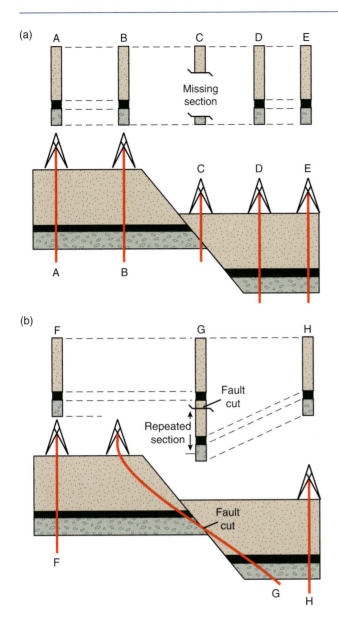

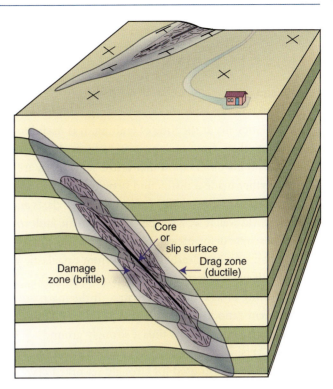

Figure 8.10 Simplified anatomy of a fault.

Figure 8.9 (a) Missing section in vertical wells (well C) always indicates normal faults (assuming constant stratigraphy). (b) Repeated section (normally associated with reverse faults) occurs where the normal fault is steeper than the intersecting wellbore (well G).

however, faults are rarely simple surfaces or zones of constant thickness. In fact, most faults are complex structures consisting of a number of structural elements that may be hard to predict. Because of the variations in expression along, as well as between, faults, it is not easy to come up with a simple and general description of a fault. In most cases it makes sense to distinguish between the central **fault core** or slip surface and the surrounding volume of brittlely deformed wallrock known as the **fault damage zone**, as illustrated in Figure 8.10.

The fault core can vary from a simple slip surface with a less than millimeter-thick cataclastic zone through a zone of several slip surfaces to an intensely sheared zone up to several meters wide where only remnants of the primary rock structures are preserved. In crystalline rocks, the fault core can consist of practically non-cohesive **fault gouge**, where clay minerals have formed at the expense of feldspar and other primary minerals. In other cases, hard and flinty **cataclasites** constitute the fault core, particularly for faults formed in the lower part of the brittle upper crust. Various types of **breccias**, cohesive or non-cohesive, are also found in fault cores. In extreme cases, friction causes crystalline rocks to melt locally and temporarily, creating a glassy fault rock known as **pseudotachylyte**. The classification of fault rocks is shown in Box 8.1.

In soft, sedimentary rocks, fault cores typically consist of non-cohesive smeared-out layers. In some cases, soft layers such as clay and silt may be smeared out to a continuous membrane which, if continuous in three dimensions, may greatly reduce the ability of fluids to cross the fault. In general, the thickness of the fault core shows a positive increase with fault throw, although variations are great even along a single fault within the same lithology.

The damage zone is characterized by a density of brittle deformation structures that is higher than the background level. It envelops the fault core, which means

BOX 8.1 | FAULT ROCKS

When fault movements alter the original rock sufficiently it is turned into a brittle fault rock. There are several types of fault rocks, depending on lithology, confining pressure (depth), temperature, fluid pressure, kinematics etc. at the time of faulting. It is useful to distinguish between different types of fault rocks, and to separate them from mylonitic rocks formed in the plastic regime. Sibson (1977) suggested a classification based on his observation that brittle fault rocks are generally non-foliated, while mylonites are well foliated. He further made a distinction between cohesive and non-cohesive fault rocks. Further subclassification was done based on the relative amounts of large clasts and fine-grained matrix. Sibson's classification is descriptive and works well if we also add that cataclastic fault rocks may show a foliation in some cases. Its relationship to microscopic deformation mechanism is also clear, since mylonites, which result from plastic deformation mechanisms, are clearly separated from cataclastic rocks in the lower part of the diagram.

Fault breccia is an unconsolidated fault rock consisting of less than 30% matrix. If the matrix-fragment ratio is higher, the rock is called a **fault gouge**. A fault gouge is thus a strongly ground-down version of the original rock, but the term is sometimes also used for strongly reworked clay or shale in the core of faults in sedimentary sequences. These unconsolidated fault rocks form in the upper part of the brittle crust. They are conduits of fluid flow in non-porous rocks, but contribute to fault sealing in faulted porous rocks.

Pseudotachylyte consists of dark glass or microcrystalline, dense material. It forms by localized melting of the wall rock during frictional sliding. Pseudotachylyte can show injection veins into the sidewall, chilled margins, inclusions of the host rock and glass structures. It typically occurs as mm- to cm-wide zones that make sharp boundaries with the host rock. Pseudotachylytes form in the upper part of the crust, but can form at large crustal depths in dry parts of the lower crust.

Crush breccias are characterized by their large fragments. They all have less than 10% matrix and are cohesive and hard rocks. The fragments are glued together by cement (typically quartz or calcite) and/or by microfragments of mineral that have been crushed during faulting.

	Non-foliated	Foliated	
Incohesive	Fault breccia (>30% visible fragments)		
	Fault gouge (<30% visible fragments)	Foliated gouge	
Cohesive	Pseudotachylyte		
	Crush breccia (fragments > 5 mm)		<10%
	Fine crush breccia (fragments 1-5 mm)		
	Crush microbreccia (fragments < 1 mm)		
	Cataclasites — Grain size reduction by cataclastic mechanisms — Protocataclasite	*Mylonite series* — Grain size reduction by plastic def. mechanisms — Protomylonite	10–50%
	Cataclasite	Mylonite	50–90%
	Ultracataclasite	Ultramylonite	>90%
		Grain size increase by recrystalliz. — Blastomylonite	

(% Matrix)

Pseudotachylyte injection veins in protomylonitic gneiss, Heimefrontfjella, Antarctica.

Continued

BOX 8.1 | (CONT.)

Cataclasites are distinguished from crush breccias by their lower fragment–matrix ratio. The matrix consists of crushed and ground-down microfragments that form a cohesive and often flinty rock. It takes a certain temperature for the matrix to end up flinty, and most cataclasites are thought to form at 5 km depth or more.

Mylonites, which are not really fault rocks although loosely referred to as such by Sibson, are subdivided based on the amount of large, original grains and recrystallized matrix. Mylonites are well foliated and commonly also lineated and show abundant evidence of plastic deformation mechanisms rather than frictional sliding and grain crushing. They form at greater depths and temperatures than cataclasites and other fault rocks; above 300 °C for quartz-rich rocks. The end-member of the mylonite series, blastomylonite, is a mylonite that has recrystallized after the deformation has ceased (postkinematic recrystallization). It therefore shows equant and strain-free grains of approximately equal size under the microscope, with the mylonitic foliation still preserved in hand samples. Plastic deformation and mylonites are treated further in Chapters 10 and 15.

Figure 8.11 Damage zone in the footwall to a normal fault with 150–200 m throw. The footwall damage zone is characterized by a frequency diagram with data collected along the profile line. A fault lens is seen in the upper part of the fault. Entrada Sandstone near Moab, Utah.

The width of the damage zone can vary from layer to layer, but, as with the fault core, there is a positive correlation between fault displacement and damage zone thickness (Figure 8.12a). Logarithmic diagrams such as shown in Figure 8.12 are widely used in fault analysis, and straight lines in such diagrams indicate a constant relation between the two plotted parameters. In particular, for data that plot along one of the straight lines in this figure, the ratio between fault displacement D and damage zone thickness DT is the same for any fault size, and the distance between adjacent lines in this figure represents one order of magnitude. Much of the data in Figure 8.12a plot around or above the line $D = DT$, meaning that the fault displacement is close to or somewhat larger than the damage zone thickness, at least for faults with displacements up to 100 meters. We could use this diagram to estimate throw from damage zone width or vice versa, but the large spread of data (over two orders of magnitude) gives a highly significant uncertainty.

A similar relationship exists between fault core thickness (CT) and fault displacement (Figure 8.12b). This relationship is constrained by the straight lines $D = 1000CT$ and $D = 10CT$, meaning that the fault core is statistically around 1/100 of the fault displacement for faults with displacements up to 100 meters.

Layers are commonly deflected (folded) around faults, particularly in faulted sedimentary rocks. The classic term for this behavior is **drag**, which should be used

that it is found in the tip zone as well as on each side of the core. Structures that are found in the damage zone include deformation bands, shear fractures, tensile fractures and stylolites, and Figure 8.11 shows an example of how such small-scale structures (deformation bands) only occur close to the fault core, in this case defining a footwall damage zone width of around 15 meters.

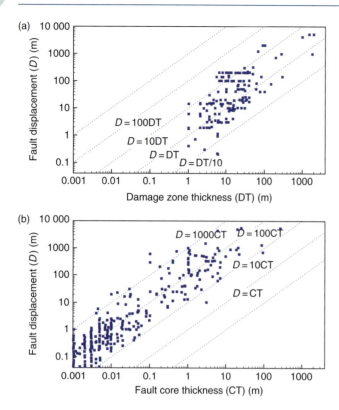

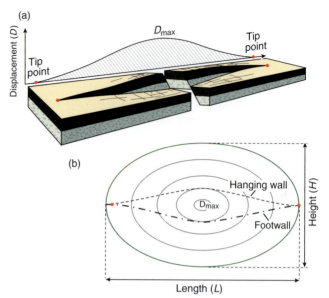

Figure 8.12 (a) Damage zone thickness (DT) (one side of the fault) plotted against displacement (D) for faults in siliciclastic sedimentary rocks. (b) Similar plot for fault core thickness (CT). Note logarithmic axes. Data from several sources.

Figure 8.13 (a) Schematic illustration of an ideal, isolated fault. The displacement profile indicates maximum displacement near the center. (b) The fault plane with displacement contours. Stippled lines are the hanging-wall and footwall cutoff lines and the distance between them indicates the dip separation.

as a purely descriptive or geometric term. The drag zone can be wider or narrower than the damage zone, and can be completely absent. The distinction between the damage zone and the drag zone is that drag is an expression of ductile fault-related strain, while the damage zone is by definition restricted to brittle deformation. They are both part of the total strain zone associated with faults. In general, soft rocks develop more drag than stiff rocks.

8.3 Displacement distribution

It is sometimes possible to map displacement variations along a fault in the field in the horizontal or vertical direction. In both directions faults tend to show a maximum displacement in the central part of the fault trace, gradually decreasing toward the tips, as illustrated in Figure 8.13. The shape of the displacement profile may vary from linear to bell-shaped or elliptic. Displacement profiles are sometimes classified into those that have a well-defined central maximum (peak type) and those that have a wide, central part of fairly constant displacement (plateau type). Examples are shown in Figure 8.14.

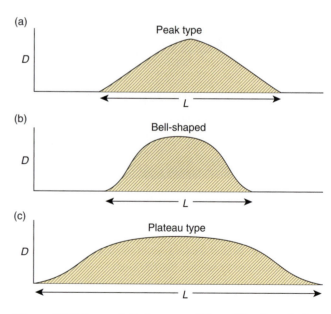

Figure 8.14 Types of displacement (D) profiles along faults.

Single faults generally show a gradual increase in displacement from the tip line toward a central point.

It may be hard to collect enough displacement data from a single fault to obtain a good picture of the

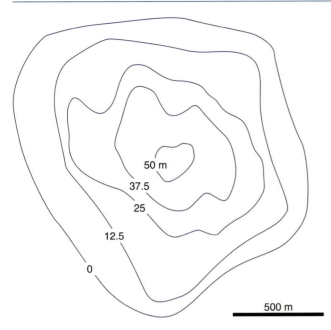

50 m
37.5
25
12.5
0

500 m

Figure 8.15 Displacement contours for a fault interpreted from high-resolution seismic data from the Gulf of Mexico. Modified from Childs *et al.* (2003).

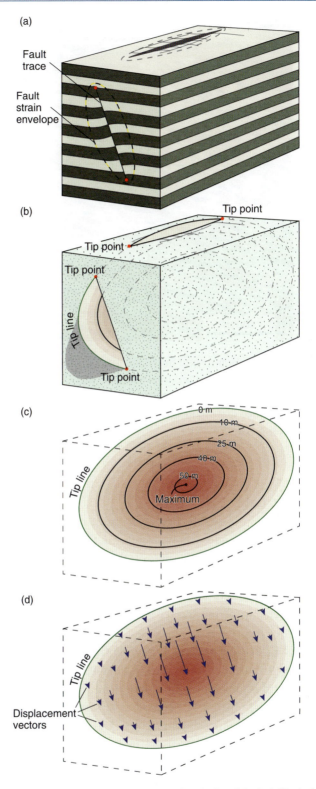

Figure 8.16 Geometric aspects of an isolated fault (elliptical fault model). (a) The fault trace is the intersection between the fault surface and an arbitrary surface (outcrop, seismic line). The endpoints of the fault trace are called tip points (b). They lie on the tip line, which is the zero-displacement line that outlines the fault (c). Displacement increases towards the center of the fault. This can be expressed in terms of displacement contours (c) or displacement vectors (d).

displacement distribution on the fault surface. However, data sets from coalmines and high-quality 3-D seismic data have made it possible to contour displacement on a number of faults. Figure 8.15 is an example of such studies, and shows the general pattern that the displacement is generally greatest in the central part of a single, isolated fault, gradually decreasing toward the tip line – a conclusion that is consistent with the field observations mentioned in the previous paragraph. Hence, these observations support the idealized model shown in Figure 8.16, where an isolated fault has an elliptical tip line and elliptical displacement contours. A geometrically similar elliptical model can be applied to extension fractures, where the displacement vectors are perpendicular to the fracture. It should be emphasized that the simple elliptical fault model is meant to describe an isolated fault in an isotropic medium. In most cases fault growth is complicated by fault interaction and mechanical layering, which causes deviations from this simple model.

8.4 Identifying faults in an oil field setting

It is crucial to collect and correctly interpret information about faults in petroleum exploration and production. Here we will review some sources of data that provide key information about faults in an oil field in an extensional setting; the same principles can be applied in contractional and strike-slip regimes.

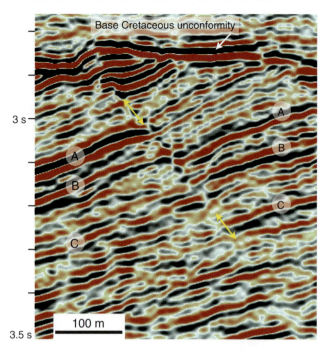

Base Cretaceous unconformity

3 s

A
A
B
B
C
C

100 m

3.5 s

Figure 8.17 A fault imaged in seismic data (arrows). There are no clear reflections from the fault itself. The dip separation is identified by the discontinuity of reflectors (cutoffs). Three-dimensional seismic from the Visund Field, North Sea.

Seismic data

Interpretation of seismic data is the most common way of identifying and mapping faults in the subsurface. Identification of faults depends on the presence of mappable seismic reflectors. Discontinuous reflectors indicate fault locations, and the dip separation is estimated by correlation of seismic reflectors across the fault, as shown in Figure 8.17.

Seismic data have a limit in resolution that varies from data set to data set, but it is usually difficult or impossible to identify faults with throw less than 15–20 meters, even on high-quality 3-D seismic data sets. Faults that fall below seismic resolution are generally referred to as **subseismic faults**, as discussed in Box 8.2. Complications along faults, such as fault lenses and fault branching, may be difficult to resolve on seismic data alone. Where available, well information is used to constrain the seismic interpretation.

As illustrated in Figure 8.18, a 3-D data set represents a cube of data where faults and reflectors can be studied and interpreted in any direction, including horizontal sections (time slices). This allows the interpretation of the 3-D geometry of faults and fault populations.

BOX 8.2 | SUBSEISMIC FAULTS

Ductile or brittle? It all depends on the scale of observation. Consider large-scale drag in the hanging wall of large normal faults. When imaged on a seismic section (or from a far distance), the layers may appear continuous and the deformation can be described as being ductile. There could still be lots of subseismic faults, because what appear to be continuous layers may be signals that are smeared out to continuous reflectors during the data processing. Two different cases are shown. In one case a series of antithetic subseismic faults affect the layers, in the other the faults are synthetic with respect to the main faults. Since the faults are too small to be imaged seismically, the two seismic images are identical, but the true small-scale dip is different in the two cases. Core data and perhaps dipmeter data will reveal the true dip. If the difference between dip determined from cores and dipmeter data is significant, then it is likely that subseismic faults occur. If the dip determined from cores is the same as the seismic dip, then the deformation is microscopic, perhaps by granular flow (typical for sediments that were poorly lithified at the time of deformation).

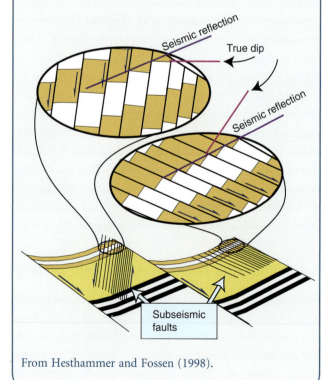

Seismic reflection
True dip
Seismic reflection

Subseismic faults

From Hesthammer and Fossen (1998).

Basin-scale faults and regional fault arrays are more commonly mapped by means of regional 2-D lines. Some 2-D lines are deep seismic lines that image the deep parts of the crust and the upper mantle (Figure 1.6). Deep seismic lines show large faults that penetrate the upper crust and sometimes the entire crust as they pass into deep shear zones.

Fault cut and well log correlation

Faults along the wellbore are typically identified by means of stratigraphic correlation. As shown in Figure 8.9, reverse and normal faults cause repetition and omission of stratigraphic sections, respectively. Knowledge of the

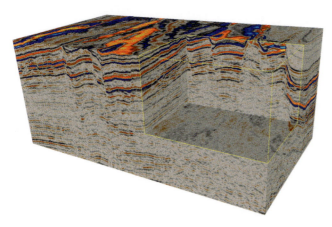

Figure 8.18 Faults as they may appear in a 3-D seismic data cube. Data from the Barents Sea.

stratigraphy from other wells in the area forms the basis for this type of fault identification.

The size of faults identifiable by this method depends on the characteristic stratigraphic markers, the number of wells, distance to other wells in the area, sedimentary facies variations and the orientation of the well. The identification of the **fault cut** (fault location in the well-bore) also depends on characteristic signatures on the well logs. Cores are generally not available, and standard logs such as gamma-ray logs, density logs, neutron logs and resistivity logs are used for stratigraphic well correl-ations. Figure 8.19 shows an example of how faults with as little as 6 meters of stratigraphic separation are detect-able in parts of the North Sea Brent Group, where the density of wells is high. In this example the fault was confirmed by core inspection.

Dipmeter data and borehole images

Microresistivity is measured continuously along the wellbore by the three or more (usually 16) electrodes of a dipmeter tool. The responses from the different electrodes are correlated around the borehole in narrow depth intervals to fit a plane. The planes are generally bedding or lamination, but may also represent deform-ation bands or fractures.

Orientations (usually given by dip and dip azimuth) are plotted in dipmeter diagrams. For structural analysis, separating dip azimuth and dip into individual plots

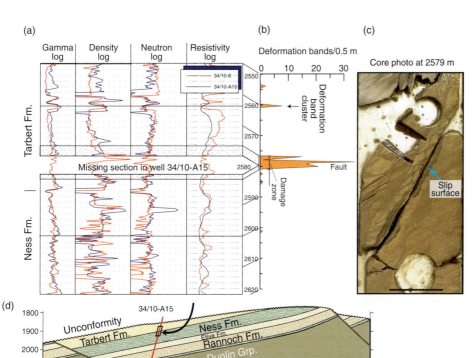

Figure 8.19 Fault separation detected by log correlation with a neighboring well. The complete log in well 8 is shown in red, and the correlation gives a missing section of ~6 m in well A15. The damage zone (orange) was estimated from core inspection to be a few meters wide. Based on Fossen and Hestshammer (2000).

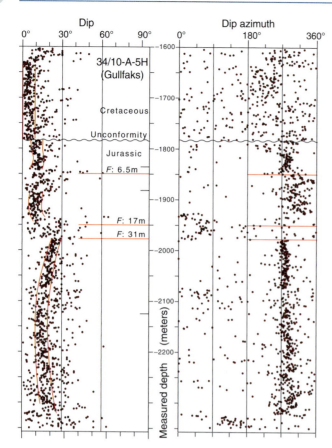

Figure 8.20 Dipmeter data from the Gullfaks Field, North Sea, where dip and dip azimuth (dip direction) are plotted against the depth measured along the wellbore. Faults identified by stratigraphic correlation (missing section) are indicated.

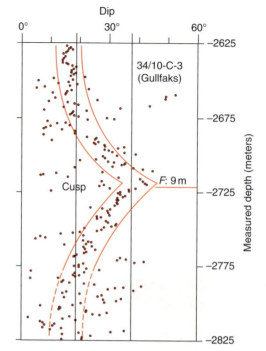

Figure 8.21 Dipmeter data (dip against depth) showing a classic cusp geometry related to drag around a minor fault. Data from the Gullfaks Field, North Sea.

and compressing the vertical scale can be advantageous, as shown in Figure 8.20. Faults can then be identified in at least three different ways. The first is represented by sudden changes in dip or dip azimuth. These occur where a fault separates two blocks in which the layering is differently oriented – a fairly common situation across many faults. An example is seen between the 17- and 31-meter faults in Figure 8.20.

Another characteristic feature is the presence of local intervals with rapid but progressive changes in dip and/or dip azimuth. Such anomalies are known as **cusps** (Figure 8.21). Cusp-shaped dip patterns indicate fault-related drag in many cases. Stratigraphic log correlation indicates a 9-meter missing section in the example shown in Figure 8.21.

The third characteristic is the appearance of anomalous orientations related to fractures or deformation bands in the damage zone or from the main slip surface itself. The high dip values at the locations of the faults in Figure 8.20 may be such examples.

It is now common to create a (almost) continuous microresistivity image of the borehole based on resistivity data from more sophisticated tools, such as the **FMI** (Formation MicroImager). This tool measures microresistivity by means of a few hundred electrodes. The result is a continuous image of the wall that is reminiscent of an actual picture of the rock. Such images are analyzed at workstations where bedding and structural features are interpreted.

Drill core information

Only a small percentage of the drilled section in a reservoir will be cored, and only rarely are faults represented in the drill core material. Drillers are reluctant to cut cores across faults because of the risk of jamming and potential pressure problems. Furthermore, some cored fault rocks may be so non-cohesive that they fall apart to form what is known as **rubble zones**. However, successfully cored faults and damage zones represent valuable information. Such samples allow for microscopic studies and permeability measurements. Furthermore, the width and nature of the damage zone, and sometimes even the fault core, can be estimated. Figure 8.22 shows an example of a core through a fault with a 6-meter missing

Figure 8.22 One-meter-long core section across a minor (6 m missing section) fault in the Gullfaks Field. Holes are from plugs sampled for permeability analysis. See also Figure 8.19.

section. The central slip surface (very thin core) and deformation bands in the damage zone are visible.

The orientation of faults and fractures in cores can be measured if the core is oriented. Usually it is not, and its orientation must then be reconstructed based on knowledge of bedding orientation from dipmeter data or seismic data. This can only be done if bedding is non-horizontal.

8.5 The birth and growth of faults

Fault formation in non-porous rocks

Faults in rocks with low or no porosity somehow grow from small shear fractures. However, this cannot happen directly from a single shear fracture, since shear fractures cannot expand in their own planes. Instead they will curve and form wing cracks or related cracks across which there is tension (Box 7.3). Experiments show that a phase of intense microfracturing occurs prior to fracture initiation or propagation. Once the density of microfractures reaches a critical level, the main fracture expands by linkage of favorably oriented microfractures. The zone of microfractures (and mesofractures) ahead of the fracture tip zone is called the **frictional breakdown zone** or the **process zone**.

For a fault to develop, a number of small shear fractures, tensile fractures and hybrid fractures must form and connect. The incipient fault surface is irregular, leading to grinding and microfracturing of the walls. A thin core of brecciated or crushed rock typically forms. During fault growth, new fractures form in the walls next to the fault core. Hence, most faults have a well-defined **core** of intense cataclastic deformation and a surrounding **damage zone** of less intense fracturing.

Fault formation and growth is a complicated process involving a frontal process zone where microfractures form and eventually connect.

Natural rocks are not isotropic, and in many cases faults form along preexisting mesoscopic weaknesses in the rock. Such weaknesses can be layer interfaces or dikes, but the structures that are most likely to be activated as faults are joints (and, of course, preexisting faults). Joints tend to be very weak planar structures with little or no cohesion. Joints may also form surfaces many tens of meters or more in length and/or height, because as extension structures they have had the freedom to expand in their own plane. Faults formed by faulting of joints inherit some of the features of the original joints. If a fault forms by frictional sliding on a single, extensive joint, the initial fault tends to be a sharp slip surface with almost no fault core and with (almost) no damage zone. If slip accumulates, however, the fault outgrows the joint and links with other joints in the vicinity of its tip zone. The damage zone then thickens, and the fault core may grow.

Fault formation by joint reactivation requires less stress, causes less off-fault damage (narrower damage zone) and may result in a lower displacement gradient along the fault.

Fault formation in porous rocks

In highly porous rocks and sediments, fault growth follows a somewhat different path. Pore space gives the grains a unique opportunity to reorganize. If the grains in a sandstone are weakly cemented together, then the grains will reorganize by rotation and frictional grain boundary sliding (translation) during deformation. In other cases grains can also break internally. In either case, the deformation is likely to localize into narrow zones or bands to form structures known as **deformation bands**. Different types of deformation bands are discussed in the previous chapter (Section 7.8).

Field observations, as well as experimental and numerical work, show that deformation proceeds by sequential formation of new deformation bands adjacent to the initial one (Figure 8.23). This means that at some point it becomes easier to form a new band next to the existing one than to keep shearing the primary band. The result is a **deformation band zone**, and this development is commonly explained in terms of strain hardening. Strain hardening is thought to be related to the loss of porosity in the band and is most pronounced where grains are crushed (cataclastic bands). Note the difference between process zones in non-porous and highly porous rocks: The process zone in non-porous rocks weakens the host rock and increases porosity by the formation of cracks. In high-porosity rocks the deformation bands in the process zone in many cases harden the rock and reduce porosity.

Once a certain number of deformation bands have accumulated in the deformation band zone, porosity is sufficiently reduced that a slip surface can form and grow. Slip surfaces nucleate in small patches that propagate, link up, and ultimately form through-going slip surfaces. Mechanically, slip surfaces are weak structures that

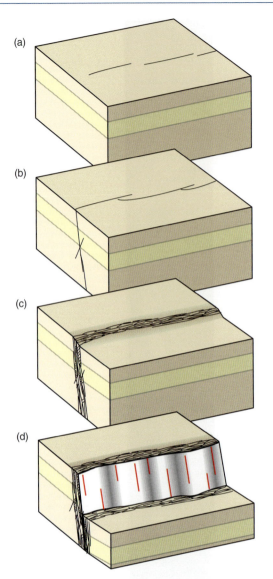

Figure 8.23 The general model for fault formation in porous sandstone, as proposed by Aydin and Johnson (1978). (a) Individual deformation bands, (b) linking of bands, (c) the formation of a deformation band zone, and (d) faulting of the zone.

relatively quickly can accumulate meters of slip or more. Through-going slip surfaces are commonly associated with a thin (millimeter-thick) zone of ultracataclasite, which may be considered as the local fault core.

Faults in highly porous rocks form in precursory deformation band zones.

The damage zone

The growth of deformation bands and/or ordinary fractures prior to the formation of a through-going slip surface has implications for our understanding of the damage zone. The moment the slip surface (fault) forms, the enclosing zone of already existing structures will become the damage zone. Once the fault is established, the process zone in front of the fault tip moves ahead of the fault tip as the fault expands, leaving behind a zone that becomes the initial fault damage zone (Figure 8.24). In a porous rock, this zone is likely to consist of deformation bands. Because faults form in porous rocks by faulting of a deformation band zone as shown in Figure 8.23, the length of the deformation band process zone tends to be longer than the process zone seen in many non-porous rocks. This is particularly true if the deformation bands are cataclastic, in which case the process zone can be several hundred meters long.

If the structures of the damage zone form ahead of a propagating fault tip, then the damage zone should be slightly older than the associated slip surface. A consequence of this assumption would be that the width and strain of a damage zone are independent of fault displacement. Empirical data (Figure 8.12) show that this is not the case, even though the fault (slip surface) represents the weakest part of the rock and is therefore prone to reactivation without the creation of more sidewall damage. The reason is simply that faults are not perfectly planar structures, nor do they expand within a perfect plane. Faults are irregular at many scales because the rocks that they grow in are both heterogeneous and anisotropic. For example, faults may bend as they meet a different lithologic layer or as they link with other faults (Figure 8.25). Figure 8.26 shows an example of how damage may generate in the vicinity of a fault bend, in this case along with a gentle fault-bend fold.

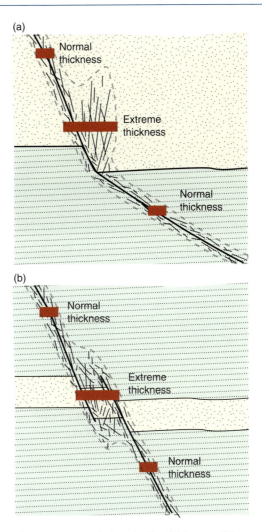

Figure 8.25 Variations in the thickness of damage zones related to (a) change in dip and (b) linkage. In these situations, minor structures are added to the damage zone until the fault cuts through the complex zone along a straighter path.

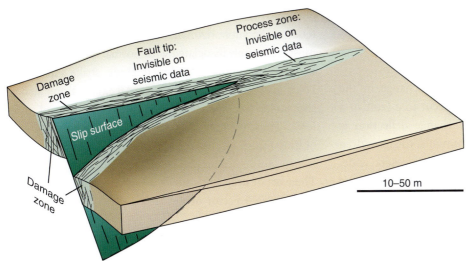

Figure 8.24 A fault is contained within a damage zone, which means that there is a (process) zone ahead of the tip where the rock is "processed" prior to fault propagation. The process zone may potentially contribute to compartmentalization of petroleum reservoirs. From Fossen *et al.* (2007).

Figure 8.26 Hanging-wall rollover (fault-bend fold) related to bend in the main fault. The damage zone is unusually wide due to the complications posed by the fault bend. Synthetic and antithetic shear bands are separated by color. Matulla Formation, Wadi Matulla, Sinai. The offset of the fault exceeds the 4 m cliff height.

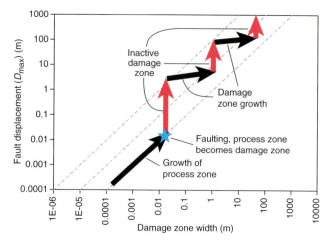

Figure 8.27 Schematic illustration of how a damage zone can grow periodically. The first stage is the growth of the process zone. Once the fault forms, the process zone becomes the damage zone and slip occurs smoothly for a while (red arrow) until complications lock the fault, which causes renewed growth of the damage zone. This repeats itself as the fault grows. The result is considerable scatter of fault data in fault displacement–damage zone width diagrams.

The structures in the damage zone form both prior to, during and after the local formation of the slip surface (fault).

If the fault is temporarily or locally planar and smooth, then there may be periods during which displacement accumulates without any deformation of the wall rocks, i.e. without any widening of the damage zone. However, at locations of fault linkage or fault bends, wall damage may also occur during fault growth, which leads to a local widening of the damage zone. Eventually, the fault may find a more planar way through zones of complications, and the damage zone becomes inactive again. Thus, the growth of damage zones may be temporal and local (Figure 8.27), contributing to the scatter seen in Figure 8.12.

The ductile drag zone

Drag is best defined as any systematic change in the orientation of layers or markers adjacent to a fault in a way that makes it clear that the change deflection is genetically related to the fault. Commonly, the term drag describes zones some meters or tens of meters wide. However, hanging-wall synclines related to normal faults in continental rift basins can extend several hundred meters into the hanging wall. Similarly, large-scale rollovers (reverse drag), up to several kilometers long on the hanging-wall side are associated with large listric faults.

Drag is folding of layers around a fault by means of brittle deformation mechanisms, directly related to the formation and/or growth of the fault.

Drag is seen in layers that are soft enough to deform ductilely in the upper, brittle part of the crust, most commonly in faulted sedimentary sequences. Although drag is commonly limited to a few-meters-wide zone along the fault (Figure 8.28), it may also be large enough to be imaged on seismic data, which in Figure 8.29 is documented by dipmeter data.

Drag can form in any tectonic regime. The kinematic requirement is that the angle between the slip vector of the fault and the layering is not too small. Because layering tends to be subhorizontal in sedimentary rocks, drag is most commonly associated with normal and reverse faults and less commonly developed along strike-slip faults. Folds also develop in subhorizontal layers along strike-slip faults (see Figure 18.10), but these are not drag folds. Thus, we may want to add another characteristic of drag folds:

The axes of drag folds make a high angle to the displacement vector of the fault.

For dip-slip faults and subhorizontal layers that means subhorizontal fold axes.

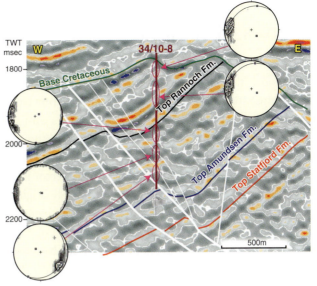

Figure 8.29 Dipmeter data collected along a vertical North Sea well. Stereonets show dip azimuth of layering in selected intervals. A change from west- to east-dipping layers is consistent with the normal drag portrayed by the seismic reflectors. Also note that the drag zone widens upward, consistent with the trishear model.

There are two geometrically different types of drag: normal drag and reverse drag. **Normal drag** is the shear zone-like geometry where layers flex toward parallelism with the fault. This is the geometry shown in Figures 8.28–32. Normal drag involves displacement, so that the total offset is the sum of the ductile normal drag and the discrete fault displacement. **Reverse drag** is used for the usually larger-scale rollover structures that occur on the hanging-wall side of listric normal faults. In this case the layers are concave in the slip direction. Both normal and reverse drag occur along faults, depending on the local fault geometry.

It was originally thought that drag was the result of friction along the fault during fault growth, but the term now includes bending of the layers prior to fault formation. The latter model seems to fit many examples of drag. This model is similar to that of damage zone development: in both cases layers are deformed in the wake of the fault tip. The difference is that drag folding is ductile down to a certain scale, commonly that of a hand sample.

The geometry of a drag fold contains information about how it formed. A drag fold in which the layers have the same geometry along the fault, with constant width of drag zone dip isogons (see Chapter 11) running parallel to the fault trace, can be modeled by simple

Figure 8.28 Drag of layers of siltstone and sandstone along vertical fault located along the left margin of the picture. Colorado National Monument, USA.

BOX 8.3 | TRISHEAR

The trishear method, published by Eric Erslev in 1991, models the ductile deformation in a triangular fault propagation fold area in front of a propagating fault tip. The hanging-wall side of the triangle moves with a constant velocity above a fixed footwall, and the apical angle of the triangle is chosen. The triangle is symmetric with respect to the fault and the velocity vector is identical for all points located along any ray originating at the fault tip. Across the zone the velocity increases toward the hanging wall, Also, the direction of the velocity vector changes gradually from the hanging wall toward the footwall, as shown in the figure. We choose the apical angle of the trishear zone, the fault dip and amount of slip, and the propagation-to-slip ratio (P/S ratio). P/S determines how rapidly the fault tip propagates relative to the slip on the fault. $P/S = 0$ means that we fix the triangular zone to the footwall, while $P/S = 1$ fixes the zone to the hanging wall. In most cases $P/S > 1$ is realistic. Trial and error will give the geometry that most closely matches your field observations or seismically imaged structure. With a trishear program we can model the ductile deformation in front of a propagating fault tip and the resulting drag along the fault in a surprisingly realistic manner. Playing with Rich Allmendinger's freely available Fault/Fold program is recommended for understanding trishear and its parameters.

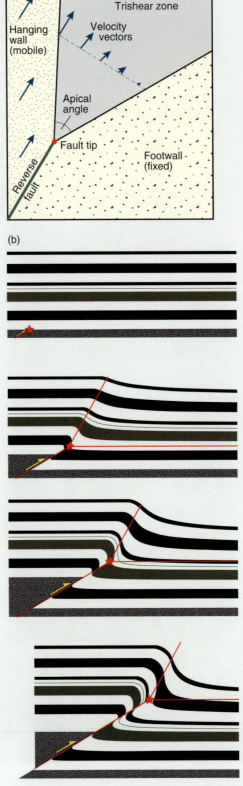

Trishear model of a reverse fault affecting an overlying sedimentary sequence. $P/S = 1.5$. Star indicates the fault tip at each stage.

shear. In this case we have a simple shear zone with a central discontinuity.

In other cases the drag zone is upward-widening, and a different kinematic model must be applied. A popular model is called **trishear**. In this model, which is more closely discussed in Box 8.3, strain is distributed in a triangular or fan-shaped zone of active deformation ahead of the fault tip (Figure 8.30). This zone moves through the rock as the fault propagates, and no further

folding occurs once the fault has cut through the layers. The width of the triangular deformation zone varies from case to case, but in all cases the drag zone widens upsection. This model seems to work particularly well in places of reactivated basement faults that grow into overlying sedimentary strata. Many examples of such structures are found in the uplifts on the Colorado Plateau and in the Rocky Mountains foreland in Wyoming and Colorado, where the fold structures are commonly referred to as **forced folds**.

Folds that form ahead of a propagating fault tip are called **fault propagation folds**. Thus, many drag folds are faulted fault propagation folds. However, drag can also form or become accentuated in the walls of an already existing fault. Just like the damage zone, fault drag can develop due to locking of the fault at fault bends, fault linkage and other complications that can increase the friction along faults. The effect of non-planar fault geometry is discussed in Chapter 20, and the development of normal drag between two overlapping fault segments is illustrated in Figure 8.31. The latter mechanism can take drag to the point where the rotated layer, which typically consists of clay or shale, forms a **smear** along the fault.

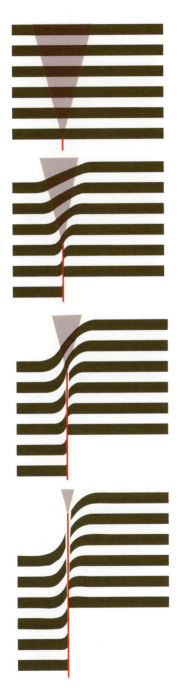

Figure 8.30 Trishear modeling of normal drag development. In this fault propagation fold model the drag zone widens upward.

(a)

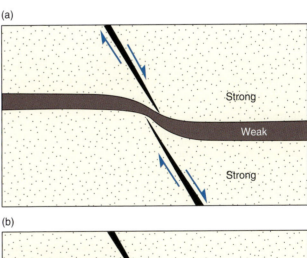

(b)

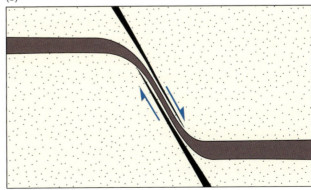

Figure 8.31 Normal dragging of mechanically weak layer (e.g. clay) between two overlapping fault segments.

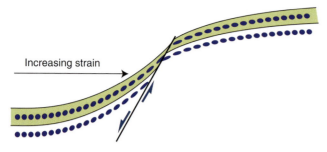

Figure 8.32 Strain ellipses portraying the relation between layer orientation and strain. Generated by means of the trishear program FaultFold. (R. Almendinger 2003).

Drag can form both ahead of the fault tip and in the walls of an active fault.

Drag, deformation mechanisms and the damage zone

Drag can occur by granular flow, particularly in poorly lithified sediments. Granular flow leaves little or no trace of the deformation except for the rotation of layering or modification of sedimentary structures. In consolidated sedimentary rocks grains may start to fracture, and the mechanism becomes distributed cataclastic flow. The mechanisms are the same as those that operate in the different deformation bands discussed in the previous chapter, but the deformation during drag folding is less localized and strain is generally lower. There is, however, a strain gradient toward the fault, as shown in Figure 8.32.

Fractures or deformation bands may occur in drag zones. In such cases the density of fractures or deformation bands increases toward the fault, as shown in Figure 8.33. The appearance of mesoscopically mappable fractures or deformation bands indicates that we are in the damage zone. Where drag folds are well developed, the drag zone tends to be wider than the damage zone, although the opposite situation also occurs.

Some faults, particularly in metamorphic rocks, show drag-like folding of the layering similar to that shown in Figure 8.28. A closer examination of many such "drag" folds reveals that they are controlled by plastic deformation mechanisms and are thus shear zones around faults that could have formed in a variety of ways, generally in the zone of brittle–plastic transition. We generally do not consider such plastic fold structures as drag folds, although the similarity can be striking.

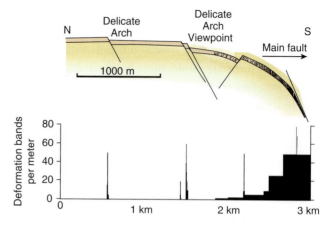

Figure 8.33 Folding of layers (rollover or reverse drag) adjacent to a major fault (not shown) accommodated by deformation band formation. Note the relationship between bed rotation and density of deformation bands. Example from Arches National Park, based on Antonellini and Aydin (1994).

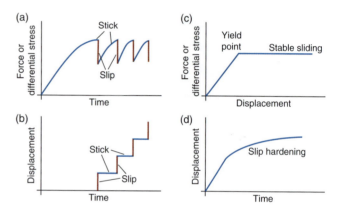

Figure 8.34 Idealized graphs illustrating the difference between stick-slip and stable sliding: (a) and (b) stick-slip graphs, (c) ideal stable sliding and (d) stable sliding with slip hardening.

Fault growth and seismicity

Once a fault surface is established it will represent a mechanically weak structure that is likely to fail again during renewed stress build-ups. Faults grow by two mechanisms. The most common one is called **stick-slip**, where slip accumulates at very sudden seismic slip events, separated by periods of no slip (Figure 8.34). Stress builds up between the slip events until it exceeds the frictional resistance of the fault. This is the model used to understand earthquakes, where each slip event causes an earthquake whose magnitude is related to the amount of energy released during the stress drop. In terms of strain, this is related to the amount of elastic strain that is released as the fault moves.

The other way for faults to accumulate slip is by **stable sliding** or **aseismic slip**. Ideally, displacement accumulates at a constant rate during stable sliding (Figure 8.34c). Some laboratory experiments show that a gradually increasing force is needed for slip to continue. This effect is called **slip hardening** (Figure 8.34d) and is related to damage of the slip surface during deformation.

Several factors control whether fault displacement accumulates gradually or by sudden slip events. Rock experiments indicate that stable sliding is more likely when the normal stress across the fault is small, which means that stable sliding is more common in the uppermost part of the brittle crust than deeper down. Low-angle faults along overpressured layers in a sedimentary sequence would also be likely to experience stable sliding even at depths of several kilometers because overpressure reduces the effective normal stress across the fault (Section 7.7).

Lithology is another important factor: porous sediments and sedimentary rocks are more likely to deform by stable sliding than are low-porosity crystalline rocks. In particular, stick-slip is favored in low-porosity quartz-rich siliceous rocks, while clay promotes stable sliding. Clay-bearing incohesive fault gouge in the fault core has some of the same effect as claystone: thick and continuous zones of clay gouge promote stable sliding. The fact that gouges (or rather slip surfaces along gouge zones) tend to represent pathways for fluid flow may add to their ability to slide in a stable fashion.

Close to the brittle–plastic transition, elevated temperatures introduce plastic deformation mechanisms that also promote stable sliding. Stick-slip deformation is of minor importance in the plastic regime, which for granitic rocks means temperatures above ~300 °C.

In summary, we could say that in the very top of the crust (upper kilometer or two), earthquakes are expected to be rare because of low normal stresses, weak and unconsolidated fault cores (gouge) and, at least in sedimentary basins, weak and porous rocks in general. Below this depth one would expect abundant seismic or stick-slip activity until the brittle–plastic transition is reached. This is exactly what earthquake data indicate, and the zone is called the **seismogenic zone** (Figure 8.35).

Typically, a single fault in the seismogenic zone shows evidence of both stick-slip and stable sliding. While seismic events may be responsible for the majority of total displacement accumulated over time, slow, **aseismic** "creep" is found to occur between **seismic** events. There

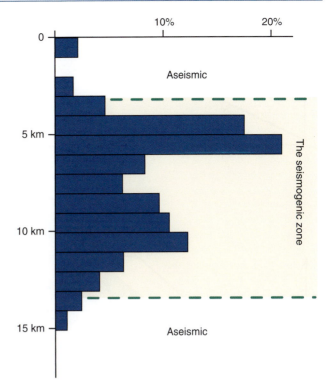

Figure 8.35 Distribution of 630 earthquakes in the crust beneath Parkfield, California. The distribution is characteristic for the continental crust away from subduction zones. Data from Marone and Scholz (1988).

is also a need for small **postseismic** adjustments that may or may not be seismic. Seismic means sudden energy release and displacement accumulations by means of earthquakes. Aseismic means gradual displacement accumulation without the generation of earthquakes.

Fault slip and displacement accumulation are commonly discussed in terms of seismicity and seismic slip behavior. It is important to realize that a single earthquake is unlikely to add more than a few meters of displacement. A quake of magnitude 6.5–6.9 that activates a 15–20 km long fault adds no more than one meter of maximum displacement to a fault. Only the largest earthquakes can generate offsets of 10–15 m. This has a very important implication:

A fault with a kilometer displacement must be the product of hundreds of earthquakes.

It is worth noting that the accumulation of such displacements would take thousands or millions of years, depending on the local displacement rate. Throw rates for faults can be found by dating sedimentary layers that are offset and measuring their displacements. Average

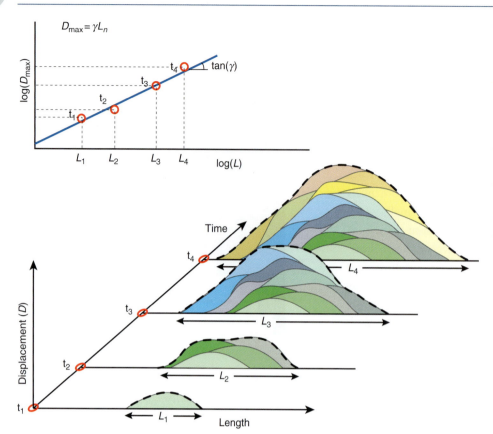

Figure 8.36 Schematic illustration of displacement accumulation through repeated slip events (earthquakes). Each event results in up to a few meters of displacement. In this model a bell-shaped cumulative displacement profile emerges which resembles that of a single slip event. The result of this model is a straight line in a logarithmic length–displacement diagram.

displacement rates of around 1–10 mm/year have been published for major faults in tectonically active areas.

Large faults tend to slip along just a limited portion of the total fault surface. The total displacement distribution for a large fault is therefore the sum of displacements contributed by individual slip events (earthquakes) (Figure 8.36). While it seems clear that single slip events produce more or less elliptical displacement contours, similar to those shown in Figure 8.15, the finite displacement distribution from a large number of slip events (earthquakes) is more difficult to predict or understand. The **characteristic earthquake model** assumes that each slip event is equal to the others in terms of slip distribution and rupture length. However, the location of the displacement maximum is shifted for each slip event. The **variable slip model** predicts that both the amount of slip and the rupture length vary from event to event, while the **uniform slip model** considers the slip at a given point to be the same in each slip event (the area varies). We will not go into the details of these models here, but simply state that displacement accumulation results in a displacement maximum near the middle of the fault, gradually tapering off toward the tips, as shown in Figure 8.13.

8.6 Growth of fault populations

Faults grow from microfractures or deformation band zones and accumulate displacement over time as deformation proceeds. Moreover, faults tend to nucleate at many different places as a region is critically stressed, for instance during rifting, and we refer to such groups of faults as **fault populations**. In general, many faults in a population soon become inactive and thus remain small. Others reach an intermediate stage before dying, while a few grow into long faults with large displacements. Hence, a fault population is always dominated by small faults, while additional long faults develop as strain accumulates.

Faults are unlikely to grow as individual structures over a long period of time. As they grow, they are likely to interfere with nearby faults. In this way two faults can join to form a single and much longer fault. Growth by linkage is a very common mechanism that creates some of the most interesting and important structures in faulted regions (Figures 8.37 and 8.38).

Fault linkage and relay structures

In a population where faults grow in length and height, faults and their surrounding stress and strain fields will

Figure 8.37 Extension fracture population along the edge of a paved road. Each of the fractures has grown from microfractures and they have reached a variety of sizes. The pavement is now more or less saturated with fractures, implying that additional strain will be accommodated by coalescence of existing fractures rather than by the nucleation of new ones.

locally interfere. Let us consider two faults whose tips approach each other during growth. Before the tips have reached each other (but after their strain fields have started to interfere) the faults are said to **underlap** (Figure 8.39a). Once the tips have passed each other, the faults are **overlapping** (Figure 8.39b, c). Under- and overlapping faults are said to be **soft linked** as long as they are not in direct physical contact. Eventually the faults may link up to form a **hard link** (Figure 8.39d).

Underlapping faults "feel" the presence of a neighboring fault tip in the sense that the energy required to keep the deformation going increases. The propagation rate of the fault tips in the area of underlap is thus reduced, which causes the local displacement gradient to increase. This results in asymmetric displacement profiles, where

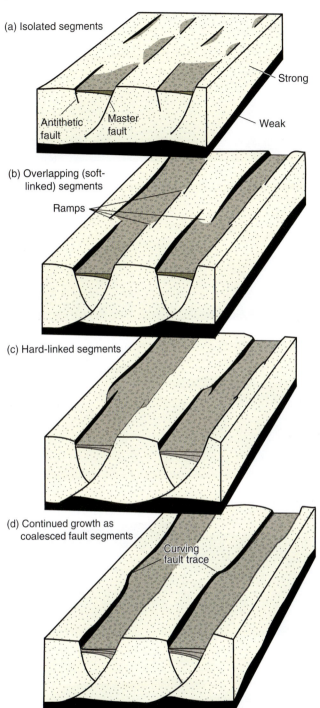

Figure 8.38 Simplified model for the development of a portion of a fault population in Canyonlands, Utah. The faults develop from isolated fractures into long faults through the formation and destruction of relay ramps. Based on Trudgill and Cartwright (1994).

the maximum is shifted toward the overlapping tip (Figure 8.40; t_1).

This asymmetric displacement distribution becomes more pronounced as the faults overlap and the layers in

(a)

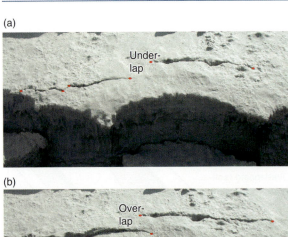

(b)

(c)

(d)

(e)

Figure 8.39 The development of curved fault systems in unconsolidated sand. Two isolated fractures (a) overlap (b, c) to form a relay ramp that eventually becomes breached (d, e). Faults were initiated by splashing water on the beach sand of a Colorado lake. The width of sand shown in each picture is *c.* 50–60 cm.

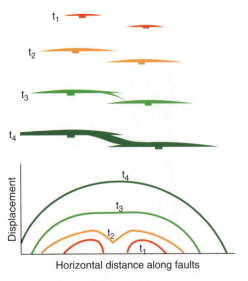

Figure 8.40 Illustration of the change in displacement along two faults that overlap and coalesce. The upper part shows the two segments in map view at four different stages of growth (t_1–t_4). The lower part shows the displacement profile at the four different stages.

the overlap zone become folded. The folding is a result of ductile displacement transfer (relay) from one fault to the other and is directly related to the high displacement gradients in the overlapping tip zones. If the fault interference occurs perpendicular to the slip direction, which for normal and reverse faults means in the horizontal direction, and if the layering is subhorizontal, then the folding is well expressed in the form of a ramp-like fold. The fold itself is called a **relay ramp** and the entire structure is known as a **relay structure** (Figures 8.39c and 8.41), and Box 8.4 explains why such structures are of particular interest to petroleum geologists.

The ramp is a fold that may contain extension fractures, shear fractures, deformation bands and/or minor faults depending on the mechanical rock properties at the time of deformation. Eventually the ramp will break to form a **breached relay ramp**. The two faults are then directly connected and associated with an abnormally wide damage zone.

Upon breaching there will be a displacement minimum at the location of the relay structure. The total displacement curve along the fault will therefore show two maxima, one on each side of the relay structure. As the deformation proceeds and the fault accumulates displacement, the displacement profile will approach that of a single fault, with a single, central maximum (Figure 8.40; t_4). However, the link is still characterized by the wide damage zone and a step in the horizontal

Figure 8.41 Relay ramp formed between two overlapping fault segments in Arches National Park, Utah. There is a higher density of deformation bands within the ramp than away from the ramp.

BOX 8.4 | RELAY RAMPS IN PETROLEUM RESERVOIRS

Relay ramps can be important structures in a petroleum reservoir. In the context of exploration, ramps can cause communication (migration of oil) across a fault that is elsewhere sealing. During production, relay ramps may represent pathways for water, oil or gas that cause pressure communication between otherwise isolated fault blocks. Relay ramps may contain abundant subseismic structures, depending on their stage of maturity. They are generally associated with damage that could cause problems to wells placed in or near the relay structure. Both breached and intact ramps are easily interpreted as sudden bends in the fault trace. The kinks can represent the interpreter's smoothing of an intact ramp around or below seismic resolution, or a breached ramp. Unbreached and barely breached ramps may show a displacement minimum in the ramp area that may be detectable from seismic data.

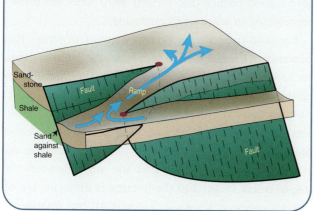

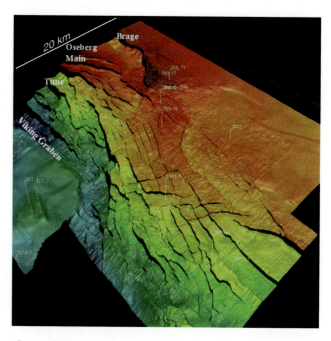

Figure 8.42 Normal fault population on the east flank of the Viking Graben, northern North Sea, illustrated at the base Cretaceous level (warm colors indicate shallow depths). The fault population shows various stages of fault linkage.

fault trace. If the mapping is based on poorly constrained outcrop information or seismic data (which always have a resolution issue), a sudden change in strike may be the only indication of a breached ramp. Such steps, seen at many locations in the seismic interpretation shown in Figure 8.42, are therefore very important as they may hint at the locations of both breached and intact relay ramps.

Bends and jogs of faults in map view are very common on many scales. Figure 8.39 shows the development of a non-planar fault in sand. The final fault geometry can

Figure 8.43 Curved fault systems (white dashed lines) in the northern North Sea basin (base Cretaceous level). Note similarities with Figures 8.39 and 8.44.

be seen to be the result of interaction between individual fault segments through the creation and breaching of relay ramps. The curved fault pattern seen in map view is very similar to that displayed by much larger faults, such as the northern North Sea faults shown in Figure 8.43 and the Wasatch Fault in Utah (Figure 8.44), so it seems likely that these large faults formed by fault linkage as portrayed in Figure 8.39. Assuming that this analogy holds, we are now able to interpret a deformation history from the geometry of a dead fault system.

It is clear that ramps come in any size and stage of development. It is also important to understand that relay ramps and overlap zones are formed and destroyed continuously during the growth of a fault population.

Fault growth by linkage involves the formation and destruction of relay structures, deviations from the ideal elliptical displacement distribution and generation of wide damage zones and fault bends at locations of linkage.

Fault linkage in the slip direction

Faults grow in the directions both normal and parallel to the slip direction and therefore interfere in both the vertical and horizontal directions (Figure 8.45). In the previous section we looked at the horizontal interaction of normal fault tips. Now we will look at the interference of normal fault tips in the vertical plane, i.e. parallel to the slip vector. Fault linkage is most commonly recognized and described in map view, but this is simply because they are easier to observe in map view. Tall vertical sections are less common than long horizontal

Figure 8.44 The Wasatch fault zone near Salt Lake City, Utah, crudely indicated by white dashed line. Note the curved fault geometry, indicating a history of segment linkage. Sketches of various stages of plaster extension experiment indicate how such fault zones can form.

exposures, and the continuity of good seismic reflectors in the horizontal direction makes it easier to map relays in map view than in the vertical direction. It takes a whole package of good reflectors to identify and map vertical overlap zones on seismic sections. There is therefore a good chance that vertical relay zones are under-represented in seismic interpretations.

Faults initiate after a certain amount of strain that depends on the mechanical properties (Young's modulus, etc.). As strong rock layers start to fracture, weak rocks continue to accumulate elastic and ductile deformation. As these fractures grow into faults, they will interfere and connect. In many cases sandstones become faulted before shales. The process is similar to that occurring in map view, except that the angle between the displacement vector and the layering is different. We do not get ramps as shown in Figure 8.41, but rotation of

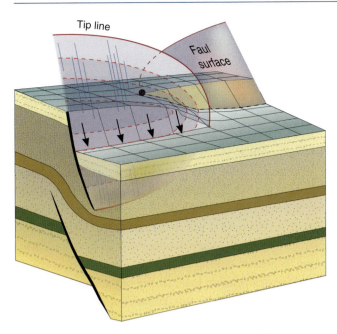

Figure 8.47 Overlapping faults where a shale layer is caught and smeared in the overlap zone. Moab, Utah.

Figure 8.45 Faults interfere in both the horizontal and vertical directions as they grow. In both cases displacement is transferred from one fault to the other, and layers between the overlapping tips tend to fold into ramps or drag folds. Modified from Rykkelid and Fossen (1992).

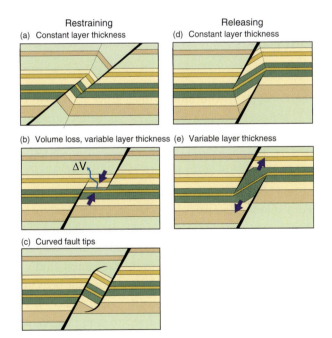

Figure 8.46 Different types of vertical overlap zones (horizontal layering). (a) Contractional or restraining type where constant layer thickness implies marked reverse drag. (b) Contraction compensated for by local dilation. (c) Restraining zone where the fault tips bend towards each other. Extensional or releasing zones with constant (d) and variable (e) layer thickness give normal drag. Based on Rykkelid and Fossen (2002).

layers as shown in Figure 8.46. The rotation depends on fault geometry and how the faults interfere.

Restraining overlap zones (Figure 8.46a–c) are, in this connection, overlap zones with shortening in the displacement direction. In principle, volume reduction may accommodate the deformation within the zone. More commonly, however, the layers within the overlap zone rotate as shown in Figure 8.46c.

Releasing overlap zones (Figure 8.46d, e) are zones where the fault arrangement and sense of displacement cause stretching within the overlap zone. Weak layers such as shale or clay layers are rotated within releasing overlap zones. If the overlap zone is narrow, such weak layers can be smeared along the fault zone (Figures 8.31 and 8.47). Field observations show that this is a common mechanism for the formation of clay smear in sedimentary sequences, but usually on too small a scale to be detected from seismic data. Such structures may cause faults to be sealing with respect to fluid flow, which can have important implications in petroleum or groundwater reservoirs.

The role of lithology

Layering or **mechanical stratigraphy** is important as fault populations develop in layered rocks. The timing of fault formation in the different layers is one aspect, and the way that faults form (ordinary fracturing versus faulting of deformation band zones) is another. Mechanical stratigraphy simply implies that the rock consists of layers that respond mechanically differently to stress, i.e. they have different strengths and different Young's moduli (E). In simple terms, some layers, such as clay

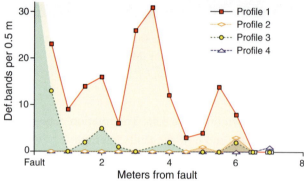

Figure 8.48 Distribution of deformation bands in the footwall to a fault with 6 m displacement. The frequency is considerably higher in the clean, highly porous sandstone (Profile 1) than in the more fine-grained layers (Profiles 2 and 3). It is also lower in the thin sandstone layer (Profile 4). San Rafael Desert, Utah.

The moment a fracture is constrained, its area increases only by layer-parallel growth, and its displacement/length (D/L) ratio becomes lower than what it was during its unconstrained growth history. In simple terms, this is because displacement scales with fracture area, and since the fracture area only increases along its length, the length has to increase at a faster rate relative to displacement.

Eventually, if the fracture keeps accumulating displacement, it will break through the bounding interface and expand into the overlying/underlying layers (Figure 8.49c). The D–L relationship will then return to its original trend (Figure 8.49d). The same development is seen for deformation bands in sandstone–shale sequences. At a critical point the deformation band cluster in the sandstone is cut by a slip surface (fault) that extends into the over and/or underlying shale.

D–L relations during fault growth

The maximum displacement (D_{max}) along a fault is a function of the fault's eccentricity (ellipticity: length/height or L/H ratio) and the strength (driving stress) of the rock (Figure 7.20). L is commonly plotted against D_{max} as shown in Figure 8.50. Straight lines in such logarithmic diagrams indicate an exponential or **power-law** relation between D and L that can be expressed as

$$D_{max} = \gamma L^n$$

Faults or other discontinuities that grow such that D and L are proportional, i.e. $D_{max} = \gamma L$, define straight, diagonal lines in the logarithmic diagram with slope γ ($n = 1$). Field data seem to plot along diagonal lines, although with a considerable amount of scatter (Figure 8.50). Some structures, including joints, veins, igneous dikes, cataclastic deformation bands and compaction bands show lower slopes ($n \sim 0.5$). This may be related to their sensitivity to mechanical stratigraphy, as discussed in the previous section. Mechanical stratigraphy occurs at different scales, from meter-scale beds up to the thickness of the entire brittle crust. Hence, it is thought that the effect repeats itself to some degree at different scales, causing some of the scatter seen in Figure 8.50. Another reason for the scatter is the growth of faults by linkage, as discussed above. The relation between D and L can be of interest in several cases, for instance where the displacement is known from well information or seismic data, and the total length of the fault is to be estimated. However, the scatter of data makes predictions uncertain.

or shale, can accommodate a considerable amount of ductile strain, while other layers, such as limestone or cemented sandstone, fracture at much lower amounts of strain. The result is that, in a layered sequence, fractures or deformation bands initiate in certain layers, while adjacent layers are unaffected or less affected by such brittle structures (Figure 8.48).

So long as the deformation band or fracture grows within a homogeneous layer, a proportionality exists between maximum displacement, length and height. This is illustrated in Figure 8.49a, where the fracture (or deformation band) has not yet (or just barely) reached the upper and lower boundaries of the layer. Once the fracture touches the layer boundaries (Figure 8.49b) it is called a **vertically constrained fracture**, and the fracture will only expand in the horizontal direction. This means that the fracture gets longer and longer while its height remains constant, i.e. its eccentricity increases. In fact, its shape is likely to become more rectangular than elliptical.

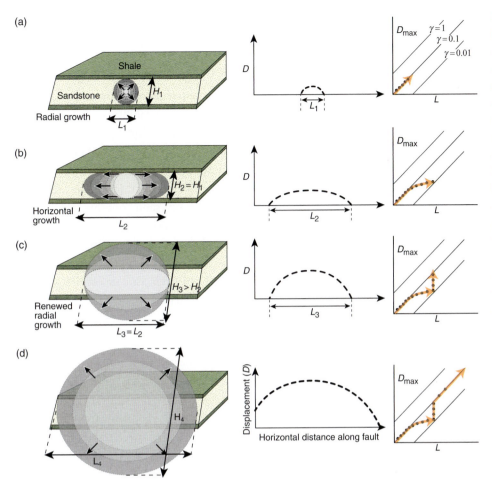

Figure 8.49 Growth of a fault in a layered sequence, with displacement profile and displacement–length evolution shown to the right (logarithmic axes). The fault nucleates in the sandstone layer (a) with a normal displacement profile and expands horizontally when hitting the upper and lower boundaries (b). A relatively long or plateau-shaped displacement profile evolves. At some point the fault breaks through the under/overlying layers (c) and starts growing in the vertical direction again. The displacement profile regains a normal shape. Such lithologic influence on fault growth causes scatter in D–L diagrams.

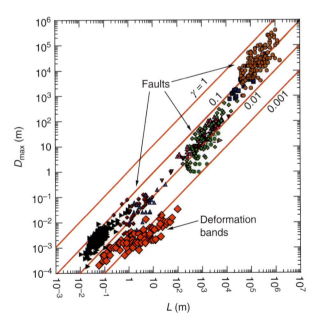

Figure 8.50 Displacement–length diagram for faults and cataclastic deformation bands. Faults from a number of localities and settings are plotted. The deformation bands show a clear deviation from the general trend in that they are longer than predicted from their displacement. Modified from Schultz and Fossen (2002).

8.7 Faults, communication and sealing properties

Faults may affect fluid flow in different ways. Faults in non-porous or low-porosity rocks are generally conduits of fluids. In particular, the fracture-filled damage zone is a good target when drilling for water in such rocks. The fault core is commonly less permeable because of its content of secondary clay minerals (fault gouge).

In highly porous rocks, faults more commonly act as baffles to fluid flow. In a petroleum reservoir setting it is important to distinguish between their effect over geologic time and their role during production. Some faults are sealing over geologic time (millions of years) and can stop and trap considerable columns of oil and gas. Other faults that are not sealing over geologic time may still baffle fluid flow during production of an oil or gas field (over days or years). The ability of faults to affect fluid flow is commonly referred to as fault **transmissibility** or **transmissivity**. Fault transmissibility is influenced by the nature of the damage zone, but is mostly controlled by the thickness and properties of the fault core.

> Faults tend to increase permeability in non-porous rocks, while they commonly reduce permeability in porous rocks.

Juxtaposition

The lithological contact relations along a fault are essential when it comes to its effect on fluid flow in a porous reservoir, and several cases are shown in Figure 8.51. Where sand is completely juxtaposed against shale, the fault is sealing regardless of the properties of the fault itself. This type of seal is called a **juxtaposition seal** (1 in Figure 8.51). However, where sand is juxtaposed against sand without any clay or shale smeared between the sand (stone) beds, the transmissibility of the fault is solely controlled by the physical properties of the fault core and the fault damage zone. These properties are again controlled by the amount of smearing of fine-grained material along the fault, the fault core thickness, the deformation mechanisms within the core and in deformation bands and other structures surrounding the core. Sand–sand juxtaposition occurs when the offset is smaller than the thickness of the sand layer. Any seal resulting from brittle deformation or cementation and dissolution along the fault in this case is referred to as a **self juxtaposed seal** (2 in Figure 8.51). When the fault displacement is larger than the sand thickness, two different sand layers can be juxtaposed. If they are stratigraphically separated by a shale the shale may be smeared

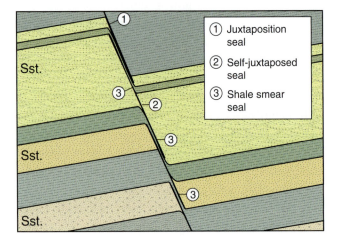

Figure 8.51 Principal sketch showing different contact relations along a fault. Yellowish layers are reservoir sandstone, while greenish layers are impermeable shale. Three principal types of seals are shown. Note that for the fault to be sealing, the membrane must be continuous also in the third direction for as far as the two sands are in contact.

out to form an impermeable membrane, thereby forming a **shale smear seal** (3 in Figure 8.51).

Cataclasis

Cataclasis in the fault core reduces grain size and therefore reduces porosity and permeability. Cataclasis is promoted by deep (>1 km) burial depths, low phyllosilicate content, well-sorted grains and low pore-fluid pressure. Cataclasis can create a cataclasite or ultracataclasite so dense that it will stop fluid flow across the fault even if there is high-permeability sandstone on each side of the fault. Cataclasis also occurs in deformation bands in the damage zone, and the more deformation bands, the larger their effect on fault transmissibility during production. The sealing effect of cataclastic deformation bands in the damage zone is probably negligible in most cases.

Diagenetic effects

Diagenetic changes that occur after or during the faulting process can change the mechanical and petrophysical properties of the fault rock significantly in some cases. The most important change is probably caused by dissolution and precipitation of quartz, which can turn the fault rock into a non-permeable 'quartzite'. Quartz dissolution and cementation is a problem at temperatures above ~90 °C (3 km). In many basins, such as the North Sea, quartz cementation occurred much later than the faulting as the sediments were buried during post-rift subsidence. Quartz and other minerals may be preferentially deposited in faults because of the reactant surfaces that form during faulting due to the scratching and breaking of grains. In addition, fluids can easily move along faults in some cases, increasing the flux of silica-bearing fluids. Calcite cementation in faults is also fairly common, but is thought to form less continuous structures.

Clay and shale smear

The fine-grained nature of clay and shale causes them to have very small pore spaces and pore throats (connections between pores), which therefore effectively stop or hinder fluids flowing through them. Clay minerals can form secondarily along faults in almost any rock type, but in most sedimentary rocks the primary source of clay and shale is shale layers in the sedimentary sequence itself. Shale layers become incorporated into the fault core during fault movements (Figure 8.51), and the process where clay or shale is smeared out into a more or less continuous membrane is simply called **smearing**.

Sand-sand juxtaposition

Smear

2 m

Figure 8.52 Minor fault in fluvial sandstone–shale layers of the Cretaceous Castlegate Formation, Salina, Utah. A centimeter-thick membrane of clay-rich material fed from shales in the hanging wall seals most of the section. Note that the uppermost part of the fault has no shale membrane, probably because of lack of shale layers at that level.

Field observations and experimental results alike show that clay and shale are likely to become smeared during faulting into a fault-core membrane separating the hanging wall and footwall. Figure 8.52 shows an example where a centimeter-thick shale membrane separates highly porous sandstones of reservoir quality. If such a membrane represents a physical barrier to fluid flow it is called a **seal**, and the fault is a **sealing fault**. A membrane must be continuous over a certain critical area of the fault to effectively seal a structure. The more clay or shale involved in a faulted sedimentary sequence, the larger the chance of smearing and sealing. A common mechanism is the smearing of clay or shale between two **vertically overlapping fault segments**, as shown in Figure 8.47. A less common mechanism is injection, where abnormally high pressures in clay layers cause **clay injection** along the fault core. **Clay abrasion**, where clay is tectonically eroded from clay-rich

layers along the fault and incorporated into the fault core, is a third mechanism that contributes to clay smearing.

Smearing of clay or shale along a fault may result in sealing faults in a fluid reservoir.

The likelihood of smearing increases with increasing amount of clay, i.e. with the number of clay layers and their cumulative thickness, and decreases with increasing fault displacement. The more displacement, the higher the probability that the seal is discontinuous and the fault is leaky. If the discontinuity of the seal is local and small, it may still reduce the flow rate across the fault.

There is a need to put numbers on the clay smearing potential of faults. In the simple case where there is a single clay or shale layer and a single fault (Figure 8.53a), then the **shale smear factor** (SSF) gives the ratio between fault throw T and the thickness of the shale or clay layer (Δz):

$$\text{SSF} = \frac{T}{\Delta z} \qquad (8.1)$$

This puts a number to the local probability of smear, and the number will vary along the fault as displacement and perhaps also layer thickness change. For faults with offsets of tens of meters or more an SSF ≤ 4 is considered to indicate a continuous smear and thus a sealing fault, while smaller faults are less predictable. Obviously, there are other factors that influence the sealing capacity of a fault, so this method should be used with some care.

In cases where there is more than one source of clay or shale the combined contribution must be considered. This can be done at any point on the fault by summing the shale bed thicknesses of the stratigraphy that has passed that point and calculating the percentage of clay or shale in the slipped interval. Divided by the fault throw (T) we get what is known as the **shale gouge ratio** (SGR):

$$\text{SGR} = \frac{\sum \Delta z}{T} \cdot 100\% \qquad (8.2)$$

Alternatively, if the clay minerals are more distributed within the sequence we should use the sum of the volume fractions from all units in the interval instead of the sum of shale layer thicknesses. A high SGR-value indicates high sealing probability. There is a good chance that the

(a)

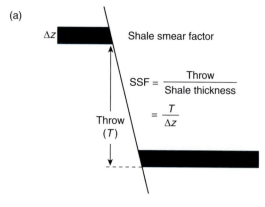

(b)

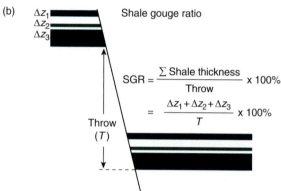

(c)

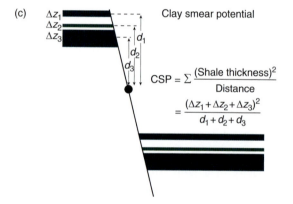

Figure 8.53 Three algorithms for estimating the likelihood of smear on a fault.

fault is sealing when the SGR-value exceeds 20%, and the probability increases with the SGR. The probability will also depend on the mechanical properties of the shale or clay layers at the time of deformation. In general, shallow burial promotes smearing while faulting at deep burial depths involves a higher risk of a leaky seal.

Clay smear can also be characterized by estimating the **clay smear potential** (CSP). CSP relates to how far a clay or shale layer can be smeared before it breaks and becomes discontinuous:

$$CSP = \sum \frac{\Delta z^2}{d} \tag{8.3}$$

where d is the distance from the source (clay) layer and Δz represents the individual thickness of each clay or shale layer.

Coal can also be smeared along faults if it is mixed with clay. However, pure coal normally behaves brittlely when faulted. Even sand can be smeared along faults when in a poorly consolidated state. Sand smearing at shallow depths may improve communication across faults, but it is rarer than shale smear.

The shale smear factor and similar algorithms help us to estimate the sealing potential of a fault, but do not catch all the complexities and variations that exist within and along natural faults.

Juxtaposition and triangle diagrams

Lithology and displacement are, as emphasized above, important factors in the estimation of the sealing properties of faults. Because both lithology and displacement vary on a fault, different portions of the fault have different sealing properties or sealing potentials. Considering a reservoir with porous reservoir sandstones interbedded with impermeable shale we can have juxtaposition seal, shale smear seal and self juxtaposed seal along the same fault, with variations occurring both across (Figure 8.51) and along stratigraphy.

(a)   (b)

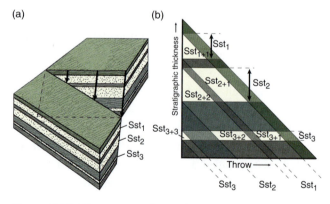

Figure 8.54 The concept of triangle diagram construction. A synthetic fault with a linearly increasing displacement is considered (a). In the block diagram the layers on the upthrown side are horizontal while they are dipping on the opposite side. This is carried over to the triangular diagram to the right (b). Because the displacement increases from zero on the left-hand side, different lithologic contact relations occur in different parts of the diagram. Areas of sand–sand and sand–clay contact can easily be found.

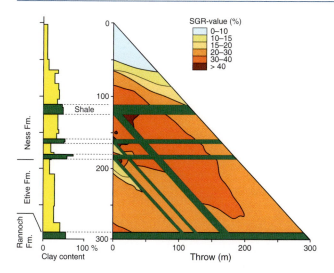

Figure 8.55 SGR-values can be added to the triangle diagram, as for this example from the North Sea Brent Group (stratigraphy shown to the left). The SGR-value is calculated for different points in the diagram and contoured. High SGR-values mean high sealing probability. Based on Høyland Kleppe (2003).

Juxtaposition as a function of stratigraphy and displacement can be visualized in a **triangle diagram** (Figure 8.54). The local stratigraphy is plotted along the vertical axis and extended parallel to the other two sides so that they gradually separate. This is in effect similar to what happens along a fault, where layer separation increases with increasing displacement. In the diagram, horizontal stratigraphic layers represent the hanging wall layers, while the dipping layers represent the footwall. Layer separation and fault displacement increase as we move to the right in the diagram, and we can read off the contact relations for any given value of fault displacement.

The different lithologies (layers) in the diagram are colored. For a sand–shale sequence different colors indicate sand–sand, sand–shale and shale–shale juxtaposition. Furthermore, for each point in the diagram one or more of the parameters SGR, SSF and CSP can be calculated and the triangular diagram can be colored to illustrate the variation in, for example, SGR (Figure 8.55).

Summary

While faults may look simple when portrayed as lines on geologic maps and interpreted seismic sections, more detailed considerations reveal that they are complicated and composite structures. Although our understanding of faults and related structures has increased significantly over the last couple of decades, much research remains to be done before we reach the point where we can predict or model their geometries and properties based on input such as tectonic regime, lithology and burial depth. Some key points and review questions from this chapter are presented here:

- Faults consist of a central core, which is a high-strain zone dominated by fault rock and slip surface(s).

- The damage zone is a low-strain zone around the core formed during the formation and evolution of the fault.

- Both the damage zone and the fault core thickness tend to increase with increasing fault displacement, but the relationship is complicated and not related to gradual widening of the two during growth.

- Faults grow from small fractures to map-scale structures by accumulating displacement from successive earthquakes.

- Faults accumulate displacement as they grow in length and height.

- Ideally, fault displacement increases from the tip line toward the centre of the fault.

- During growth, faults tend to interact and link up.

- Fault linkage is a gradual process from underlapping via overlapping to hardlinking of fault segments.

- Fault relay structures are areas of complications and involve much small-scale deformation (damage zones).

- Drag can form by faulting of a fault-propagation fold.

- The geometry of drag folds can be used to determine the sense of displacement on a fault.

- Smearing of clay or shale along faults can cause the fault to be sealing.

- A sealing fault stops fluids from flowing from the hanging wall to the footwall or vice versa.

Review questions

1. What is the difference between shear fractures and faults?

2. Why do normal faults tend to be steeper than reverse faults?

3. What is the main differences between a mylonite and a cataclasite?

4. A vertical well is drilled through a stratigraphic section twice (repeated section). What type of fault can we infer, and why can we not explain this by folding?

5. Why would the damage zone grow during faulting?

6. Would the damage zone be visible on good 3-D seismic data?

7. How can dipmeter data help identifying faults?

8. Is fault sealing good or bad in terms of petroleum exploration and production?

E-MODULE

The e-learning module called *Faults* is recommended for this chapter.

FURTHER READING

Petroleum oriented

Aydin, A., 2000, Fractures, faults, and hydrocarbon entrapment, migration, and flow. *Marine and Petroleum Geology* **17**: 797–814.

Cerveny, K., Davies, R., Dudley, G., Kaufman, P., Knipe, R. J. and Krantz, B., 2004, Reducing uncertainty with fault-seal analysis. *Oilfield Review* **16**, 38–51.

Yielding, G., Walsh, J. and Watterson, J., 1992, The prediction of small-scale faulting in reservoirs. *First Break* **10**: 449–460.

Displacement and growth rates

Barnett, J. A. M., Mortimer, J., Rippon, J. H., Walsh, J. J. and Watterson, J., 1987, Displacement geometry in the volume containing a single normal fault. *American Association of Petroleum Geologists Bulletin* **71**: 925–937.

Ferill, D. A. and Morris, A. P., 2001, Displacement gradient and deformation in normal fault systems. *Journal of Structural Geology* **23**: 619–638.

Hull, J., 1988, Thickness–displacement relationships for deformation zones. *Journal of Structural Geology* **4**: 431–435.

Morewood, N. C. and Roberts, G. P., 2002, Surface observations of active normal fault propagation: implications for growth. *Journal of the Geological Society* **159**: 263–272.

Roberts, G. P. and Michetti, A. M., 2004, Spatial and temporal variations in growth rates along active normal fault systems: an example from The Lazio–Abruzzo Apennines, central Italy. *Journal of Structural Geology* **26**: 339–376.

Walsh, J. J. and Watterson, J., 1989, Displacement gradients on fault surfaces. *Journal of Structural Geology* **11**: 307–316.

Walsh, J. J. and Watterson, J., 1991, Geometric and kinematic coherence and scale effects in normal fault systems. In A. M. Roberts, G. Yielding and B. Freeman (Eds.), *The Geometry of Normal Faults*. Special Publication **56**, London: Geological Society, pp. 193–203.

Yielding, G., Walsh, J. and Watterson, J., 1992,
The prediction of small-scale faulting in reservoirs.
First Break **10**: 449–460.

Fault geometry and linkage

Benedicto, A., Schultz, R. A. and Soliva, R., 2003, Layer
thickness and the shape of faults. *Geophysical Research
Letters* **30**: 2076.

Caine, J. S., Evans, J. P. and Forster, C. B., 1996, Fault
zone architecture and permeability structure. *Geology*
24: 1025–1028.

Childs, C., Watterson, J. and Walsh, J. J., 1995, Fault overlap
zones within developing normal fault systems. *Journal of
the Geological Society* **152**: 535–549.

Childs, C., Manzocchi, T., Walsh, J. J., Bonson, C. G.,
Nicol, A. and Schöpfer, P. L., 2009, A geometric model
of fault zone and fault rock thickness variations.
Journal of Structural Geology **31**: 117–127.

Peacock, D. C. P. and Sanderson, D. J., 1994, Geometry and
development of relay ramps in normal fault systems.
American Association of Petroleum Geologists Bulletin
78: 147–165.

Damage zones

Kim, Y.-S., Peacock, D. C. P. and Sanderson, D. J., 2004,
Fault damage zones. *Journal of Structural Geology*
26: 503–517.

Shipton, Z. K. and Cowie, P., 2003, A conceptual model
for the origin of fault damage zone structures in
high-porosity sandstone. *Journal of Structural Geology*
25: 333–344.

Wibberley, C. A. J., Yielding, G. and Di Toro, G., 2008,
Recent advances in the understanding of fault zone
internal structure: a review. In C. A. J. Wibberley,
W. Kurz, J. Imber, R. E. Holdsworth and C. Collettini
(Eds.), *The Internal Structure of Fault Zones:
Implications for Mechanical and Fluid-Flow Properties*.
Special Publication **299**, London: Geological Society,
pp. 5–33.

Fault sealing and smear

Færseth, R. B., Johnsen, E. and Sperrevik, S., 2007,
Methodology for risking fault seal capacity: Implications
of fault zone architecture. *American Association of
Petroleum Geologists Bulletin* **91**: 1231–1246.

Hesthammer, J., Bjørkum, P. A. and Watts, L. I., 2002,
The effect of temperature on sealing capacity of faults in
sandstone reservoirs. *American Association of Petroleum
Geologists Bulletin* **86**: 1733–1751.

Knipe, R. J., 1992, Faulting processes and fault seal.
In R. M. Larsen, H. Brekke, B. T. Larsen and E. Talleraas
(Eds.), *Structural and Tectonic Modelling and its
Application to Petroleum Geology*. NPF Special
Publication, Amsterdam: Elsevier, pp. 325–342.

Manzocchi, T., Walsh, J. J. and Yielding, G., 1999, Fault
transmissibility multipliers for flow simulation models.
Petroleum Geoscience **5**: 53–63.

Fault strain

King, G. and Cisternas, A., 1991, Do little things matter?
Nature, **351**: 350.

Marrett, R. and Allmendinger, R. W., 1992, Amount of
extension on "small" faults: an example from the Viking
Graben. *Geology* **20**: 47–50.

Reches, Z., 1978, Analysis of faulting in three-dimensional
strain field. *Tectonophysics* **47**: 109–129.

Dipmeter data, drag and fault-propagation folding

Bengtson, C. A., 1981, Statistical curvature analysis
techniques for structural interpretation of dipmeter
data. *American Association of Petroleum Geologists*
65: 312–332.

Erslev, E. A., 1991, Trishear fault-propagation folding.
Geology **19**: 617–620.

Fault rocks

Sibson, R., 1977, Fault rocks and fault mechanisms.
Journal of the Geological Society **133**: 191–213.

Snoke, A. W., Tullis, J. and Todd, V. R., 1998, *Fault-related
Rocks: A Photographic Atlas*. Princeton: Princeton
University Press.

Terminology

Peacock, D. C. P., Knipe, R. J. and Sanderson, D. J., 2000,
Glossary of normal faults. *Journal of Structural Geology*
22: 291–305.

Chapter 9

Kinematics and paleostress in the brittle regime

In the previous chapters we have indicated that a close relationship exists between stress and faulting, for example according to Anderson's tectonic stress regimes. It should therefore be possible to say something about the stress field at the time of faulting and fracturing, based on the orientation and nature of the faults and fractures. This is referred to as paleostress analysis, a field that is hampered by several assumptions. However, many paleostress analyses yield reasonable results, as can be verified by independent information. The fundamental input to paleostress analysis is kinematic observations of fault structures made in the field. Relevant structures and the fundamentals of paleostress analysis in the brittle regime are briefly presented in this chapter.

9.1 Kinematic criteria

The true finite displacement vector on a fault surface can be found directly where a point in the hanging wall can be connected to an originally neighboring point in the footwall. Such points can be faulted fold hinges or other recognizable linear structures that intersect with the fault surface.

Unfortunately, such points are rarely found. In most cases we are pleased if we can correlate layers or seismic reflectors from one side to the other. If the fault surface is exposed in the field we would use the lineation on the fault surface to estimate the orientation and length of the displacement vector. The assumption is typically made that the lineation on the fault surface represents the displacement direction. It may, however, be that the lineation only reveals the last part of the deformation history, i.e. the last slip event(s), and that lineations related to earlier slip events have been obscured or obliterated. Careful searching for multiple, overprinting lineations is therefore called for when collecting fault slip data in the field.

A lineation on a slip surface may represent only the last of several slip events, and does not have to be parallel to the finite (total) displacement vector.

In other cases it is not possible to correlate stratigraphy or marker horizons from one side of the fault to the other. We then have no information about the size of the fault (the width of the damage zone and fault core may give us a hint), and we do not even know whether the fault movement was reverse or normal, sinistral or dextral. The lineation on the fault surface is useful, but we need to look for kinematic criteria in order to determine the sense of slip. Such criteria exist, although many of them tend to be ambiguous. Hence, as many kinematic criteria as possible must be combined for kinematic fault analysis.

Mineral growth and stylolites

Fault surfaces are never perfectly planar structures, and useful kinematic structures may form where geometric irregularities occur. Where an irregularity causes local contraction, such as the contractional bend in Figure 9.1, contractional structures such as stylolites (Figure 9.2). and "cleavage" may be found. Differently oriented irregularities, such as the left bend in Figure 9.1, may cause extension and opening of voids where mineral growth takes place. A study of the relation between fault surface geometry and the occurrence of contractional

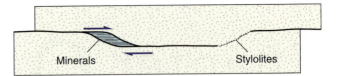

Figure 9.1 Irregularities along a fault create steps where mineral growth or shortening structures (stylolites) can form. The localization of such structures relative to the local fault geometry gives reliable information about the sense of slip.

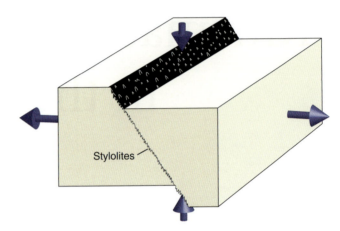

Figure 9.2 Faults with a component of shortening across the fault surface can result in pressure solution and stylolite formation in limestone and marble. Linear stylolitic structures sometimes form, called slickolites. Slickolites form a lineation that parallels the movement direction quite precisely.

versus extensional structures may reveal the sense of slip with a high degree of confidence.

Subsidiary fractures

Small fractures developed along a fault or slip surface may show geometric arrangements that carry information about the sense of slip on the fault. These small fractures have been given different names depending on their orientations and kinematics. Figure 9.3e shows these different categories of fractures in a section perpendicular to the main slip surface (M) and parallel to the slip direction. **T-fractures** are the name often used for small extension fractures in this setting. They may be open, but are more commonly mineralized with quartz or carbonates and do not show striations. The orientation of T-fractures with respect to the main or average slip surface or **M-surface** characterizes the sense of slip. T-fractures typically dip around 45° in the slip direction with respect to a horizontally oriented M-surface (Figure 9.3 a, c).

A set of shear fractures, known as **P-fractures**, is sometimes seen to dip in the opposite direction. With M still being horizontal, these make low angles to M and

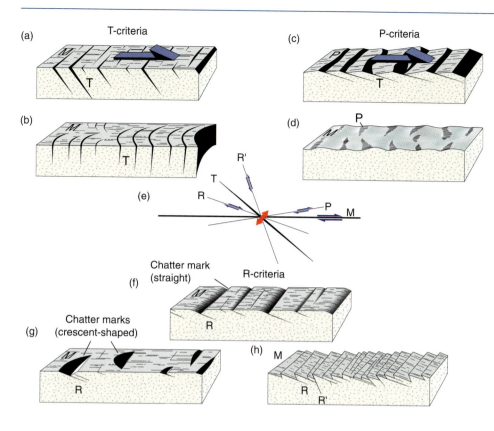

Figure 9.3 Kinematic criteria along a dextral fault with subordinate fractures or irregularities. The general nomenclature for fractures in a shear system is used (R, R′, P, T and M fractures). R, Riedel shears; P, shear fractures; T, extension fractures; M, average slip surface (fault). Identification of the type of subordinate fracture helps interpret the movement on the fault. Based on Petit (1987).

kinematically correspond to low-angle "reverse" or "thrust faults". In this setting Riedel shear fractures or **R-fractures** represent low-angle normal "faults" while **R′-fractures** correspond to antithetic reverse faults that make a high angle to M. Riedel fractures tend to be more common than R′ and P, but they all exist and their local kinematics as well as their orientation with respect to M reveal the sense of movement on the main structure. T-structures are perhaps the most reliable of these, because they are easy to distinguish from the various shear fractures.

The French geologist Jean-Pierre Petit separated the various structures that one may see along fault surfaces by T, P and R-criteria, where the letters T, P and R indicate the dominant subordinate fracture element of the structure. **T-criteria** comprise extension fractures (T) that intersect the striated fault slip surface (M). In cross-section T and M tend to form an acute angle of intersection pointing toward the slip direction, as shown in Figure 9.3a. The intersections may also occur as curved structures pointing toward the direction of slip (Figure 9.3b). Such structures resemble glacial chatter marks that form when flowing ice plucks pieces off fairly massive bedrock such as quartzite or granite.

P-criteria are dominated by P-fractures and may occur together with T-fractures (Figure 9.3c). P-surfaces may be polished and striated, and they are characterized by their low angle to M. Situations where undulations of the main slip surface create a systematic pattern with striations on the side facing the movement of the opposite wall (contractional side) and no striations on the lee (extensional) side (Figure 9.3d) can also be regarded as P-criteria.

R-criteria, which form the most commonly used type of kinematic criterion, are based on the acute angle between R and M (Figure 9.3f, g). The lines of intersection between R and M show a high (close to 90°) angle to the striae on M, and such straight (Figure 9.3f) or curved (Figure 9.3g) linear structures on the slip surface are commonly referred to as **chatter marks**. Faults with small offsets may not have developed a through-going slip or M-surface, in which case the fault may be defined by en-echelon arranged R-fractures (Figure 9.3h), and sometimes also R′-fractures. The R-fractures are then closely arranged and striations are not well developed.

Ploughing, mineral growth and slickensides

Asperities or relatively hard objects (rock fragments, pebbles or strong mineral grains) on one side of a fault surface may mechanically plough grooves or **striations**

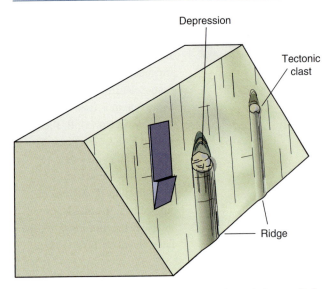

Figure 9.4 Pointed asperities or tectonic clasts sitting on fault surfaces may form depressions on the lee side and ridges on the opposite side. Such lineations are ridge-in-groove lineations.

into the opposing wall. Material in front of the object is being pushed aside, while a crescent-shaped opening occurs on the lee side, typically filled with material from the opposite wall. A ridge (or groove, depending on the point of view) is sometimes found ahead of the object, as illustrated in Figure 9.4, and such lineations are sometimes referred to as **ridge-in-groove lineations**.

The crescent-shaped opening soon evolves into a groove whose length ideally corresponds to the movement of the hard object relative to the rest of the wall. At least that's what we would like to think. But there are many examples of centimeter-scale offsets causing decimeter- or even meter-long striations. Hence, there must be other mechanisms at work than just physical carving. One explanation is that some striations are **corrugations** that did not form purely as frictional grooves. These are linear structures that formed at the initial stage of fracture growth and may be polished and striated as slip accumulates. The result may be long and well-developed striae on slip surfaces that have very small (cm- or even mm-scale) offsets. Hence, small-scale fault corrugation structures do *not* necessarily reflect the amount of slip on the surface. Striations are typically found on (but not restricted to) polished slip surfaces called **slickensides**, where the striations are known as **slickenlines**.

In addition to frictional striations, minerals may crystallize on the lee side of asperities (irregularities), thus revealing the sense of slip. When minerals crystallize as fibers, the orientation of the fibers tends to be close to the

slip direction. It is commonly found that minerals precipitated on fault slip surfaces are subsequently affected by renewed slip. It is therefore common to see slickenlines developed on deformed mineral fill along fault surfaces. Lineations related to fractures are further discussed in Chapter 13.

9.2 Stress from faults

Fault observations used in paleostress analysis include the local strike and dip of the fault surface, the orientation of the lineation (usually given by its rake or pitch) and the sense of movement. When such data are collected for a fault population it is in principle possible to obtain information not only about strain but also about the orientation of the principal stresses (σ_1, σ_2 and σ_3) and their relative magnitudes (i.e. the shape of the stress ellipsoid). Paleostress methods hinge on several assumptions. The most basic is that the faults under consideration formed in the same stress field. Others are that the rocks are fairly homogeneous, that the strain is relatively low and that the structures have not rotated significantly since they initiated.

Groups of small faults from a limited area (an outcrop or a quarry) can be analyzed to reconstruct the local paleostress field.

Conjugate sets of faults

A simple case of deformation is **plane strain** expressed in terms of conjugate fault systems. As shown in Figure 9.5 for the three main tectonic regimes, a conjugate system has two sets of oppositely dipping faults where the lineations are perpendicular to the line of fault intersection. The sense of slip is complementary on the two sets, and the angle between the two sets should be constant. According to the Coulomb fracture criterion, which tells us that faults or shear fractures form at an angle to σ_1 that is controlled by the internal friction of the rock, the angle between the two sets should be consistent with the mechanical properties of the rock (and the pore fluid pressure) at the time of deformation. More importantly, conjugate sets should develop symmetrically about the principal stresses for both normal, reverse and strike-slip faults (Figure 9.5). According to the Andersonian theory of faulting (Section 5.6), the slip direction or lineation would be in the dip direction (reverse and normal faults) or in the strike direction (strike-slip faults). Hence,

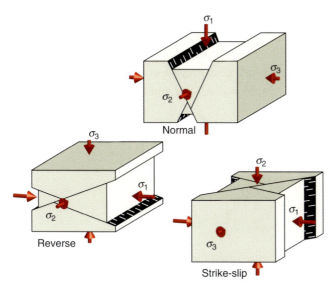

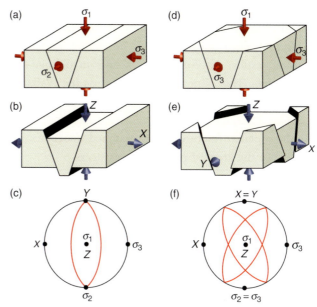

Figure 9.5 Conjugate shear fractures and their relation to stress according to Anderson's conditions for planar strain. The acute angle is bisected by σ_1 and the two fractures intersect along σ_2.

Figure 9.6 A single pair of dip-slip conjugate fault sets (a–c) form in plane strain, while multiple sets imply 3-D strain (d–f). The pattern shown in (e) is called orthorhombic from its high degree of symmetry. In these idealized cases the principal stresses (red arrows) and strains are parallel.

conjugate faults and their striations reveal the orientations of the principal stresses.

The maximum principal stress axis bisects the acute angle of conjugate faults.

In some cases we may find that the principal stresses deviate somewhat from the horizontal/vertical directions predicted by Anderson, which may indicate that the area has been tilted since the time of deformation or that the stress field was rotated or refracted through this volume of rock. As mentioned in Chapter 6, such deviations can occur near large and weak faults or joint zones.

Two conjugate sets of faults indicate plane strain (Figure 9.6a, b), and the strain axes are found by plotting the fault sets in a stereonet (Figure 9.6c). The orientation of the principal stress axes can be inferred based on the assumption of isotropic rheology, constantly oriented stress field and negligible rotation of the fault blocks. In fact, we can draw a parallel between conjugate sets observed in rock mechanics experiments and their angular relation to the principal stresses.

In other cases fault systems may contain more than two conjugate sets. A simple case is shown in Figure 9.6d, e, where two genetically related pairs of fractures coexist. This arrangement is referred to as **orthorhombic** based on its symmetry elements. Again, the strain axes are easily found from stereoplots, and the stress axes can be inferred (Figure 9.6f).

Complex fault populations

Conjugate fault sets are found in rocks that have experienced a single phase of brittle deformation. In many cases, fracture patterns show evidence of repeated activation under changing stress conditions and commonly more complicated arrangements than a simple conjugate pattern. If we have a fracture or fault population that is complex in the sense that it contains elements with a wide range of orientations, how would this fracture population respond to a new stress field?

Clearly some faults will slip while others will not. We looked at a similar example in Figure 7.22, where a weak foliation was reactivated for a certain range of orientations. Now we are looking at a population of weak elements, and we also want to know the direction and sense of slip. Prediction is not trivial, but there are some important statements that can be made.

The most obvious assumption that can be made is that faults or fractures oriented perpendicular to one of the three principal stresses will not slip. This is obvious since there is no shear component in this situation ($\sigma_s = 0$). For any other orientation $\sigma_s \neq 0$ and frictional sliding (slip) will occur if σ_s exceeds the frictional resistance against slip.

It also seems reasonable to make the assumption that slip on a surface will occur in the direction of maximum resolved shear stress. If the largest shear stress is in the dip direction we will get normal or reverse faults. If the maximum shear stress vector is horizontal, strike-slip faulting results. Any other case results in oblique-slip movements. This assumption is known as the **Wallace–Bott hypothesis**:

Slip on a planar fracture can be assumed to occur parallel to the greatest resolved shear stress.

The hypothesis implies that the faults are planar, fault blocks are rigid, block rotations are negligible, and the faults were activated during a single phase of deformation under a uniform stress field. Clearly, these are idealized conditions. For example, intersecting faults have the potential to locally perturb the stress field. On the other hand, empirical observations and numerical modeling suggest that the deviations are relatively small in most cases, suggesting that the Wallace–Bott hypothesis is a reasonable one. Building on these simple assumptions we can measure a fault population with respect to fault orientation, lineation orientation and sense of slip, and use these data to calculate the orientation and relative size of the stress axes. The methods used for this purpose are known as **stress inversion** or **fault slip inversion techniques**.

Paleostress from fault slip inversion

While the absolute values of the principal stresses are unachievable in most cases, their relative magnitude, i.e. the shape of the stress ellipsoid, can be estimated from fault population data. For this purpose we use the stress ratio

$$\phi = (\sigma_2 - \sigma_3)/(\sigma_1 - \sigma_3) \tag{9.1}$$

where $0 \leq \phi \leq 1$. The ratio ϕ is also called R (note that R is defined as $1 - \phi$ by some authors). $\phi = 0$ for a prolate stress ellipsoid, where $\sigma_2 = \sigma_3$ (uniaxial compression). $\phi = 1$ implies that $\sigma_1 = \sigma_2$, and the stress ellipsoid is oblate (uniaxial tension). We use the ratio ϕ to express the stress tensor, which in the coordinate system defined by the principal stresses is

$$\begin{bmatrix} \sigma_1 & 0 & 0 \\ 0 & \sigma_2 & 0 \\ 0 & 0 & \sigma_3 \end{bmatrix} \tag{9.2}$$

An isotropic stress component can be added to the stress tensor, and the components can be multiplied by a constant without changing the orientation and shape of the stress ellipsoid. The stress tensor that contains information about the shape and orientation of the stress ellipsoid, but not the absolute magnitude of the principal stresses, is known as the **reduced stress tensor**. This tensor is here expressed in terms of two constants k and l, where: $k = l/(\sigma_1 - \sigma_3)$ and $l = -\sigma_3$

$$\begin{bmatrix} \sigma_1 + l & 0 & 0 \\ 0 & \sigma_2 + l & 0 \\ 0 & 0 & \sigma_3 + l \end{bmatrix} \begin{bmatrix} k & 0 & 0 \\ 0 & k & 0 \\ 0 & 0 & k \end{bmatrix}$$
$$= \begin{bmatrix} 1 & 0 & 0 \\ 0 & \phi & 0 \\ 0 & 0 & 0 \end{bmatrix} \tag{9.3}$$

Adding the constant l to each principal stress is the same as adding an isotropic stress. Furthermore, multiplying the axes by the constant k means contracting or inflating the stress ellipsoid while maintaining its shape.

While the full stress tensor contains six unknowns, the reduced tensor contains only four unknowns, represented by ϕ and the orientation of the principal stresses. In other words, the reduced stress tensor gives us the orientation and shape of the stress ellipsoid. While we have discussed how stress influences slip on fractures (by causing slip in the direction of the maximum resolved shear stress), the inverse problem where the (reduced) stress tensor is calculated from slip data is more relevant to paleostress analysis.

By inversion of fault slip data we mean reconstructing the orientation and shape of the stress ellipsoid based on measured fault slip data.

Because the reduced stress tensor has four unknowns we need data from at least four different fault surfaces to find the tensor. The tensor that fits all the data is the tensor we are looking for. However, because of measuring errors, local stress deflections, rotations etc., we are searching for the tensor that best fits the fault slip data. We therefore collect data from more than four slip surfaces, usually 10–20 or more. This gives many more equations than unknowns, and a statistical model is applied that minimizes the errors.

The calculations are done by a stress inversion program, of which several are available. The results are presented in stereoplots that show the orientations of the principal stress axes, and in dimensionless Mohr diagrams that reflect the relative sizes of the principal stresses (Figure 9.7). During the calculation of the reduced stress tensor there may be remaining data that do not fit in. These may be treated separately to see if they together

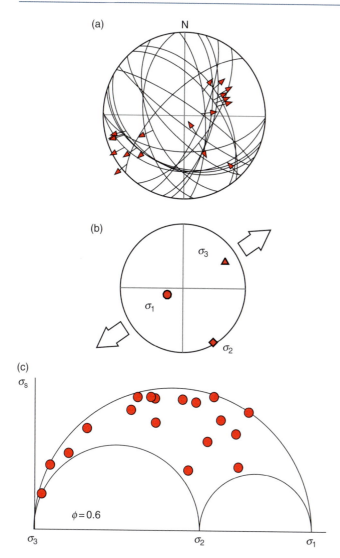

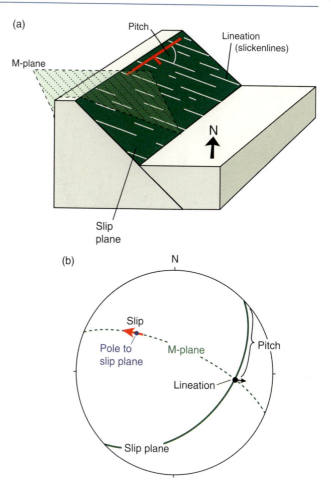

Figure 9.8 (a) Schematic illustration of pitch as measured on a fault surface. (b) The pitch plotted in a stereonet (lower hemisphere, equal area). This projection also shows how the tangent-lineation is found for a known slip plane (as shown in (a)) and its lineation (striation) by plotting a plane (the movement or M-plane) that contains the lineation and the pole to the slip plane. The tangent-lineation (red arrow) is the arrow drawn tangential to the M-plane at the pole to the slip plane. Its direction describes the movement of the footwall relative to the hanging wall; in the present case a normal movement, where the footwall moves up to the west.

Figure 9.7 (a) Actual fault data, shown by great circles (fault surfaces) and arrows (lineations) on an equal-area lower-hemisphere plot. (b) Principal stresses found from stress inversion. (c) The data plotted in the dimensionless Mohr diagram, indicating the relative values of the principal stresses.

define a second tensor that may relate to a separate tectonic event. A different and usually safer way to distinguish between subpopulations is to use field criteria such as cross-cutting relations and characteristic mineral phases on slip surfaces.

A geometric way of extracting stress from fault slip data is to construct **tangent-lineation diagrams**. As shown in Figure 9.8, this is done by plotting each fault plane and its lineation, the M-plane, which contains the pole to the fault plane and the lineation, and by drawing an arrow tangent to the M-plane at this pole. The direction of the arrow reflects the movement of the footwall relative to the hanging wall. When differently oriented

slip surfaces are plotted, such as the example presented in Figure 9.9, the resulting pattern may reveal both the orientations and the relative magnitudes (ϕ) of the principal stress axes (Figure 9.10). Software, such as FaultKin, can be used to plot the data collected in the field, but it is important that we understand what such programs do with our data.

Clearly, many data points from faults with a variety of orientations are required to obtain a good result. An example is shown in Figure 9.9, and it is clear from this plot that the distribution of data is not consistent enough to constrain the relative stress magnitudes – a relatively

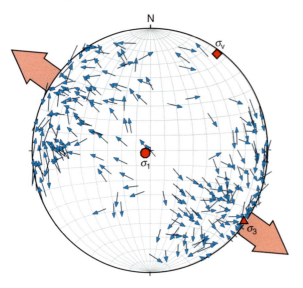

Figure 9.9 Tangent-lineation diagram for fault data from basement gneisses west of Bergen, Norway. σ_1 can be defined, but the ϕ-value, which in principle is found by comparing with the next figure, is more difficult to define in this case. This is quite common and is partly due to limited variation in fault orientations.

common situation when working with real data sets, especially data sets from polyphasally deformed rocks.

Paleostress analyses need to be treated with care for several reasons. One is the fact that they heavily depend on our ability to identify fault populations formed under a stress field that was constant during the history of faulting. Another is that the stress field needs to be uniform within the locality under investigation. We know that fault interaction, as well as any mechanical layering, tends to perturb the stress field, so this assumption is likely to be an approximation only. Outputs from paleostress analyses have to be interpreted in the light of these facts and additional locality-specific circumstances.

9.3 A kinematic approach to fault slip data

The paleostress method based on inversion of fault slip data relates stress and slip through the Wallace–Bott hypothesis and relies on assumptions that clearly over-simplify the actual situation in the crust. For this reason, some of us prefer a purely kinematic approach where strain or strain rate rather than stress are treated.

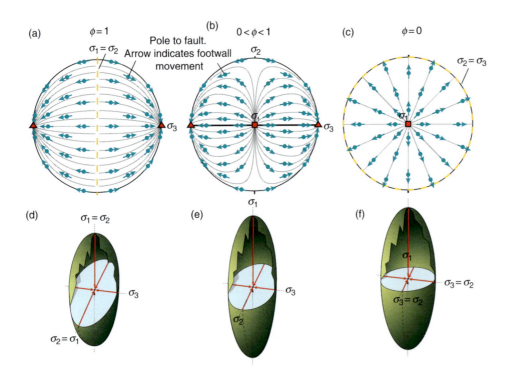

Figure 9.10 (a–c) Stereographic projection of poles to faults (points). Arrows indicate footwall versus hanging wall movements. The pattern emerges by considering faults with different orientations, and the assumption is made that the movement is parallel to the direction of maximum shear stress on each surface. The pattern depends on the ratio between the principal stresses, expressed by ϕ. By plotting field data in these diagrams one can obtain an estimate for the orientation of the principal stresses and ϕ. The geometry of the strain ellipsoid is shown (d–f). Based on Twiss and Moores (2007).

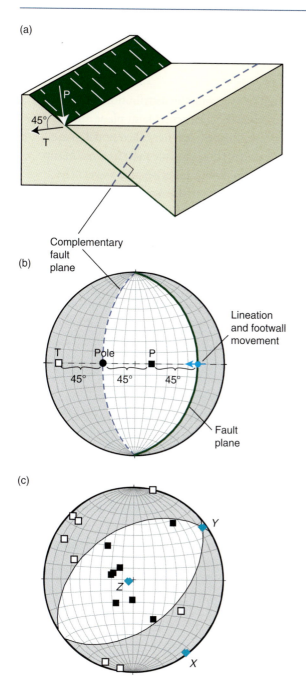

Graphically the kinematic method consists of plotting the fault planes and corresponding slip-related lineations in a stereoplot, as shown in Figure 9.11. Based on the orientation and sense of slip, P- and T-axes are found for each fault plane, somewhat similar to focal mechanism solutions treated in Box 9.1. These axes are symmetry axes of the contractional and extensional quadrants. They are perpendicular to each other and are contained in the plane that is perpendicular to the fault plane. The fault plane also contains the lineation (slip direction). Furthermore, P and T make 45° to the fault plane (Figure 9.11). The position of T and P depends on the sense of fault movement (normal, reverse, sinistral or dextral).

A pair of P- and T-axes is found for each fault in the population. For fault populations with a distribution in fault orientations we end up with a distribution of P- and T-axes. Ideally, they can be separated by two orthogonal surfaces, which are fitted to the plot. Y (the intermediate strain axis) is represented by the line of intersection between the planes, while Z and X are located in the middle of the P- and T-fields, respectively (Figure 9.11c). This gives the orientation of the paleo-strain axes – the magnitude can only be approached by adding information about displacement and area of each fault included in the analysis. The kinematic method is quickly done by means of a computer program, and in most cases it can be shown that the strain axes found by this method are very similar to the stress axes found by inversion of slip data as described in the previous section. The kinematic method also relies on a uniform strain field and on a correct grouping of slip information generated during different phases of deformation.

Figure 9.11 Kinematic analysis of fault data. (a) A simple normal fault. (b) Plot of the fault, the lineation and the sense of movement. The complementary shear plane oriented 90° to the fault is shown. The T- and P-axes bisect these two planes. (c) Data from 16 faults with different orientations. Each of the faults is plotted as illustrated in (b). Ideally, the P- and T-axes will plot within two sectors separated by two mutually orthogonal planes. If they do not, then the conditions are not fulfilled; they may for example have formed in two different stress fields. The few data are a minimum, but indicate a NW–SE stretching and vertical shortening, i.e. an extensional regime.

9.4 Contractional and extensional structures

While analyses of fault slip data are concerned with shear fractures and faults, contractional and extensional structures are also useful for stress or strain rate considerations. Contractional structures, also known as anticracks, are found as solution seams or stylolites in some (mostly) brittlely deformed rocks. In general, such structures represent small strains with little or no rotation. A close connection between stress and strain axes can therefore be assumed. The same is the case with low-strain extension structures such as veins and joints.

BOX 9.1 | FOCAL MECHANISMS AND STRESS

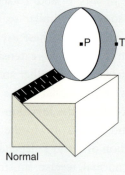

Normal

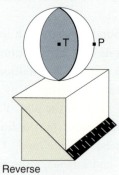

Reverse

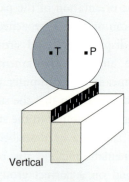

Vertical

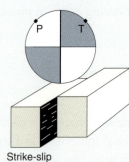

Strike-slip

A so-called **fault-plane solution** is found by mapping the distribution of P- or S-waves around the hypocenter of an earthquake. The actual fault plane and its complementary, orthogonal theoretical shear plane (together called nodal planes) are plotted as stereographic projections based on the principle that the sense of slip (normal, reverse etc.) on a fault controls the distribution of seismic waves. The nodal planes are found by using observations of the first P-wave movement at various seismic stations and deciding whether they are compressive or tensile. The planes are plotted as stereographic projections, and the result is the well-known "beach-ball"-style projections shown to the left. Information about the first arrival (compressive P or tensile T) from several seismologic observatories is used to constrain the orientations of the nodal planes and their senses of movement, known as the focal mechanism. The quadrants are separated by the fault plane and the complementary nodal plane, and the P- and T-axes are plotted in the middle of the quadrants. As shown to the left, different "beach balls" indicate different **focal mechanisms** or sense of fault movement.

Which of the two nodal planes actually represents the fault plane is unknown from seismic data alone, unless aftershock analysis is involved. Only the P- and T-axes are known. P and T are *not* identical to σ_1 and σ_3, although this assumption was commonly made previously. With a crust full of weak structures, many quakes must be expected to result from reactivation of preexisting fractures. There is therefore no precise relationship between fault orientation and the stress field. However, we know that σ_1 must lie in the P-field, and σ_3 in the T-field. Thus, observations from multiple earthquakes on variously oriented faults give a distribution of data that increases our chances of a good estimate of the stress axes. In principle we can use stress inversion to estimate the principal stresses from such data.

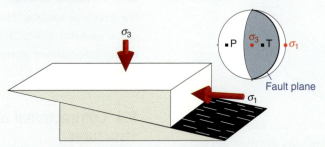

Example where the principal stresses and the P- and T-axes are not identical. A low-angle thrust fault is activated with a vertical σ_3 and a horizontal σ_1. Since P and T always bisect the nodal planes the principal stresses and the P- and T-axes will only coincide if the fault is oriented at 45° to σ_1.

Paleostress indicators must record small strains that do not involve much rotation to ensure that strain and stress axes can be correlated.

While contractional structures tend to form perpendicular to σ_1, extensional fractures form perpendicular or at least at a high angle to σ_3. This makes joints and veins the structures that give the quickest and simplest connection between stress and strain. The combination of contraction and extension structures is particularly valuable, and when they occur in faulted rocks these structures can be used in or compared to the results of stress-inversion methods.

Non-rotated joints, veins and dikes are structures that immediately define the orientation of σ_3 at the time of initiation.

A related kinematic method involves the reconstruction of walls displaced during dike intrusion. Magma overpressure during intrusion may cause fracturing of the surrounding rock, into which magma flows to form dikes. In the presence of a significant tectonic stress, the fractures and thus the dikes are oriented perpendicular to σ_3. In most cases, however, the rock contains preexisting fractures that will be filled with magma.

Fractures with an orientation perpendicular to σ_3 will preferentially open perpendicular to the walls, but if we

Figure 9.13 Left-stepping dike intrusion, well suited for stress determination. The corners were connected prior to the intrusion, and the extension direction, which can be assumed to be close to the minimum principal stress, is found by connecting the corners. Permian dike along the shoulder of the North Sea rift system.

(a)

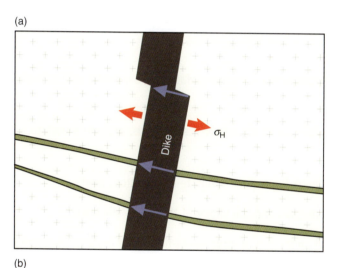

(b)

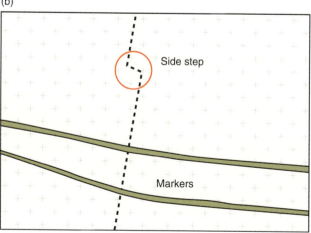

Figure 9.12 Reconstruction of a horizontal section (a) to the situation prior to dike intrusion (b). Arrows can be drawn between points that once were neighbors to find displacement vectors. For small strains these vectors approximate σ_H. From Fossen (1998).

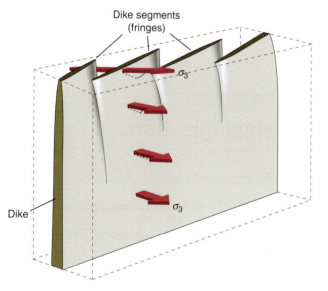

Figure 9.14 En-echelon dike system that is connected at depth – one of Anderson's popular interpretations published in 1951. Anderson interpreted this phenomenon as the result of an upward rotation of σ_3 towards the surface during magma intrusion.

want to find the orientation of σ_3 we have to look for points on each side of the dike that once were adjacent to each other. Such points can be corners where dikes step sideways (Figure 9.12), in which case displacement vectors can be constructed by connecting separated corners, as shown in Figure 9.13. Structures that were separated by dikes can also be used, such as veins, older dikes, fold hinges and steeply dipping layers. In these cases we observe strain (the local extension direction), but because

of the small displacements usually involved it is reasonable to correlate the horizontal extension direction with σ_H. Assuming that the vertical stress is a principal stress axis we can further assume that $\sigma_H = \sigma_3$. It is the local stress that is found, and it is not uncommon to find that dikes rotate in the vertical direction into a fringe-like geometry such as shown in Figure 9.14. Such geometries indicate that the σ_H changes direction in the vertical direction, similar to fringe zones around fractures (Figure 7.26).

Summary

We have looked briefly at ways to determine sense of slip on faults and to use this and other brittle structures to get information about paleostress. In all cases, we should be aware of the fact that stress can only be deduced from strain patterns and never observed directly. Some assumption must always be made to go from strain to stress. It is therefore wise to take precautions when extracting paleostress information from deformed rocks.

Paleostress and the paleostress method are not treated in much detail in this chapter, so the interested reader is encouraged to explore the literature listed after these key points and review questions:

- Several kinematic criteria can help determine the sense of displacement on a slip surface. We should be able to make a sketch of the most important ones.

- Joints and veins are extension structures that initiate perpendicular to the smallest stress axis, and are therefore useful paleostress indicators.

- Conjugate faults are another useful set of structures that allows for a quick estimate of the stress field.

- Inversion of slip data gives the orientation of the principal stresses and the shape of the stress ellipsoid.

- Complex fault slip datasets may contain slip surfaces formed under two or more stress fields that must be separated to obtain useful results.

Review questions

1. Why must we be careful when interpreting lineations as displacement indicators?

2. What are the premises for successful paleostress analysis?

3. What is meant by the expression "slip inversion"?

4. What are conjugate faults, and what stress information do they give?

5. What does the Wallace–Bott hypothesis postulate?

6. What is the reduced stress tensor?

E-MODULE

 The section called *Brittle structures* in the *Kinematic indicators* e-learning module is recommended for this chapter.

FURTHER READING

General

Anderson, E. M., 1951, *The Dynamics of Faulting*. Edinburgh: Oliver & Boyd.

Sense of slip

Petit, J.-P., 1987, Criteria for the sense of movement on fault surfaces in brittle rocks. *Journal of Structural Geology* **9**: 597–608.

Stress and strain from fault populations

Angelier, J., 1994, Fault slip analysis and palaeostress reconstruction. In P. L. Hancock (Ed.), *Continental Deformation*. Oxford: Pergamon Press, pp. 53–100.

Cashman, P. H. and Ellis, M. A., 1994, Fault interaction may generate multiple slip vectors on a single fault surface. *Geology* **22**: 1123–1126.

Etchecopar, A., Vasseur, G. and Daignieres, M., 1981, An inverse problem in microtectonics for the determination of stress tensors from fault striation analysis. *Journal of Structural Geology* **3**: 51–65.

Lisle, R. J., Orfie, T. O., Arlegui, L., Liesa, C. and Srivastava, D. C., 2006, Favoured states of palaeostress in the Earth's crust: evidence from fault-slip data. *Journal of Structural Geology* **28**: 1051–1066.

Marrett, R. and Allmendinger, R. W., 1990. Kinematic analysis of fault-slip data. *Journal of Structural Geology* **12**: 973–986.

Stress from extension and contraction structures

Dunne, W. M. and Hancock, P. L., 1994, Paleostress analysis of small-scale brittle structures. In P. L. Hancock (Ed.), *Continental Deformation*. Oxford: Pergamon Press, pp. 101–120.

Fry, N., 2001, Stress space: striated faults, deformation twins, and their constraints on paleostress. *Journal of Structural Geology* **23**: 1–9.

Jolly, R. J. H. and Sanderson, D. J., 1995, Variation in the form and distribution of dykes in the Mull swarm, Scotland. *Journal of Structural Geology* **17**: 1543–1557.

The relationship between stress and strain

Marrett, R. and Peacock, D. C. P., 1999, Strain and stress. *Journal of Structural Geology* **21**: 1057–1063.

Twiss, R. J. and Unruh, J. R., 1998, Analysis of fault slip inversions: do they constrain stress or strain rate? *Journal of Geophysical Research* **103** (B6): 12,205–12,222.

Watterson, J., 1999, The future of failure: stress or strain? *Journal of Structural Geology* **21**: 939–948.

Chapter 10

Deformation at the microscale

In most of this book we study structures observable in thin sections, outcrops, maps and satellite photos. However, it is very useful and interesting to also take a closer look at the processes and mechanisms that take place from the grain scale down to that of atoms. This is a range that is more difficult to approach, especially the atomic scale, but a basic understanding is important and forms a foundation for a good understanding of mesoscale structures. The most important distinction is between brittle and plastic deformation mechanisms. Brittle deformation is sudden and violent: atomic lattices are forcefully torn apart and the lattice structure is forever damaged and weakened. Mechanisms in the plastic regime are more complicated and sluggish. Several factors influence the atomic-scale response of a crystal to stress, but temperature is the single most important factor: high temperature promotes plastic deformation mechanisms and microstructures. In this chapter we will briefly review brittle deformation mechanisms before focusing on the fundamentals of microscale plastic deformation of rocks and crystals.

10.1 Deformation mechanisms and microstructures

When strain accumulates in a deforming rock, certain **deformation processes** occur at the microscale that allow the rock to change its internal structure, shape or volume. The processes involved may vary – we have already looked at brittle processes in Chapter 7, and in the plastic regime there are other and different processes (Table 10.1). If these microscale processes lead to a change in shape or volume of a rock we call them **deformation mechanisms**:

Strain is accommodated through the activation of one or more microscale deformation mechanisms.

Most of the structures that reveal the type of mechanism at work are microscopic, and we therefore call them **microstructures**. They range in size from the atomic scale to the scale of grain aggregates. **Intracrystalline** deformation occurs within individual mineral grains. The smallest deformation structures of this kind actually occur on the atomic scale and can only be studied with the aid of the electron microscope. Larger-scale intracrystalline microstructures can be studied under the optical microscope, and encompass features such as grain fracturing, deformation twinning and (plastic) deformation bands.

When deformation mechanisms produce microstructures that affect more than one grain, such as grain boundary sliding or fracturing of mineral aggregates, we have **intercrystalline** deformation. Intercrystalline deformation is particularly common during brittle deformation.

The two expressions deformation mechanisms and deformation processes are closely related and are often used interchangeably. However, it is sometimes useful to distinguish between the two. Some define deformation mechanisms as processes that lead to strain. There are other microscopic changes that can occur that do not lead to strain, even though they may still be related to deformation. These are still deformation processes, and include rotation recrystallization, grain boundary migration (explained later in this chapter) and, in some cases, rigid rotation. Furthermore, two or more processes can combine to form a (composite) deformation mechanism. Note that the term deformation process is also used in a more general sense in other fields of structural geology.

Table 10.1 An overview of deformation processes in the brittle and plastic regimes.

| BRITTLE DEFORMATION | Fracturing | |
	Frictional sliding	
BRITTLE FLOW	Granular flow	Frictional sliding / Rolling
	Cataclastic flow	Grain fracturing
PLASTIC FLOW	Diffusion	Wet diffusion / Grain-boundary diffusion / Volume diffusion
	Crystal plasticity	Twinning / Dislocation creep

10.2 Brittle versus plastic deformation mechanisms

Brittle mechanisms dominate the upper crust while plastic mechanisms are increasingly common as pressure and, particularly, temperature increase with depth. However, brittle mechanisms can occur in deep parts of the lithosphere, and plastic mechanisms do occur at or near the surface in some cases. The reason is that not only temperature and pressure control deformation mechanisms, but also the rheology of the deforming mineral(s), availability of fluids and strain rate. While brittle deformation mechanisms can completely dominate deformation of a granitic rock in the upper crust, the transition from completely brittle to perfectly crystal-plastic deformation is gradual. There is a wide range in physical conditions or crustal depths where brittle and plastic mechanisms coexist. For a granitic rock, for example, quartz and feldspar respond differently to stress, particularly for the 300–500 °C window. Because most rocks consist of more than one mineral, and different minerals have different brittle–plastic transition windows, the brittle–plastic transition may be a kilometer-thick transition zone even for a single rock type. We therefore find that brittle and plastic deformation mechanisms commonly coexist in the same sample even when deformed at a constant depth during a single phase of deformation.

The controlling deformation mechanism determines whether the deformation belongs to the brittle or plastic regime.

Brittle deformation in an overall plastic setting is mostly restricted to intergranular fracturing. This is again related to the different ways different minerals respond to stress. Fracturing of a stiff mineral, such as garnet or feldspar, in a matrix of quartz is a typical example. The fractures are restricted to the garnet or feldspar grains, while plastic deformation mechanisms make quartz flow plastically.

10.3 Brittle deformation mechanisms

The characteristic feature of brittle deformation is fracturing and frictional sliding. As pointed out in Section 7.1, a distinction is drawn between **intergranular fracturing**, **intragranular fracturing**, **frictional sliding** on fractures and grain boundaries, and **grain rotation**. The combination of these deformation mechanisms is called **cataclastic flow**. Note that intragranular and intracrystalline are equivalent expressions, except that intragranular is used for granular media such as sand and sandstone, while intracrystalline is used about crystalline rocks where the porosity is (almost) zero.

The deformation mechanisms at very shallow depths are different for porous sediments and (almost) non-porous, crystalline rocks. Deformation of an unconsolidated sand or weakly consolidated sandstone buried at less than ~1 km depth is governed by two mechanisms: **grain rolling** (grain rotation) and **frictional grain boundary sliding**. The process involving these mechanisms is referred to by either of the equivalent terms **particulate flow** and **granular flow**. This is intergranular deformation in the sense that there is no permanent internal deformation of the grains.

Granular flow characterizes the deformation of highly porous sediments deforming in a shearing mode or in response to vertical loading (compaction). This is explored in a field of research known as **soil mechanics**, relevant to slope stability issues and other soil-oriented engineering problems. If the stresses across grain contacts become high enough, grains in a highly porous sediment or sedimentary rock will fracture. The fractures are confined to individual grains and are therefore intragranular microfractures. Under certain circumstances (low pore pressure and small grain contact areas) microfractures may form close to the surface, commonly by chipping off small flakes of the grains. This type of microfracturing is referred to as **spalling** or **flaking** (Figure 10.1). At higher confining pressures, corresponding to depths in excess of ~1 km, fractures split the grains into more evenly sized parts, and this mechanism is sometimes described as **transgranular fracturing**. Some use the term transgranular as a substitute for intergranular, i.e. about fractures that cut across *several* grains,

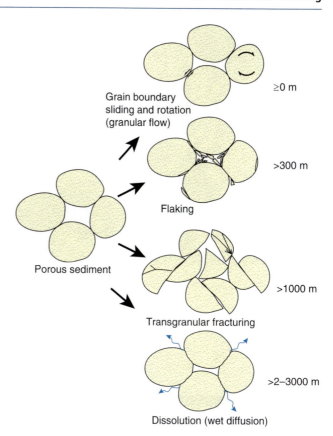

Figure 10.1 Deformation mechanisms operative at shallow depths. Very approximate depths are indicated.

so some care should be exercised when using these terms. Once fractured, the grains reorganize themselves by frictional sliding and rotation, leading to porosity reduction.

Grains in non-porous or low-porosity rocks develop fractures that may already be intergranular at the initial stages of deformation. As more and more fractures form, frictional sliding along these fractures together with rotation of grains may be widespread enough that we can characterize the deformation by the term cataclastic flow. From this point on the rock becomes crushed into a gouge, breccia or cataclasite (Box 8.1). In general, cataclasis in low-porosity rocks involves dilation and increased permeability as fractures form and open. Incohesive gouge and breccia dominate the uppermost few kilometers of the crust, while cohesive breccia and cataclasite are more common from roughly 3–5 to 10–12 km depth. In the rest of this chapter we will look at crystal-plastic deformation mechanisms.

10.4 Mechanical twinning

Stress can result in mechanical bending or kinking of the crystal lattice of some minerals, even at very low temperatures. Plagioclase feldspar and calcite (Figure 10.2)

Figure 10.2 Deformation twins in an aggregate of calcite crystals.

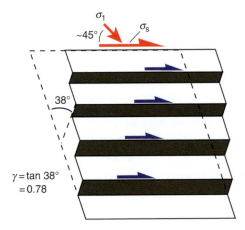

Figure 10.3 Deformation (glide) twins in a calcite crystal. Stress is ideally at 45° to the shear (glide) plane. Dark lamellae have been sheared (simple shear).

a shear stress just above a critical value of $\sim$10 MPa for so-called e-twins (the most common of several calcite twin systems) – a value that seems to be more or less independent of normal stress, temperature and strain rate. This means that calcite can deform by twinning at any level in the crust, even at the surface (and by hand for demonstration purpose), as long as the critical stress is reached. Low-temperature ($<$200 °C) calcite twins are typically thinner and straighter than those formed at higher temperatures.

The ideal orientation for a deformation twin plane is where the shear stress is at its peak, which we know from Chapter 2 to be at 45° to σ_1. This relationship between the orientation of well-developed twins and σ_1 can be used to estimate stress. Optical c-axes for the twinned portions of calcite crystals can be found using a universal stage or goniometer. Then the statistically determined orientation of the c-axis from a number of observations in the same thin section gives the approximate orientation of σ_1. A simple method is shown in Figure 10.4. More sophisticated methods also exist.

Mechanical twinning accumulates small shear strains and elongations of just a few percent. The amount of shear strain associated with a single kink in the crystal lattice is restricted to a fixed angle (38° for calcite), the angle that turns the twin plane into a mirror plane. The shear strain in the twinned portion of a crystal is $\tan(38°) = 0.78$. Depending on how much of the crystal

are common examples. These intracrystalline kink structures are expressions of strain and are also known as **deformation twins** and the process as **mechanical twinning**. Mechanical twinning does not involve breaking of the crystal lattice, and is therefore considered a plastic deformation mechanism. The structures must be distinguished from twins formed during crystal growth (growth twins) and cooling (transformation twins). Because deformation twins are found in only a few of our common minerals, they are easily distinguished. One useful criterion is that mechanical twins tend to taper out to form an interfingering pattern.

Calcite commonly shows a type of deformation twin formed by **twin gliding**. Twin gliding involves simple shear along the twin plane, as illustrated in Figure 10.3. The kinked and unkinked parts are mirror images of one another about this plane. Calcite twin gliding occurs at

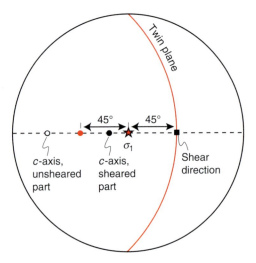

Figure 10.4 Finding the orientation of σ_1 from the measured orientation of the twin plane and c-axes of twinned and untwinned lamellae. The two c-axes lie on a great circle that also contains σ_1 and the pole to the twin plane (red). Ideally the angle between them should be 45°. When several grains are plotted, the orientation of σ_1 can be found statistically.

is twinned, the total shear strain accommodated by the grain is around half of this value.

Mechanical twinning of calcite is a low-strain and low-temperature plastic deformation mechanism that stores information about the stress field at the time of deformation.

When a twin has formed, further strain is accommodated by the formation of new twins. In aggregates of twinned calcite grains the different grains will be sheared according to their crystallographic orientation. Those favorably oriented relative to the X-axis of the strain ellipse are more sheared than other grains. Hence, the orientation of the strain ellipse can also be found. The classic textbook example of how patterns of stress or small strain can be mapped over large regions is the study done on calcite twinning in the carbonates of the Appalachian foreland region. Stress attributed to the Alleghany orogeny in the Appalachian Mountains can be mapped almost 100 km into the North American continent from the actual thrust front on the east coast. However, calcite twinning also has applications on a small scale, for example in the mapping of stress and strain associated with folded layers.

10.5 Crystal defects

The atomic lattice of any mineral grain, deformed or not, contains a significant number of defects. This means that the crystal has energy stored in the lattice. The more defects, the higher the stored energy.

There are two main types of defects. Some are known as **point defects**, represented by either vacancies or, less importantly, impurities in the form of extra atoms in the lattice (Figure 10.5). The point defect of interest to us is the one represented by a missing atom. Movement of vacancies is called **diffusion** (Figures 10.6 and 10.7).

The other type of defect is **line defects**, generally referred to as **dislocations**. A dislocation is a mobile line defect that contributes to intracrystalline deformation by a mechanism called **slip**. Slip implies movement of a dislocation front within a plane. A slip plane is usually the plane in a crystal that has the highest density of atoms. There are also some called **plane defects**, which include structures such as grain boundaries, subgrain boundaries and twin planes. These defects are discussed in Section 10.6.

When a crystal is deformed by plastic deformation, the dislocation density increases. Deformation adds energy to the crystal, and a high density of defects implies that

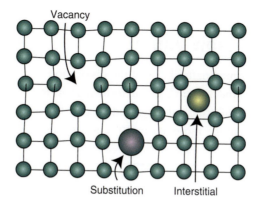

Figure 10.5 Point defects in a crystal lattice include vacancies (holes), substitutional impurities, and interstitial impurities. Vacancies represent the most important point defect in crystal-plastic flow.

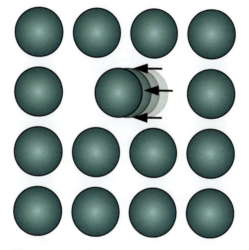

Figure 10.6 Migration of vacancies through an atomic lattice is called diffusion. A hole moves as it is replaced by a neighboring atom.

the crystal is in a high-energy state. A low-energy state is more stable, and there is thus a thermodynamic drive to reduce the number of crystal defects. Both the building up of defects such as dislocations and the reduction of them occur by the movement of defects within the atomic lattice. Such movements are not "painless". It takes energy to move dislocations around, and the movement will occur in the crystallographic plane or direction where dislocation movements involve the least energy.

Diffusion creep

Migration of vacancies in crystallographic lattices (Figure 10.6) is called **diffusion mass transfer**, usually referred to simply as **diffusion** or **diffusion creep**. Diffusion of vacancies through crystals is known as **volume diffusion** or **Nabarro–Herring creep** (Figure 10.7). The rate is not

BOX 10.1 | HOW MANY?

In an undeformed natural crystal the dislocation density is around $10^6/cm^2$. In a deformed grain the density is several orders of magnitude higher (an order of magnitude more means 10 times as many).

BOX 10.2 | DEFORMATION MECHANISM MAPS

The different deformation mechanisms that are operative in a deforming mineral under various physical conditions can be expressed by means of a deformation mechanism map. Stress–temperature maps have been constructed, and the diagrams (maps) are contoured for a range of strain rates. Deformation mechanism maps show the range for which each deformation mechanism dominates. They are partly based on experimental data that have been extrapolated into geologically realistic strain rates and temperatures, and partly on theoretical considerations. The example shown here is for quartz. Realistic natural strain rates are indicated in yellow (10^{-12}–10^{-15} s^{-1}). Note that this and similar maps are hampered by many uncertainties and limited data availability.

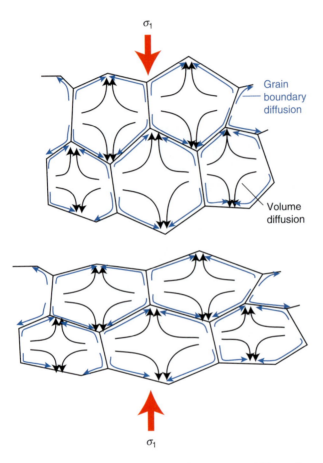

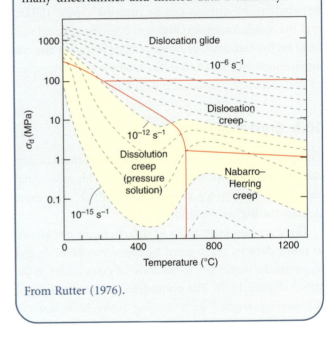

From Rutter (1976).

Figure 10.7 Diffusion in a mineral can occur within grains by means of volume diffusion, or along grain boundaries by means of grain boundary diffusion. In both cases vacancies move toward high-stress sites so that the minerals accumulate strain over time.

overly high, perhaps a few centimeters per million years. However, at some point vacancies will reach a grain boundary and disappear. Because vacancies migrate towards sites of maximum stress the crystals acquire a shape fabric or strain. During this process the crystal will gain regularity and turn into a more perfect crystal. Volume diffusion requires a lot of energy, so the migration rate is highly temperature dependent: high temperatures give high vibrations in the lattice, which increases the

rate. For this reason, volume diffusion is important only in the lower part of the crust and in the mantle.

In other cases, migration of vacancies occurs preferentially along grain boundaries. This type of diffusion is known as **grain boundary diffusion** or **Coble creep**. Coble creep is a bit less energy demanding than Nabarro–Herring creep, and is more important in the deformation of the plastic crust. In both types of diffusion, mineral grains change shape, and over time this change adds to mesoscopic strain that can be seen in a hand sample or in outcrop. It is interesting to note that the grain size is important, particularly for Nabarro–Herring creep: the smaller the grains, the higher the strain rate.

Pressure solution (or dissolution) is another important diffusion process. It bears similarities to Coble creep, and geometrically and mathematically it can be treated in the same way. However, in the case of pressure solution, diffusion occurs along a thin film of fluid along grain boundaries. A better name for pressure solution is therefore **wet diffusion**. Wet diffusion can occur at very low (even diagenetic) temperatures. During such diffusion the mineral is dissolved and the ions are carried with the fluid to be precipitated some other place. This mechanism is chemically controlled but also strongly affected by stress. Dissolution is significantly quicker where stress is high, particularly at surfaces oriented perpendicular to σ_1, while precipitation is favored on surfaces at high angles to σ_3. In porous rocks, wet diffusion at grain contacts is promoted by the stress concentrations there (Figure 7.36).

Precipitation can also occur far away in a different part of the rock or in a completely different rock layer or unit. Rocks exposed to wet diffusion experience a volume reduction. Wet diffusion is the main mechanism in what is known as **chemical compaction**, for example of sand that is undergoing lithification. Quartz sand(stone) experiences wet solution at temperatures above ~90 °C (Figure 10.8). The mechanism reduces the pore space in sandstones and not only makes the sandstone stronger and more cohesive, but also reduces porosity and permeability in clastic reservoir rocks. Wet diffusion is also common in limestones, where pressure solution seams known as **stylolites** form.

Volume diffusion: vacancies move through crystals (temperature and stress controlled).

Grain boundary diffusion: vacancies move along grain boundaries (temperature and stress controlled).

Pressure solution: ions move in fluid films and pore fluid (chemically and stress controlled).

Grain boundary sliding accommodated by dry or wet diffusion can occur at high temperatures when diffusion is quick enough to modify the shapes of the grains as they slide along each other. In contrast to frictional sliding in the brittle regime, diffusion-accommodated sliding along the grain boundaries is frictionless and no voids open during the deformation. This deformation mechanism, which is characterized by relatively rapid strain rates at low differential stress, occurs in fine-grained rocks in the mantle and lower crust, typically after a phase

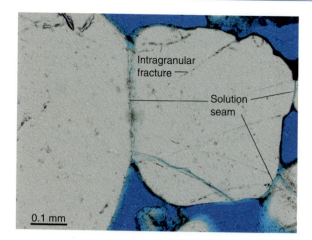

Figure 10.8 Pressure solution at grain contacts in Nubian Sandstone (Sinai). Also note intragranular fractures.

of grain-size reduction by dynamic recrystallization (dislocation creep).

Small grain sizes favor diffusion because of short distances to grain boundaries (short diffusion paths).

Rocks deformed by diffusion-driven grain boundary sliding are always fine-grained and can accommodate large strains without developing any preferred grain shape fabric. The type of deformation process sometimes referred to as **superplastic creep** or **superplasticity** is dominated by grain boundary sliding and fine grain size.

Dislocations and dislocation creep

A dislocation is a mobile line defect that contributes to intracrystalline deformation by a mechanism called **slip**. Slip implies movement of a dislocation front within a **slip plane**, as shown in Figure 10.9, and should not be confused with frictional slip associated with brittle deformation and fault slippage. Slip planes are relatively weak crystallographic directions controlled by the atomic structure, and are usually the plane(s) in a crystal that have the highest density of atoms. Minerals have one or more such directions that can be variably activated, depending on temperature and state of stress. Both the amount of differential stress and the orientation of the stress field is of importance, as the critically resolved shear stress on any given slip system must be high enough for the system to be activated.

Mica has only one slip plane, while quartz has as many as four. One of the slip planes in quartz is in the basal plane (normal to the crystallographic c-axis) and is activated at low metamorphic grade (300–400 °C). The activation of basal slip for quartz produces a strong preferred orientation

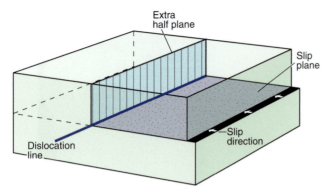

Figure 10.9 Block diagram representing the concept of dislocation line, slip plane, extra half plane and slip direction for an edge dislocation. Compare with the next two figures. Based on Hobbs *et al.* (1976).

of quartz *c*-axes that when measured optically (the *c*-axis and optical axis of quartz are coincident) can reflect the deformation mechanism as well as the kinematic framework of the deformation. For high shear strains, the quartz *c*-axes are oriented at a high angle to the foliation.

Dislocations are too small to be seen under the optical microscope, but can be identified by means of the transmission electron microscope (TEM) (Figure 10.10a and b). Studies of dislocations have shown that there are two different types of dislocations. The simplest one is the **edge dislocation**, which is the edge of an extra half-plane in the crystal lattice (Fig. 10.11a). The edge dislocation is the line drawn in Figure 10.11a, and it moves in the horizontal plane, which is the slip plane. The dislocation line or end of the extra half-space is thus perpendicular to its slip direction.

The other type is the **screw dislocation**, where the dislocation line is oriented parallel to the slip direction. The slip plane is vertical in Figure 10.11b. Screw dislocations are a bit like tearing a piece of paper. The motions of these two types of dislocations through a crystal are therefore somewhat different, and they may join forces to make composite dislocations that contain elements of both kinds.

The formation, motion and destruction of dislocations in a crystal are all contained in the term **dislocation creep**. Only a small volume around the line defect is being deformed at any time during dislocation creep. The process by which edge dislocations move is called **dislocation glide** and is shown in Figure 10.12. Eventually the dislocation has slipped through the crystal, and the offset is complete. This is different from brittle microfracturing, where the entire grain is cut almost instantaneously.

Dislocation creep allows the deformation to take place at much lower differential stress than that required for

(a)

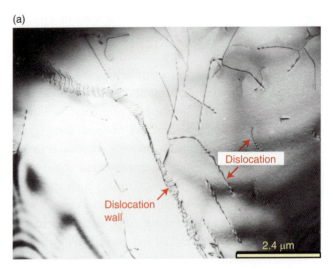

(b)

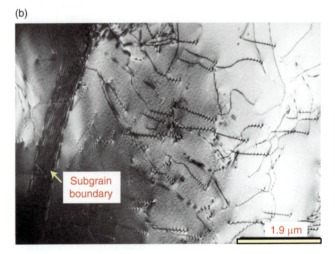

Figure 10.10 Electron microscope (TEM) photo of dislocations in deformed quartz from a deformed conglomerate in the Swedish Caledonides (Lisle, 1984). (a) Low-density area where several dislocations have free terminations. A dislocation wall is seen. (b) Somewhat higher dislocation density. The thick zone to the left is a subgrain boundary. Dislocation creep was an important mechanism in this case, together with some grain boundary sliding and pressure solution. The dislocation density indicates a deviatoric stress of 30–60 MPa. Photo: Martyn Drury.

brittle fracturing. This is why rocks do not fracture if dislocation creep is active, and the reason why the strength of the crust decreases downward as we enter the brittle–plastic transition (Box 10.3).

There is another difference between brittle fracturing and deformation by dislocation movement, namely that dislocations leave no trace:

Dislocation movements do not damage or weaken the mineral.

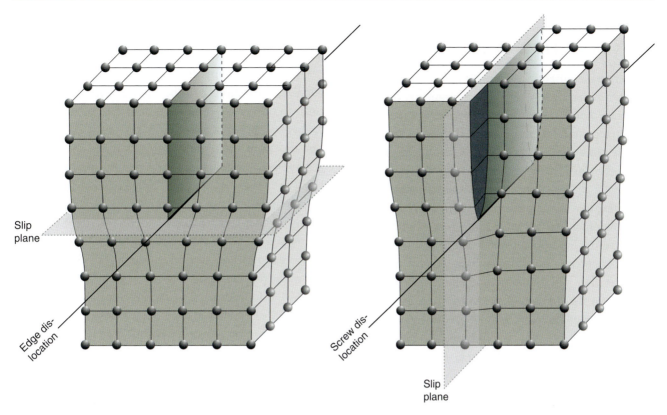

Figure 10.11 The two types of dislocations. An edge dislocation (left) occurs where an extra half-space of atoms interrupts the lattice, while a screw dislocation (right) involves twisting of the lattice.

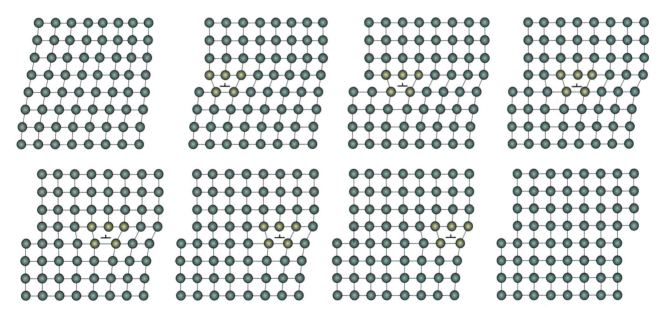

Figure 10.12 Formation and movement of an edge dislocation through a crystal lattice. The process can be compared to the way a caterpillar moves, as only bonds along the dislocation line (which is perpendicular to the page) are broken at the same time. Thus, the energy that it takes to move the dislocation is kept at a low level. Instantaneous fracturing of the whole crystal would require much higher energy.

Once the dislocation has slipped through the crystal, the crystal is completely healed with respect to this particular imperfection. The imperfection is gone and no weakness is introduced, which there would be where fracturing has occurred. Hence, dislocation movements do not reduce the internal strength of crystals.

When a crystal is strained, the dislocation density increases. We therefore refer to this energy as **strain**

BOX 10.3 | FLOW LAWS

Flow laws are useful for estimating the strength of the lithosphere. Flow laws and experimental data indicate that the crust becomes weaker as pressure and temperature increase within the plastic regime. At the same time we know that there is a temperature limit for plastic deformation. At lower temperatures frictional gliding controls the crustal strength, meaning that the upper crustal strength is controlled by how much stress it takes to form or reactivate a fracture. This regime is governed by Byerlee's law from Chapter 7. We therefore have to combine Byerlee's law and flow laws to obtain a realistic model for the strength of the entire crust. The point of intersection between Byerlee's law and the appropriate plastic flow law indicates the brittle–plastic transition (see also Figure 6.18). In reality, this intersection is not sharp, but a gradual transition.

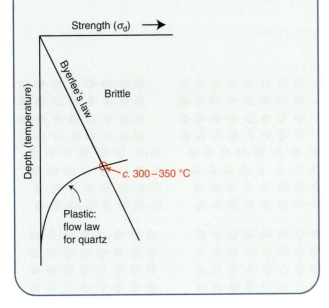

energy. We could also say that differential stress creates new dislocations at the grain boundaries, and these dislocations will move through the crystal. Hence differential stress adds energy to the crystal:

A high density of defects implies that the crystal is in a high-energy state.

A low-energy state is more stable, and there is thus a thermodynamic drive to reduce the number of crystal defects. Slip of defects through the atomic lattice requires

energy. The movement therefore occurs in the crystallographic plane and direction where dislocation movements require the least energy.

Gliding or slipping dislocations may encounter interstitials, substitutions or other dislocations on their way through the crystal. If the dislocation has too little energy to bypass the obstacles it will get stuck. **Dislocation pile-ups** form where multiple dislocations entangle and accumulate. Bypassing obstacles requires dislocations to change slip planes by a process called **cross-slip**. This may be possible for screw dislocations, but edge dislocations "jump" to another slip plane by the process known as **climb**. Climb and cross-slip require energy, which in this context means temperature. As a rule of thumb, climb and cross-slip occur for temperatures in excess of 300 °C for quartz and 500 °C for feldspar. Below these temperatures it is hard to move dislocations, and we quickly leave crystal-plasticity and enter the frictional or brittle regime. In general:

Dislocation glide is most important where temperature is too low for volume diffusion and it is not wet enough for wet diffusion to occur.

Flow laws

Dislocation movement depends not only on temperature (T), but also on differential stress (σ_d) and the activation energy (E^*) involved. These three variables can be related to the strain rate (σ_d) by means of a flow law. This law, which is a constitutive equation because it relates stress to strain rate, also depends on the deformation mechanism.

Flow laws relate stress to strain rate and depend on the dominating deformation mechanism, which again depends on the mineral and temperature.

For dislocation glide, i.e. where temperature is too low for dislocations to climb over lattice obstacles, the flow law is:

$$\dot{e} = A \exp(\sigma_d) \exp(-E^*/RT) \qquad (10.1)$$

where A is an empirically determined material constant, R is the gas constant and T is the temperature in K. For higher temperatures, where dislocation creep dominates, the law becomes:

$$\dot{e} = A(\sigma_d)^n \exp(-E^*/RT) \qquad (10.2)$$

Note that stress is the main variable in this formula. Differential stress is raised to the nth power and dislocation creep, where dislocations can both glide and climb, is therefore called **power-law creep**. Typical values of n lie between 3 and 5 for power-law creep. This is the most widely used flow law, applicable to many crustal and even mantle-level settings.

When temperature is really high (or grain size is very small), the flow law for diffusion is applied:

$$\dot{e} = A(\sigma_d)\exp(-E^*/RT) \qquad (10.3)$$

This formula is identical to the one for dislocation creep (Equation 10.2), with $n = 1$. This points to the linear relation between strain rate and stress, typical for perfectly (Newtonian) viscous deformation.

10.6 From the atomic scale to microstructures

Atomic-scale deformation structures such as dislocations can only be studied by means of electron microscopy at 10 000–100 000 times magnification. However, the effects of these structures and related mechanisms can be seen under the optical microscope. They are referred to as **microstructures** and carry information about temperature, state of stress and rheological properties at the time of deformation. Microstructures and microtextures such as recovery and recrystallization can be seen under the optical microscope. However, keep in mind that the controlling mechanisms are the atomic-scale ones discussed above.

Recovery

Processes such as dislocation creep reduce the internal energy of a mineral grain by moving dislocations to the grain boundary or collecting them in zones within the grain. Dislocations can organize themselves into what are known as **dislocation walls** (Figures 10.10 and 10.13). Such walls are visible in thin sections if they contain sufficient dislocations. What makes them visible is the change in their crystallographic orientation across the walls. For minerals such as quartz, the two sides of the wall show slightly different extinction angles. Thus, **undulose extinction** is characteristic of dislocation walls in mineral grains. Elongated grain-internal zones with slightly different extinction, as shown in Figure 10.14, are known in the (older) literature as **deformation bands**. Such deformation bands have nothing to do with mesoscopic deformation bands in deformed sandstones (Chapter 7), and unless the meaning is clear from the context, the name should be reserved for tabular strain discontinuities in porous rocks. Where dislocations migrate further and arrange themselves into more well-

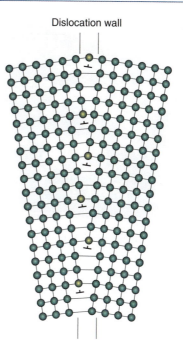

Dislocation wall

Figure 10.13 Simple dislocation wall composed of edge dislocations. Dislocation walls separate parts of a crystal with slightly different lattice orientation, such as the boundary between neighboring subgrains.

1 mm

Figure 10.14 Deformation bands in quartz crystals, characterized by undulose extinction. Quartz pebble in the deformed conglomerate shown in Box 3.1.

defined networks that outline small patches with few or no dislocations, we have a process called **subgrain formation**. Subgrains are polygonal patches of a mineral grain that are slightly (usually less than 5°) misoriented with respect to their neighbors or host grain. Subgrain formation (Figures 10.15 and 10.16) is an advanced stage in the process called **recovery**, where deformed grains can reduce their stored energy by the removal or rearrangement of dislocations.

Figure 10.15 Subgrains and deformation bands in quartz. A large grain is breaking down, forming a core of relict quartz with a mantle of subgrains and new grains (core–mantle structure). Quartz band in sheared phyllite, Scandinavian Caledonides.

Figure 10.17 Recrystallized quartz bands in metarhyolite. Note the even size and strain-free nature of the recrystallized grains. Grain boundaries are more irregular than what would be expected for static recrystallization, and are therefore interpreted as dynamic.

(a) Subgrain rotation recrystallization

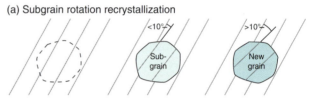

(b) Grain boundary migration (bulge nucleation)

Figure 10.16 Gradual evolution of subgrains at the tail of a larger quartz grain. Note faint shadows of subgrains as they rotate out of alignment with the host grain. Heimefrontfjella, Antarctica.

Figure 10.18 (a) Illustration of recrystallization by means of subgrain rotation. (b) Bulging, resulting from migration of a grain boundary into a more strained grain (with more dislocations).

> Recovery comprises all processes that move, cancel out and order dislocations into walls that separate portions of the original grain with slightly different crystallographic orientations.

Recrystallization

If recovery continues so that the dislocations still present in subgrains are removed and the grains become strain-free with little or no undulose extinction, then the mineral has recrystallized (Figure 10.17). This type of recrystallization, where subgrains rotate until they classify as a separate

grain (by definition more than 10° relative to neighboring grains) is known as **subgrain rotation recrystallization** (Figure 10.18a). Subgrain rotation requires that dislocations are free to move and climb relatively freely, a process favored by elevated temperatures and known as climb-accommodated dislocation creep.

Minerals can also recrystallize by the migration of grain boundaries, a process known as **grain boundary migration** or **migration recrystallization**. In general, grain boundary migration is driven by differences in strain energy, and grains with high dislocation density have higher energy than a neighboring strain-free (dislocation-free) grain. Along the boundary between two grains there will be a slight movement of atoms

(a)

(b)

Figure 10.19 Dynamic recrystallization in a greenschist-facies shear zone. The new grains are oblique to the main foliation because they have only experienced the last part of the non-coaxial deformation. The middle grain in (a) is a feldspar porphyroclast. (b) is a close-up view of part of (a).

in the more strained grain to fit the lattice of the strain-free grain. In this sense the grain boundary migrates into the grain with high dislocation density. One variant of this process, illustrated in Figure 10.18b, is referred to as **bulging**. Nucleation of new strain-free grains that expand in a strained grain is also described. Both types of boundary migration are stimulated by temperature and differences in dislocation density.

In many cases recrystallization in the crust is seen to be a combination of grain boundary migration and subgrain rotation. Recrystallized grains tend to be bigger than related subgrains and in many cases the grain boundaries are straighter.

Recrystallization is the process whereby strained and dislocation-rich grains are replaced by unstrained grains with few or no dislocations.

Recrystallization that occurs as the rock is being deformed (under differential stress) is referred to as **dynamic recrystallization** (Figure 10.19). Rocks can also recrystallize after the deformation has come to a halt. This process is called **static recrystallization** or **annealing**. Static recrystallization tends to produce larger and more equant grains, typically forming a polygonal pattern. Grains that undergo dynamic recrystallization are continuously recrystallizing under the influence of tectonic stress. New dislocations will form in the grains, seen under the microscope as undulatory extinction. In addition, dynamically recrystallized grains will soon become strained with a preferred orientation that depends on the sense of shear. There is always competition between continuous deformation by crystal-plastic deformation mechanisms and temperature-stimulated recovery by recrystallization during dynamic recrystallization. The higher the temperature, the faster the recrystallization.

Dislocation accumulation is counteracted by recrystallization, which involves formation or migration of grain boundaries.

One of the characteristics of recrystallized rocks is the **pinning** effect of non-recrystallizing minerals, such as small mica grains in quartzite or quartz-rich mylonites. Such minerals hinder grain boundary migration and cause the recrystallized rock to have an uneven or smaller grain size.

Stress and grain size

Recrystallization is driven by differences in dislocation density across grain boundaries, which again depend on differential stress. This implies that the dislocation density in a deformed rock can tell us something about the differential stress at the time of deformation. Hence, the size of subgrains and dynamically recrystallized grains is related to the differential stress during deformation.

Using grain size to estimate paleostress in dynamically recrystallized rocks is a tool referred to as a **paleopiezometer** ("piezo" is derived from the Greek word for pressure). In general, the average grain size goes down with increasing differential stress and strain rate. The temperature is also of interest: low temperature implies higher stress and, according to Figure 10.20, a smaller grain size. This is consistent with the general observation that greenschist facies mylonites have lower grain size than amphibolite facies mylonites. Estimated stress values from naturally deformed rocks lie between a few MPa for high-temperature mylonites and up to ~100 MPa for low-temperature mylonites (i.e. mylonites formed close to the brittle–plastic transition).

The method is based on the assumption that the deformation is stable. In this context this means that the average grain size is constant for subgrains and recrystallized grains, independent of the duration of the deformation. As reflected by the different curves in Figure 10.20, there is considerable uncertainty involved in using this method. There is uncertainty in defining the average grain size, different types of recrystallization mechanisms may give different stress–grain size relations, and the influence of fluids should be accounted for. Static (postkinematic) growth of minerals that interfere with dynamic recrystallization textures is yet another source of error. Nevertheless, this is the only way of quantifying stress in the middle and lower crust that we know of today.

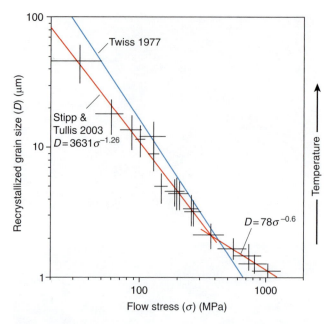

Figure 10.20 Grain size plotted against differential stress for quartz. Experimentally derived data by Stipp and Tullis (2003) are shown together with the two curves that best fit their data. A theoretically estimated curve (Twiss 1977) is shown for comparison.

Summary

Plastic deformation at the microscale is the foundation for all plastic deformation structures that can be observed in hand samples, outcrops, maps, profiles etc. Folds, plastic shear zones, mylonite zones and similar structures are all results of dislocation creep and diffusion. Keep this in mind as we move on to look at such structures in the next chapters. First go through these summary points and make sure you can address the review questions below:

- Brittle deformation mechanisms involve frictional sliding and breaking of crystal lattice and atomic bonds. Plastic deformation is healing and produces or leaves no flaws.

- Plastic deformation can occur by twinning, different types of diffusion and dislocation creep.

- Minerals can recrystallize during deformation (synkinematic or dynamic recrystallization) or after deformation (postkinematic or static recrystallization).

- Dynamic recrystallization competes against dislocation formation and straining of grains.

- The stress required to drive dislocation motion decreases with increasing temperature.

- The size of recrystallized grains is related to differential stress, and can to some extent be used to estimate paleostress.

- Recrystallization occurs by concentration of dislocations along existing or new grain boundaries so that dislocation-free domains (new grains) emerge.

- Undulatory extinction in quartz indicates the presence of dislocations (strain).

Review questions

1. What is the difference between a slip plane in a plastically deforming crystal and a slip plane associated with brittle faulting?

2. What are the main principal differences between brittle and plastic deformation?

3. Why is intracrystalline fracturing so common in brittle deformation of highly porous rocks?

4. Name two plastic deformation mechanisms that can operate at shallow crustal depths.

5. What is meant by the term dislocation creep and how does it differ from diffusion?

6. What deformation mechanism is particularly active in fine-grained rocks at high temperatures (in the lower crust and the mantle)?

7. What information can we get from dynamically deformed quartz that disappears during static recrystallization?

8. What is the difference between recrystallization by subgrain rotation and grain boundary migration?

E-MODULE

The e-learning module called *Plastic deformation* is recommended for this chapter.

FURTHER READING

General

de Meer, S., Drury, M., Bresser, J. H. P. and Pennock, G. M., 2002, Current issues and new developments in deformation mechanisms, rheology and tectonics. In S. de Meer, M. R. Drury, J. H. P. de Bresser and G. M. Pennock (Eds.), *Deformation Mechanisms, Rheology and Tectonics: Current Status and Future Developments.* Special Publication **200**, London: Geological Society, pp. 1–27.

Karato, S.-I., 2008, *Deformation of Earth Materials: An Introduction to the Rheology of Solid Earth.* Cambridge: Cambridge University Press.

Knipe, R. J., 1989, Deformation mechanisms: recognition from natural tectonites. *Journal of Structural Geology* **11**: 127–146.

Passchier, C. W. and Trouw, R. A. J., 2006, *Microtectonics.* Berlin: Springer Verlag.

Dislocation creep

Hirth, G. and Tullis, J., 1992, Dislocation creep regimes in quartz aggregates. *Journal of Structural Geology* **14**: 145–159.

Flow laws

Carter, N. L. and Tsenn, M. C., 1987, Flow properties of continental lithosphere. *Tectonophysics* **136**: 27–63.

Schmid, S. M., 1982, Microfabric studies as indicators of deformation mechanisms and flow laws operative in mountain building. In K. J. Hsü (Ed.), *Mountain Building Processes.* London: Academic Press, pp. 95–110.

Grain size piezometers

Shimizu, I., 2007, Theories and applicability of grain size piezometers: the role of dynamic recrystallization mechanisms. *Journal of Structural Geology* **30**: 899–917.

Stipp, M. and Tullis, J., 2003, The recrystallized grain size piezometer for quartz. *Journal of Geophysical Research* **30**: doi:10:1029/2003GL018444.

Pictures of structures

Snoke, A. W., Tullis, J. and Todd, V. R., 1998, *Fault-related Rocks: A Photographic Atlas.* Princeton: Princeton University Press.

Chapter 11

Folds and folding

Folds are eye-catching and visually attractive structures that can form in practically any rock type, tectonic setting and depth. For these reasons they have been recognized, admired and explored since long before geology became a science (Leonardo da Vinci discussed them some 500 years ago, and Nicholas Steno in 1669). Our understanding of folds and folding has changed over time, and the fundament of what is today called modern fold theory was more or less consolidated in the 1950s and 1960s. Folds, whether observed on the micro-, meso- or macroscale, are clearly some of our most important windows into local and regional deformation histories of the past. Their geometry and expression carry important information about the type of deformation, kinematics and tectonics of an area. Besides, they can be of great economic importance, both as oil traps and in the search for and exploitation of ores and other mineral resources. In this chapter we will first look at the geometric aspects of folds and then pay attention to the processes and mechanisms at work during folding of rock layers.

11.1 Geometric description

It is fascinating to watch folds form and develop in the laboratory, and we can learn much about folds and folding by performing controlled physical experiments and numerical simulations. However, modeling must always be rooted in observations of naturally folded rocks, so geometric analysis of folds formed in different settings and rock types is fundamental. Geometric analysis is important not only in order to understand how various types of folds form, but also when considering such things as hydrocarbon traps and folded ores in the subsurface. There is a wealth of descriptive expressions in use, because folds come in all shapes and sizes. Hence we will start this chapter by going through the basic jargon related to folds and fold geometry.

Shape and orientation

Folds are best studied in sections perpendicular to the folded layering, or perpendicular to what is defined as the axial surface, as shown in Figure 11.1. Unless indicated, we will assume that this is the section of observation in this chapter. In general, folds are made up of a **hinge** that connects two usually differently oriented **limbs**. The hinge may be sharp and abrupt, but more commonly the curvature of the hinge is gradual, and a **hinge zone** is defined. A spectrum of hinge shapes exists, from the pointed hinges of **kink bands** and **chevron folds** (sharp and angular folds) to the well-rounded hinges of **concentric folds** (Figure 11.2). Classification of folds relative to hinge curvature is referred to as **bluntness**.

The shape of folds can also be compared to mathematical functions, in which case we can apply terms such as **amplitude** and **wavelength**. Folds do not necessarily show the regularity of mathematical functions as we know

(a) Kink band
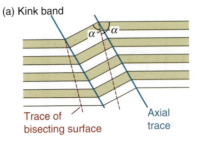

Trace of bisecting surface

Axial trace

(b) Chevron folds
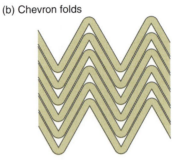

(c) Concentric folds
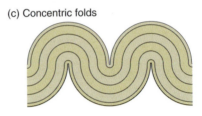

(d) Box fold
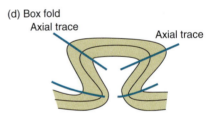

Axial trace

Axial trace

Figure 11.2 (a) Kink band, where the bisecting surface, i.e. the surface dividing the interlimb angle in two, is different from the axial surface. (b) Chevron folds (harmonic). (c) Concentric folds, where the arcs are circular. (d) Box folds, showing two sets of axial surfaces.

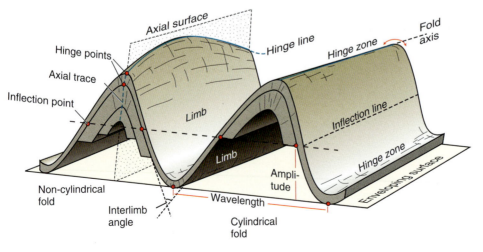

Figure 11.1 Geometric aspects of folds.

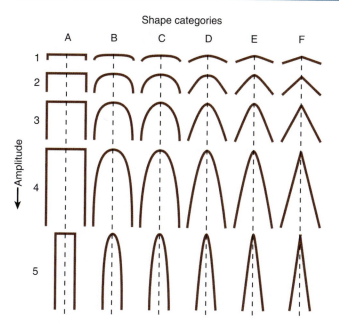

Shape categories

A   B   C   D   E   F

Figure 11.3 Fold classification based on shape. From Hudleston (1973).

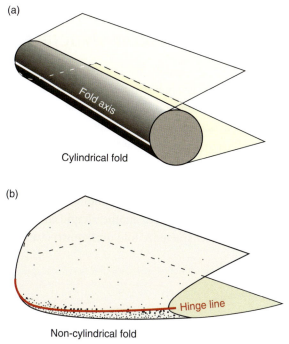

(a) Fold axis / Cylindrical fold

(b) Hinge line / Non-cylindrical fold

Figure 11.4 Cylindrical and non-cylindrical fold geometries.

them from classes of elementary algebra. Nevertheless, simple harmonic analysis (Fourier transformation) has been applied in the description of fold shape, where a mathematical function is fitted to a given folded surface. The form of the Fourier transformation useful to geologists is

$$f(x) = b_1 \sin x + b_3 \sin 3x + b_5 \sin 5x \dots \quad (11.1)$$

This series converges rapidly, so it is sufficient to consider only the first coefficients, b_1 and b_3, in the description of natural folds. Based on this method, Peter Hudleston prepared the visual classification system for fold shape shown in Figure 11.3.

In multilayered rocks, folds may be repeated with similar shape in the direction of the axial trace, as seen in Figure 11.2a–c. Such folds are called **harmonic**. If the folds differ in wavelength and shape along the axial trace or die out in this direction they are said to be **disharmonic**.

The point of maximum curvature of a folded layer is located in the center of the hinge zone and is called the **hinge point** (Figure 11.1). Hinge points are connected in three dimensions by a **hinge line**. The hinge line is commonly found to be curved, but where it appears as a straight line it is called the **fold axis**.

This takes us to an important element of fold geometry called **cylindricity**. Folds with straight hinge lines are **cylindrical**. A cylindrical fold can be viewed as a partly unwrapped cylinder where the axis of the cylinder defines the fold axis (Figure 11.4a). At some scale all folds are non-cylindrical, since they have to start and end

somewhere, or transfer strain to neighboring folds (Box 11.1), but the degree of cylindricity varies from fold to fold. Hence, a portion of a fold may appear cylindrical as observed at outcrop (Figure 11.5), even though some curvature of the axis must exist on a larger scale.

Cylindricity has important implications that can be taken advantage of. The most important one is that the poles to a cylindrically folded layer define a great circle, and the pole (π-axis) to that great circle defines the fold axis (Figure 11.6a). When great circles are plotted instead of poles, the great circles to a cylindrically folded layer will cross at a common point representing the fold axis, in this case referred to as the β-axis (Figure 11.6b). This method can be very useful when mapping folded layers in the field, but it also works for other cylindrical structures, such as corrugated fault surfaces.

Another convenient property of cylindrical folds is that they can be projected linearly, for example from the surface to a profile. Cylindricity is therefore commonly assumed when projecting mapped structures in an area onto cross-sections, particularly in the early 1900 mapping of the Alps by Swiss geologists such as Emile Argand and Albert Heim. Since the validity of such projections relies on the actual cylindricity of the projected structures, the uncertainty increases with projection distance.

The **axial surface**, or **axial plane** when approximately planar, connects the hinge lines of two or more folded surfaces. The **axial trace** of a fold is the line of intersection

BOX 11.1 | FOLD OVERLAP STRUCTURES

Individual folds can overlap and interfere. Just like faults, they initiate as small structures and interact through the formation of overlap or relay structures. Fold overlap structures were first mapped in thrust and fold belts, particularly in the Canadian Rocky Mountains, and many of the fundamental principles of fault overlap structures come from the study of fold populations. Fold overlaps are zones where strain is transferred from one fold to another. Rapid changes in fold amplitude characterize fold overlap structures.

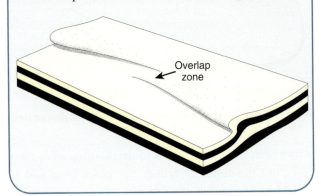

Figure 11.5 Cylindrically folded granitic dike in amphibolite.

of the axial surface with the surface of observation, typically the surface of an outcrop or a geologic section. The axial trace connects hinge points on this surface. Note that the axial surface does not necessarily bisect the limbs (Figure 11.2a). It is also possible to have two sets of axial surfaces developed, which is the case with so-called **box folds**, which are also called **conjugate folds** from the characteristic conjugate sets of axial surfaces (Figure 11.2d). In other cases, folds show axial surfaces with variable orientations, and such folds are called **polyclinal**.

The orientation of a fold is described by the orientation of its axial surface and hinge line. These two parameters can be plotted against each other, as done in Figure 11.7, and names have been assigned to different fold orientations. Commonly used terms are **upright** folds (vertical axial plane and horizontal hinge line) and **recumbent** folds (horizontal axial plane and hinge line).

Most of the folds shown in Figure 11.7 are **antiforms**. An antiform is a structure where the limbs dip down and away from the hinge zone, whereas a **synform** is the opposite, trough-like shape (Figure 11.8b, c). Where a stratigraphy is given, an antiform is called an **anticline** where the rock layers get younger away from the axial surface of the fold (Figures 11.8e and 1.6). Similarly, a

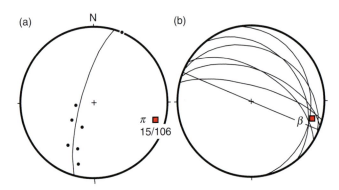

Figure 11.6 Measurements of bedding around a folded conglomerate layer. (a) Poles to bedding plot along a great circle. The pole to this great circle (π-axis $= 15/106$) represents the fold axis. (b) The same data plotted as great circles. For a perfectly cylindrical fold the great circles should intersect at the point (β-axis $= 15/106$) representing the fold axis. Data from the fold shown in Box 3.2.

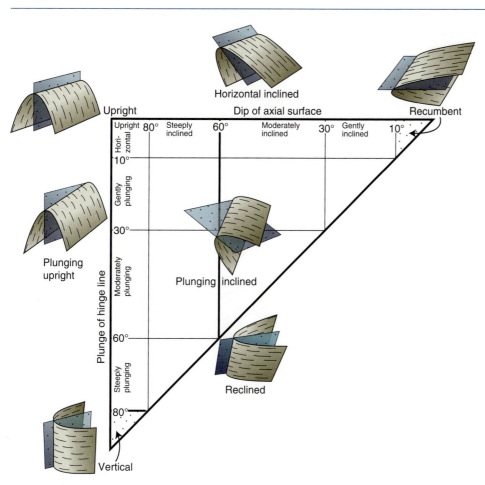

Figure 11.7 Classification of folds based on the orientation of the hinge line and the axial surface. Based on Fleuty (1964).

syncline is a trough-shaped fold where layers get younger toward the axial surface (Figure 11.8d). Returning to Figure 11.7, we can have upright or plunging synforms as well as antiforms. We can even have recumbent synclines and anticlines, because their definitions are related to stratigraphy and younging direction. However, the terms recumbent and vertical antiforms and synforms have no meaning.

Imagine a tight to isoclinal recumbent fold being refolded during a later tectonic phase. We now have a set of secondary synforms and antiforms. The younging direction across their respective axial surfaces will depend on whether we are on the inverted or upright limb of the recumbent fold, as shown in Figure 11.8h. We now need two new terms, synformal anticline and antiformal syncline, to separate the two cases (Figure 11.8f, g). A **synformal anticline** is an anticline because the strata get younger away from its axial surface. At the same time, it has the shape of a synform, i.e. it is synformal. Similarly, an **antiformal syncline** is a syncline because of the stratigraphic younging direction, but it has the shape of an antiform. Technically, a synformal anticline is the same as an anticline turned upside down, and an antiformal syncline looks like an inverted syncline. Confused?

Remember that these terms only apply when mapping in polyfolded stratigraphic layers, typically in orogenic belts.

As already stated, most folds are non-cylindrical to some extent. A non-cylindrical upright antiform is sometimes said to be **doubly plunging**. Large doubly plunging antiforms can form attractive traps of oil and gas – in fact they form some of the world's largest hydrocarbon traps. When the non-cylindricity is pronounced, the antiform turns into a **dome**, which is similar to a cereal bowl turned upside-down (or Yosemite's Half Dome made whole). Domes are classic hydrocarbon traps, for example above salt structures, and geoscientists commonly talk about such traps as having a **four-way dip closure**. Correspondingly, a strongly non-cylindrical synform is in fold terminology called a **basin** (the cereal bowl right-way up).

A **monoclinal fold** is a sub-cylindrical fold with only one inclined limb (Figure 11.8a). Monoclinal folds (or just monoclines for short) are commonly found as map-scale structures related to reactivation of, or differential compaction across, underlying faults or salt structures (Figure 1.6).

In addition to orientation and stratigraphic relations, folds are commonly described or classified according to **tightness**. Tightness is characterized by the opening or

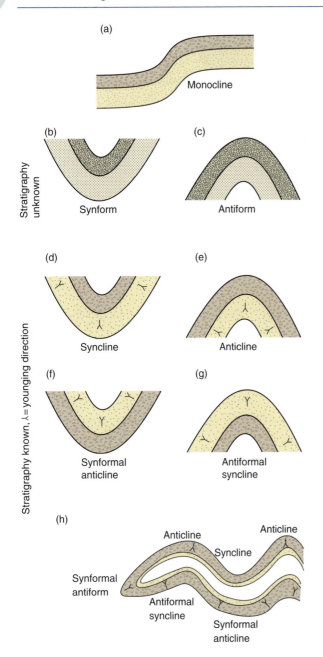

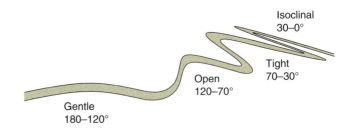

Figure 11.9 Fold classification based on interlimb angle.

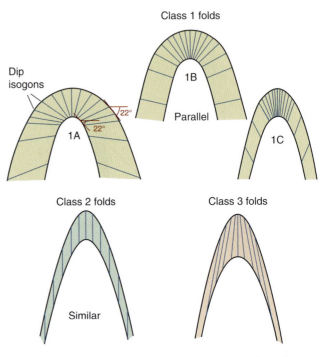

Figure 11.10 Ramsay's (1967) classification based on dip isogons. Dip isogons are lines connecting points of identical dip for vertically oriented folds.

Figure 11.8 Basic fold shapes. The bottom figure illustrates how various types of syn- and antiforms may occur in a refolded fold.

interlimb angle, which is the angle enclosed by its two limbs. Based on this angle, folds are separated into gentle, open, tight and isoclinal (Figure 11.9). Tightness generally reflects the amount of strain involved during the folding.

Folds usually come in groups or systems, and although folds may be quite non-systematic, neighboring folds tend to show a common style, especially where they occur in rows or trains. In these cases they can, akin to mathematical functions, be described in terms of wavelength, amplitude, inflection point and a reference surface called the **enveloping surface**. The enveloping surface is the surface tangent to individual hinges along a folded

layer, as shown in Figure 11.1. Note that the enveloping surface does not generally connect the hinge lines, although it does so for symmetric folds.

Dip isogons

Some folds have layers that maintain their thickness through the fold, while others show thickened limbs or hinges. These, and related features, were explored by the British geologist John Ramsay, who classified folds geometrically by means of **dip isogons**. By orienting the fold so that its axial trace becomes vertical, lines or dip isogons can be drawn between points of equal dip on the outer and inner boundaries of a folded layer. Dip isogons portray the difference between the two boundaries and thus the changes in layer thickness. Based on dip isogons, folds can be classified into the three main types shown in Figure 11.10:

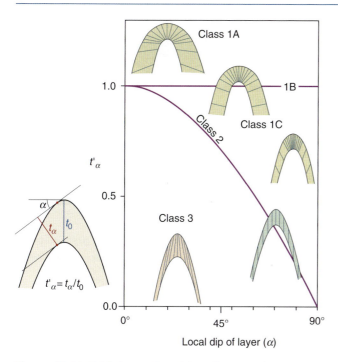

Figure 11.11 Fold classes plotted in a diagram where normalized layer thickness is plotted against dip of the folded surface. t'_α is the local layer thickness divided by the layer thickness in the limb, according to Ramsay (1967).

Class 1: Dip isogons converge toward the inner arc, which is tighter than the outer arc.

Class 2 (**similar folds**, also called **shear folds**): Dip isogons parallel the axial trace. The shapes of the inner and outer arcs are identical.

Class 3: Dip isogons diverge toward the inner arc, which is more open than the outer arc.

Class 1 folds are further subdivided into classes 1A, 1B and 1C. 1A folds are characterized by thinned hinge zones, while 1B folds, also called **parallel folds** and, if circle-shaped, **concentric folds** (Figure 11.2c), have constant layer thickness. Class 1C folds have slightly thinned limbs. Class 2 and, particularly, Class 3 folds have even thinner limbs and more thickened hinges. Among these classes, Class 1B (parallel) and 2 (similar) geometries stand out because they are easy to construct and easy to identify in the field.

One way of plotting folds according to the dip isogon classification is shown in Figure 11.11, where folds are considered as upright structures (vertical axial planes), so that the dip of the limb (α) increases in each direction from 0° at the hinge point. The vertical parameter in this figure, t'_α, is the normalized version of the orthogonal thickness, which is denoted t_α in Figure 11.11. This is the thickness measured orthogonal to the layer at one of the two corresponding points of equal dip on each arc (red points in Figure 11.11). For a Class 1B fold $t'_\alpha = t_\alpha$

regardless of the location on the folded layer, and each fold limb will plot along the horizontal $t'_\alpha = 1$ line. Hence, plotting measurements from a single folded layer will give a series of points that define two lines (one for each side of the hinge point) in Figure 11.11.

Symmetry and order

Folds can be symmetric or asymmetric in cross-section. A fold is perfectly symmetric if, when looking at a cross-section perpendicular to the axial surface, the two sides of the axial trace are mirror images of one another. This implies that the two limbs are of equal length. The chevron folds and concentric folds shown in Figure 11.2 are examples of symmetric folds.

If we extend this concept to three dimensions, the axial plane becomes a mirror plane, and the most symmetric folds that we can think of have two other mirror planes perpendicular to the axial plane. This is the requirement of **orthorhombic** symmetry. For symmetric folds the bisecting surface coincides with the axial plane. Hence, the kink band shown in Figure 11.2a is not symmetric. In fact, this is how we distinguish between kink bands and chevron folds: chevron folds are symmetric while kink bands have one long and one short limb. This leaves us with one symmetry (mirror) plane only, the one perpendicular to the axial surface, and the symmetry is said to be **monoclinic**.

Symmetric folds are sometimes called **M-folds**, while asymmetric folds are referred to as **S-folds** and **Z-folds**, as shown in Figure 11.12. Distinguishing between S- and Z-folds may be confusing to some of us, but Z-folds have short limbs that appear to have been rotated clockwise with respect to their long limbs. Z-folds thus mimic the letter Z when considering the short limb and its two adjacent long limbs. S-folds imply a counter-clockwise rotation, and resemble the letter S (this has nothing to do with the difference in angularity between S and Z). Interestingly, S-folds become Z-folds when viewed from the opposite direction. Plunging folds are usually evaluated when looking down-plunge, while viewing direction must be specified for folds with horizontal axes.

Fold systems consisting of folds with a consistent asymmetry are said to have a **vergence**. The vergence can be specified, and the vergence direction is given by the sense of displacement of the upper limb relative to the lower one (Figure 11.13). We can also relate it to the clockwise rotation of the inclined short limb in Figure 11.13, where a clockwise rotation implies a right-directed vergence.

Fold vergence is important in structural analysis in several ways. Large folds tend to have smaller folds occurring

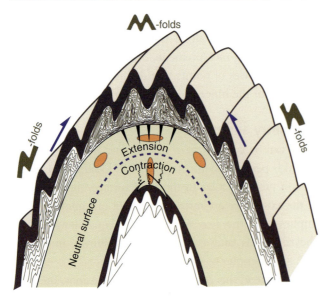

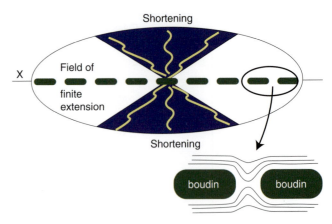

Figure 11.14 Fold vergence in relation to the strain ellipsoid for coaxial deformation, Note that folds can also occur between boudins in the field of finite extension.

Figure 11.12 Z-, M- and S-folds may be related to lower-order folds, in which case they provide information about the geometry of the large-scale fold.

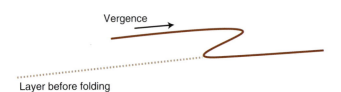

Figure 11.13 The concept of fold vergence. This fold is right-verging and a Z-fold according to the clockwise rotation of the short limb.

in their limbs and hinge zones, as shown in Figure 11.12. The largest folds are called the first-order folds, while smaller associated folds are second- and higher-order folds. The latter are also called **parasitic folds**. First-order folds can be of any size, but where they are map scale we are likely to observe only second- or higher-order folds in outcrops. If a fold system represents parasitic (second-order) folds on a first-order synformal or antiformal structure, then their asymmetry or vergence indicates their position on the large-scale structure. As shown in Figure 11.12, parasitic folds have a vergence directed toward the hinge zone. This relationship between parasitic and lower-order folds can be extremely useful for mapping out fold structures that are too large to be observed in individual outcrops.

The vergence of asymmetric fold trains in shear zones is generally unrelated to lower-order folds and can give information about the sense of shear of the zone. Such kinematic analysis requires that the section of observation contains the shear vector and should be used together with independent kinematic indicators (see Chapter 15).

The (a)symmetry of folds may also reflect strain and the orientation of the strain ellipse. In general, layers that are parallel to ISA$_3$ (the fastest shortening direction, see Chapter 2) will develop symmetric folds. For coaxial strain this is straightforward (Figure 11.14), but for simple shear and other non-coaxial deformations things get somewhat more complicated, since layers rotate through the position of ISA$_3$ during the deformation.

Fold asymmetry may relate to position on a lower-order fold, sense of shear or orientation of the folded layer relative to the strain ellipse.

11.2 Folding: mechanisms and processes

Every geologist mapping or describing folds in the field probably has the same question in mind: how did these structures actually form? As geologists we tend to look for a simple history or mechanism that can explain our observations reasonably well. Folding is no exception, and there are different approaches and process-related terms. One approach is to consider the way force or stress acts on a layered rock, which leads to the three-fold classification and terminology shown in Figure 11.15. Other terms are related to how the layer(s) react to force and stress, for instance whether layers fold by layer-parallel shearing, orthogonal flexure or some other mechanism that is controlled by rock rheology. Still other classes of folding, such as kinking and chevron folding, are related to fold geometry. For this reason, several different fold mechanisms are defined, and many of them overlap in definition. This is why terms such as buckling, kink

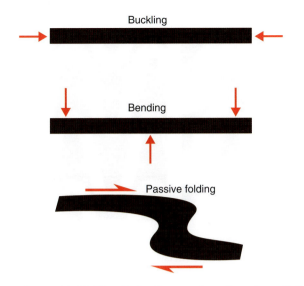

Figure 11.15 The relation between how force is applied and fold mechanisms.

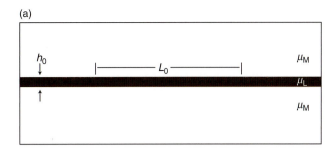

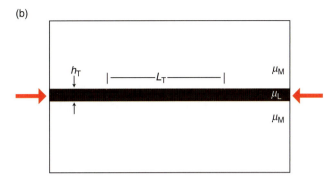

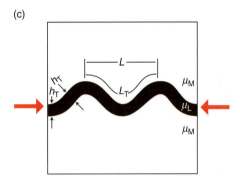

Figure 11.16 Buckling of a single layer. L_0 is the original length that is changed into L_T after initial shortening (a, b) while μ_L and μ_M are layer and matrix viscosities, respectively. h_0 is the original layer thickness (a), which increases to h_T during the initial thickening phase (b). L is the wavelength while L_T is the arc length. Based on Hudleston (1986).

folding and bending can be confusing when discussed in terms of mechanisms such as flexural slip and simple shear. In summary, we are dealing with differences in orientation of stress axes relative to the layering, kinematics, and mechanical and rheological properties, and thus mechanisms that emphasize different aspects of folding.

The most important distinction between the ways folds form probably lies in whether the layering responds actively or passively to the imposed strain field. We will start out by considering **active folding** (buckling), where the competence or viscosity contrast between the folding layer and its host rock is important. We will then look at **passive folding**, where layers are simply passive markers with no rheological influence, and then consider **bending**, where forces are applied across the layering (Figure 11.15). The following sections will then discuss models known as **flexural folding** mechanisms (flexural slip, flexural shear and orthogonal flexure), which can contribute to both active folding and bending. Finally, we will discuss **kinking** and the formation of **chevron folds**.

Active folding or buckling (Class 1B folds)

Active folding or **buckling** is a fold process that can initiate when a layer is shortened parallel to the layering, as shown schematically in Figure 11.16. Folds such as the ones seen in Figure 11.17 appear to have formed in response to layer-parallel shortening. A contrast in viscosity is required for buckling to occur, with the folding layer being more competent than the host rock (matrix). The result of buckling is rounded folds, typically parallel and with more or less sinusoidal shape.

Figure 11.17 Two folded layers of different thickness. The upper and thinner one shows a smaller dominant wavelength than the lower one.

Buckling occurs when a competent layer in a less competent matrix is shortened parallel to the length of the layer.

If an isotropic rock layer has perfectly planar and parallel boundaries and is perfectly parallel with a constantly oriented σ_1 or ISA_1, then it will shorten without folding even though there is a significant viscosity contrast between the layer and the host rock. However, if there are small irregularities on the layer interfaces, then these irregularities can grow to form buckle folds with a size and shape that depend on the thickness of the folded layer and its viscosity contrast with its surroundings.

Buckling or active folding implies that there is layer parallel shortening and a viscosity contrast involved, and also irregularities on which folds can nucleate.

Buckling of single, competent layers in a less competent matrix (Figure 11.16) is relatively easy to study in the laboratory and has also been explored numerically. Single-layer folds formed by buckling have the following characteristics:

- The fold wavelength–thickness ratio (L/h) is constant for each folded layer if the material is mechanically homogeneous and if they were deformed under the same physical conditions. Such folds are often called **periodic folds**. If the layer thickness varies, then the wavelength is changed accordingly (Figure 11.17).

- The effect of the folding disappears rapidly (about the distance corresponding to one wavelength) away from the folded layer.

- The folds in the competent layer approximate Class 1B folds (constant layer thickness). If there are two or more folded competent layers then the incompetent layers in between are folded into Class 1A and Class 3 folds (Figure 11.18). Cusp (pointed) hinges point to the more competent layers.

- The outer part of the competent layer is stretched while the inner part is shortened. The two parts are typically separated by a **neutral surface** (Figure 11.19). Note that layer-parallel shortening, which always takes place prior to folding, can reduce or eliminate the outer extensional zone.

- The normal to the axial surface or axial cleavage indicates the direction of maximum shortening (Z).

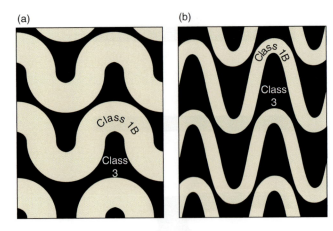

Figure 11.18 Alternating Class 1B and 3 folds are commonly seen in folded layers. Competent layers exhibit Class 1B geometry.

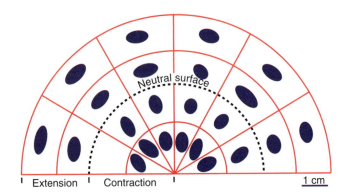

Figure 11.19 Strain distribution in the hinge zone of a folded limestone layer in shale. Outer-arc stretching is separated from inner-arc shortening by a neutral surface. From Hudleston and Holst (1984).

If the layers are Newtonian viscous, and disregarding any layer-parallel shortening, then the relation between wavelength and thickness is given by

$$L_d/h = 2\pi(\mu_L/6\mu_M)^{1/3} \tag{11.2}$$

μ_L and μ_M are the viscosities of the competent layer and the matrix, respectively, while L_d is the dominant wavelength and h the layer thickness. Experiments and theory show that homogeneous shortening (T) occurs initially, together with the growth of irregularities into very gentle and long-amplitude fold structures. When the most accentuated folds achieve opening angles around 160–150°, the role of layer-parallel shortening decays. From that point on the folds grow without any significant increase in layer thickness. Equation 11.2 can be expanded to include layer-parallel thickening:

$$L_{dT}/h_T = 2\pi(\mu_L/6\mu_M(T+1)T^2)^{1/3} \tag{11.3}$$

L_{dT} is here the revised expression of the dominant wavelength, while h_T is the thickness when layer-parallel shortening (thickening) is taken into account. The factor T is identical to the strain ratio X/Z, or $(1 + e_1)/(1 + e_3)$.

The viscosity contrast $\mu_L/6\mu_M$ can be estimated (formulas not shown here) by measuring the average length of the folded layer over one wavelength and h_T for a fold population. In addition, the layer-parallel shortening T in the competent layers must be estimated.

Buckling has been modeled under the assumption of linear or Newtonian viscosity (Equation 6.23). It is likely that most rocks show non-linear rheological behavior during plastic deformation, which has consequences for the buckling process. A power-law rheology is then assumed (Equation 6.24), where the exponent $n > 1$. The higher the n-exponent, the quicker the fold growth and the less the layer parallel shortening T. Many natural folds show low T-values, and, together with low L/h ratios ($L/h < 10$), this indicates a non-linear rheology. However, the differences between the results from viscous and power-law rheology models are not great.

Buckle folds are most easily recognized as single competent layers, but can also occur where several competent layers occur in parallel. L_d/h is significantly less for multilayer than for single layer buckling. Where two thin layers are close they will behave more like a single layer whose thickness is the sum of the two thin layers, as seen from the experimental results shown in Figure 11.20. Where we have alternating thick and thin layers, the thin layers will start to develop folds first (Figure 11.21a, b). At some point the thick layers will start to fold (with longer wavelength) and take control over the further development. The result is relatively large folds controlled by

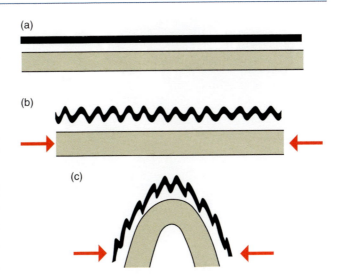

Figure 11.21 Illustration of how folding initiates in thin layers. Once the thicker layer starts to fold, the smaller folds in the thin layer become parasitic and asymmetric due to flexural flow.

thick layers together with small, second-order folds formed earlier in the process (Figure 11.21c). An example of multilayer folding where the wavelength is controlled by a package of layers is shown in Figure 11.22.

Several mechanisms can be involved during buckling. The simplest ones can collectively be termed **flexural folding** and are separated into orthogonal flexure, flexural slip and flexural flow. In addition there is always the possibility of having volume change, particularly in the hinge zone. We will briefly review these idealized models after a look at two other models of folding known as passive folding and bending.

Passive folding (Class 2 folds)

Passive folding is typical for rocks where passive flow occurs, i.e. where the layering exerts no mechanical influence on the folding. In these cases the layering only serves as a visual expression of strain with no mechanical or competence contrast to neighboring layers. Such layers are called **passive layers**. Perfectly passive folds produced by simple shear are Class 2 (similar) folds, and passive folds that are associated with simple shear, or at least a significant component of simple shear, are called **shear folds** (Figure 11.23a).

Passive folds generated by simple shearing are perfectly similar folds.

Passive folds of perfect Class 2 geometry can easily be generated by differentially shearing a card deck. Drawing lines perpendicular to the cards prior to shearing helps visualize the fold. However, the formation of passive folds is not restricted to simple shear. Passive folds can form in

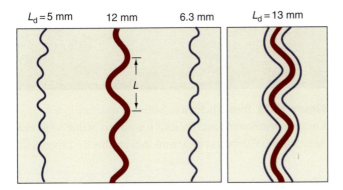

$L_d = 5$ mm   12 mm   6.3 mm $L_d = 13$ mm

L

Figure 11.20 Folding of multilayered rocks. Far-apart layers act as individual layers (left). The closer they get, the more they behave as a single layer with thickness larger than that of the thickest of the individual layers. Based on experiments by Currie *et al.* (1962).

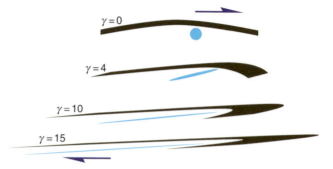

(a) Simple shear passive folding

$\gamma = 0$

$\gamma = 4$

$\gamma = 10$

$\gamma = 15$

(b) Pure shear passive folding

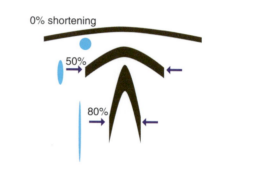

0% shortening

50%

80%

Figure 11.23 Formation of Class 2 folds by (a) simple shearing and (b) pure shearing of a gently curved layer. No viscosity contrast is involved, meaning that the folds can be regarded as passive.

Figure 11.22 Buckled multilayers. Note how the largest folds affect the entire layer package.

response to any kind of ductile strain, for instance subsimple shear, transpression (Chapter 18) and even coaxial strain (Figure 11.23b). Hence simple shear is only one of an infinite spectrum of kinematic models that can produce passive folds.

Passive folding produces harmonic folds where the layering plays no mechanical role and therefore no influence on the fold shape.

Examples of passive folding are found where passive layers enter shear zones or otherwise are affected by heterogeneous strain. Drag folds along faults (Chapter 8) are examples typical for the brittle regime, although many layered sequences contain beds of quite different mechanical properties so that slip occurs between layers (see flexural slip below). Passive folds are frequently found in mylonite zones, particularly in monomineralic rocks such as quartzite (Figure 11.24), marble and salt.

Figure 11.24 Passive harmonic folding of quartzite in a Caledonian mylonite zone. The similar geometry of this Z-fold and its setting in a Caledonian shear zone indicate that it is a shear fold.

Bending

Bending occurs when forces act across layers at a high angle (Figure 11.25), unlike buckle folds where the main force acts parallel to a layer. This is also the case for passive folding, and the two are closely related. However, bending

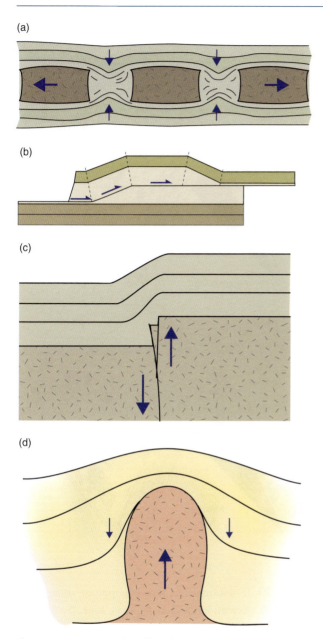

Figure 11.25 Examples of bending in various settings and scales: (a) between boudins; (b) above thrust ramps; (c) above reactivated faults; and (d) above shallow intrusions or salt diapirs.

is generally thought of as something that is more directly forced upon the layers by geometries and kinematics of the bounding rock units. Several aspects of bending have been studied in great detail by engineers because of its importance in the field of construction engineering, such as in horizontal beams supported by vertical pillars.

Bending occurs when forces act across layers, and may involve more than one mechanism.

Classic geologic results of bending are the **forced folds** created in sedimentary layers blanketing faulted rigid

basement blocks (Figure 11.25c). The displacement is forced upon the sediments by fault movement on a preexisting basement fault, and the sediments are soft enough to respond by monoclinal folding until at some critical point they rupture and the fault starts propagating up-section. Such structures are particularly well exposed in the Colorado Plateau–Rocky Mountains area, where numerous Laramide-related uplifts have created such structures.

Bending as such is a boundary condition- or external load-related model, not a strain model, particularly not when a free surface is involved such as during forced folding mentioned above (Figure 11.25c). In other words, there are many ways that folding and strain can accumulate internally in a fold during bending.

An obvious response to bending is deformation by simple shear, in which case we are back to passive folding. The simple shear passive folding model may work if we have a wide fault zone underneath the fold or if the fold is very narrow. In most cases the fold widens upward, telling us that we have to modify the simple shear model. In this case trishear comes in handy. Trishear distributes shear in a triangular zone ahead of a propagating fold, and seems to work very well for several mapped examples.

Still, trishear cannot explain all features seen in many forced folds. Field studies show evidence of bedding-parallel slip or shear. This is manifested by striations on weak bedding-parallel surfaces or by bedding-parallel deformation bands. We will discuss this mechanism below as flexural slip. Also the related flexural mechanism described below as orthogonal flexure can result from bending loads.

There are many other examples of bending. One is **fault-bend folds**, for instance where thrust sheets are passively bent as they move over a ramp structure (Figure 11.25b) (see Chapter 16). Such folds are commonly modeled as kink folds, again related to flexural slip. They may also be modeled by means of simple shear, which is commonly done for fault-bend folds formed above non-planar (e.g. listric) faults (see Chapter 20).

Differential compaction, where a sedimentary sequence compacts more in one area than in another due to different degrees of compaction of the underlying layers, is also a type of bending. This is common across the crests of major fault blocks in postrift-sequences in sedimentary basins, but can also occur along salt diapirs (19.5a) and shallow intrusions. Folds formed by differential compaction are gentle.

Forceful intrusion of magma or salt can also bend roof layers, as shown in 19.21. Again the strain accumulation mechanism may vary, with flexural slip being a common constituent.

Figure 11.26 Passive folding of layers between boudins.

In the plastic regime, bending is less common because of the high ductility of all or most parts of the deforming rocks. However, bending is frequently associated with rigid **boudins** (Figure 11.25a and 11.26).

Flexural slip and flexural flow (Class 1B)

Flexural slip implies slip along layer interfaces or very thin layers during folding (Figure 11.27). It is one of three kinematic models of folding (the others being flexural flow and orthogonal flexure) that maintains bed thickness and thus produces Class 1B or parallel folds. Simple flexural slip experiments can be performed simply by folding double sandwiches with jelly. The sandwich maintains its thickness even though slip occurs between the pieces of bread, until the fold becomes too tight. It is a prerequisite for flexural slip that the deforming medium is layered or has a strong mechanical anisotropy.

In nature, the anisotropy could be mica-rich thin layers in a quartzite or mylonite, or thin shale layers between thicker sandstone or limestone beds in sedimentary rocks. Flexural slip can occur in the middle crust where plastic deformation mechanisms would be involved, but is perhaps more common where sedimentary strata are folded in the upper crustal brittle regime. In the latter case, bedding surfaces act like faults, and slickenlines (red lines in Figure 11.27a) will sometimes develop on slipping surfaces.

Maximum slip occurs at the inflection points and dies out toward the hinge line, where it is zero. The sense of slip is opposite on each limb, and the slip is consistent relative to the hinges, where sense of slip changes. Relative slip on the convex side of a flexural slip fold is always toward the fold hinge, whereas on the concave side slip is opposite.

Slickenlines on folded weak layers and constant bed thickness reveal flexural slip.

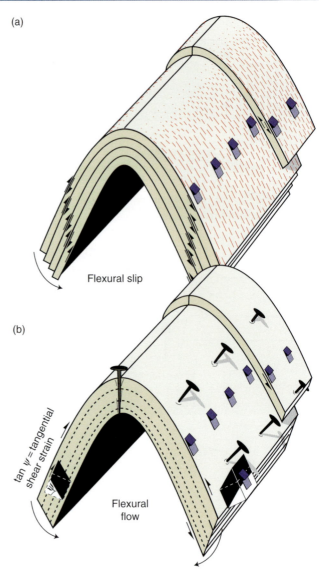

Figure 11.27 (a) Flexural slip, showing opposite sense of slip on each limb, decreasing towards the hinge zone. (b) Flexural flow, where fold limbs are being sheared. Ideally, layer thickness is preserved in both models.

In cases where strain is more evenly distributed in the limbs in the form of shear strain, as is more commonly the case in the plastic regime, flexural slip turns into the closely related mechanism called **flexural shear** or **flexural flow**. Flexural flow experiments are conveniently done by bending a soft paperback book or a deck of cards (remember to draw circles for strain markers). During this process slip occurs between individual paper sheets. If we put strain markers on our paperback, we would see that strain is zero in the hinge zone and increasing down the limbs. This is so because the shear strain is directly related to the orientation (rotation) of the layers, as shown in Figure 11.27b: the higher the rotation, the higher the shear strain.

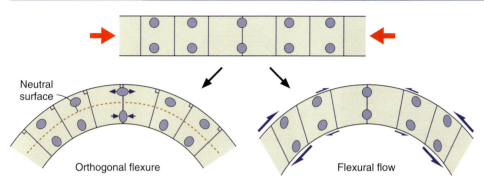

Figure 11.28 Layer-parallel shortening resulting in orthogonal flexure and flexural flow. Note what happens to the originally orthogonal lines. Strain ellipses are indicated.

For originally horizontal layers folded into an upright fold, shear strain is directly related to dip ($\gamma = \tan($layer dip$)$), and the sense of shear is opposite on each side of the axial trace (Figure 11.27b). This results in a characteristic strain distribution in the fold. For example, the neutral surface separating extension from contraction, typical for many buckle folds, is not found in pure flexural-flow folds. Flexural flow produces identical strain in the inner and outer part of a fold, but strain increases away from the hinge. Note that evidence for a combination of orthogonal flexuring (see below) and some flexural flow or slip is commonly found in buckle folds, in which case a neutral surface may well exist.

Pure flexural folds have no neutral surface, and strain increases away from the hinge zone.

Pure flexural folds are perfect Class 1B folds. We can estimate the amount of layer-parallel shortening for such folds by measuring the length of any one of the folded layers. This layer has maintained its original length because it was the shear plane throughout the folding history. Constant layer length and thickness are assumptions that simplify restoration of cross-sections (Chapter 20).

Orthogonal flexure (also Class 1B)

Orthogonal flexure, also called tangential longitudinal strain, is a deformation type with its own specific conditions:

All lines originally orthogonal to the layering remain so throughout the deformation history.

The result is stretching of the outer part and shortening of the inner part of the folded layer. The long axis of the strain ellipse is therefore orthogonal to bedding in the inner part of the layer and parallel to bedding in the outer part, as shown in the folded limestone layer in Figure 11.19.

Figure 11.28 shows a comparison between flexural flow and orthogonal flexure. Orthogonal flexure and flexural flow have in common that they produce parallel (Class 1B) folds. But the two models produce quite different strain patterns: The neutral (no strain) surface separating the outer-extended and inner-contracted part of the folded layer does not exist in flexural flow, where strain is identical across the fold along dip isogons. During the folding history, the neutral surface moves inward toward the core of the fold, which can result in contraction structures overprinted by extension structures.

Orthogonal flexure produces parallel folds with a neutral surface.

Pure orthogonal flexure is only possible for open folds. When folds get tighter, the conditions for orthogonal flexure become harder and harder to maintain, and flexural slip or flow will gradually take over. Evidence for orthogonal flexure is typically found in stiff, competent layers that resist ductile deformation. Some have simplified the definition of orthogonal flexure to a mechanism resulting in outer-arc contraction and inner-arc extension. By getting rid of the requirement of orthogonality, the model becomes more general and embraces many more natural examples.

Kinking and chevron folding

Kink bands are common in well-laminated and anisotropic rocks rich in phyllosilicate minerals, and some field occurrences are shown in Figure 11.29. Kink bands are centimeter- to decimeter-wide zones or bands with sharp boundaries across which the foliation is abruptly rotated. Wider zones are sometimes referred to as kink folds. Kink bands and kink folds are characterized by their strong asymmetry and their Class 2 fold geometry. They are closely related to **chevron folds**, which also are Class 2 folds, but differ in terms of symmetry. Both are relatively low-temperature (low metamorphic

Figure 11.29 (a) Conjugate kink bands in mylonitized anorthosite gabbro, Bergen, Norway. (b) Kink folds related to Laramide thrusting in north Wyoming (Dead Indian Summit) (c) Kink-like folds in oceanic sediments in Oman.

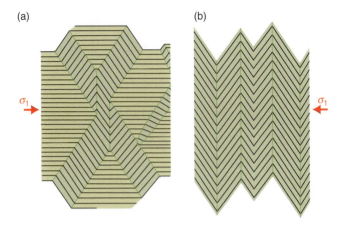

Figure 11.30 (a) The orientation of σ_1 can be determined from the orientation of conjugate sets of kink bands. (b) Continued kink band growth can produce chevron folds.

grade) deformation structures where there is a significant mechanical anisotropy represented by lamination or repeated competent–incompetent layers, and both imply layer shortening.

Classic kink bands have very angular hinges and lack even the narrow hinge zone found in the outer arc of chevron folds. There is another important difference between the two. While chevron folds initiate with their axial surface perpendicular to the shortening direction, kink bands form oblique to this direction, typically in conjugate pairs.

When conjugate sets of low-strain kink bands are observed, such as the examples portrayed in Figure 11.29a, b, σ_1 or ISA_1 is commonly assumed to bisect the sets, as shown in Figure 11.30. As stated before, going from strain to stress is not straightforward, but the smaller the strain the better the correlation. When a single set of kink bands occurs, we know that σ_1 is oblique to the band, but its precise orientation is unknown because kink bands may rotate during progressive deformation. In addition we still do not understand kink band formation in detail, and there seem to be several mechanisms that apply.

Kink folds generated by bending do not directly reveal the orientation of stress. Such kink folds have orientations that are controlled by the local geometries of ramps or fault bends. Hence, in such cases the bisecting axis between two kink zones does not in general represent σ_1 or ISA_1. See Chapters 16 and 20 for examples of such structures.

Experiments have shown that conjugate sets can nicely merge to form chevron folds if strain is high enough (around 50%) (Figure 11.30). However, 50% shortening is not commonly achieved by kinking in naturally deformed rocks, so this way of forming chevron folds may not be the most common one after all. Classic chevron folds with beds on the centimeter scale are more likely to form by flexural slip of multilayered rocks during layer-parallel shortening, as illustrated in Figure 11.31. The typical setting is where competent beds are separated by thin incompetent layers, for instance quartzite or chert separated by shale or phyllite. Flexural slip then occurs between the competent layers, which are strained only in the thin hinge zones. Just like buckle folds, the hinges have to stretch in the outer arc and shorten in their inner parts. Figure 11.32 shows an example of this, where extension veins have

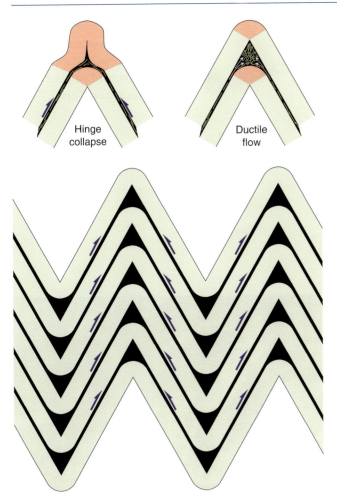

Figure 11.32 Extension fractures (veins) in the outer, stretched arc of folded competent layers. Chevron folds, Varanger, northern Norway.

Figure 11.31 Chevron folds forming by the flexural slip mechanism imply a space problem in the hinge zone that is resolved by ductile flow of the incompetent (dark) layers or collapse of the competent layers in the hinge zone. Strained parts of competent layers are marked in red, showing that layer thickness is maintained on the limbs.

formed in the outer arc and (less obvious) contractional structures dominate the inner arc. Furthermore, geometric problems in the hinge zone require flow of the incompetent rock into the hinge, or alternatively inward hinge collapse of the competent bed as seen in Figures 11.31 and 11.33. Hinge collapse is particularly common in relatively thick competent layers that occur between thinner ones. Another way of resolving hinge compatibility problems is by reverse faulting, as seen in Figure 11.33.

11.3 Fold interference patterns and refolded folds

In areas affected by two or more deformation phases, a secondary set of folds may be superimposed on earlier folds. Folds modified by a later fold phase are known as

Figure 11.33 Hinge compatibility problems solved by ductile hinge collapse and reverse faulting. Detail from Figure 16.23 (glacio-tectonic detachment folds). The height of the wall is about 13 m.

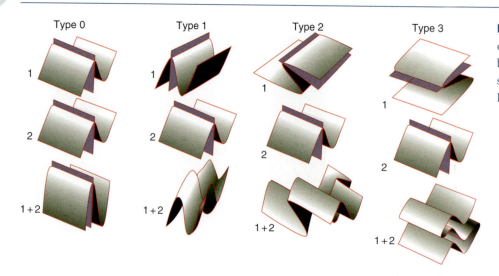

Figure 11.34 Principal types of fold superposition (1 + 2, bottom), formed by the superposition of system 2 on 1. Based on Ramsay (1967).

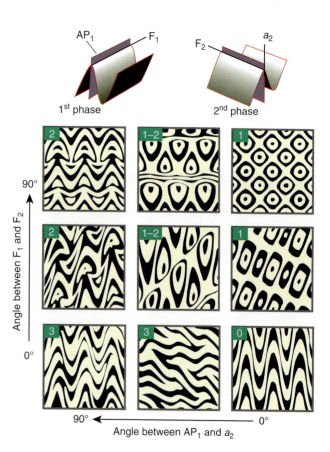

Figure 11.35 Fold interference patterns of cylindrical folds, classified according to the relative orientations of fold axes (vertical direction) and the angle between the first axial plane and the direction a_2 indicated in the upper part of the figure. The patterns are numbered 0–3, according to Ramsay (1967).

Figure 11.36 Type 1–2 interference pattern in folded quartz schist.

distinguished between four main patterns, based on the relative orientation of axial planes and fold axes (Figures 11.34 and 11.35). Type 1 is the classic dome-and-basin structure, Type 2 is the so-called boomerang type (Figure 11.36), and Type 3 has been described as the hook-shaped type. There is also a Type 0 pattern defined by two identical, but temporally separate fold systems. The result of Type 0 interference is simply a tighter fold structure.

The four patterns shown in Figure 11.34 represent end-members in a spectrum of possible interference patterns, as indicated in Figure 11.35. Note that their appearance in outcrops also depends on the section through the fold structures as well as the folding mechanism, although the patterns shown in Figure 11.35 are qualitatively useful for a range of orientations. In many cases it is possible to perform simple unfolding exercises to reconstruct the geometry of the first set of fold structures.

refolded folds and the resulting patterns are referred to as **fold interference patterns**. We may find simple or complex fold interference patterns, depending on the relative orientations of the two fold sets. John Ramsay

Interference patterns by definition arise from the overprinting of a second phase of deformation on an earlier set of deformation structures, and in most cases it is possible to determine their age relationship based on overprinting relations. A typical overprinting relation is one axial plane cleavage being crenulated by the other (and therefore later) one. The occurrence of such relationships, and fold interference patterns in general, is traditionally taken as evidence of polyphasal deformation. However, it is important to understand that some fold interference patterns can result from a single period of non-steady-state flow, where the orientation of the ISA locally or regionally changes during the course of the deformation, or where folds and foliations are rotated internally, for example in a shear zone, during deformation. In particular, Type 1 patterns can be the result of a single phase of heterogeneous non-coaxial deformation, or of amplification of preexisting irregularities, as shown in Figure 11.37. Extremely non-cylindrical folds form in high-strain shear zones or slump zones and are often called **sheath folds**. In general it is useful to be able to recognize the geometric relations between different fold phases by use of the patterns depicted in Figure 11.35.

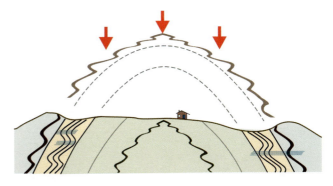

Figure 11.38 Christmas-tree folding due to vertical shortening after formation of the first-order antiform. A subhorizontal axial plane cleavage is typical. Compare to Figure 11.14

The patterns discussed above and in Figures 11.34 and 11.35 involve two deformations, each producing folds of comparable size. There are other cases where the amplitude and wavelength of the two fold sets are very different, and mapping is required to sort out their relationship. A case is shown in Figure 11.38, where secondary folds are small compared to the primary fold structures that they are superimposed on. Compare these folds with those shown in Figure 11.12, and you will see a significant difference in vergence, which calls for an alternative interpretation. How should we explain the picture shown in Figure 11.38? A simple explanation is that the large-scale fold was later affected by vertical shortening, initiating layer-parallel shear on each limb. This can happen if a gravitational collapse-type deformation occurs toward the end of or after orogenic and fold-forming deformation, in which case small asymmetric folds form that are unrelated to the large-scale fold in the way shown in Figure 11.12. Instead, their asymmetry is related to the orientation of the preexisting layering, as shown in Figure 11.38. The pattern resembles a Christmas tree, and one of my lecturers of the past (Donald Ramsay) called them **Christmas-tree folds**. Such folds are common in parts of the Caledonian orogen, which went through a late phase of gravitational-influenced collapse.

11.4 Folds in shear zones

In high-strain shear zones or mylonite zones, folds form and grow continuously during shearing (Figure 11.39). Folding can occur if layers initially lie in the contractional field or where layers are rotated into this field due to irregularities in the zone. Such folds can be regarded as passive if the competence contrast between layers was negligible during folding, or may have an active component if a viscosity contrast exists. Whether a

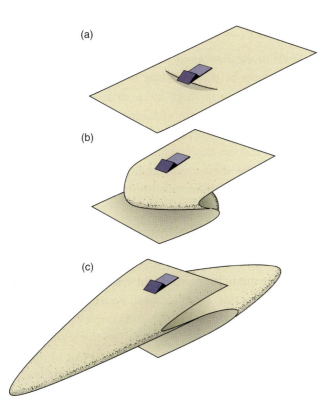

Figure 11.37 Development of sheath folds (highly non-cylindrical folds) by amplification of a preexisting irregularity. Note that it takes high shear strains to form sheath folds by simple shear. Shear strain increases from (a) to (c).

(a)

(b)

Figure 11.39 (a) Early-stage folds in Caledonian shear zone. Hinge lines make a high angle to the lineation and transport direction (arrow). (b) More mature stage of folding in the same shear zone. Hinge lines are highly curved and refolded, oriented both parallel, oblique and orthogonal to the transport direction (arrow).

fold is active or passive can be addressed by means of geometric analysis, since passive folds are Class 2 and active folds are Classes 1 and 3 (Figures 11.10 and 11.11).

At high strains the foliation in a shear zone will, in theory, be almost parallel to the shear plane. It will still be in the extensional field, but so close to the shear plane that just a modest perturbation of the layering can make it enter the contractional field. The result is a family of folds that verge in accordance with the sense of shear. This sort of rotation

can occur around tectonic lenses or heterogeneities (see Figure 15.13) or by a slight rotation of the ISA due to a change in the stress field or rotation of the shear zone.

If a fold hinge line lies in the shear plane it will remain in this position without any rotation. In general, however, hinge lines will initiate at an angle to the shear plane and rotate toward parallelism with the shear direction. Open folds with hinge lines that make a high angle to the transport direction (Figure 11.37a) are therefore thought to have experienced less shearing than those that are tighter with hinge lines closer to the transport direction (Figure 11.37c). This rotational behavior of fold hinges in non-coaxial flow has been used to explain why many high-strain shear zones contain folds with hinges subparallel to the lineation. Extensive rotation of hinges may result in sheath folds, where the hinge line is subparallel to the lineation for the most part, except for their noses where the hinge line is strongly curved (Figure 11.37c). Sections through sheath folds resemble Type 1 fold interference patterns, but are formed during a single event of shearing (although not necessarily during a perfect simple shearing history).

> Fold hinges can bend into very non-cylindrical folds during a single event of shearing, forming geometries similar to Type 1 interference patterns.

Folds in shear zones can also form in other ways. The foliation may be perpendicular to the shear zone walls, which is the case in many steep strike-slip shear zones (Chapter 18). Folds form with hinge lines oblique to the zone. The obliquity depends on the vorticity number of the zone and on the exact orientation of the layering relative to the shear zone. Examples of such folds are found along the San Andreas Fault in California.

11.5 Folding at shallow crustal depths

While many or most of the folds discussed above and seen in nature form in the plastic regime, folds also form in the brittle upper crust. Folds can form even at or close to the surface, as shown in Box 11.2. Examples are gravity-controlled soft-sediment deformation, for example on unstable delta slopes or continental slopes, or in the toe zone of landslides. Folds formed in such environments are commonly found to be strongly non-cylindrical.

Fluidization or liquefaction of sediments shortly after burial can create contorted beds and folds related to **mud diapirs** and **sand intrusions**. The triggering mechanisms may be gravity loading, dipping layering, earthquakes and

BOX 11.2 | DUCTILE FOLDING OF SEDIMENTS

Most folds are the result of plastic flow in the middle and lower crust, but folding can also occur at shallower depth. Folds may form where gravitational instabilities or dewatering cause deformation of unconsolidated sediments. Folding of buried porous sediments occurs easily if the pore fluid pressure becomes abnormally high (overpressure). In extreme cases the sediment is liquefied and acts more like a fluid until the fluid pressure normalizes. Folds in sediments are often found to be very non-cylindrical, somewhat similar to non-cylindrical folds found in some mylonite zones. They also show a range in fold style, from gentle to isoclinal and upright to recumbent.

Soft sediment folds have several characteristics. One is that they generally lack the axial planar cleavage so commonly associated with folds formed under metamorphic conditions. The upper picture shows a tight fold formed shortly after deposition of the sandstone, but no evidence of cleavage is visible. Another characteristic is that soft sediment folds tend to be confined to distinct stratigraphic levels where the layers above and below are untouched by the folding, as seen in the lower picture, where horizontal beds occur a few meters above the fold. Also, sharp discontinuities (fractures) and cataclastic deformation are absent.

Upper: Folded cross-sets in Neoproterozoic fluvial sandstone, Finnmark, Norway, formed prior to consolidation. Lower: Folded sandstone–shale interface in the Dewey Bridge Member near Moab, Utah, overlain by completely planar layers at the base of the Slick Rock Member.

even meteor impact events. On a larger scale, salt diapirs bend and fold layers as they move toward the surface. Tectonic folding at shallow levels includes the formation of **fault propagation folds** ahead of propagating fault tips (see Chapter 16), for instance in accretionary prisms. Thrust- and salt-related shallow fold structures are normally affected by erosion and show variations in layer thickness, hence their nature and timing can commonly be explored from the sedimentary record.

Deflections of layers along a fault, known as **drag** and **smear,** are other examples of fault-related folds that can form at very shallow depths. Hanging-wall folds or **rollovers** controlled by fault geometry are yet other examples (Chapters 8, and 20).

Summary

Simple models of folding, such as flexural slip and simple shear, are attractive and quite useful at times. Nevertheless, we should be aware of the possibility that natural folds have gone through a growth history where different mechanisms have been active at the same time or during different parts of the deformation history. We can use strain distribution and fold geometry to try to discover the dominant mechanism, and from field observations and experimental results we can to some extent predict both mechanism and geometry if we have knowledge of such factors as competence contrasts, layer thickness, mineralogy and anisotropy. Use the following summary points and review questions to test your knowledge of folds and folding:

- Folds form and occur at all levels in the crust.

- Buckle folds and chevron folds imply layer-parallel shortening.

- Shear folds involve an increase in layer length and not necessarily shortening perpendicular to the axial surface.

- Thick layers produce longer-wavelength folds than thin layers.

- No fold is perfectly cylindrical, and the most non-cylindrical folds form in shear zones.

- The asymmetry of parasitic folds indicates the geometry of the higher-order fold that they are related to.

- Small asymmetric folds do not have to be related to higher-order structures, for example in shear zones, where they may indicate sense of shear.

- Folds forming in soft sediments tend to lack axial plane cleavage and are confined to certain stratigraphic levels.

Review questions

1. Make a sketch map of a dome, a basin, an upright synform and a plunging upright synformal anticline.

2. How could dome-and-basin patterns form?

3. How can we distinguish between flexural flow and orthogonal flexure?

4. How would we identify a buckle fold?

5. A shear fold?

6. What is the difference between a parallel and a similar fold?

7. What fold types are parallel?

8. Where can we expect to find similar folds?

9. In what settings do monoclines typically form?

10. If we compress a thin layer and a thick layer so that they start buckling, which layer do you think starts to buckle first?

11. Which one do you think thickens the most prior to folding?

12. Which of the two layers would form the largest folds?

13. What conditions give more or less concentric and parallel folds?

14. How can dip isogons help us distinguish between buckle folds and shear folds?

15. Why do we get asymmetric folds on the limbs of lower-order folds?

E-MODULE

 The e-learning module called *Folding* is recommended for this chapter.

FURTHER READING

General

Donath, F. A. and Parker, R. B., 1964, Folds and folding. *Geological Society of America Bulletin* **75**: 45–62.

Hudleston, P. J., 1986, Extracting information from folds in rocks. *Journal of Geological Education* **34**: 237–245.

Ramsay, J. G. and Huber, M. I., 1987, *The Techniques of Modern Structural Geology. 2: Folds and Fractures*. London: Academic Press.

Folds in shear zones

Bell, T. H. and Hammond, R. E., 1984, On the internal geometry of mylonite zones. *Journal of Geology* **92**: 667–686.

Cobbold, P. R. and Quinquis, H., 1980, Development of sheath folds in shear regions. *Journal of Structural Geology* **2**: 119–126.

Fossen, H. and Holst, T. B., 1995. Northwest-verging folds and the northwestward movement of the Caledonian Jotun Nappe, Norway. *Journal of Structural Geology* **17**: 1–16.

Harris, L. B., Koyi, H. A. and Fossen, H., 2002, Mechanisms for folding of high-grade rocks in extensional tectonic settings. *Earth-Sciences Review* **59**: 163–210.

Krabbendam, M. and Leslie, A. G., 1996, Folds with vergence opposite to the sense of shear. *Journal of Structural Geology* **18**: 777–781.

Platt, J. P., 1983, Progressive refolding in ductile shear zones. *Journal of Structural Geology* **5**: 619–622.

Skjernaa, L., 1989, Tubular folds and sheath folds: definitions and conceptual models for their development, with examples from the Grapesvare area, northern Sweden. *Journal of Structural Geology* **11**: 689–703.

Vollmer, F. W., 1988, A computer model of sheath-nappes formed during crustal shear in the Western Gneiss Region, central Norwegian Caledonides. *Journal of Structural Geology* **10**: 735–745.

Buckling

Biot, M. A., 1961, Theory of folding of stratified viscoelastic media and its implications in tectonics and orogenesis. *Geological Society of America Bulletin* **72**: 1595–1620.

Hudleston, P. and Lan, L., 1993, Information from fold shapes. *Journal of Structural Geology* **15**: 253–264.

Sherwin, J.-A. and Chapple, W. M., 1968, Wavelengths of single layer folds: a comparison between theory and observation. *American Journal of Science* **266**: 167–179.

Fold geometry

Bell, A. M., 1981, Vergence: an evaluation. *Journal of Structural Geology* **3**: 197–202.

Stabler, C. L., 1968, Simplified fourier analysis of fold shapes. *Tectonophysics* **6**: 343–350.

Mechanisms and processes

Bobillo-Ares, N. C., Bastida, F. and Aller, J., 2000, On tangential longitudinal strain folding. *Tectonophysics* **319**: 53–68.

Hudleston, P. J., Treagus, S. H. and Lan, L., 1996, Flexural flow folding: does it occur in nature? *Geology* **24**: 203–206.

Ramsay, J. G., 1974, Development of chevron folds. *Geological Society of America Bulletin* **85**: 1741–1754.

Tanner, P. W. G., 1989, The flexural-slip mechanism. *Journal of Structural Geology* **11**: 635–655.

Strain in folds

Holst, T. B. and Fossen, H., 1987, Strain distribution in a fold in the West Norwegian Caledonides. *Journal of Structural Geology* **9**: 915–924.

Hudleston, P. J. and Holst, T. B., 1984, Strain analysis and fold shape in a limestone layer and implications for layer rheology. *Tectonophysics* **106**: 321–347.

Ramberg, H., 1963, Strain distribution and geometry of folds. *Bulletin of the Geological Institution of the University of Uppsala* **42**: 1–20.

Roberts, D. and Strömgård, K.-E., 1972, A comparison of natural and experimental strain patterns around fold hinge zones. *Tectonophysics* **14**: 105–120.

Folds in extensional settings

Chauvet, A. and M. Séranne, 1994, Extension-parallel folding in the Scandinavian Caledonides: implications for late-orogenic processes. *Tectonophysics* **238**: 31–54.

Fletcher, J. M., Bartley, J. M., Martin, M. W., Glazner, A. F. and Walker, J. D., 1995, Large-magnitude continental extension: an example from the central Mojave metamorphic core complex. *Geological Society of America Bulletin* **107**: 1468–1483.

Chapter 12

Foliation and cleavage

Foliation and cleavage are terms for penetrative tectonic planar structures in rocks. Tectonic foliations go hand in hand with folds and lineations and form the most common type of structure encountered in metamorphic rocks, and their wide occurrence makes them particularly important for deciphering the deformation history of rocks. Primary foliations such as bedding are needed to initiate buckle folds and to observe folds in rocks in general. And where strain markers are absent, tectonic foliations provide us with useful and widespread strain information, since most foliations are associated with perpendicular shortening. Cleavage and foliations also create slates and schists that are of significant economic importance all over the world. In this chapter we will introduce basic terminology and discuss how and under what conditions different types of foliations initiate. Foliations associated with shear zones are covered in Chapter 15.

12.1 Basic concepts

Fabric

In structural geology the term **fabric** is used to describe penetrative and distributive components of rock masses (Figure 12.1). It can be composed of platy or elongate minerals with a preferred orientation. Examples include mica flakes in a mica schist or actinolite needles in an actinolite schist.

A fabric is built of minerals and mineral aggregates with a preferred orientation that penetrate the rock at the microscopic to centimeter spacing scale.

The fact that a fabric consists of penetrative elements means that minerals restricted to a fracture surface do not form a fabric even if they are perfectly aligned. The distance between the elements that constitute a fabric is typically less than about a decimeter, which excludes sets of faults or small shear zones as fabric elements.

A variety of objects in rocks, such as minerals, mineral aggregates, conglomerate pebbles etc., can be arranged in different ways and thereby give rise to different kinds of fabrics. It is useful to make a distinction between random fabrics, linear fabrics and planar fabrics. A **linear fabric** is characterized by elongate elements with a preferred orientation. A **planar fabric** contains tabular or platy minerals or other "flat" objects with a common orientation. A planar fabric need not be planar in the mathematical sense – the planar structures or elements are commonly seen to bend around rigid objects or may be affected by subsequent folding. In fact, curviplanar fabric would be a more appropriate term in some cases. A **random fabric** is one where its elements show no preferred orientation. Completely random fabrics may not be very common in rocks, but more or less random fabrics occur in some undeformed sedimentary rocks (clasts, ooids) and igneous rocks (phenocrysts). However, clasts and phenocrysts can obtain a preferred orientation during sedimentation or crystallization. Such **primary fabrics**, or more specifically, sedimentary and magmatic fabrics, must be accounted for when considering the role of tectonic deformation in a rock. Recognition of primary fabric is required to interpret deformation fabrics, and to make sure that we do not confuse primary structures with those formed during rock deformation.

Almost any rock, whether magmatic, sedimentary or metamorphic, shows a fabric, but fabrics are particularly well developed in strongly deformed metamorphic rocks

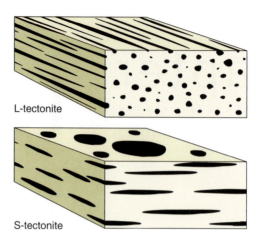

L-tectonite

S-tectonite

Figure 12.1 Fabric is a configuration of objects penetrating the rock. Linear objects form L-fabrics (top) while planar objects constitute S-fabrics (bottom). The rocks are known as L- and S-tectonites, respectively.

referred to as tectonites. In such rocks, **tectonic fabrics** are named according to the shape and organization of the fabric elements. As illustrated in Figure 12.1, rocks that show a marked linear fabric are called **L-tectonites**, and those showing a pronounced planar fabric are called **S-tectonites**. **LS-tectonites** is the term used for deformed rocks that contain both a linear and a planar fabric. A close connection between fabric and the shape of the strain ellipsoid has been suggested for strongly deformed metamorphic rocks: L-fabrics indicate constrictional strain, LS-fabrics relate to plane strain, and S-fabrics are the signature of flattening strain.

Foliation

Foliation (derived from the Latin word *folium*, meaning leaf) is generally used for any fabric-forming planar or curviplanar structure in a metamorphic rock, but may also include primary sedimentary bedding or magmatic layering. Some geologists prefer to reserve the term foliation for planar structures formed by tectonic strain, but it is now common to include depositional bedding and other primary planar structures in the definition of this term. It is important to make a clear distinction between the **primary foliations** that form during the deposition of sediments and formation of magmatic rocks, and **secondary foliations** such as axial plane cleavages in metamorphic rocks. Primary foliations are bedding in sedimentary rocks, flow banding in lavas and magmatic layering in intrusive rocks. Secondary foliations are products of stress and strain and most are **tectonic foliations** because they form in response to tectonic stress. The most important example of a non-tectonic secondary foliation is one resulting from compaction. In structural

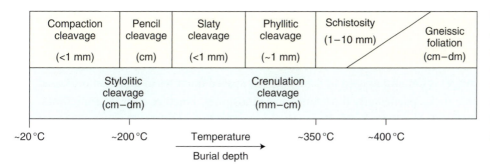

Figure 12.2 Schematic overview of important cleavage and foliation types, arranged according to burial depth or temperature and with an indication of the spacing of foliation domains. Temperatures indicated are very approximate.

geology we tend to restrict the term foliation to planar structures formed by deformation, and a tentative classification scheme for such structures is shown in Figure 12.2.

A tectonic foliation is a planar structure formed by tectonic processes, and includes cleavages, schistosity and mylonitic foliations.

A foliation is a fabric, which means that a single surface, such as a fault or a joint, will not represent a foliation. Even if parallel fractures are distributed throughout a hand sample they may lack the **cohesion** that is a second characteristic of foliations. Although foliated rocks typically split along the foliation, force is required to overcome that cohesion.

A foliated rock is by definition cohesive, although rocks may split preferentially along the foliation.

A foliation may be defined by zones of different grain sizes, flattened objects (e.g. conglomerate pebbles), recrystallized tabular grains with a uniform orientation, platy minerals arranged into millimeter-thick zones or domains, densely distributed cohesive microfractures and microfolds (crenulations).

Cleavage

The term **cleavage** refers to the ability of a rock to split or cleave into more or less parallel surfaces. Cleavage is a (large) subgroup of foliation – not all foliated rocks split preferentially along the foliation. Cleavage is found in very low-grade (lower greenschist facies and below) and barely metamorphic rocks, and in micaceous gneisses or schists in the form of late-stage crenulation cleavage.

Both cleavage and foliation are penetrative at the scale of a hand sample, but the spacing of the planar elements varies. If the distance is greater than 1 mm and distinguishable in the hand sample as individual surfaces or zones, the cleavage is designated a **spaced cleavage** (left

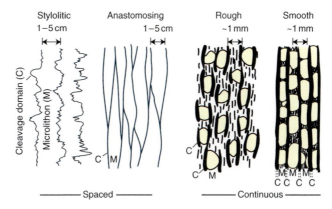

Figure 12.3 Disjunctive cleavage types. Stylolitic (limestones) and anastomosing (sandstones) cleavages are usually spaced, while continuous cleavages in more fine-grained rocks are separated into rough and smooth variants, where the rough cleavage can develop into the smooth version. All disjunctive cleavages are domainal, and the cleavage domains (C) are separated by undeformed rock called microlithons (M).

side of Fig. 12.3, also see Box 12.1). The structure is a **continuous cleavage** if the spacing of the planar elements is 1 mm or less (right side of Figure 12.3). Naturally, rocks with a continuous foliation split into thinner slices or flakes than do rocks with spaced foliations.

Rock cleavage should not be confused with mineral cleavage, which is the tendency of a mineral to break along specific crystallographic planes.

12.2 Relative age terminology

The Austrian geologist Bruno Sander introduced, around 1930, the designation S for foliations. Foliations in a given outcrop or area that can be shown to be of different ages are designated S_n, where the subscript n indicates the relative age. Primary foliations, such as bedding or the magmatic layering in Figure 12.4, are referred to as S_0. The oldest secondary foliation is designated S_1, the second one S_2 and so on. Foliations are commonly

BOX 12.1 | FRACTURE CLEAVAGE

Fracture cleavage (spaced cleavage) is sometimes used to denote closely spaced fractures in unmetamorphosed or very low grade metamorphic rocks, particularly limestones and sandstones. Fracture cleavage sometimes forms by fracturing of preexisting cleavage planes. Ordinary cleavage forms by contraction perpendicular to the cleavage planes. Fracture cleavage, as defined here, involves extension across the cleavage. It is therefore better classified as a small-scale fracture array in many cases.

Shear fractures can also occur closely and systematically enough that they resemble cleavage. This type of "cleavage" is very restricted, typically to fault cores and fault damage zones. There are additional uses of the term fracture cleavage in the literature, and it is wise to avoid this term to avoid confusion.

Slaty cleavage (left-dipping) turning into a spaced disjunctive cleavage (vertical) in the very fine-grained sandstone above. The disjunctive cleavage could be termed a fracture cleavage because of the fracturing along the pressure-solution seams.

Right-dipping subparallel fractures in sandstone near Temple Mountain, Utah, resembling a cleavage in the unmetamorphosed Navajo Sandstone.

related to folds, in which case the designations for folds (F) and foliations (S) share suffixes. For example, folds related to the first foliation (S_1) in a rock are referred to as F_1 etc. We have already used this terminology in Figure 11.35, where we named fold axes of two different phases of folding F_1 and F_2, respectively. Similarly, consecutive deformation phases are named D_1, D_2 and so on.

12.3 Cleavage development

There are many types of cleavages and a rich terminology is available. To efficiently deal with cleavages and foliations it is useful to keep an eye on crustal depth and lithology. **Crustal depth** is related to temperature (and pressure), and with increasing temperature we first obtain increasing mobility of minerals and at yet higher temperatures the possibility that minerals will recrystallize. Around

350–375 °C we leave the realm of cleavage and enter that of schistosity and mylonitic foliations. **Lithology** and mineralogy are important because different minerals react differently to stress and temperature. Phyllosilicates are particularly important in cleavage development. In general, if there are no phyllosilicates in the rock there will not be a very strong cleavage or schistosity. Cleavage formation in calcareous rocks is controlled by the mobility of carbonate and the easy formation of stylolites.

Cleavage is the low-temperature version of foliation and is best developed in rocks with abundant platy minerals.

We will here consider the most common types of foliation that develop due to deformation during prograde metamorphism, i.e. as a rock is being buried to progressively greater depths.

Figure 12.4 Two generations of foliations in metagabbro in a Caledonian ophiolite fragment. The primary magmatic layering (S_0) has been reworked into a shear-related foliation (S_1) during Caledonian shearing.

Compaction cleavage

The first secondary foliation forming in sedimentary rocks is related to their compaction history. Reorientation of mineral grains and collapse of pore space result in accentuation and reworking of the primary foliation (bedding). For a clay or claystone, the result is a shale with a marked **compaction cleavage** (Figure 12.5a). In this process there is also dissolution going on, and in some quartzites we can find pressure solution seams. Such structures are much more common in limestone, in which dissolution of carbonate produces subhorizontal and irregular seams where quartz or carbonate has been dissolved and where clay and other residual minerals are concentrated. The seams are stylolites or pressure solution seams, and the foliation can be called a **stylolitic cleavage** (Figure 12.3, left), which is a type of (**pressure**) **solution cleavage**.

The spacing of the seams in calcareous rocks is usually several centimeters, and the cleavage is therefore a widely spaced cleavage. In fact, the stylolitic surfaces may be too far apart to define a cleavage. In contrast, the compaction cleavage in shale is recognizable under the microscope and therefore a continuous cleavage. These non-tectonic cleavages are usually regarded as S_0 foliations.

Early tectonic development and disjunctive cleavage

A tectonic foliation commonly results when a sedimentary rock is exposed to tectonic stress that leads to progressive horizontal shortening of sedimentary beds – a condition typical of the foreland regions of orogenic belts. In limestone and some sandstones, the first tectonic foliation

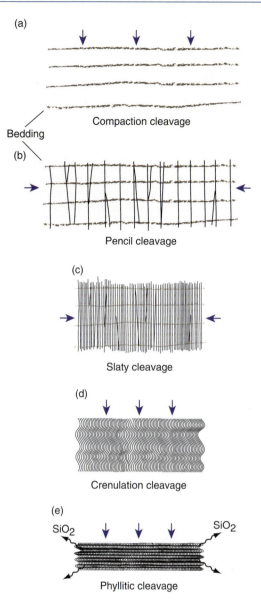

Figure 12.5 Theoretical cleavage development in a mudstone.

to form is a pressure solution cleavage that typically is stylolitic (toothlike, showing a zigzag suture in cross-section). If σ_1 is horizontal, a vertical pressure solution cleavage forms that will make a high angle to S_0 and previously formed compaction-related stylolites.

Pressure solution is also important when shales are exposed to tectonic stress. In this case extensive dissolution of quartz causes concentration and reorientation of clay minerals. At some point the secondary cleavage will be as pronounced as the primary one, and clay minerals will be equally well oriented along S_1 and S_0. The shale will now fracture along both S_1 and S_0 into pencil-shaped fragments, which explains why the cleavage is known as **pencil cleavage** (Figure 12.5b and Figure 12.6). See Box 12.2 for a discussion of the formation of

Figure 12.6 Pencil cleavage in shale in the Caledonian foreland fold-and-thrust belt near Oslo.

such cleavage. Pencil cleavage also occurs where two tectonic cleavages develop in the same rock due to local or regional changes in the stress field. Such purely tectonic pencil cleavage is associated with some thrust ramps, formed close in time during the same phase of deformation.

If the tectonic shortening persists, it will eventually dominate over the compactional cleavage. More and more clay grains become reoriented into a vertical orientation as quartz grains are being dissolved and removed, somewhat similar to a collapsing house of cards. Microfolding of the clay grains may also occur. The result is a continuous cleavage that totally dominates the structure and texture of the rock. The rock is now a slate and its foliation is known as **slaty cleavage**, and we have reached stage (c) in Figure 12.5.

The formation of slaty cleavage occurs while the metamorphic grade is very low, so that recrystallization of clay minerals into new mica grains appears to have just started. A close look at a well-developed slaty cleavage reveals that a change has taken place in terms of mineral distribution. There are now domains dominated by quartz and feldspar, known as **QF-domains**, that separate **M-domains** rich in phyllosilicate minerals. The letters Q, F and M relate to quartz, feldspar and mica, and we need a microscope to discern the individual domains, which are considerably thinner than 1 mm. The QF-domains are typically lozenge- or lens-shaped while the M-domains form narrower, enveloping zones. As shown below, many types of cleavages and foliations show such domainal structures, and they can all be referred to as **domainal cleavages**.

The term **disjunctive cleavage** (Figure 12.3) is commonly used about early tectonic domainal cleavage in previously unfoliated rocks such as mudstones, sandstones and limestones. This term implies that the cleavage cuts across, rather than crenulating (folding), preexisting foliations.

BOX 12.2 | PENCIL CLEAVAGE AND STRAIN

The development of pencil cleavage can be represented in the Flinn plot as shown below. Step I involves compaction during burial. The progressive flattening involved causes an initial path along the horizontal axis. Compaction is completed at stage 1. Horizontal tectonic shortening (Step II) then transforms the strain ellipse into a prolate geometry (stage 2). Pencil structures are now formed. If the deformation continues (Step III), then the Z-axis becomes vertical and the vertical tectonic cleavage starts to dominate. Strain becomes perfectly oblate again at stage 3 and further deformation (Step IV) causes vertical stretching.

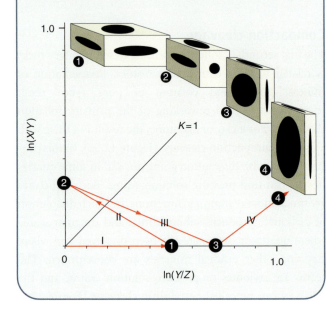

It was once thought that slaty cleavage formed by physical grain rotation. We now know that so-called wet diffusion or pressure solution is what chiefly produces the domainal structure that characterizes slaty cleavage. Grains of quartz and feldspar are dissolved perpendicular to the orientation of the cleavage and achieve lensoid shapes (disk shapes in three dimensions). Where this happens, phyllosilicates are concentrated and M-domains form. The importance of dissolution or pressure solution allows us to use the term (pressure) solution cleavage also for slaty cleavage.

Cleavage forms through grain rotation, growth of minerals with a preferred direction and, most importantly, wet diffusion (pressure solution) of the most solvent minerals in the rock.

Wet diffusion implies that dissolved minerals diffuse away through a very thin film of fluid located along grain boundaries. The material is precipitated in so-called pressure shadows of larger and more rigid grains in the QF-domains or becomes transported out of the rock. In fact, very significant amounts of matter appear to have left most slate belts, which represent interesting aspects in terms of thermodynamics and fluid flow through the upper crust.

Greenschist facies: from cleavage to schistosity

New phyllosilicate minerals grow at the expense of clay minerals in shales and slates when they enter the field of greenschist facies metamorphism. A phyllite forms and the cleavage changes into a **phyllitic cleavage** (Figures 12.5e and 12.7). The new mica minerals grow with their basal plane more or less perpendicular to the Z-axis of the strain

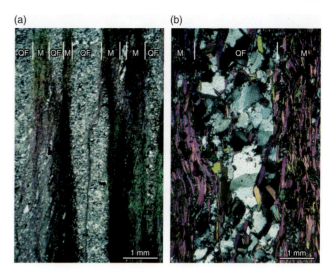

(a)   (b)

Figure 12.8 (a) Phyllitic cleavage in lower greenschist facies phyllite. (b) This cleavage formed higher into the greenschist facies, showing very well-developed QF- (middle) and M-domains and coarser grain size.

ellipsoid, and more or less perpendicular to σ_1. The newly formed mica grains are thus parallel and a phyllitic cleavage is established. The cleavage is still a continuous one, and the development of QF- and M-domains is more pronounced than for slaty cleavage. The domainal cleavage becomes better developed because dissolution (wet diffusion) becomes more efficient as greenschist facies temperatures are reached (Figure 12.8).

When original claystone reaches upper greenschist facies and perhaps lower amphibolite facies, the mica grains grow larger and become easily visible in a hand sample. At the same time, the foliation becomes less planar, wrapping around quartz–feldspar aggregates and strong metamorphic minerals such as garnet, kyanite and amphibole. The foliation is no longer called a cleavage but a **schistosity**, and the rock is a **schist**.

Schistosity is also found in quartz-rich rocks such as quartz schists and sheared granites. Here the M- and QF-domains are on the millimeter or even centimeter scale and they appear more regular and planar than for micaschists. This is why quartz schists and sheared granites split so easily into slabs that can be used for various building purposes. In summary, while wet diffusion (solution) and grain reorientation dominate the formation of slaty cleavage, recrystallization is more important during the formation of schistosity.

Secondary tectonic cleavage (crenulation cleavage)

An already established tectonic foliation can be affected by a later cleavage (S_2 or higher) if the orientation of the ISA

Figure 12.7 Phyllitic cleavage bears similarities to crenulation cleavage when viewed under the microscope. The difference is that phyllitic crenulations are microscopic and thus invisible to the naked eye. In this case a phyllitic cleavage dies out towards a folded competent lamina.

Figure 12.9 Asymmetric crenulation cleavage affecting a mylonitic foliation. The cleavage is discrete in the middle, micaceous layer, while it is zonal and less well developed in the more quartz-rich adjacent layers.

(see Chapter 2) changes locally or regionally at some point during the deformation, or if a later cleavage-forming deformation phase occurs. Because cleavages tend to form perpendicular to the maximum shortening direction (X), a new cleavage will form that overprints the preexisting one. In many cases this occurs by folding the previous foliation into a series of microfolds, in which case the cleavage is called a **crenulation cleavage**. Hence, a crenulation cleavage is a series of microfolds at the centimeter scale or less with parallel axial surfaces (Figure 12.9). Depending on the angle between the existing foliation and the secondary stress field, the crenulation cleavage will be symmetric or asymmetric. A **symmetric crenulation cleavage** has limbs of equal length, while an **asymmetric crenulation cleavage** is composed of small, asymmetric folds with S- or Z-geometry.

Crenulation cleavage through which the earlier foliation can be traced continuously is known as **zonal crenulation cleavage**. In the opposite case, where there is a sharp discontinuity between QF- and M-domains, the cleavage is called a **discrete crenulation cleavage**. The M-domains here are thinner than the QF-domains and mimic microfaults. Discrete and zonal crenulation cleavages can grade into each other within a single outcrop.

Crenulation cleavage is restricted to lithologies with a preexisting well-developed foliation that at least partly is defined by phyllosilicate minerals. It is commonly seen in micaceous layers while absent in neighboring mica-poor layers. The domainal thickness of the affected foliation is connected with the wavelength of the new crenulation cleavage: thicker domains produce longer crenulation wavelengths. This is the same relationship between layer thickness and wavelength that was explored in Chapter 11,

Figure 12.10 Crenulation cleavage affecting the phyllitic cleavage shown in Figure 12.8a. The crenulation cleavage is seen to be axial planar to decimeter-scale folds. Photo: Ø. J. Jansen.

where the viscosity contrast was shown to be important. A close connection is also seen between crenulation cleavage and folding, as discussed in the next section (Figure 12.10).

We can find any stage of crenulation cleavage development, from faint crenulation of foliations to intense cleavage development where recrystallization and pressure solution have resulted in a pronounced domainal QF-M-structure. In the latter case the original foliation can be almost obliterated in a hand sample although usually observable under the microscope. Progressive evolution of crenulation cleavage is accompanied by progressive shortening across the cleavage, and eventually a crenulation cleavage can transform into a phyllitic foliation.

12.4 Cleavage, folds and strain

Axial plane cleavage

A close geometric relation between cleavages and folds is seen in most cases, for example in the fold shown in

Figure 12.11, and it is clear that the two types of structures commonly form simultaneously. Geometrically a cleavage splits the fold more or less along the axial surface, particularly near the hinge zone. Where a cleavage parallels the axial surface, the cleavage is called **axial plane cleavage**.

Many cleavages are axial planar and thus represent an important link between tectonic foliation and folds in deformed rocks.

Interestingly, if we study cleavage–fold relations in detail we may see a difference in orientation between the axial plane and the cleavage. In fact, the orientation of the cleavage may be seen to vary from layer to layer. The variations occur across layers of contrasting competence or viscosity and the phenomenon is called **cleavage refraction**. The higher the contrast in competency, the more pronounced the refraction.

Cleavage can also form a variety of patterns in the hinge zone. Figure 12.12 shows a drawing of the cleavage pattern developed in incompetent shale layers between sandstone beds in the Caledonian nappes in Finnmark, Norway. In general, the cleavage is perpendicular to the shortening axis, i.e. it represents the XY-plane of the strain ellipsoid. Hence, the pattern seen in the hinge zone of the shale layers becomes particularly interesting. In the upper part the cleavage (and XY-plane) is perpendicular to the bedding, which is consistent with layer-parallel shortening in the competent layer above. In the lower part of the shale the layer-parallel cleavage conforms to outer arc stretching in the sandstone layer below. In between there is a place where the cleavage vanishes, and that point is called a **neutral point** (Figure 12.12). The neutral surface

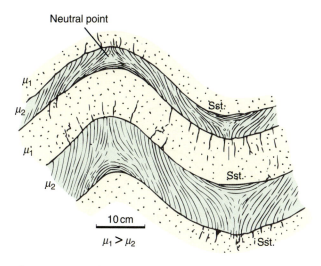

Figure 12.12 Folded sand–shale layers. Non-planar cleavage in the shale (green) reflects variations in the orientation of the local *XY*-plane. Note that there is no cleavage at the neutral points. Compare with the next figure. From Roberts and Strömgård (1972).

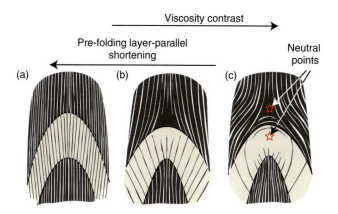

Figure 12.13 Different cleavage patterns in and around the hinge zones of folded competent layers. The amount of layer-parallel shortening prior to the folding is important, together with the viscosity contrast. Based on Ramsay and Huber (1987).

found in classic buckle folds (see Chapter 11) is missing because of the influence of flexural shear. Even in the competent layers a component of flexural shear would reduce the neutral surface to a neutral point (Figure 12.13c).

The pattern shown in Figure 12.12 is not always observed. In fact, it is more common to see a continuous cleavage through alternating competent and incompetent layers, as shown in Figure 12.13a, or to see the cleavage being developed in the incompetent layer only. The final cleavage pattern is linked to the early deformation history, because the final cleavage pattern is sensitive to the amount of layer-parallel shortening prior to cleavage formation. More specifically, the cleavage will be continuous if it is already well developed at the onset of folding. The result is

Figure 12.11 Discrete cleavage in phyllite, axial planar to mesoscopic folds. Joma area, Central Norwegian Caledonides.

(a)

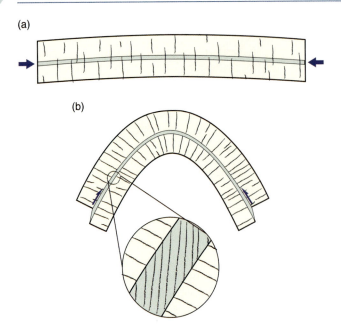

(b)

Figure 12.14 Cleavage refraction caused by localized shear in incompetent layers on fold limbs (flexural shear).

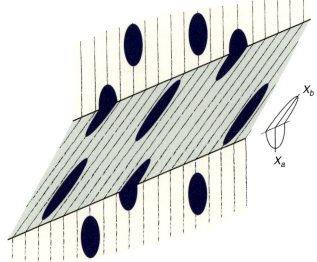

Figure 12.15 Strain associated with cleavage refraction (assuming no slip along layer interfaces). The strain must be compatible across the interfaces, i.e. strain ellipses must fit together as shown. In this situation, the only deformation possible is simple shear and/or volume change across the layering.

illustrated in Figure 12.13a. If, on the other hand, the layer starts to fold before the cleavage is established, then the pattern shown in Figure 12.13c is more likely to emerge.

Field investigations show that cleavage refraction occurs when layers of contrasting competence are folded. We can explain this observation by a difference in shear strain (γ) between competent and incompetent layers (Figure 12.14). In this flexural shear model the strain ellipse at the interface between a competent and incompetent layer is shared by the strain ellipsoids in both layers, as shown in Figure 12.15.

Cleavage refraction reveals layer competence contrasts, location on the fold structure and is influenced by pre-folding layer-parallel shortening and flexural shear during folding.

The angular relation between cleavage and layering is related to the position on the overall fold structure, as shown in Figure 12.16. Regardless of its tendency to show refraction, cleavage has a fairly consistent orientation across a fold while the attitude of the folded layers changes systematically. As shown in Figure 12.16a, b the fold geometry can easily be constructed (qualitatively) from observations of the cleavage–layering relation from only a single locality. Clearly this method assumes a genetic relationship between the cleavage and the folding. Later cleavage that is unrelated to the folding can also occur in a polydeformed area, in which case the angular relations tell nothing about the larger-scale fold geometry (Figure 12.16c).

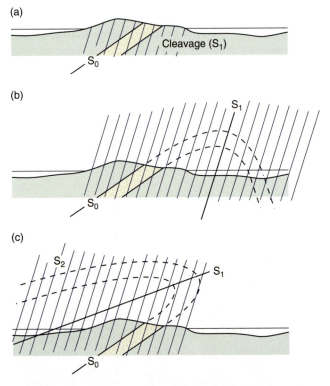

Figure 12.16 (a) Mapped cleavage-layering relation in a deformed area. (b) Interpretation of a large-scale fold based on the assumption that the cleavage is related to such a fold. If the fold has a completely different geometry (c), then the fold and the cleavage must have formed at different times.

Strain

Cleavage formed at low metamorphic grades, i.e. slaty cleavage and in part crenulation cleavage and phyllitic cleavage, is generally assumed to represent the *XY* or flattening plane. This is clearly the case where the deformation history is coaxial (Figure 12.17a), and the cleavage variations in the hinge zone shown in Figure 12.12 are an example of its usefulness. But it is also true for non-coaxial deformations, such as simple shear (Figure 12.17b). In simple shear the foliation rotates toward the shear plane, and the foliation in the shear zone tracks the *XY*-plane through a shear zone – a very useful feature that is discussed further in Chapter 15. As always there are exceptions that we must look out for, for example where the foliation starts to slip or where it wraps around rigid lenses or other rigid objects.

The shape of the strain ellipsoid may be more difficult to extract, but we know from the concentration of insoluble minerals in the M-cleavage domains that volume loss (dilation) is an important component of cleavage-related strain. At low temperature and shallow burial depths compaction cleavage in sediments involves a 30–40% reduction in volume. Pressure solution-dominated cleavages in limestones may involve even larger losses of volume, and rocks with a well-developed slaty cleavage typically show evidence of impressive 50–75% shortening across the cleavage (Figure 12.15). How do we know?

Fortunately, strain markers such as the reduction spots in sedimentary rocks (reduction spheres in three dimensions) can be preserved. When deformed, these reduction spots change from circles to ellipses that represent sections through the local strain ellipsoid (Figure 3.1). Fossils with known original shape or geometry can also be used to estimate strain in cleaved (meta)sedimentary rocks.

Most cleavages approximate the *XY*-plane of the strain ellipsoid and are characterized by substantial volume loss.

Shortening perpendicular to the cleavage may or may not be compensated by stretching within the plane of cleavage. If pressure solution (wet diffusion) is prominent and not balanced by precipitation, then the shortening may be uncompensated for in the *XY*-plane. In this case the deformation is anisotropic volume loss, and the strain ellipse becomes oblate. Significant loss of volume has been documented by a number of studies (Figure 12.18), which shows that diffusion of material out of the rock is an important mechanism during cleavage formation. Alternatively, the dissolved material would be precipitated in the QF-domains, but in most cases such local precipitation seems to be of minor importance. While physical compaction may be important in the formation of compaction cleavage in porous rocks, wet diffusion becomes more important as temperature increases.

Pressure solution (wet diffusion) is largely responsible for the oblate strain ellipsoids associated with cleavage formation.

(a)

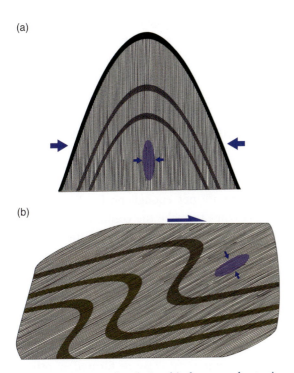

(b)

Figure 12.17 Simple relationship between shortening direction (arrows), strain ellipse and cleavage for (a) upright folds, for instance in slate belts, and (b) shear zones.

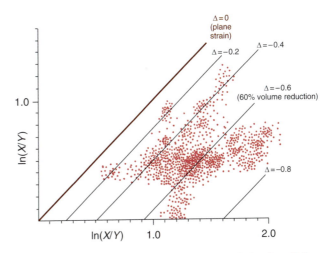

Figure 12.18 Strain data from slates with well-developed slaty cleavage. The data fall completely within the field of flattening. Δ is the volume loss factor from Chapter 2. Data from Ramsay and Woods (1973).

(a)

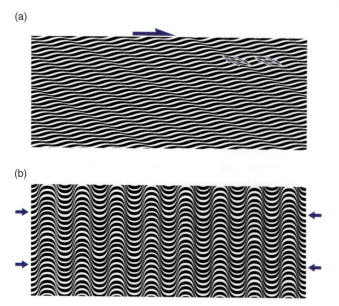

(b)

Figure 12.19 (a) Shear bands (extensional crenulation cleavage), where the lamination is primarily offset by shearing. (b) Symmetric crenulation cleavage formed by layer-parallel shortening. Note that ordinary crenulation cleavage can also be asymmetric.

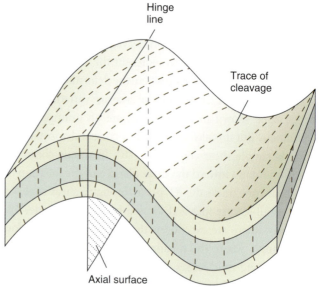

Figure 12.20 Cleavage transecting the axial surface of a transected fold.

Phyllitic cleavage and the foliation in schists are not always the result of coaxial shortening, but can also be influenced by, or form as a result of, simple shearing. During non-coaxial shearing, so-called **shear bands** that may look similar to axial plane cleavages can develop (Figure 12.19). Shear bands are also called extensional crenulation cleavage, and are small shear zones in rocks that extend the layering or foliation that they deform. Shear bands do not relate directly to the strain ellipsoid but provide important information about the sense of shear, as discussed in Chapter 15. It may be difficult to distinguish between ordinary crenulation cleavage and shear bands, but shear bands are not directly related to folding and therefore not an axial plane cleavage. They may also show a striation related to localized shear or slip. It may be wise to restrict the term cleavage to foliations formed by a significant component of shortening across the structure.

Transected folds

In most cases the theoretical axial surface has more or less the same orientation as the axial plane cleavage. Less commonly there is a marked difference between the orientations of cleavage and axial surface (Figure 12.20), even where the folding and the cleavage are genetically related and where there is no refraction due to rheologically contrasting layers. In such cases the cleavage transects not only the axial surface but also the fold hinge (Figure 12.20). Such cleavage is called **transecting cleavage**, and the folds are referred to as **transected folds**.

The extensive slate belts of Wales and the Scottish Southern Uplands exhibit classic examples of transected folds. The obliquity is systematic in these regions, with the cleavage being rotated clockwise relative to the fold axis, as shown in Figure 12.20. The systematic mismatch between the axial planes and cleavage in the British Caledonides has been explained by sinistral transpression (see Chapter 18). In such a tectonic regime, folds and cleavages will rotate during the deformation and a slight time difference between the onsets of folding and cleavage formation can result in transected folds. However, transected folds can also form during coaxial deformation in the case of non-steady-state flow, i.e. where the ISA rotate relative to the deforming rock.

12.5 Foliations in quartzites, gneisses and mylonite zones

Mica-bearing quartzites develop a foliation known as **schistosity**. There are not enough phyllosilicate grains in impure quartzites to make the continuous cleavages seen in slates and phyllites, or the wavy foliation seen in micaschists, but the impure quartzite or **quartz schist** will split neatly into centimeter- to decimeter-thick slabs typically used for pavements or other building purposes. The splitting is mainly due to parallel phyllosilicates that impose an anisotropy on the rock.

While wet diffusion governs cleavage formation, other crystal-plastic deformation mechanisms become more important during the formation of schistosity.

A pure quartzite will never develop a well-defined schistosity because there are no phyllosilicate minerals to become reoriented and concentrated. Instead, a shape fabric formed by flattening of individual quartz grains can give the rock an anisotropy, but not as pronounced as in a quartz schist. Pure quartzites preserve a **quartzitic banding** from variations in grain size or color variations, commonly portraying tight to isoclinal similar folds where strain is high. Otherwise, foliation may only be expressed at the microscopic scale, where flattened plastically deformed grains are observed.

Granites and other quartzofeldspathic rocks poor in phyllosilicate minerals do not develop cleavage as easily as most other rocks because micas are few and much time and energy is required to reorganize them into a cleavage. However, at low temperatures and when water is present, feldspar alteration to micas may generate foliated rocks. Some magmatic rocks also have a high feldspar/quartz ratio. Feldspars require considerably higher temperature than quartz to deform plastically and lack the mobility of quartz in wet diffusion.

This does not mean that there is no foliation in deformed magmatic rocks. In some cases we can see a spaced cleavage in coarse-grained magmatic rocks. The spacing between the mica-enriched domains may well be several centimeters. At high strains granites and other magmatic rocks will develop a foliation from reorientation and flattening of minerals and mineral aggregates. In many cases this process can be related to non-coaxial deformation, in contrast to the coaxial deformations that seem to be responsible for cleavages in many slates and phyllites. The tectonic foliation in deformed magmatic rocks, such as seen in Figure 12.21, is called **schistosity** or **gneissic banding** rather than cleavage.

Gneissic banding commonly results from the type of process illustrated in Figure 12.22, where earlier structures, including dikes, veins and foliations, are flattened and

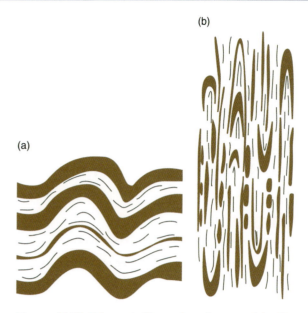

Figure 12.22 Schematic illustration of transposition by horizontal shortening and vertical extension. Both coaxial and non-coaxial strain can produce this result. The result (b) is a banded rock with intrafolial folds and isolated fold hinges. Note how the dominant foliation changes (transposes) from horizontal (a) to vertical (b).

Figure 12.23 Mylonitic foliation formed by shear-related plastic grain size reduction of a coarse-grained granitic rock.

Figure 12.21 Gneissic banding, formed during shearing of a heterogeneous intrusive complex. High strains are required to reach this stage of transposition.

rotated into almost complete parallelism. This process is called transposition, in which case the planar structure can be called **transposition foliation** consisting of **transposed layering**. Both coaxial and non-coaxial deformation can result in transposed foliation if strain is high.

Even if there are no preexisting planar structures in our magmatic rock, high strain, and particularly high non-coaxial strain, may result in **mylonitic foliation**. Mylonitic foliations are related to transposed layering and gneissic banding, but the distance between the foliation domains is smaller, typically on the millimeter or centimeter scale (Figure 12.23). This is related to the high strains involved; high strains flatten objects and thin layers. Mylonite zones are typically found in shear zones or thrust zones that involve large (kilometer scale or more) displacements. Mylonites and mylonitic shear zone structures are discussed in more detail in Chapter 15.

Summary

Foliations are common in metamorphic rocks and are often used in the definition of the different types of metamorphic rocks. Without foliations it would be difficult to do proper structural analysis. In this chapter we have looked at the different types of foliations and how they may form. There are many mechanisms and relations that we have not explored in depth in this chapter, and there are aspects of foliation development that need further research. Some of the most important things to remember from this chapter are listed here:

- In simple terms, foliations form in metamorphic rocks while fractures form at shallower crustal depths.
- Most foliations form perpendicular to or at a high angle to the shortening direction.
- Foliations give us important strain information where regular strain markers are absent.
- Different types of foliations reflect variations in lithology and temperature or depth of burial during deformation.
- Cleavages are commonly found to closely define the axial surface of related folds.
- Schistosity commonly forms during shearing and does not have to be related to folding.
- Pressure solution removes large volumes of rock during cleavage formation, but not during the formation of schistosity and mylonitic foliations.

Review questions

1. What is the difference between primary, secondary and tectonic foliations?
2. How are primary foliations recognized in deformed and metamorphosed sedimentary rocks?
3. What separates a fracture zone from a cleavage or foliation?
4. What type of strain can produce transected folds?
5. What favors the formation of continuous (dense) cleavage?
6. What is the most important mechanism for cleavage formation?
7. What is meant by QF- and M-domains?
8. What is cleavage refraction and how can we explain it?

9. How can we use the angle between cleavage and a pre-cleavage foliation to predict large-scale fold geometry?

10. Why is cleavage always associated with flattening strain (oblate strain ellipsoids)?

11. What is the difference between shear bands (extensional crenulation cleavage) and ordinary crenulation cleavage?

E-MODULE

 The e-learning module called *Foliation and cleavage* is recommended for this chapter.

FURTHER READING

General

Dieterich, J. H., 1969, Origin of cleavage in folded rocks. *American Journal of Science* **267**: 155–165.

Ramsay, J. G. and Huber, M. I., 1983, *The Techniques of Modern Structural Geology. 1: Strain Analysis.* London: Academic Press.

Volume loss and strain

Beutner, E. and Charles, E., 1985, Large volume loss during cleavage formation, Hamburg sequence, Pennsylvania. *Geology* **13**: 803–805.

Goldstein, A., Knight, J. and Kimball, K., 1999, Deformed graptolites, finite strain and volume loss during cleavage formation in rocks of the taconic slate belt, New York and Vermont, U.S.A. *Journal of Structural Geology* **20**: 1769–1782.

Mancktelow, N. S., 1994, On volume change and mass transport during the development of crenulation cleavage. *Journal of Structural Geology* **16**: 1217–1231.

Ramsay, J. G. and Woods, D. S., 1973, The geometric effects of volume change during deformation processes. *Tectonophysics* **16**: 263–277.

Robin, P. -Y., 1979, Theory of metamorphic segregation and related processes. *Geochimica et Cosmochimica Acta* **43**: 1587–1600.

Slaty cleavage

Siddans, A. W. B., 1972, Slaty cleavage: a review of research since 1815. *Earth-Science Reviews* **8**: 205–212.

Sorby, H. C., 1853, On the origin of slaty cleavage. *Edinburgh New Philosophical Journal* **55**: 137–148.

Wood, D. S., 1974, Current views of the development of slaty cleavage. *Annual Reviews of Earth Science* **2**: 1–35.

Crenulation cleavage

Cosgrove, J. W., 1976, The formation of crenulation cleavage. *Journal of the Geological Society* **132**: 155–178.

Gray, D. R. and Durney, D. W., 1979, Crenulation cleavage differentiation: implications of solution-deposition processes. *Journal of Structural Geology* **1**: 73–80.

Hanmer, S. K., 1979, The role of discrete heterogeneities and linear fabrics in the formation of crenulations. *Journal of Structural Geology* **1**: 81–91.

Swager, N., 1985, Solution transfer, mechanical rotation and kink-band boundary migration during crenulation-cleavage development. *Journal of Structural Geology* **7**: 421–429.

Worley, B., Powell, R. and Wilson, C. J. L., 1997, Crenulation cleavage formation: evolving diffusion, deformation and equilibration mechanisms with increasing metamorphic grade. *Journal of Structural Geology* **19**: 1121–1135.

Pencil cleavage

Engelder, T. and Geiser, P., 1979, The relationship between pencil cleavage and lateral shortening within the Devonian section of the Appalachian Plateau, New York. *Geology* **7**: 460–464.

Cleavage refraction

Treagus, S. H., 1983, A theory of finite strain variation through contrasting layers, and its bearing on cleavage refraction. *Journal of Structural Geology* **5**: 351–368.

Treagus, S. H., 1988, Strain refraction in layered systems. *Journal of Structural Geology* **10**: 517–527.

Cleavage-transected folds

Johnson, T. E., 1991, Nomenclature and geometric classification of cleavage-transected folds. *Journal of Structural Geology* **13**: 261–274.

Chapter 13

Lineations

Linear structures go hand in hand with planar structures in deformed rocks, where they are mesoscopic structures pointing in a specific direction. We have already looked at the role of lineations that are found on slip surfaces and how they can reveal paleostress and kinematics. Lineations are even more common in metamorphic rocks, where they tend to be closely associated with strain and transport or shear directions. In this chapter we will sort out the different types of lineations that are commonly encountered in deformed rocks and discuss their origins and implications.

13.1 Basic terminology

Lineation is a term used to describe linear elements that occur in a rock, such as the linear structures seen in the gneiss portrayed in Figure 13.1. A large number of non-tectonic or **primary linear structures** occur in both undeformed and deformed rocks. Ropy lava, flow lineations and columns in columnar basalts in igneous rocks, and long axes of aligned non-spherical pebbles, groove marks and aligned fossils in sedimentary rocks are some examples. In our context we are concerned with linear structures resulting from deformation, although primary structures may also be involved, such as in the formation of S_0–S_1 intersection lineations (see below).

A lineation is a fabric element in which one dimension is considerably longer than the other two.

The fabric elements constituting **tectonic linear structures** include elongated physical objects, such as strained mineral aggregates or conglomerate pebbles, lines of intersection between two sets of planar structures, and geometrically defined linear features such as fold hinge lines and crenulation axes. A distinction is made between **penetrative lineations**, which build up a linear fabric or L-fabric, **surface lineations**, which are restricted to a surface (e.g. slickenlines), and non-physical, **geometric lineations** such as fold axes and intersection lineations.

The term lineation should not be confused with the related term **lineament**, which is used for linear features at the scales of topographic maps, aerial photos, satellite images or digital terrain models. Most lineaments are planar structures such as fractures and foliations that intersect the surface of the Earth.

13.2 Lineations related to plastic deformation

Penetrative lineations are found almost exclusively in rocks deformed in the plastic regime. Where the lineation forms the dominating fabric element so that the S-fabric is weak or absent, the rock is classified as an **L-tectonite**. It can be seen from rocks with strain markers that most L-tectonites plot in the constrictional field of the Flinn diagram, i.e. $X \gg Y \geq Z$ (Figure 2.14). A balanced combination of a foliation (S-fabric) and a penetrative lineation (L-fabric) is more common, and such a rock is referred to as an **LS-tectonite**. LS-tectonites tend to plot close to the diagonal in the Flinn diagram. **S-tectonites**, which have no or just a hint of linear fabric, typically plot in the flattening field of the Flinn diagram.

Mineral lineations

A penetrative linear fabric is typically made up of aligned prismatic minerals such as amphibole needles in an amphibolite, or elongated minerals and mineral aggregates such as quartz–feldspar aggregates in gneiss. Mineral lineations can form by several processes:

Minerals and mineral aggregates can form a linear fabric by means of recrystallization, dissolution/ precipitation or rigid rotation.

Physical **rotation** of rigid prismatic minerals in a soft matrix can in some cases occur during deformation. An example is amphibole or epidote crystals in micaschist, where statically grown amphiboles become aligned in zones of localized deformation. In most cases the competence contrast between elongate-shaped minerals and their

Figure 13.1 Lineation in gneiss.

matrix is not high enough for rotation to be important. Instead, **synkinematic recrystallization** by means of plastic deformation mechanisms or a **dissolution/precipitation** process reshapes minerals and mineral aggregates. In addition, precipitation of quartz in pressure shadows or **strain shadows** is a common way to facilitate growth of minerals or mineral aggregates in a preferred direction. Even crushing or **cataclasis** of brittle minerals and mineral aggregates enclosed in a ductile matrix can reshape mineral aggregates to linear fabric elements.

Cataclasis, pressure solution and recrystallization all contribute to change the shape of minerals and mineral aggregates during deformation.

In a homogeneously strained rock, if a mineral aggregate had a spherical shape at the onset of deformation, its shape after deformation would represent the strain ellipsoid. In most cases the original shape is unknown so that the final shape only gives us a qualitative impression of the shape of the strain ellipsoid. Nevertheless, deformed mineral aggregates in gneisses have been used for strain analysis, although the difference in viscosity between the aggregates and their surroundings may add uncertainty to the results. In cases where the initial shape is known and the competence contrast is small, such analyses are particularly useful.

Deformed conglomerates or oolites are examples of rocks where linear shapes can quantitatively be related to strain (see Chapter 4). These and other lineations defined by the shape of deformed objects are named **stretching lineations** (Figure 13.2) and the related fabric is called a **shape fabric**. Stretching lineations and shape fabrics are extremely common in plastically deformed rocks such as gneisses.

Figure 13.2 Stretching lineation in quartzite conglomerate. The long axes of the pebbles are plunging to the right. The Bergsdalen Nappes, West Norway Caledonides.

Stretching of minerals and mineral aggregates into a penetrative stretching lineation forms the most common type of lineation in deformed metamorphic rocks.

Quartz, calcite and some other common minerals can grow well-aligned fibrous crystals that define linear elements. Such mineral lineations are referred to as **mineral fiber lineations**. Fibers may grow in the instantaneous stretching direction (ISA_1), but may also grow perpendicular to the face of the opening walls. Besides, once formed they may rotate away from this direction as a result of progressive deformation. In addition to their occurrence in veins in retrograde metamorphic rocks and unmetamorphosed sedimentary rocks such as overpressured mudstones, they are commonly found in strain shadows of porphyroclasts in metamorphic rocks during low-grade metamorphism. Fibers do not form if pressure and temperature get too high, and seldom above middle greenschist-facies conditions. This rather restricted occurrence makes fiber lineations less common and also less penetrative than many other lineation types.

Rodding describes elongated mineral aggregates that are easily distinguished from the rest of the rock. Quartz rods are common in micaschists and gneisses where striped quartz objects occur as rods or cigars in the host rock. Rods are often considered as stretching lineations, but are commonly influenced by other structure-forming processes. They may represent isolated fold hinges, or be related to boudinage or mullion structures (see below), or to deformed veins with an originally elongated geometry.

Intersection lineations

Many deformed rocks host more than one set of planar structures. A combination of bedding and cleavage is a common example. In most cases such planar structures intersect, and the line of intersection is regarded as an **intersection lineation**. Where the first tectonic cleavage (S_1) cuts the primary layering or bedding (S_0), the resulting intersection lineation (L_1) appears on the bedding planes, as shown in Figure 13.3. Intersection lineations formed by the intersection of two tectonic foliations are also common. In most cases intersection lineations are related to folding, with the lineation running parallel to the axial trace and the hinge line. Note that for transected folds (Figure 12.20) there will be an angle between the intersection lineation and the axial trace.

In some deformed rocks an intersection lineation appears only locally. In most cases, however, their frequency and

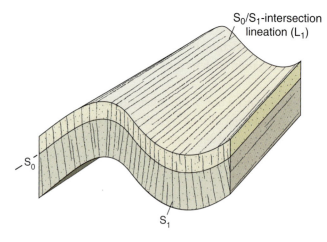

S₀/S₁-intersection lineation (L₁)

Figure 13.3 Intersection lineations appearing on bedding or foliation surfaces that are intersected by a later foliation.

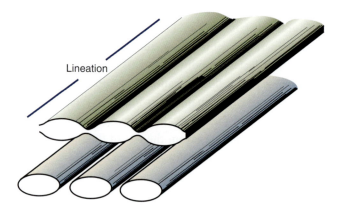

Lineation

Figure 13.4 Cylindrical pinch-and-swell structures (above) and boudins (below) represent linear elements in many deformed rocks.

distribution are large enough that the lineation can be considered penetrative. Like other lineations, intersection lineations may be folded about later folds, showing their use in tracing the deformation history of a rock.

Fold axes and crenulation lineations

Fold axes are generally regarded as linear structures, despite being theoretical lines related to the geometrical shape of the folded surface. Some rocks have a high enough density of parallel fold axes that they constitute a fabric. This is often the case with phyllosilicate-rich metamorphic rocks, where small-scale folds or crenulations constitute a **crenulation lineation**. Crenulation lineations are thus composed of numerous millimeter-to centimeter-scale fold hinges of low-amplitude folds. They are commonly seen in multiply deformed phyllites, schists and in micaceous layers in quartz-schists, mylonites and gneisses. Crenulation lineations are closely

associated with intersection lineations but are different in that they are comprised of fold hinges identifiable to the naked eye.

During folding of layered rocks crenulation cleavages and crenulation lineations form at an early stage, while larger folds form later on during the same process. It is therefore of interest to compare the orientation of early crenulation lineations with related but slightly younger fold axes. If there is a difference in orientation this could be related to how layers rotated during deformation. We are already familiar with the concept of transected folds, where the lineation makes an angle to the hinge line (Figure 12.20).

Boudinage

Boudins are competent rock layers that have been stretched into segments (Figure 13.4). Individual boudins are commonly much longer in one dimension than the other two and thus define a lineation. Such linear boudins form where the X-axis of the strain ellipsoid is significantly larger than Y. Chocolate-tablet boudins (see next chapter) can form when $X \approx Y$.

When occurring in folded layers, boudins typically appear on the limbs of the fold with their long axes oriented in the direction of the fold axis (Figure 13.5). In general, boudinage structures are most easily recognized in sections perpendicular to the long axes of the boudins. Because of this fact they may be difficult to recognize as linear features in deformed rocks. It is also true that boudins are restricted to competent layers and therefore more restricted in occurrence than most other lineations.

Mullions

Mullion is the name that structural geologists use for linear deformation structures that are restricted to the

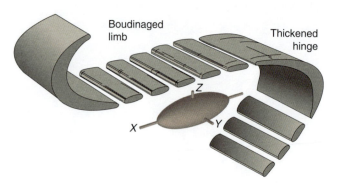

Boudinaged limb

Thickened hinge

Figure 13.5 Common connection between folding and boudinage. The fold hinges are thickened while the limbs are extended and boudinaged. The strain ellipse is indicated.

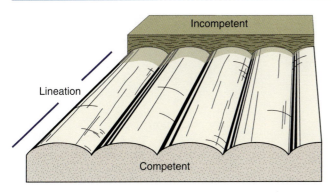

Figure 13.6 Mullion structures form lineations at the interface between rocks of significantly different competence (viscosity).

Figure 13.7 Two perpendicular sets of mineral lineations on a fault surface in serpentinite. The two directions indicate movements under two different stress fields. Leka Ophiolite, Scandinavian Caledonides.

Figure 13.8 Fiber lineation (talk) in extension fracture where the fibers have grown perpendicular to the fracture walls as the fracture opened.

interface between a competent and an incompetent rock. The term mullion has been used in several different ways in the literature, ranging from striations on fault surfaces (fault mullions) to layer-interface structures formed during layer-parallel extension as well as contraction. We will relate the term to layer-interface structures where the viscosity contrast is significant. In such cases the cusp shapes of mullions always point into the more competent rock, i.e. the one with the higher viscosity at the time of deformation (Figure 13.6). Such mullions are closely related to buckle folds in the sense that their formation is predicted by a contrast in viscosity, they form by layer-parallel shortening, and their characteristic wavelength is related to the viscosity contrast. But they differ from buckle folds in having shorter wavelengths and they are restricted to a single layer interface. A common place to find mullion structures in metamorphic rocks is at the boundary between quartzite and phyllite or micaschist. Mullions also occur on the surface of quartz pods in micaschists.

Pencil structures

The formation of **pencil structures** occurs as a result of discrete interference between compaction cleavage and a subsequent tectonic cleavage, or between two equally developed tectonic cleavages, as discussed in Chapter 12. Pencil structures have a preferred orientation and form a lineation in unmetamorphosed and very low-grade metamorphic rocks.

13.3 Lineations in the brittle regime

Some lineations occur only on fracture surfaces. They are not fabric-forming elements and are more characteristic of the brittle regime in the upper crust. These lineations form by mineral growth in extension fractures, as striations carved on the walls of shear fractures and faults, by intersections between fractures and by fracture curvatures that form early during fracturing.

Mineral lineations in the brittle regime tend to be restricted to **fiber lineations**, where minerals have grown in a preferred direction on fractures (Figure 13.7). The growth of minerals on fractures usually requires that the fractures open to some extent, either as true extension fractures or as shear fractures with a component of extension. Furthermore, the minerals must grow in a preferred direction for a lineation to be defined. Minerals such as quartz, antigorite, actinolite, gypsum and anhydrite may appear fibrous on fractures.

Mineral fibers are found in many extensional or Mode I fractures, as shown in Figure 13.8. The orientation of the fibers is commonly taken to represent the extension

(a)

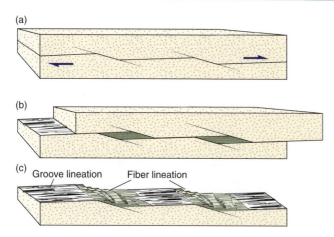

(b)

(c) Groove lineation   Fiber lineation

Figure 13.9 Formation of fiber lineation in irregular shear fractures. (a) Early stage. (b) Final stage. In (c) the upper wall is removed for inspection. Groove lineations (striations) are found on surfaces that have not opened during faulting.

Figure 13.10 Slickensides with slickenlines, formed by cataclastic shearing of epidote on a post-Caledonian normal fault in the Precambrian of West Norway.

Figure 13.11 Fault grooves on striated fault surface separating limestone from sandstone. Also see Figure 9.4. Moab Fault, Utah.

direction. Curved fibers are sometimes seen, implying that the extension direction has changed during the course of deformation, or that shear has occurred during or after the formation of the fibers.

Even though extension is involved in the formation of fibrous mineral lineations, this does not imply that such lineations are restricted to extension fractures. Because of the irregular shapes of most shear fractures, a component of extension may occur in extensional stepovers, and minerals may grow as walls separate (Figure 13.9). The mineral fibers then grow on the lee side of steps or other irregularities, precipitated from fluids circulating on the fracture network. Thus, the sense of slip is detectable from the relation between fiber growth and fracture geometry.

Non-planar shear fractures can contain extensional (pull-apart) segments where fibers can grow and form a local lineation.

Striations or **slickenlines** are lineations found on shear fractures and form by physical abrasion of hanging-wall objects into the footwall or vice versa (Figure 13.10). The smooth and striated slip surface itself is called a **slickenside**. Slickensides tend to be shiny, polished surfaces coated by a ≤1 mm thick layer of crushed, cohesive fault rock. Hard objects or asperities can carve out linear tracks or grooves known as **fault grooves** (Figure 13.11). The term **groove lineation** can be used for this type of slickenlines.

Such mechanically formed slickensides may show similarities with glacially striated surfaces. Close examination

of many slickensided slip surfaces reveals that they are formed on mineral fill or that they actually are fiber lineations.

There are two principal types of slickenlines: those that form by mechanical abrasion (striations) and those formed by fibrous growth (slip fiber lineations).

Minerals may grow during the movement history of a fault, and it is common to find a combination of fiber lineations and striations. Such lineations may be of the same or different ages, and sometimes two or more different sets of lineations of different minerals exist that may or may not be striated due to mechanical abrasion.

Figure 13.12 Lineations in a zone of deformation bands. The lineation points down dip and is an expression of the cigar-shaped geometry of undeformed volumes of rock between the deformation bands. Compass for scale. Entrada Sandstone, San Rafael Desert, Utah.

Geometric striae relate to the irregular or corrugated shape of a slip surface. Such irregularities may have a preferred orientation or axis in the slip direction and appear as lineations on an exposed wall. A special type of geometric striae is the cigar-shape seen on the walls of deformation band cluster zones, as portrayed in Figure 13.12. Geometric striae and physical striae or slickenlines commonly coexist.

Intersection lineations are found on fractures where the main slip plane is intersected by secondary fractures such as Riedel fractures or tensile fractures. The lines of intersection typically (but not necessarily) form a high angle to the slip direction, in marked contrast to striae and mineral lineations that tend to parallel the slip direction.

Finally, we mention a type of lineation that is typical for some deformed limestones. It forms perpendicular to well-developed pressure solution seams where shortening occurs across the fracture surface and is composed of tubular structures known as **slickolites**. These structures tend to point in the direction of contraction and slickolites are thus a kind of lineation that is kinematically different from the other lineations discussed above.

13.4 Lineations and kinematics

Geologists have always had a feeling that linear structures indicate the kinematic or movement pattern during deformation, in one way or another. We will discuss this here, first in the brittle regime of the upper crust and then in the plastic regime.

Fault-related kinematics

Lineations are important structures for understanding the sense of movement on individual slip surfaces as well as the kinematics of fault populations. Fiber lineations, slickenlines and so-called geometric lineations give good indications of the movement, although a reliable determination of the sense of movement may require additional information.

Slickenlines, striae and several other lineations associated with faults parallel the movement direction, but do not in themselves reveal the sense of shear.

To distinguish normal from reverse movements or sinistral from dextral slip we need detailed information about fracture morphology relative to mineral growth, second-order fracture geometry, correlation of markers across the slip surface or observation of drag adjacent to the fault. The traditional method of "feeling" the sense of slip from moving your hand along the slip surface has been proven by many field geologists to be an unreliable method. Instead we should closely examine any irregularities and structures on and around the slip surface to try to understand their geometry and formation, and based on that evaluate the sense of slip. This requires that we understand the difference and implications of extension fractures and different types of shear fractures forming during faulting, and have diagrams such as Figure 9.3 in mind.

It is common to find two or more sets of linear structures on a single slip surface (Figure 13.7). Different sets record different movements at different times, implying that the stress field changed between different slip events or that local geometric complications perturbed the stress field during faulting. In some cases the growth history and composition of minerals on such fractures can be used to reveal the relative timing of movement events.

Lineations and kinematic axes in the plastic regime

Lineations are common structures in plastic shear zones and mylonite zones, including extensional shear zones, strike-slip zones and mylonites associated with thrust nappes. Recognized as important kinematic structures as early as the nineteenth century, the discussion whether such lineations form parallel or perpendicular to the general movement direction persisted into the mid-1900s.

Around 1950, the Austrian geologist Bruno Sander set a new scene with his focus on coordinate systems and symmetry. He introduced the concept of **kinematic axes** by defining three orthogonal axes, *a*, *b* and *c*, with the *a*-axis representing the transport direction and *c* being perpendicular to the shear plane (Figure 13.13). The issue was: do lineations and fold axes parallel the kinematic *a*- or *b*-axis?

Sander and his school insisted that lineations represent the *b*-direction, i.e. not the transport direction. Others, such as the British geologist E. M. Anderson and the Norwegian geologist Anders Kvale, opposed Sander's view. In the Caledonides the stretching lineation (e.g. Figure 1.11) was found to parallel the regional transport direction. The lineation was also parallel to hinges of tight folds. Anderson, Kvale and others thus concluded that the lineation in most cases is parallel to the *a*-axis. This is also the view that soon became generally accepted.

The concept of kinematic axes was soon abandoned and replaced by **strain axes** and their angular relations with the shear zone walls or other suitable planes of reference. For simple shear, the stretching lineation lies in Sander's *a–c* plane, or in the *XZ* plane of the strain ellipsoid as it rotates from its initial position at 45° to the shear plane toward the shear direction (i.e. the kinematic *a*-axis; Figure 13.13). The lineation thereby has an orientation that depends on strain. For very high strains the stretching lineation approximates the *a*-axis. For a thrust or penetratively deformed thrust nappe, this means that the trend of the lineation indicates the direction of transport. The actual sense of transport or sense of shear is not taken into account here and cannot be resolved by studies of lineations alone, but requires studies of asymmetric structures discussed in Chapter 15.

Stretching lineations commonly indicate the direction of transport when projected onto the shear plane.

Transport directions are somewhat less meaningful when dealing with coaxial deformations such as pure shear, but even in these cases a characteristic pattern of transport can be predicted. Just think of a soft thrust nappe collapsing coaxially above a rigid basement. A consistently arranged stretching lineation is then expected, and for perfect pure shear this lineation will be parallel to the transport direction. Hence, the lineation becomes a perfect *a*-lineation in Sander's nomenclature.

For other types of deformation, such as transtension, the lineation can be found to be oblique to the shear direction no matter how intense the deformation may be (see Chapter 18). In fact, it may even change orientation in quite dramatic ways. Other general deformations, such as oblique combinations of pure and simple shear (where the shear plane and the pure shear flattening plane are oblique), can give a whole range of lineation orientations. The relationship between lineations and the transport direction may become quite complicated, but it is important to realize that the implications of a lineation depend on the type of deformation, be it is simple shear, pure shear, or some combination of the two.

Map-scale linear patterns are also related to strain and kinematics. An example is the radial lineation patterns generated by three-dimensional gravitational collapse, as shown in Figure 16.29. Observations of such lineation patterns suggest that the deformation is far from plane. However, there are many places where lineations are parallel over wide areas, and strain data can be used to find out if these formed during plane strain (with no transport perpendicular to the lineation) or not. If the

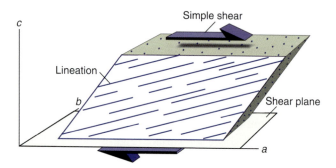

Figure 13.13 The stretching lineation and its relation to Sander's kinematic axes for simple shear. The lineation rotates toward the kinematic *a*-axis (the shear or transport direction) by increasing strain.

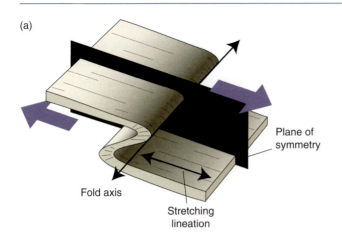

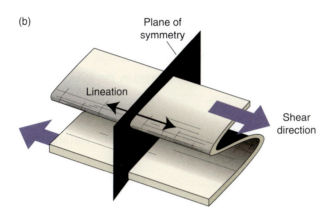

Figure 13.14 Folds and symmetry. Folds may have axes that are parallel as well as perpendicular to the stretching lineation. Using the concept of symmetry to estimate the transport direction is clearly not a reliable approach when dealing with folds.

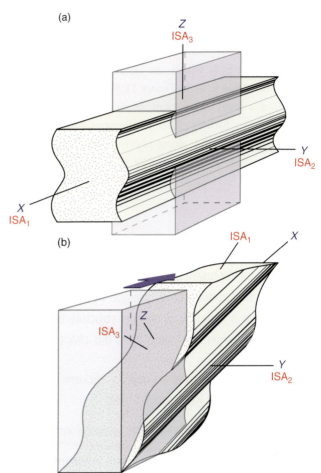

Figure 13.15 Folds and stretching lineations formed under pure shear (top) and simple shear (bottom). In both cases the fold axes and stretching lineation become subparallel (perfectly parallel for pure shear, subparallel for simple shear). Based on Rykkelid (1987).

strain ellipsoid is close to plane (plotting along the diagonal of the Flinn diagram) we have a good reason to assume plane strain with the lineation indicating the transport direction. If strain is far from plane a more complex relationship may exist between transport direction and lineations. In general, however, the assumption that lineations indicate the transport direction is a useful working hypothesis.

Fold hinges and related types of lineations

Stretching lineations are often the easiest type of lineation to interpret when transport direction is the issue. Fold axes, crenulation lineations and intersection lineations can be more ambiguous. Such lineations may be results of a later, superimposed deformation that has nothing to do with the main movement picture. In other cases they are the last expression of a progressive deformation phase, and in this case they give useful information

about the kinematic picture. This is sometimes the case in mylonites, where the foliation is continually folded and new cleavages are formed repeatedly during the process of mylonitization.

Fold axes and intersection lineations are genetically different from stretching lineations and have a much more complicated relationship with the transport direction. In some cases they make a high angle to the stretching and mineral lineation (Figure 13.14a), in other cases they are oblique or parallel (Figure 13.14b). It is not uncommon in high-strain zones to find fold axes and intersection lineations to be more or less parallel to the stretching and mineral lineation (Figure 13.14b and 13.15).

Fold axes that parallel other lineations were still considered a problem in the middle of the twentieth century. Today we have a set of reasonable explanations for parallel fold axes and other lineations. It is, for example, clear that fold axes may rotate toward parallelism with the shear direction. This may be a matter of going from Figure 13.15 a to b, which can occur if the hinge line is curved and strain is high (Figure 11.39). It can also occur if there is a local component of strike-slip movement. Folds can also form parallel to the stretching or shear direction, for instance along highly elongated lenses of more competent material. Figure 13.15a shows a model where vertical layering in a horizontal high-strain zone is affected by pure shear (a) and simple shear (b). In (a) folds form parallel to X (the stretching direction) without any rotation. In (b) folds also form parallel to X and the stretching lineation, but rotate toward the horizontal transport direction with increasing strain.

Folds can form with axes at any angle to the transport direction and typically rotate toward this direction as strain accumulates and folds tighten.

Summary

Lineations are important because of their close relationship to kinematics or movement directions and, together with the foliation(s) that contain them, provide significant information about the deformation history. Boudinage, often considered as a lineation, is another structure that deserves closer attention, and will be treated in the following chapter. Some important points from this chapter:

- Lineations can be stretched objects, slickenlines on slip surfaces, lines of intersection between planar surfaces and non-physical lines such as fold axes.

- Each of these types of lineations must be interpreted separately.

- Lineations in the brittle regime form in extension fractures, veins and on shear fractures.

- In the plastic regime lineations form penetrative fabrics alone (L-fabrics) or together with foliations (LS-fabrics).

- Stretching of minerals and mineral aggregates gives stretching lineations.

- A very pronounced stretching lineation may indicate prolate strain.

- Many lineations are closely related to strain or slip and therefore to kinematics.

- Using a lineation as an indicator of transport direction (movement direction) requires that we understand how the lineation formed.

- Stretching lineations are commonly favored when estimating the transport directions in orogenic belts, while intersection lineations and fold hinges are less reliable.

Review questions

1. What makes mullions and buckle folds similar?

2. What types of lineations form in the brittle regime?

3. What lineations mark the X-axis of the strain ellipsoid?

4. What type of lineation can be related to ISA_1?

5. How do striations or slickenlines relate to kinematics?

6. What is the difference between crenulation lineations and intersection lineations?

7. How do boudins relate to the strain ellipsoid?

8. How do mineral lineations that form in the brittle and plastic regime differ?

9. Why must we have a strain model when considering the relationship between stretching lineation and transport direction?

10. How are the various lineations defined and how do they develop?

E-MODULE

The e-learning module called *Lineations* is recommended for this chapter.

FURTHER READING

General

Cloos, E., 1946, *Lineation*. Memoir 18, Boulder: Geological Society of America.

Sander, B., 1930, *Gefügekunde der Gesteine*. Vienna: Springer-Verlag.

Sander, B., 1948, *Einführung in die Gefügekunde der Geologischen Körper*, Vol. 1. Vienna: Springer-Verlag.

Sander, B., 1950, *Einführung in die Gefügekunde der Geologischen Körper*, Vol. 2. Vienna: Springer-Verlag.

Turner, F. J. and Weiss, L. E., 1963, *Structural Analysis of Metamorphic Tectonites*. New York: McGraw-Hill.

Stretching lineations

McLelland, J. M., 1984, The origin of ribbon lineation within the southern Adirondacks, U.S.A. *Journal of Structural Geology*, **6**: 147–157.

Transport direction

Ellis, M. A. and Watkinson, A. J., 1987, Orogen-parallel extension and oblique tectonics: the relation between stretching lineations and relative plate motions. *Geology*, **15**: 1022–1026.

Kvale, A., 1953, Linear structures and their relation to movements in the Caledonides of Scandinavia and Scotland. *Quaternary Journal of the Geological Society*, **109**: 51–73.

Lin, S. and Williams, P. F., 1992, The geometrical relationship between the stretching lineation and the movement direction of shear zones. *Journal of Structural Geology*, **14**: 491–498.

Petit, J.-P., 1987, Criteria for the sense of movement on fault surfaces in brittle rocks. *Journal of Structural Geology*, **9**: 597–608.

Ridley, J., 1986, Parallel stretching lineations and fold axes oblique to a shear displacement direction: a model and observations. *Journal of Structural Geology*, **8**: 647–653.

Shackleton, R. M. and Ries, A. C., 1984, The relation between regionally consistent stretching lineations and plate motions. *Journal of Structural Geology*: **6**, 111–117.

Rotation of lineations

Sanderson, D. J., 1973, The development of fold axes oblique to the regional trend. *Tectonophysics*, **15**: 55–70.

Skjernaa, L., 1980, Rotation and deformation of randomly oriented planar and linear structures in progressive simple shear. *Journal of Structural Geology*, **2**: 101–109.

Williams, G. D., 1978, Rotation of contemporary folds into the X direction during overthrust processes in Laksefjord, Finnmark. *Tectonophysics*, **48**: 29–40.

Chapter 14
Boudinage

In the plastic regime, layers tend to fold when shortened, particularly if there is a viscosity contrast between individual layers. In this chapter we will look at how layers that are being stretched can part into pieces known as boudins. Classic boudins represent the counterpart to buckle folds and provide solid evidence of layer-parallel extension that is preserved even if the layers experience later shortening and folding. Just like folds, boudins form in different ways and provide us with different types of information that are well worth our attention.

14.1 Boudinage and pinch-and-swell structures

The term boudins, which comes from the French word for sausage, was first used in 1908 by Max Lohest for shortened boudins or mullions in Bastogne in Belgium. Before that, boudins and pinch-and-swell structures had been described several times. The exact definition and meaning of the terms boudins and boudinage have changed through much of the twentieth century, but there is now consensus that **boudins** are extensional structures formed by layer-parallel extension, while **boudinage** is the process that leads to the formation of boudins from originally continuous layers.

Classic boudins form where single competent layers are extended into separate pieces through plastic, brittle or a combination of plastic and brittle deformation mechanisms (Figure 14.1). The boudinaged layer is located in a rock matrix that deforms plastically. As indicated in Figure 14.1, boudins are separated by brittle extension fractures (left side) or by shear fractures that may be symmetric or asymmetric (middle and right side of Figure 14.1). Instead of fractures, boudins may also be separated by narrow ductile shear zones that are confined to the boudinaged layer.

Boudins are more or less regularly shaped and spaced fragments formed by stretching of competent layers or foliations.

There are also examples of regularly spaced areas of thinning in many extended competent layers without the separation into isolated fragments or boudins. Such structures are called **pinch-and-swell structures** and the process by which they form is known as **necking**. Pinch-and-swell structures are structures where the boudin-like elements are barely connected, as shown in Figure 14.2b. Both regular boudins and pinch-and-swell structures are controlled by temperature, strain rate and viscosity contrast or a well-developed foliation. High temperature promotes plastic deformation mechanisms also in the most competent layer. High-viscosity contrast and strain rate promote fracturing of competent layers. These parameters will also affect the geometry of boudins.

14.2 Geometry, viscosity and strain

A boudin has **thickness** and **width**, and boudins have a measurable **separation** (Figure 14.2a). Experiments show that thick layers develop wider boudins than thin layers, just as thick buckled layers show longer wavelengths than thin layers (Figure 11.20). During boudinage a competent layer is broken up into more and more boudins until a characteristic width/thickness or **aspect ratio** is reached.

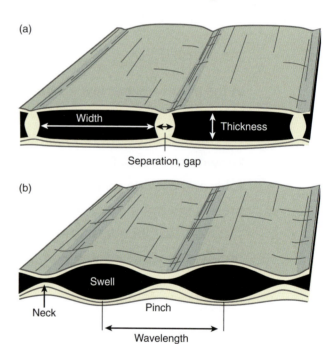

Figure 14.2 Descriptive terminology of boudins and pinch-and-swell structures.

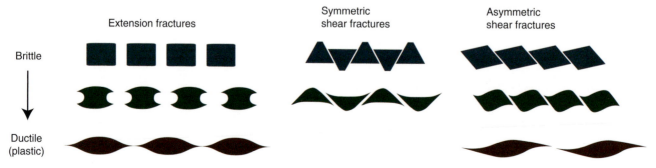

Figure 14.1 The geometry of boudins is largely controlled by whether boudins are separated by extension or shear fractures, and the influence of plastic versus brittle deformation mechanisms. Asymmetric boudins may indicate non-coaxial deformation.

(a)

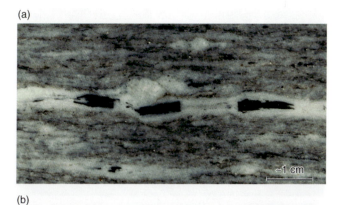

(b)

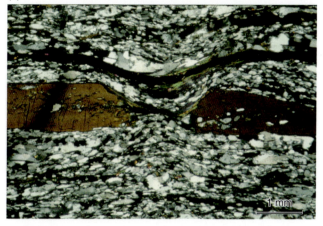

Figure 14.3 Boudinage on the microscale. Top: boudinaged Swiss amphibole crystals as seen in hand sample. Crystals are 2 mm thick. Bottom: boudinaged Norwegian amphiboles as seen in thin section. This amphibole is 1 mm thick.

Typical characteristic aspect ratios fall in the range 2–4 (Figures 14.3 and 14.4), and further layer-parallel extension will only increase the separation of the boudins. In some ways we can compare the aspect ratio of boudins to the characteristic wavelength of buckle folds.

The separation is the distance between the boudins. Unlike the aspect ratio, the separation is independent of the viscosity contrast between the boudinaged layer and its adjacent layers or matrix. Instead it depends mostly on strain, more specifically on the amount of layer-parallel extension. Separation is therefore more variable than the aspect ratio of boudins.

Hans Ramberg extended competent layers in a less competent matrix in the laboratory. He found that the competent layers developed into boudins or formed pinch-and-swell structures. He also realized that the strongest or most competent layers formed the most rectangular boudins (Figure 14.5a, b). Less competent layers developed pinch-and-swell structures (Figure 14.6). Ramberg's experiments tell us that the shape of boudins reflects the viscosity contrast in the rock during deformation. Well-rounded or pulled-out corners, such as those seen in Figure 14.7, indicate that the margins of the boudins deformed predominantly by plastic deformation mechanisms, similar to the neighboring layer. In some cases we can see plastic deformation of one margin only, as shown in Figure 14.5d. Experiments tell us that the

Figure 14.4 Rectangular boudins, formed by stretching of a granitic dike in metasediments. Cross-cutting relations are preserved between the dike margins and the foliation, showing that the dike rotated anticlockwise relative to the foliation. Hydrothermal deposition of quartz in the gaps between the boudins is a common feature in boudinaged felsic rocks. Photo: E. Rykkelid.

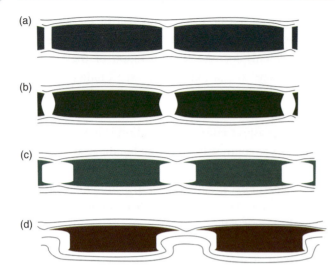

Figure 14.7 Plastic deformation of the encircled margins of boudins in the Proterozoic gneisses of southwest Norway correlate with the type shown in Figure 14.5c.

Figure 14.5 Different types of boudin geometry, all involving extensional fracturing. (a) Rectangular, (b) weak plastic component, (c) marked plastic component along margins, (d) pronounced plastic component along upper margin, almost absent along lower margin, indicating that the viscosity contrast is higher along the lower margin.

Figure 14.8 Boudinaged granitic layer in gneiss. The asymmetric boudins have fairly similar, but not identical lengths. According to theory, boudinage continues until all of the boudins have a length shorter than some critical value. Sinistral sense of shear.

Rectangular boudins imply higher competence contrast and more brittle deformation than barrel-shaped boudins.

Figure 14.6 Boudinage structures in amphibolite layers (metamorphosed basaltic dikes) in quartz schist, northern Norway Caledonides. The deformation is strongly influenced by plastic deformation. Photo: Steffen Bergh.

viscosity contrast is different along the two margins, being least where the plastic deformation (and rounded corners) occurs.

Numerical modeling has shown that the stress concentration is largest at the corners of boudins. This explains why the corners are the first parts of the boudins to be deformed. Where this stress concentration results in deformation of the boudins, **barrel-shaped boudins** (Figures 14.5c and 14.7) or **fish-mouth boudins** (Figure 14.5d) may result.

Ramberg and later workers carried out theoretical analyses that indicate that the tensile stress causing boudinage increases from the corners toward the central part of the boudin margin. This means that long boudins have higher tensile stress in their central parts than short boudins. The subdivision of early-formed boudins thus continues until the tensile stress in the middle of the boudin falls below a critical value, which occurs when the boudin becomes shorter than some critical length. The boudin length is below the critical length when the tensile stress in the boudin is smaller than the tensile strength of the material. This model also explains why individual boudins from the same layer tend to show a certain variation in length (Figure 14.8).

Pinch-and-swell structures have even larger similarities with buckle folds than boudins. In cases where the viscosity contrast is significant, the mathematical relationship between the dominant wavelength (L_d) and thickness (h)

can be directly applied. The relationship is given by Biot's classic formula

$$L_d/h = 2\pi(\mu_L/6\mu_M)^{1/3} \tag{14.1}$$

where L_d/h represents the length/thickness aspect ratio of the structures. Pinch-and-swell structures can also form when the viscosity contrast is small (by a mechanism known as resonance folding), but then the L_d/h values tend to fall in the range 4–6.

Buckle folds can form in both non-linear (Newtonian) and linear media. However, pinch-and-swell structures only form in media with non-Newtonian properties. This restriction is one of several reasons why folds are more abundant than pinch-and-swell and boudin structures in deformed rocks. Another is that buckling may form by amplification of initial layer irregularity during layer-parallel shortening. Such irregularity would be suppressed during layer-parallel extension/layer-perpendicular shortening, and therefore be of little help in initiating classic boudins. Folds can also form by passive mechanisms, in rocks with layers with no viscosity contrast at all. In contrast, the formation of boudins in layers without viscosity contrast is restricted to cases of well-foliated rocks in shear zones, as discussed in Section 14.4.

14.3 Asymmetric boudinage and rotation

Until this point we have regarded boudins as symmetric structures, but **asymmetric boudins** (Figures 14.1, 14.8 and 14.9) are also commonly found in deformed metamorphic rocks. Asymmetric boudins are separated by shear fractures or shear bands (small-scale shear zones) that tend to die out once they leave the boudinaged layer. If the boudinaged layer behaves in a brittle manner during deformation, then the differential stress or strength of the competent layer will determine whether shear fractures or extension fractures form. In general, extension fractures form when $(\sigma_1 - \sigma_3) < 4T_0$, where T_0 is the tensile strength of the rock layer.

Boudins separated by shear fractures are sometimes pulled apart without much rotation or shear (Figure 14.10a, b). These boudins parallel the surrounding foliation, and the shear fractures will open in an extensional or opening mode. This behavior is typically seen in migmatitic gneisses where amphibolitic layers, for example, are pulled apart and the new-formed space (volume) is filled with melt or new minerals.

Where boudins are offset by shear structures, significant rotation of the boudins occurs (Figure 14.10c, d).

Figure 14.9 Asymmetric boudins in amphibolitic gneiss. The boudins are separated by shear fractures in the amphibolite. The fractures turn into ductile shear zones outside of the boudins and rapidly die out. Sinistral sense of shear.

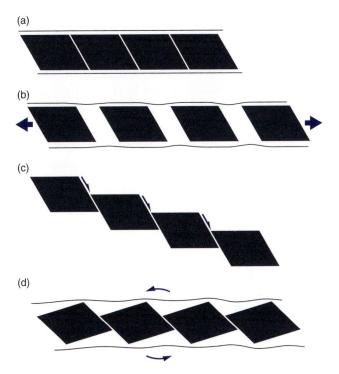

Figure 14.10 Asymmetric boudins can form by extension across shear fractures (a, b), or by a combination of shearing along shear fractures (c) and boudin rotation (d).

This process resembles the domino model for fault systems discussed in Chapter 17, which works as well for systems of asymmetric boudins as for rotated fault blocks in extensional settings (Figure 17.3). The rotation keeps the boudins aligned with the general foliation of the rock.

Rotated asymmetric boudins can also form by rotation of symmetric boudins during non-coaxial deformation.

(a)

(b)

Figure 14.11 Experimental study of the effect of boudin aspect ratio during shearing. Short boudins rotate synthetic to the sense of shear while long boudins back-rotate. Based on Hanmer (1986).

Experiments show that layer-parallel simple shear causes rotation of already existing boudins. But in what direction will the boudins rotate? In simple terms, short and competent boudins rotate as rigid objects synthetic to the direction of shear. This means that if the sense of shear is clockwise, the rotation of the boudins will also be clockwise. However, long boudins, i.e. boudins with high aspect ratios, will rotate against the direction of shear.

Whether a boudin rotates anti- or synthetically with respect to the sense of shear thus depends on the shape or aspect ratio of the boudin (Figure 14.11). However, it also depends on the type of flow (W_k), the presence or absence of a foliation in the boudins, whether slip occurs along its margins, whether extensional fractures or shear bands separate the boudins, and the viscosity or competence contrast at the time of deformation. In general, the larger the competence contrast, the more rigid the boudins, and rigid objects rotate more easily with the shear sense. If the boudins are asymmetric and separated by shear bands they tend to show antithetic back-rotation (the domino effect mentioned above). The effect of a strong foliation is discussed in the next section. Field observations show that most (but not all) boudins rotate against the sense of shear, as shown in Figure 14.12 and in Figures 14.8 and 14.9.

Short, symmetric and rigid boudins tend to rotate with the shear direction, while long boudins back-rotate.

An additional factor often overlooked is the orientation of the boudinaged layer prior to deformation. So far we have implicitly assumed that it was more or less parallel to the extension direction, which does not have to be the case. Clearly boudins can form at any orientation within the field of instantaneous stretching. Figure 14.13 shows how a layer can rotate from a high angle toward parallelism with the flattening *XY*-plane during a coaxial deformation history. Rigid objects rotate slower

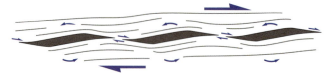

Figure 14.12 Typical sense of rotation of boudins during simple shear. The rotation is "against" the sense of shear, sometimes referred to as back-rotation.

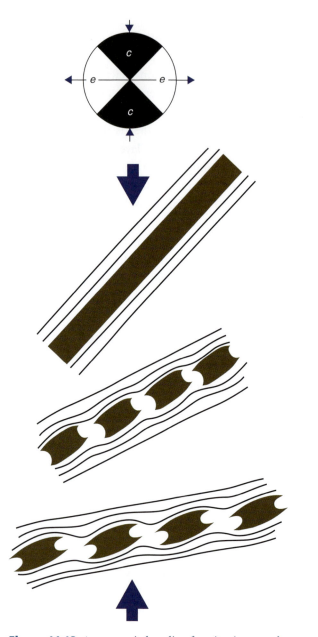

Figure 14.13 Asymmetric boudins forming in pure shear in a layer that is oblique to the plane of flattening.

than the surrounding foliation, causing an asymmetry to develop. Experiments show that the more square the boudins, the larger the difference in rotation between boudins and the foliation. If the coaxial deformation continues far enough, the difference will decrease, but the process typically stops before subparallelism is reached.

What about non-coaxial deformation? It is impossible to decide from the observation of a single occurrence of rotated boudins whether the deformation is coaxial or non-coaxial. Rotated boudins can be the result of either coaxial or non-coaxial deformation, and additional structures must be used to evaluate vorticity.

14.4 Foliation boudinage

In the discussion above we considered the boudinage of individual layers – boudinage in the classic sense. But as we know, rocks, and strongly deformed rocks in particular, are commonly multilayered or well foliated. In the somewhat similar way that layers can get close enough to interfere during buckling, densely packed layers can interfere and generate boudinage structures with a thickness that is considerably greater than that of individual layers or laminae. Whether a planar anisotropy is represented by a mesoscopically identifiable layering or a pronounced schistosity or mylonitic foliation at the microscope scale, it can generate a special type of boudinage known as **foliation boudinage**. Foliation boudins are one or several orders of magnitude thicker than the microlayering represented by the foliation, but do in many ways resemble classic boudins.

Foliation boudinage occur where there is a strong planar anisotropy (foliation) in deformed rocks.

Symmetric foliation boudins are separated by tensile fractures. The fractures are filled with quartz or other hydrothermal minerals, and/or by flow of adjacent rock layers (Figure 14.14). The foliation within the boudin is typically pinched toward the extension fracture (Figure 14.15). The fracture itself can be quite irregular, giving the impression that the foliation has been torn apart.

Asymmetric foliation boudins (Figure 14.16, see also the cover picture for this chapter) are separated by brittle shear fractures or by ductile shear bands showing relative movement along the fractures/bands. It appears that mineral fill is less common than for symmetric foliation boudins. Asymmetric foliation boudins are found on many scales. They are found as single or conjugate sets and have many similarities with the type of shear bands discussed in Chapter 15. Similar to shear bands, they are quite reliable kinematic indicators when only one set of shear fractures or shear bands is

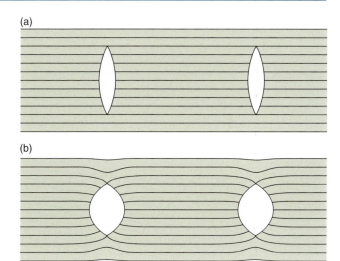

Figure 14.14 The principle of symmetric foliation boudinage. (a) Formation of tensile fractures. (b) Vertical closing and horizontal opening of the fractures. Based on Platt and Vissers (1980).

Figure 14.15 Symmetric foliation boudinage in amphibolite facies gneisses.

developed. They are more reliable than classic boudins, because no significant viscosity contrast exists between the boudins and the surrounding rock so that the effect of rigid body rotation, which promotes synthetic rotation, is eliminated. Instead, rotation is controlled by the geometrically necessary rotation of the foliation around a shear fracture or shear band of limited length, as shown in Figure 14.16.

The strong foliation that develops in many tectonites is generally considered to make homogeneous foliation–parallel extension difficult. This probably explains why foliation boudinage is common in well-foliated rocks: the foliation simply "tears", which allows for more extension. Foliation boudinage is therefore a type of structure that develops at a relatively late stage of deformation, after the foliation is well developed.

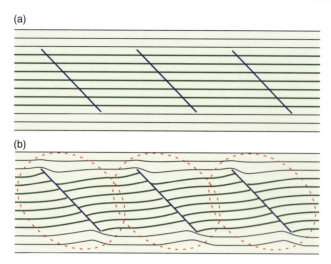

(a)

(b)

Figure 14.16 Formation of asymmetric foliation boudinage. (a) Formation of shear fracture. (b) Movement along the shear fractures (shear bands) causing rotation of the foliation between the fractures. Inspired by Platt and Vissers (1980).

14.5 Boudinage and the strain ellipse

Boudins are usually observed in sections close to the XZ-section of the finite strain ellipsoid. All of the figures shown so far represent this section. However, the third direction is also important because it contains important information about the related strain field (Figure 14.17). It may be challenging to get three-dimensional control in the field, but we should always attempt to look for evidence of strain in the Y-direction.

Boudins commonly show rod-like geometries, where sections perpendicular to the XZ-section show neither boudinage nor folding of the boudinaged layer. The boudins are arranged like railroad cross-ties or sleepers (Figure 14.17b) with no shortening or extension along their long axes. Such a geometry is consistent with plane strain, with the Y-axis oriented along the long axes of the boudins (Figure 14.17b).

If the layers are boudinaged in two directions (Figures 14.17a and 14.18) we have stretching in two directions, implying flattening strain. If the extension can be shown to be equal in the two directions, we have evidence for uniform flattening, i.e. an oblate strain ellipsoid, where $X \gg Y = Z$. This pattern of boudinage is called **chocolate tablet boudinage**.

In some cases, layers can be found to be boudinaged in one direction and folded in the other, as shown in Figure 14.17c. This tells us that there is layer-parallel shortening in one direction and layer-parallel extension in the other.

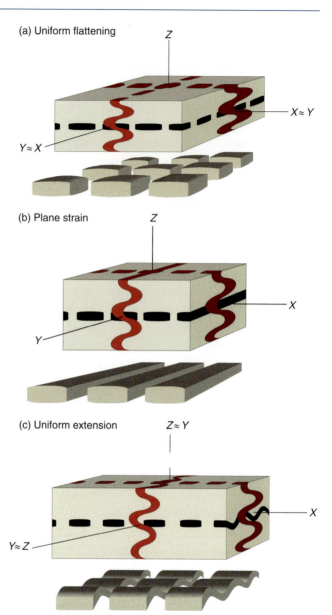

(a) Uniform flattening

(b) Plane strain

(c) Uniform extension

Figure 14.17 The relation between boudinage geometry and strain geometry. Typically, folds and boudins occur in differently oriented layers, but a single layer may show boudin structures in one section and folds in another if strain is constrictional. The 3-D layer at the base of each drawing corresponds to the black layer.

Figure 14.17c shows this in a uniaxial extension strain field, but we can get this geometry also in other fields, depending on the orientation of the layer. In fact, all the vertical layers in Fig. 14.17a–c that contain X show this geometry. Hence, it generally takes more than one boudinaged/folded layer to establish the type of strain in an outcrop.

These structures, formed when layers are simultaneously folded in one direction and boudinaged in the

Figure 14.18 Chocolate tablet boudinage, i.e. boudinage in two directions, indicating two directions of finite extension within the layer. From the Leka Ophiolite, Central Scandinavian Caledonides.

Figure 14.19 Folded boudins in a Caledonian nappe. Such structures cannot form under a single phase of deformation unless the ISA changes orientation during the deformation (non-steady state deformation).

into the shortening field around tectonic lenses or other heterogeneities in a shear zone. Hence, before using folded boudins as conclusive evidence for multiphase deformation, other possibilities must be evaluated.

14.6 Large-scale boudinage

Some seismic images of the lower crust portray conjugate sets of dipping reflectors, which are somewhat different from the patterns seen in the middle and upper crust. Such conjugate sets, which occur at the 100-meter to kilometer scale, are commonly interpreted as oppositely dipping shear zones separating mega-lenses of less-deformed rock. Seismic noise can generate similar patterns, so the interpretation of such seismic images is not always easy. However, the study of exposed pieces of lower-crustal rocks gives some support to the interpretation of large-scale boudinage structures in the lower crust. They may not be considered classic boudinage per se, but they are similar enough to be mentioned in this context. The patterns are consistent with vertical shortening and lateral extension (pure shear) and are particularly characteristic for the lower crust of continental rifts or other areas of significant extension (Figure 14.20).

Boudinage can also occur at larger scale, and it has been suggested that the middle crust can be parted into boudin-like elements (Figure 14.21). A considerable amount of extension is needed to boudinage the crust in this manner. At an even larger scale, the entire lithosphere may develop pinch-and-swell structures and boudins, as discussed in Box 14.1.

other, should not be confused with boudinaged folds or folded boudins. Boudinaged folds can form during progressive deformation when layers rotate from the field of shortening into that of extension. In contrast, folded boudins (Figure 14.19) do not form in a single, steady-state deformation because layers only rotate from the shortening field into the extensional field, and never the opposite.

Boudinaged folds commonly form during progressive deformation, but it takes two phases of deformation to form folded boudins.

However, boudinaged layers can become folded during progressive deformation where the foliation is rotated

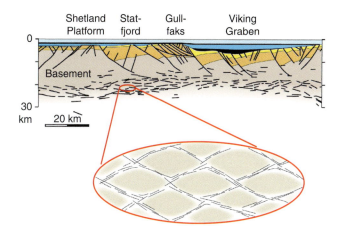

Figure 14.20 Lower crustal structures as interpreted from deep-seismic lines are sometimes interpreted as large-scale boudins and related to vertical shortening.

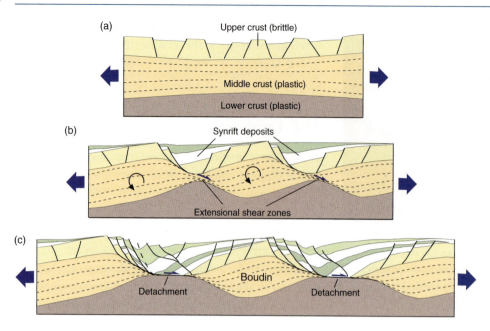

Figure 14.21 Idealized model showing how the middle crust can be extended by asymmetric boudinage. Simplified from Gartrell (1997).

BOX 14.1 | LITHOSPHERIC BOUDINAGE

Several geologic structures or aspects of such structures are scale independent, and boudinage is no exception. Classic boudinage depends on layering and differences in rheologic properties, and such differences occur also at the scale of the lithosphere/asthenosphere. Hence, when the relatively strong lithosphere is stretched during rifting, structures resembling mesoscale boudins may result, and we can talk about lithospheric necking and boudinage.

Lithospheric boudins may, akin to smaller-scale boudins, be symmetric or asymmetric, and their different geometries are controlled by the rheology of the lithosphere, such as viscosity contrasts, strain softening during the deformation history, and coupling between the crust and the underlying lithospheric mantle. For instance, modeling shows that strain softening favors asymmetric boudinage with the development of metamorphic core complexes, structures discussed in Chapter 17 (see Figure 17.8). Strong layers also play an important role, where the upper mantle is the strongest component.

Two examples are shown here. One is a section across the Mediterranean from France through the southern tip of Italy. This section shows back-arc spreading, which from the profile can be described in terms of boudinage of the crust and as a pinch-and-swell structure at the scale of the entire lithosphere. Note that the northwest-most neck is the oldest in this model because of the arc-splitting mechanism effective in back-arc settings.

The other example is one of many numerical models conducted by Ritske Huismans, where both symmetric and asymmetric lithospheric boudinage is portrayed by means of plane-strain finite element modeling. Note that the upper part of lithospheric boudin systems behaves brittlely by faulting, while the lower part deforms ductilely by means of plastic deformation mechanisms.

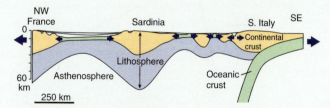

Profile across the Mediterranean, showing pinching of the lithosphere and boudinage of the crust. Based on Gueguen *et al.* (1997).

Continued

BOX 14.1 | (CONT.)

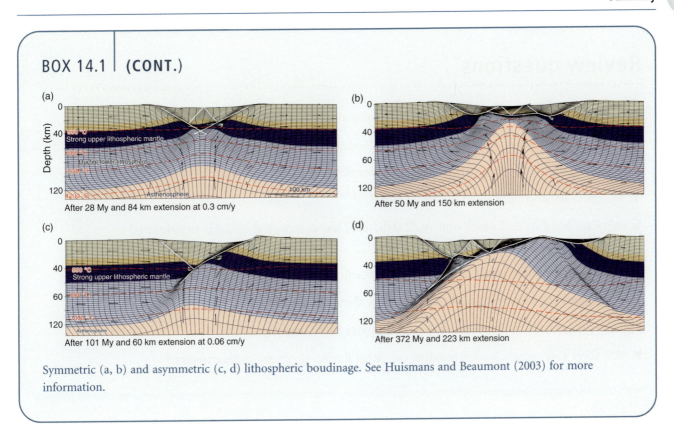

Symmetric (a, b) and asymmetric (c, d) lithospheric boudinage. See Huismans and Beaumont (2003) for more information.

Summary

Boudinage occurs as layers and foliation disrupt during layer-parallel extension. Classic boudinage is controlled by layer thickness and viscosity contrast, while foliation boudinage is less dependent on viscosity contrasts. While classic boudinage is the extensional counterpart to buckling, foliation boudinage is strongly dependent on a foliation anisotropy and hence not to be considered a direct counterpart to passive folding. Boudins are useful during strain analyses and can also be used for shear sense evaluation if original layer orientation, aspect ratio and viscosity contrast are taken into account. Important points to be aware of:

- Boudinage always indicates layer-parallel extension in the section of observation.

- Folded boudins generally indicate two phases of deformation while boudinaged folds can form during a single progressive phase.

- Short and rigid boudins rotate like rigid inclusions.

- Long boudins tend to back-rotate with respect to the sense of shearing.

- The shape of the boudin (angular, barrel or fish-mouth shaped) reflects the viscosity contrast at the time of deformation.

- Foliation boudins are commonly asymmetric, indicating sense of shear in zones of non-coaxial strain.

Review questions

1. What is the difference between classic and foliation boudinage?

2. How can a layer fold and boudinage (extend) at the same time?

3. Why do pinch-and-swell structures sometimes develop instead of boudins?

4. Where is stress concentrated in classic boudins?

5. Why are boudins and pinch-and-swell structures less common than folds in deformed metamorphic rocks?

6. Do rotated boudins imply non-coaxial deformation?

7. What does chocolate tablet boudinage tell us about the strain field?

8. Why are folded boudins indicative of two phases of deformation?

9. Are there cases where folded boudins can form during progressive deformation?

E-MODULE

 The e-learning module called *Boudinage* is recommended for this chapter.

FURTHER READING

General

Goscombe, B. D., Passchier, C. W. and Hand, M., 2004, Boudinage classification: end-member boudin types and modified boudin structures. *Journal of Structural Geology* **26**, 739–763.

Price, N. J. and Cosgrove, J. W., 1990, *Analysis of Geological Structures* (Chapter 16). Cambridge: Cambridge University Press.

Classic boudinage

Lloyd, G. E., Ferguson, C. C., 1981, Boudinage structure: some interpretations based on elastic-plastic finite element simulations. *Journal of Structural Geology* **3**: 117–128.

Ramberg, H., 1955, Natural and experimental boudinage and pinch-and-swell structures. *Journal of Geology* **63**: 512–526.

Smith, R. B., 1975, Unified theory on the onset of folding, boudinage, and mullion structure. *Geological Society of America Bulletin* **86**: 1601–1609.

Strömgård, K. E., 1973, Stress distribution during formation of boudinage and pressure shadows. *Tectonophysics* **16**: 215–248.

Boudinage and 3-D strain

Ghosh, S. K., 1988, Theory of chocolate tablet boudinage. *Journal of Structural Geology* **10**: 541–553.

Zulauf, J. and Zulauf, G., 2005, Coeval folding and boudinage in four dimensions. *Journal of Structural Geology* **27**: 1061–1068.

Folded boudins

Sengupta, S., 1983, Folding of boudinaged layers. *Journal of Structural Geology* **5**: 197–210.

Foliation boudinage

Lacassin, R., 1988, Large-scale foliation boudinage in gneisses. *Journal of Structural Geology* **10**: 643–647.

Platt, J. P. and Vissers, R. L. M., 1980, Extensional structures in anisotropic rocks. *Journal of the Geological Society* **2**: 397–410.

Lithospheric boudinage

Gueguen, E., Dogliono, C. and Fernandez, M., 1997, Lithospheric boudinage in the western Mediterranean back-arc basin. *Terra Nova* **9**: 184–187.

Huismans, R. S. and Beaumont, C., 2003, Symmetric and asymmetric lithospheric extension: Relative effects of frictional-plastic and viscous strain softening. *Journal of Geophysical Research* **108**: doi:10.1029/2002JB002026

Reston, T. J., 2007, The formation of non-volcanic rifted margins by the progressive extension of the lithosphere: the example of the West Iberian margin. In G. D. Karner, G. Manatschal and L. M. Pinheiro (Eds.), *Imaging, Mapping and Modelling Continental Lithosphere Extension and Breakup.* Special Publication **282**, London: Geological Society, pp. 77–110.

Boudinage and kinematics

Hanmer, S., 1986, Asymmetrical pull-aparts and foliation fish as kinematic indicators. *Journal of Structural Geology* **8**: 111–122.

Chapter 15

Shear zones and mylonites

Strain, and shear strain in particular, tends to localize into zones or bands. We have already looked at some types of strain localization structures, such as shear fractures and faults that form in the brittle regime. Localization also occurs in the plastic regime, where foliations and sheared markers tend to show continuity across the zone. Such classic shear zones form an important end-member in a spectrum of shear zones in which both microscale deformation mechanism and ductility vary. On the other end of this spectrum are faults with a measurable thickness. Shear zones can be many kilometers wide, but they also occur on the scale of a hand sample. We will look at shear zones and their internal structure and strain pattern in this chapter, going from a discussion of definitions via the ideal shear zone to different and more complex types of high-strain zones. The last part is devoted to kinematic structures, structures that can reveal the sense of movement in a shear zone, and shear zone growth.

15.1 What is a shear zone?

Faults and shear zones are closely related structures, and Figure 15.1 illustrates the general perception of shear zones as the deep counterpart or extension of faults. Both shear zones and faults are strain localization structures, both involve displacement parallel to the walls, and both tend to grow in width and length during displacement accumulation. Here is a fairly general and simple definition of a **shear zone**:

A shear zone is a tabular zone in which strain is notably higher than in the surrounding rock.

Some would add that a shear zone should contain at least a component of simple shear, but if we want a terminology that is consistent with that of fractures, stylolites and deformation bands (Figure 7.37), a zone of localized coaxial strain also should to be regarded as a shear zone. This opens up a classification of shear zones according to deformation type, for example **pure shear zones**, **subsimple shear zones** and **simple shear zones**.

A shear zone is bounded by two margins or **shear zone walls** that separate the shear zone from its **wall rock** (Figure 15.2). This fact allows us to apply the hanging-wall–footwall terminology from Chapter 8.

Once the margins of a shear zone are defined, we can measure its thickness, and sometimes its displacement. Displacement can be directly measured if the zone offsets a marker, and can also be estimated from internal structures in some cases, as discussed later in this chapter.

The above definition of a shear zone is wide, and embraces faults as well as classic **ductile shear zones**, where markers can be traced continuously through the zone. However, there are differences between faults and classic ductile shear zones, in terms of displacement distribution, anatomy and deformation mechanisms, that are worth noting. Let us first make the observation that any shear zone has a thickness that is significant relative to its displacement. Hence, the millimeter-thin cataclasite associated with a single frictional slip surface of several meters offset is too thin to be regarded as a shear zone. More well-developed faults tend to show a two-fold anatomy, with a central high-strain core and a low-strain damage zone (Figure 8.10). We can see from Figure 15.3 that, for a given displacement, fault cores tend to be thinner than ductile shear zones. However, the combined fault core and damage zone thickness corresponds well with the thickness of more ductile shear zones. A difference between faults and ductile shear zones lies in the distribution of strain within the zone, with strain

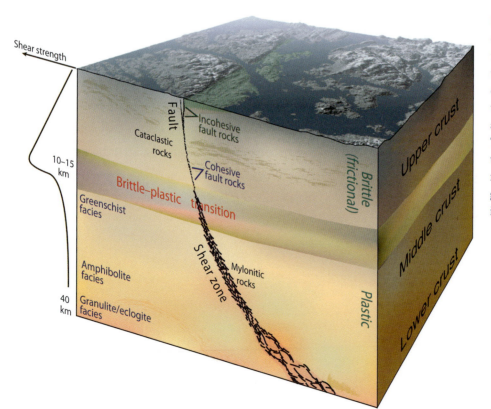

Figure 15.1 Simplified model of the connection between faults, which normally form in the upper crust, and classic ductile shear zones. The transition is gradual and known as the brittle–plastic transition. The depth depends on the temperature gradient and the mineralogy of the crust. For granitic rocks it normally occurs in the range of 10–15 km.

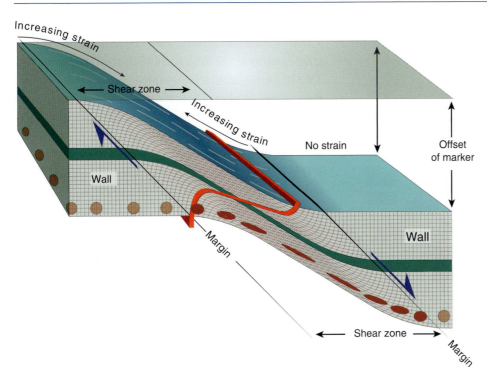

Figure 15.2 Ideal shear zone deforming a grid with two planar markers and circular strain markers. Note how the grid squares change shape and the planar markers change orientation and thickness across the zone. The strain is at its maximum in the central part of the shear zone.

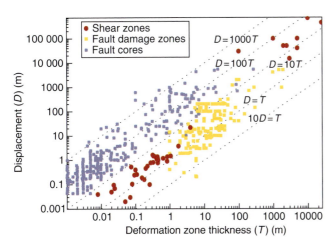

Figure 15.3 Shear zone thickness plotted against displacement for ductile shear zones dominated by plastic deformation mechanisms, faults damage zones/cores, and just fault cores (data from Figure 8.12).

variations being more gradual for ductile shear zones. Another difference is that ductile shear zones can involve both plastic and brittle deformation mechanisms, while faults are totally dominated by brittle mechanisms.

In summary, faults are non-ductile shear zones, forming a subclass of shear zones with its own characteristic features. Other subclasses of shear zones can be defined on the basis of kinematics, microscale deformation mechanisms (plastic or brittle), metamorphic grade, tectonic significance, and so on. Depending on the tectonic regime, shear zones may show normal, reverse, strike-slip or oblique shear, just like faults. Extensional and contractional shear zones tend to be low-angle ($<30°$ dips) while those dominated by strike-slip movement tend to be steep. Shear zones occur at almost any scale in any tectonic regime and form at any depth, although most commonly in the plastic regime.

Kinematic classification

Just like faults and fractures, shear zones can be classified on the basis of relative movement between the two walls, i.e. on the basis of kinematics (Figure 15.4). Although most shear zones tend to be dominated by simple shear, there may also be dilation, compaction and/or pure shear involved. Shear zones in sand and sandstones show the whole range from compaction zones (compaction bands) through simple shear zones (simple shear bands) to dilation zones (dilation bands). Pure shear is also possible in fault cores where clay and sand can more freely be extruded or injected along the fault.

Even though most shear zones may be simple shear-dominated, there is a full 2-D kinematic spectrum from compaction zones via simple shear zones to dilation zones.

Plastic shear zones can show significant deviations from simple shear. However, the ideal shear zone as seen from a strain perspective is one where the deformation is simple shear with or without additional compaction or

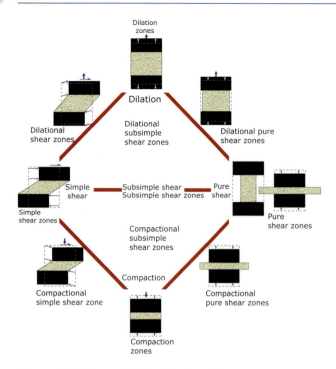

Figure 15.4 Kinematic classification of shear zones (plane strain). Note that pure shear may involve shear zone-perpendicular shortening as well as extension.

dilation. We will have a special look at the ideal shear zone before proceeding with more complex ones. Before that we need to discuss another way of classifying shear zones, which is related to the somewhat ambiguous terms brittle, plastic and ductile deformation.

Brittle versus plastic shear zones

Our general definition of shear zones at the beginning of this chapter makes no restrictions regarding microscale deformation mechanisms. Hence, a shear zone can contain some elements (mineral grains, lenses, layers) that deform plastically and others that deform brittlely at the same time, or everything may deform either plastically or brittlely. The deformation mechanism(s) activated depends on temperature, pressure, metamorphic reactions, cementation, strain rate and amounts of fluid available, in addition to the properties and distribution of rocks and minerals in the zone. It is also worth noting that deformation mechanisms can vary through the life of a shear zone due to changes in physical conditions.

In the overall brittle or frictional regime of the upper crust (Figure 15.1), brittle deformation mechanisms dominate so that deformation in the zone is by cataclastic flow. As discussed in Chapter 7, cataclastic flow involves microfracturing, frictional sliding on grain boundaries and rigid rotation of grain fragments. In sand and poorly cemented sandstone at shallow burial depths,

shear zones may develop that deform by granular flow, which involves frictional grain reorganization without grain fracturing. As long as this deformation occurs in a zone of finite width, it is possible to classify the zone as a shear zone, or shear deformation band if occurring on the scale of a hand sample. Shear zones forming predominantly by brittle deformation mechanisms are called **brittle shear zones** or **frictional shear zones**. While an individual slip surface may be too thin to be called a shear zone, more well-developed faults with cores and damage zones can be thought of as end-members in a spectrum of shear zones, even though most of us give preference to the term fault or fault core when referring to brittle discontinuities that sharply cut off structures in the wall rocks.

Deeper in the crust, in the plastic regime, the plastic deformation mechanisms discussed in Chapter 10 come into effect. Where plastic deformation mechanisms dominate we get **plastic shear zones**. In the brittle–plastic transition zone, which can be quite wide if polymineralic rocks such as granites are involved, **brittle–plastic shear zones** form in which both brittle and plastic deformation mechanisms are important. **Semi-brittle shear zone** is another commonly used term in this context, applied to brittle shear zones influenced by plastic deformation mechanisms. Most plastic shear zones contain some brittle elements, such as fractured feldspar or garnet porphyroclasts, unless the temperature is very high.

Ductile shear zones

The term ductile shear zone is a popular term, but also an ambiguous one, because some geologists use the term ductile to imply plastic deformation mechanisms. Equating ductility with plasticity is not recommended, even though many ductile shear zones are indeed plastic. Instead we relate the term ductility to continuity of originally continuous marker elements in the shear zone. A **perfectly ductile shear zone** contains no internal discontinuities, so that marker layers crossed by the shear zone can be traced continuously through the zone at the mesoscopic scale (Figure 15.2). This type of deformation is sometimes named continuous deformation or continuous strain, and most plastic and some brittle shear zones preserve continuity of passive markers. The shear zone shown in Figure 15.5 is an example of a ductile shear zone formed by brittle deformation mechanisms at near-surface conditions. Hence, a ductile shear zone can deform by means of brittle as well as plastic microscale deformation mechanisms.

Figure 15.5 Thin shear zone (disaggregation band) in the Navajo Sandstone (Arches National Park, Utah) showing continuity of sand laminae across the zone. The shear band formed prior to lithification.

Passive markers can be traced continuously through a perfectly ductile shear zone.

Field observations show that many shear zones, including those dominated by plastic deformation, show internal sharp discontinuities in the form of slip surfaces, extension fractures, veins or pressure solution seams. Such shear zones can be termed **semi-ductile**. Discontinuities are found in many plastic shear zones, but are more characteristic for brittle (frictional) shear zones.

Many faults show little or no ductility, with a very rapid transition from nearly unstrained wall rock to intensely sheared breccia, fault gouge or cataclasite in the fault core. However, other faults show fault drag along the core, and thus display a combination of continuous and discontinuous deformation. A fault-propagation fold ahead of a fault can be considered as a ductile shear zone (Figure 8.30 and Box 8.3). Where or when the fault enters the fold, the structure becomes affected by discontinuous deformation and becomes semi-ductile. Hence the deformation in a shear zone can change in space and time from being continuous (ductile) to (semi-)discontinuous.

In summary of what has been discussed in the last two sections, shear zones can be classified according to ductility (continuity of markers) and plasticity (degree of plastic

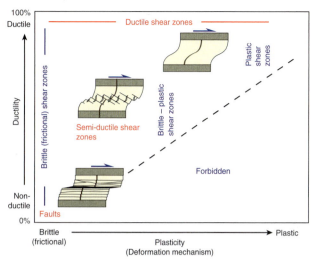

Figure 15.6 Simple classification of shear zones based on deformation mechanism (horizontal axis) and mesoscopic ductility (continuity of markers).

versus brittle deformation mechanisms). This is illustrated in Figure 15.6, which indicates that the term brittle shear zone may have two meanings, either frictional deformation at the microscale (no plasticity), the use that is preferred here, or complete discontinuity of structures (no ductility). In both understandings of the term, brittle shear zones plot in the lower left corner of the diagram shown in Figure 15.6.

15.2 The ideal plastic shear zone

The **ideal shear zone** is limited by two perfectly planar (straight in cross-section) boundaries separating it from completely undeformed wall rocks (undeformed by the shear-zone forming deformation – an earlier fabric may exist). Ideal shear zones are also ductile, so that slip surfaces or other discontinuities are non-existent. In ideal plastic shear zones, which we will focus on in this section, we can see a foliation, a lineation and evidence of strain variations that contain important information about the zone.

For closer analysis, a coordinate system is placed with the horizontal axis along the shear zone margin and parallel to the shear direction, as shown in Figure 15.7. We can now move the walls while keeping the distance between them constant. The result is a perfect simple shear (Figure 15.8b). If we wish we could also increase or decrease the distance between the walls by adding a component of dilation or compaction. This would result in a change in volume (Figure 15.8d) that can take place before, during or after the simple shearing; the end result will, in principle, be the same.

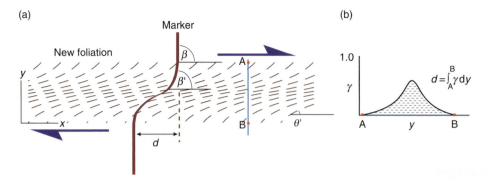

Figure 15.7 (a) Shear zone with genetically related foliation. The foliation makes ~45° with the shear zone along the margins. This angle is reduced as strain increases toward the center of the zone. θ' is the angle between the shear zone and the foliation. (b) The displacement can be found by measuring or calculating the area under a shear strain profile across the zone if the deformation is simple shear.

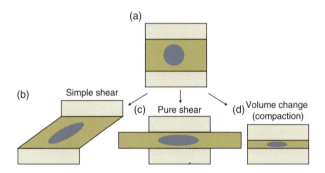

Figure 15.8 The difference between simple shear, pure shear and isotropic vertical volume change.

Ideal shear zones are perfectly ductile and involve simple shear with or without additional compaction/dilation.

Any deformation that deviates from simple shear (with or without additional compaction or dilation across the zone) requires discontinuities to form and thus contradicts the restrictions set for an ideal shear zone (Figure 15.8c). Such deviating deformation will cause extrusion of material along the zone and create structural incompatibilities. By incompatibility we mean that different parts of the shear zone do not fit together in a continuous framework, causing overlaps, gaps or discontinuities to arise. At the advanced level, tests of compatibility are necessary steps in the analysis of strain in shear zones.

Classic ideal plastic shear zones are found in more or less isotropic rocks, particularly in magmatic rocks such as the one shown in Figure 15.9. In these cases the deformation matrix for simple shear (Equation 2.16), or the one for combined simple shear and dilation, can be used to model the deformation. Using these matrices,

Figure 15.9 Ductile shear zone in greenstone where feldspar aggregates (amygdales) get progressively more strained into the shear zone (downward). White line indicates how the shear zone-related foliation rotates toward parallelism with the zone. Photo: Graham B. Baird.

shear strain, shape and orientation of the strain ellipse and displacement can be calculated at any point in the zone. Let us look at the structures forming within ideal shear zones and what they can give us.

Foliation and strain

Characteristic structures develop as displacement and strain accumulate in plastic shear zones. These structures have orientations and geometries that depend on both the type of shear zone and the amount of strain. If a shear zone develops in a fairly isotropic rock, then a faint foliation will appear at low shear strains. We can see such a faint foliation along the margins of the shear zones portrayed in both Figures 15.9 and 15.10. The angle θ

between this initial foliation and the margins will be close to 45° for a simple shear zone. In most cases it will be slightly less, because it takes a certain strain to form a visible foliation, and during that first strain increment the foliation will rotate a certain amount.

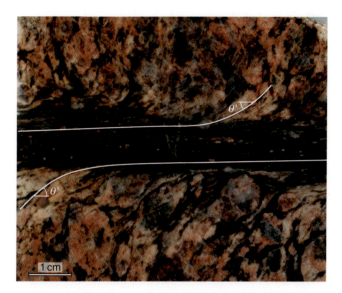

Figure 15.10 A shear zone in the Diana Syenite in Harrisville, New York State, showing profound strain increase toward the central ultramylonitic part of the zone. This is expressed by the change in orientation of the foliation and a marked decrease in grain size in the zone. Photo: Graham B. Baird.

The faint initial foliation is also seen along the low-strain shear zone margins of a well-developed shear zone, and the angle it makes with the zone (now termed θ') decreases as we trace the foliation into the more strained central parts of the zone. The foliation initiates perpendicular to the fastest shortening direction (ISA$_3$), but because simple shear is non-coaxial the foliation will constantly rotate toward parallelism with the shear plane. For low and moderate shear strains (up to 10–15) the foliation is a flattening foliation that represents the orientation of the XY-plane of the strain ellipse. This close relationship between the orientation of the shear zone foliation and strain can be very useful, and helps us to map strain in shear zones. For a simple shear zone, where θ' denotes the angle between the foliation and the shear zone, the relationship is given by the formula

$$\theta' = 0.5 \tan^{-1}(2/\gamma) \tag{15.4}$$

Note that this relationship only holds for a simple shear zone where the foliation under consideration formed during the formation of the shear zone in a previously unstrained rock, and in a section perpendicular to the foliation in the direction of shear.

In some cases the shear zone contains strain markers so that the aspect ratio $R = X/Z$ of the strain ellipse can be estimated. We can then relate the orientation of the foliation to both the shape of the strain ellipse and γ, as shown graphically in Figure 15.11.

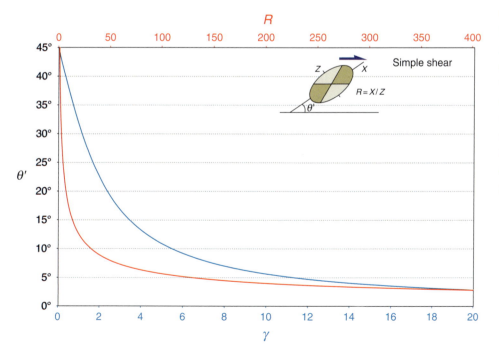

Figure 15.11 Curves showing how the angle θ' between the maximum finite strain axis (X) and the shear plane decreases with increasing strain for simple shear. Blue curve represents shear strain (γ) and red curve represent the aspect ratio $R = X/Y$ of the strain ellipse. Note different scales for γ and R (top and bottom).

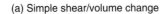
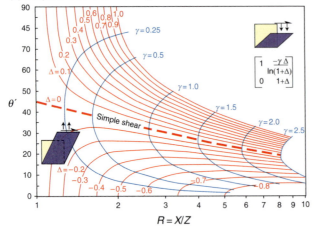

(a) Simple shear/volume change

(b) Subsimple shear

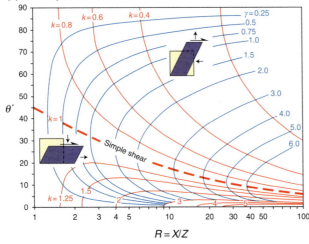

Figure 15.12 R–θ' plot for (a) simple shear and compaction and (b) simultaneous simple and pure shear (both plane strain). The simple shear curve starts at $\theta = 45°$. Note that the horizontal axes are logarithmic.

lineation indicates the X-axis of the strain ellipsoid. The stretching lineation lies in the foliation surface and thus defines the same angle θ' with the margin or shear plane as does the foliation (Figure 15.11). The projection of the lineation onto the shear plane will indicate the shear direction (Figure 15.12), which is also the flow apophysis for the simple shear.

Offset and deflection of passive markers

There is a connection between the total offset and finite strain in a shear zone. For simple shear this connection can be portrayed graphically by a shear strain profile across the zone. The length of the curve will be the thickness of the zone and the shear strain value will represent strain at any point across the zone. By integrating across the zone or by finding the area under the shear–strain curve, the total offset d is found (Figure 15.7b).

Sometimes shear zones are found to cross markers such as dikes, veins and layering (Figure 15.7a). Such markers show the offset across the shear zone directly, and the result can be compared to the strain integration method illustrated in Figure 15.7b. Remember that the strain integration method assumes that the XY-plane is represented by the foliation and that the shear zone represents simple shear deformation. A significant discrepancy between the two methods indicates that the deformation deviates from simple shear. Perhaps it can be explained by additional shortening or dilation across the zone, or perhaps other deformation types are involved. Before going into other types of deformation we will explore passive rotation of markers that cross a simple shear zone.

A line marker that is initially perpendicular to the shear zone is easy to handle, since its angle of rotation is the angular shear. More generally, if a marker originally makes an angle β to the shear zone (Figure 15.5a) then its new angle β' after the deformation depends on the type of deformation in the zone. For simple shear the relationship between these angles and the shear strain γ is given by the formula

$$\gamma = \cot \beta' - \cot \beta \qquad (15.5)$$

Remember that $\cot \beta = 1/\tan \beta$. We could also use the deformation matrix $\mathbf{D}$ to find the new orientation of markers for appropriate matrixes $\mathbf{D}$. In a section (two-dimensional analysis) the marker can be represented by a unit vector $\mathbf{l}$, and the new orientation becomes

$$\mathbf{l}' = \mathbf{D}\mathbf{l} \qquad (15.6)$$

If there is a component of compaction (or dilation) across the simple shear zone, then even the faint foliation along the margins will make an angle θ with the shear zone that is less (or more) than 45°. How much less depends on the amount of compaction. This means that we cannot use Figure 15.11 if the shear zone has thinned by compaction, but expand it with contours accounting for the compaction or dilation, as shown in Figure 15.12a.

There is a simple relationship between shear strain, the orientation of the foliation and strain in an ideal shear zone.

Lineations

A stretching lineation develops along with the foliation in plastic shear zones. Ideally, in a simple shear zone the where $\mathbf{l}'$ is the vector representing the new orientation. For simple shear $\mathbf{D}$ is the simple shear matrix from Equation 2.16. It is then easy to show that if $\mathbf{l} = (x, y)$, then its new orientation becomes $\mathbf{l}' = (x + \gamma y, y)$:

$$\mathbf{l}' = \begin{bmatrix} 1 & \gamma \\ 0 & 1 \end{bmatrix} \begin{bmatrix} x \\ y \end{bmatrix} = \begin{bmatrix} x + \gamma y \\ y \end{bmatrix} \tag{15.7}$$

For deformation types other than simple shear, different matrices apply.

One of the advantages of using matrix operations is that any line orientation (not just those in the plane perpendicular to the shear plane and along the shear direction) can easily be handled. In such three-dimensional analysis, planes are treated by means of their poles. In this case the strain takes the normal vector $\mathbf{p}$ to a new orientation represented by the vector $\mathbf{p}'$ according to the equation

$$\mathbf{p}' = \mathbf{p}\mathbf{D}^{-1} \tag{15.8}$$

Since line and plane rotations are determined by the deformation type, and hence the deformation matrix $\mathbf{D}$, we can easily model how lines and planes rotate during the deformation history by means of incremental strain. This shows that simple shear produces rotations along great circles, which differ from those caused by other deformation types (Figures 2.30 and 18.21). Passive linear and planar structures that are increasingly deflected toward the center of a shear zone can be plotted and their rotational paths may reveal important information about the deformation type. In practice such analyses are done by plotting the orientation of markers outside and inside differently strained parts of the shear zone, and then comparing to numerically modeled rotation paths using Equations 15.6 and 15.8.

Before moving to more general shear zone types we will summarize some facts about simple shear zones, compaction zones and the combination of the two.

Simple shear zones

A simple shear zone has the following characteristics:

Plane strain with $W_k = 1$.

No shortening or stretching along or normal to the zone.

ISA_1 oriented at 45° to the shear plane (walls).

The strain ellipsoid X-axis initiates at 45° to the walls and progressively rotates toward parallelism with the shear plane according to the formula $\theta' = 0.5 \tan^{-1}(2/\gamma)$.

Knowing θ', the local shear strain is given by $\gamma = 2/\tan(2\theta')$.

Passive markers at an initial angle β to the shear direction obtain a new angle β' after the deformation according to the formula $\cot \beta' = \cot(\beta) + \gamma$.

The vorticity (w) is identical to the shear strain rate: $w = \dot{\gamma}$.

Dilation/compaction zones

A compaction zone or dilation zone is a deformation zone that has experienced pure compaction or dilation perpendicular to its walls (Figure 15.8d). Such deformation zones form end-members in the spectrum of shear zone kinematics illustrated in Figure 15.4 (left-hand side) and are therefore useful to explore. Thick zones of pure compaction or dilation are rare, but millimeter- to decimeter-thick zones occur in porous rocks and are a special type of deformation band known as compaction/dilation bands. For such zones we have:

Planar and non-coaxial deformation with $W_k = 0$.

Shortening or extension perpendicular to the zone.

ISA oriented parallel and perpendicular to the zone.

X initiates and remains parallel to the zone.

Passive markers at an initial angle β with the shear vector end up at a new angle β' according to the formula:

$$\cot \beta' = \frac{\cot \beta}{1 + \Delta} \tag{15.9}$$

where Δ is the volume change across the zone (see Section 2.12).

Vorticity $w = 0$.

Dilational/compactional shear zones

Shear zones that involve a combination of simple shear and dilation or compaction are known as **dilational** and **compactional shear zones**. Such zones, where simple shearing and dilating/compacting are simultaneous, are probably common and are characterized by:

Plane strain with $0 < W_k < 1$.

Generally non-coaxial deformation.

Shortening (compaction) or extension (dilation) perpendicular to the shear zone.

No shortening or extension along the shear zone.

ISA_1 oriented obliquely to the shear zone ($<45°$ for compactional shear, $>45°$ for dilational shear) (Figure 15.12a).

X forms obliquely to the zone and rotates according to the formula

$$\tan 2\theta' = 2\frac{(1+\Delta)\Delta\gamma/\ln(1+\Delta)}{1 - [\Delta\gamma/\ln(1+\Delta)]^2 - (1+\Delta)^2} \quad (15.10)$$

For compactional shear $\theta < 45°$ and θ' is always less than for simple shear for a given shear strain. For dilational shear zones the orientation of X can show a range of values, depending on strain and amount of dilation, but always with a higher θ than simple shear for a given shear strain.

Passive markers at an initial angle β with the shear direction end up at a new angle β' according to the formula

$$\cot \beta' = \frac{\cot \beta + \Delta\gamma/\ln(1+\Delta)}{1+\Delta} \quad (15.11)$$

All of these formulas can be extracted from the deformation matrix.

15.3 Adding pure shear to a simple shear zone

Many deformation zones deviate from the conditions of ideal shear zones: walls may be non-parallel, slip surfaces or other sharp discontinuities may occur within or along the zone, walls may be deformed and displacement may vary along the zone. A shear zone that deviates from the ideal shear zone model in any of these ways is called a **general shear zone**.

A general shear zone is one that deviates from the ideal ductile shear zone.

Few natural shear zones have perfectly parallel and planar walls because of lithologic heterogeneities or variations in the internal conditions of the zone during deformation. Non-parallel and curved walls create local deviations from the simple shear flow pattern.

Sharp discontinuities or slip surfaces are seen in many shear zones. Such internal discontinuities allow for a pure shear component to occur. If the shear zone walls are undeformed, a pure shear with shortening across the shear zone will thin the zone, and material will extrude along the shear zone as shown schematically in Figure 15.8c. On the contrary, if the pure shear component involves shortening parallel to the zone, the zone will widen. Flow of rock into the widening part of the zone can then occur.

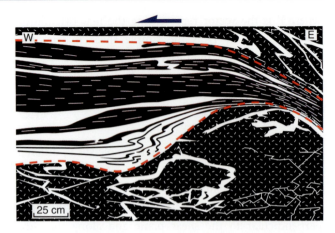

Figure 15.13 Folds formed on the lee side of a tectonic lens. Note the lee side thickening, the thinning on the opposite side and the vergence of the folds. Red stippled lines outline the shear zone. Modified from Fossen and Rykkelid (1990).

We expect to see layer thickening, imbrication and folding of layers where a shear zone widens. Local thinning of shear zones is associated with thinning of layers and possibly the formation of extensional shear bands. Figure 15.13 shows both effects: thinning of layers where the shear zone thins on the left side of the rigid lens, and thickening and folding on the lee side of the lens (central part of the figure). Furthermore, along-strike variations in coaxial deformation can lead to a shear zone of varying thickness and with internal slip surfaces of local extent. An example is shown in Figure 15.14, where a predominantly plastic shear zone changes character from completely ductile in the lowermost part to semi-ductile in the central and upper part.

If frictional slip occurs in shear zones, a slip-related lineation can form. Such lineations form on slip planes, which may or may not be parallel to the shear zone, and should not be confused with the stretching lineation illustrated in Figure 15.15.

Slip surfaces and other brittle expressions in predominantly plastic shear zones commonly form in shear zones that exhume and cool during shearing. They can also occur in the overall plastic regime of the lower crust for reasons other than changing temperature and pressure. Slip can occur where platy metamorphic minerals align to form a mechanically weak plane and where strain rate locally and temporarily increases. Narrowing shear zone walls or disturbance of the flow by a large rigid inclusion in the zone may be what it takes to increase the local shear-strain rate. External conditions forcing the shear zone to accumulate displacement at a higher rate are also possible. Finally, the role of fluids is important.

Sharp
discontinuity

Figure 15.14 Plastic shear zone with a local discontinuity along its left margin, suggesting that the deformation may deviate from simple shear.

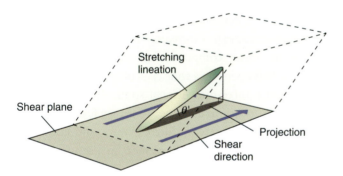

Stretching
lineation

Shear plane

θ'

Projection

Shear
direction

Figure 15.15 Illustration of the role of the stretching lineation in a plane strain shear zone. Ideally, its projection onto the shear plane indicates the shear direction. Deviations occur if there is a viscosity contrast between the linear elements and the matrix.

Dry rocks are much more prone to respond to stress in a brittle manner, so variable "wetting" of the deforming rock can control its rheologic behavior even in the lower crust. Hence, several factors can control the formation of transient sharp discontinuities in shear zones. It is also worth noting that transient formation of slip surfaces, pseudotachylyte and other brittle elements during otherwise plastic deformation implies that such structures are

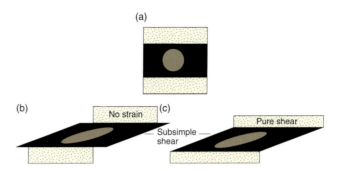

(a)

(b)   No strain   (c)   Pure shear
Subsimple
shear

Figure 15.16 A shear zone that involves coaxial strain (b) has compatibility problems along its margins that are solved if the wall rocks also deform by the same amount of coaxial strain (c).

likely to be obscured by subsequent plastic shearing and metamorphic recrystallization.

Some shear zones have walls that were coaxially strained during the shearing in the zones. In this case strain compatibility is preserved if the coaxial strain is homogeneous throughout both the shear zone walls and the shear zone (Figure 15.16c). This means that there will be no need for the discontinuities seen in Figure 15.8c and 15.16b – the walls and the zone will stretch together (Figure 15.16c).

Subsimple shear zones

It is obvious from the discussion above that a whole spectrum of plane strain shear zones exists where simple shear and pure shear are combined simultaneously in the same plane. We know this type of deformation from Chapter 2 as subsimple shear, and such shear zones are therefore called **subsimple shear zones**.

Subsimple shear zones cover the plane strain spectrum of simultaneous simple shear and pure shear.

For such zones we will see an initial angle θ between the foliation and the margin that is $>45°$ if the zone is thickening, and $<45°$ where the pure shear causes the zone to thin and extend. The close relationship between θ', strain and the type of deformation can be used to analyze natural shear zones. This is done by plotting the orientation of the foliation against strain in diagrams such as Figure 15.12b. This particular diagram is constructed for subsimple shear, so if there is additional volume change the diagram changes. In any event, data that plot off the simple shear curve must be explained by means of additional volume change or coaxial strain. To separate the two, evidence for or against volume change

(a)

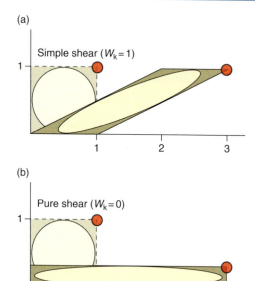

(b)

(c)

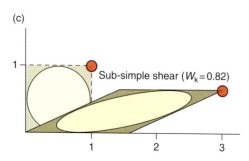

Figure 15.17 Comparison of shear-zone parallel offset and strain. A square with a circle is deformed under simple shear, pure shear and subsimple shear so that the horizontal displacement of the upper right corner is the same in all cases. The resulting strain is greatest for pure shear and least for the chosen subsimple shear. Based on Fossen and Tikoff (1997).

should be searched for. Such evidence could be stylolites, vein material, new minerals and concentrations of immobile minerals in the shear zone.

Stretching lineations in subsimple shear zones will, as for simple shear zones, reflect the shear direction when projected onto the shear plane (Figure 15.15). Note that if the simple shear component is small or zero the lineation indicates the stretching direction caused by pure shear, and can no longer be called the shear direction.

Let us take a closer look at the connection between strain and offset. Consider the upper right corner (1, 1) of a quadrant in a shear zone, as illustrated in Figure 15.17. If we apply a simple shear so that this corner is displaced to a new position (3, 1), then this offset will correspond to a strain ellipse with a particular axial ratio (Figure 15.17a). If instead we apply a pure shear to obtain the same horizontal displacement of this corner,

then we will find that the pure shear has generated a higher strain (compare the ellipses in Figure 15.17a, b). Some subsimple shear deformations require less strain than simple shear to produce the same shear-zone parallel displacement (Figure 15.17c). In fact, the subsimple shear that requires the least strain is the one with $W_k = 0.82$. This means that a subsimple shear of $W_k = 0.82$ is the most strain-efficient way to create shear-zone parallel offset.

It takes less strain to reach a given offset by means of subsimple shear than by pure or simple shear.

It is worth noting the nature of the horizontal displacement along the zone for different deformation types. For simple shear the displacement across the zone is constant along the zone. For subsimple and pure shear, however, the displacement changes along the zone.

15.4 Non-plane strain shear zones

We have so far been discussing shear zones from a plane strain perspective, in which any coaxial (pure shear) strain acts in the same plane as the simple shear. Any plane strain (without anisotropic volume change) produces strain ellipsoids that plot along the diagonal of the Flinn diagram. However, strain measurements in many shear zones show significant deviations from plane strain (Figure 15.18), implying extension or shortening related to a coaxial and/or simple shear component in the third direction (Y-axis). This opens up a wide class of general shear zones that produce three-dimensional strain. In such general shear zones, foliations and lineations may form with new orientations and rotate in different ways, and lineations do not relate to the direction of shear in a simple way.

Non-plane strain shear zones have flattening or constrictional strain geometries and lineations that may not represent the transport direction.

It is important to find out if a given shear-zone strain is plane or not. We could look for field evidence of thickness variations and flow in more than one direction. If we have strain markers, such as in the sample shown in Figure 15.9, we can section it and plot the strain data in the Flinn diagram, as done in Figure 15.18. The data in this figure come from shear zones of the type shown in Figure 15.9, and even though this shear zone may look like a simple shear zone at first glance, the strain data shown in Figure 15.18 reveal non-plane strain. This

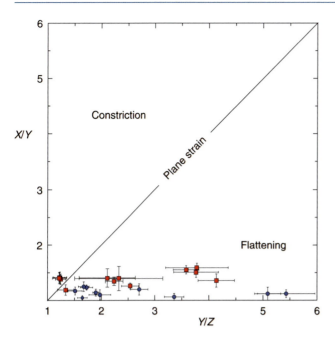

Figure 15.18 Flinn diagram with strain estimations from two shear-zone samples similar to that shown in Figure 15.9, cut parallel and perpendicular to the shear direction. The strain data clearly tell us that the two shear zones are affected by flattening and therefore deviate from simple shear. Data from Bhattacharyya and Hudleston (2001).

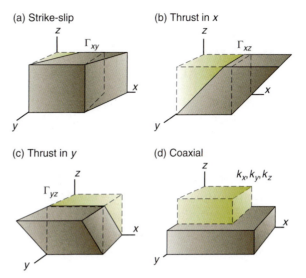

Figure 15.19 Orthogonal deformation components that can be combined into the three-dimensional strain matrix of Equation 15.12. Based on Fossen and Tikoff (1997).

means that the shear zone shown in Figure 15.9 cannot be a simple shear zone.

If we find that our shear-zone strain is three dimensional we are challenged by the fact that three-dimensional considerations open up a wealth of possible deformation types. We can still think in terms of pure and simple shear, but now we need to combine elements of simple and pure shear in more than one plane. Figure 15.19 shows how we can combine a three-dimensional coaxial strain with up to three orthogonal simple shear systems. The deformation matrix then becomes three-dimensional:

$$\begin{bmatrix} k_x & \Gamma_{xy} & \Gamma_{xz} \\ 0 & k_y & \Gamma_{yz} \\ 0 & 0 & k_z \end{bmatrix} \qquad (15.12)$$

This matrix can be used to model such deformations, but calculations generally require computer programs. There are also cases where the coaxial strain acts at an oblique angle to the shear component(s), in which case things get more complicated. The many possibilities make three-dimensional analyses of non-plane strain shear zones challenging, and we usually look for the simplest model that can explain our data. Transpression and transtension are two quite popular and closely related classes of simple

three-dimensional shear-zone deformations, discussed in the last part of Chapter 18.

15.5 Mylonites and kinematic indicators

Mylonites

In the central parts of some plastic shear zones, for example the one shown in Figure 15.10, strain can get so high that preexisting textures and structures are totally flattened and transposed. The rock becomes strongly banded and is called a **mylonite** – a word that was coined by earlier workers in the Scottish Moine Thrust Zone (see Box 15.1) The characteristics of mylonites vary with temperature, pressure, mineralogy, grain size, presence of fluids and strain rate. In general, mylonites are more fine-grained than their host rock (very apparent from Figure 15.10), with well-defined foliation and lineation.

Mylonite zones or belts can be up to several kilometers thick, particularly in Precambrian shield areas and eroded collisional orogens. In these cases the walls and shear plane may be difficult to define, and there will be many heterogeneities within the zone. Not only will finite strain vary, with lenses and slivers of less-deformed rocks, but also the deformation type will vary across the zone. The result can be a zone containing a rich variety of deformation structures such as folds, cleavages, foliations, lineations and other structures discussed later in this chapter. It is still a shear zone in the general sense of the word, but certainly more complicated and varied than ideal shear zones dealt with in the first part of this

BOX 15.1 | MYLONITE VERSUS CATACLASITE

The term **mylonite** had at least two different meanings through the last century. The word stems from the Latin word for milling or crushing into fine pieces. This was also the process that once was assumed to have formed the mylonites along the famous Scottish Moine Thrust Zone – the place where the term mylonite was first applied. With the great aid of optical and electron microscopes, we now know that mylonites are formed mainly by plastic deformation mechanisms.

The term mylonite is now being used for strongly deformed rocks that have been exposed to grain size reduction due to plastic deformation, while the related term **cataclasite** is used where cataclastic flow dominates. Some cataclasis can occur during mylonitization, for instance when feldspar is crushed in a matrix of plastically deforming quartz.

Mylonites are separated into three subgroups, depending on how much of the original groundmass is still intact (not recrystallized):

Protomylonite: <50% matrix (new grains)
Mylonite: 50–90% matrix
Ultramylonite: >90% matrix

Mylonites are common in thrusts, extensional shear zones and steep basement shear zones.

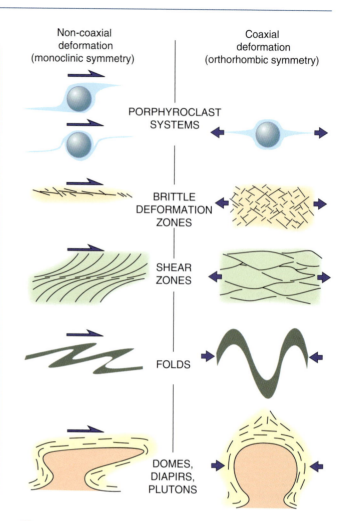

Figure 15.20 Asymmetric structures (left side) characterize non-coaxial deformations, while coaxial deformations tend to result in more symmetric structures. Based on Choukroune *et al.* (1987).

section. This makes thick mylonite zones both exciting and challenging.

Strongly sheared rocks of the kind found in high-strain shear zones have been defined in several ways through the past century, and the most common classification is shown in Boxes 8.1 and 15.1. This classification distinguishes between protomylonites, where original grains still dominate (90–50%), mylonites (50–10% original grains) and ultramylonites, which have less than 10% of the original grains intact. This change from protomylonite through mylonite to ultramylonite is recognizable in many shear zones, but is made more difficult if rocks have been exposed to previous phases of strong deformation.

Remains of the original texture or original minerals can be found as large lenses or fragments wrapped in the mylonitic foliation. Where such lenses are tens of centimeters or more in length they are called **protolithic**

lenses. Fragments of single crystals or minerals are called **porphyroclasts**. Feldspar typically forms porphyroclasts in deformed granitic rocks during greenschist facies shearing, where temperature is too low for feldspar to deform by crystalloplastic deformation mechanisms.

Kinematic indicators

Our understanding of microstructures as kinematic indicators in shear zones and mylonite zones increased considerably during the 1970s and 1980s. Understanding the connection between structural asymmetry and kinematics represented an important breakthrough in the study of strongly sheared rocks. The key point is that many mylonites contain structures that show monoclinic (low) symmetry, simply referred to as **asymmetric structures** by most geologists. The asymmetry is related to the rotational component or non-coaxiality of the deformation, or the fact that objects rotate in a preferred direction, as shown in Figure 15.20. It is

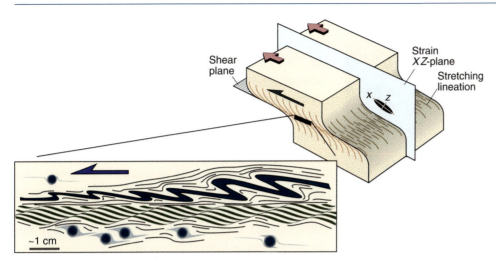

also reflected in the particle paths, which for non-coaxial deformations are asymmetric. Coaxial deformations produce structures whose geometries show a higher (ortho-rhombic) degree of symmetry, which is related to their symmetric particle paths. The connection between symmetry and coaxiality is shown schematically for a variety of structures in Figure 15.20. Structures formed in coaxial flow are thus sometimes referred to as orthorhombic.

The (a)symmetry of mylonite structures can be used to evaluate the sense of shear and sometimes also the degree of coaxiality of a mylonite zone.

It is the monoclinic structures that give information about the sense of displacement or sense of shear in mylonite zones and we will therefore focus on monoclinic or asymmetric structures in this section. In a shear zone we primarily study the section perpendicular to the foliation that contains the lineation (Fig. 15.21), although the section perpendicular to the lineation can also be of interest for evaluation of three-dimensional deformation.

Deflected markers

We have already looked at how pre-existing markers (linear or planar) become rotated into shear zones (Figure 15.2). Even if we do not see the shear zone margins, rotation of planar markers from an area of low strain to an area of high strain provides a very reliable criterion for sense of shear determination, for instance along the margins of tectonic lenses in the shear zone.

Mylonitic foliation and shear bands (S-C structures)

The foliation that develops in a shear zone is usually thought to trace the XY-plane of the strain ellipsoid.

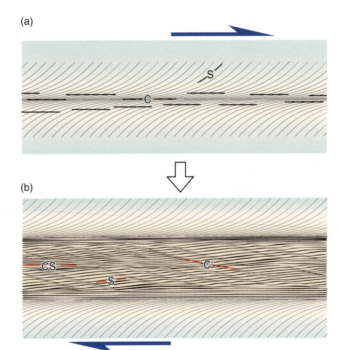

Figure 15.22 Schematic illustration of the development of S-C structures in a shear zone in a magmatic rock. (a) The new-formed foliation (S) is cut by shear surfaces (C) that parallel the shear zone margins. (b) Continued deformation rotates S into close parallelism with C, together referred to as a CS-foliation. New and oblique shear bands (C′) form and back-rotate the CS-foliation, which then is called S.

The sense of rotation of the foliation from the margin into the shear zone is generally a safe kinematic indicator.

As strain accumulates, a set of slip surfaces or shear bands commonly forms parallel to the walls of the shear zone (Figure 15.22a). These shear bands are called C (French "cisaillement" for shear, which relates to the

(a)

(b)

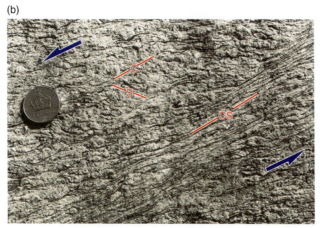

Figure 15.23 (a) Undeformed granite. (b) Sheared version of the same rock. Two sets of planar surfaces (S and C) are developed. S rotates to become subparallel with C in the zone of high strain. The combined foliation is named CS.

movement of scissors) and the foliation is named S (for schistosity or "schistosité"). C-surfaces are not really surfaces, but small-scale shear zones that affect the foliation within the main shear zone. In detail, the foliation curves into and out of the C-surfaces, and the sense of deflection shown by the curving foliation reflects the sense of shear of the entire shear zone. C-surfaces are particularly common in shear zones in magmatic rocks (Figure 15.23).

If the shear strain is high, say above 10, the angle between the shear zone and the foliation becomes indiscernible. We then have a composite foliation consisting of rotated foliation and C-surfaces. In addition, heterogeneities in the deforming rock as well as slip on micaceous elements in the foliation perturb the flow enough to break down the simple picture seen at lower strains.

Such high-strain complications promote the formation of new structures that can reveal the kinematics of

the flow. Notably, a new set of shear bands, oblique to the shear zone margins, form when strain gets high (Figure 15.22b). These shear bands are designated C′ when their obliquity to the shear zone can be demonstrated, and are particularly common in mylonites rich in platy minerals. C′ surfaces are thus similar to the Riedel shears (R) in brittle shear zones and can only be separated from shear-zone parallel C-surfaces when their orientation with respect to the shear zone boundaries is known. This information is not always available, and the structural geologists Gordon Lister and Arthur Snoke suggested in 1984 that any kind of structure that is composed of two planar structures formed during one progressive shearing event be called S-C structures. Furthermore, mylonites that show S-C structures are called **S-C mylonites**.

> S-C mylonites are made up of two sets of planar structures, a foliation (S) and shear bands (C) that obliquely transect and often back-rotate the foliation.

The angle between S and C can vary, but is typically about 25–45°. The angular relationship between the foliation and shear bands (C) (Figures 15.24 and 15.25) is a reliable shear indicator if the angular relation is consistent. The higher the degree of coaxiality, the lower the consistency, and coaxial deformation is expected to produce sets of oppositely dipping shear bands. However, a consistent angular relationship between shear bands and the general foliation is very common in mylonite zones. It is, however, important to remember that shear bands may form relatively late during the evolution of a shear zone, thus reflecting the last part of the deformation history.

Microscale foliations

Foliations that are oblique to the main foliation or the mylonitic banding occur in mineral aggregates that dynamically recrystallize during the deformation. Quartz aggregates or deformed quartz veins are typical examples. The aggregates themselves form part of the main mylonitic foliation, while the long axes of its deformed grains define an oblique foliation (Figures 15.26 and 15.27). There are two competing processes going on in the aggregates. One is the strain produced by elongation of grains. The other is the dynamic recrystallization and recovery that erases the strain recorded by those grains. The oblique foliation within the aggregates reflects only the last increment of deformation, while the main foliation is a result of the entire deformation history and therefore represents an orientation that is close to the

(a)

(b)

(c)

Figure 15.24 (a) S-C structure in protomylonitic granite, Antarctica. Note tiling of feldspar porphyroclasts. (b) Shear bands in phyllite, together with asymmetric folds. Caledonian basal décollement. (c) Asymmetric boudins in granitic gneiss. Asymmetric folds around inclusion represent quarter structure. Caledonides east of Bergen.

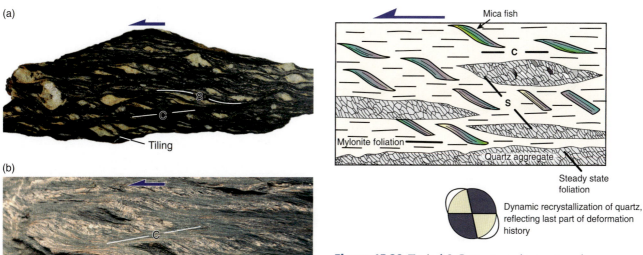

Figure 15.26 Typical S-C structures in quartz–mica-dominated mylonites.

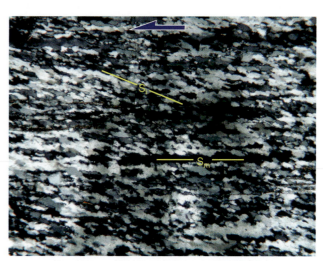

Figure 15.27 Thin section of a mylonite with a horizontal mylonitic foliation (S_m). The quartz grains are stretched in a direction S_i oblique to S_m, and the angular relations are consistent with top-to-the-left sense of shear. We could also use S-C terminology, where S_m represents C and S_i corresponds to S.

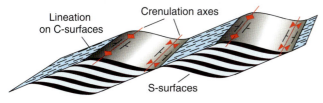

Figure 15.25 Schematic illustration of the geometry of shear band-type S-C structures. The crenulation axis is typically at a high angle to the sense of shear, which is reflected by a lineation that may appear on the shear bands (C-surfaces). Slip may or may not occur on the S-surfaces.

shear plane. The angle between the mylonitic foliation and the grain shape fabric within mineral aggregates indicates that the deformation is non-coaxial, reveals the sense of shear, and even indicates the degree of non-coaxiality or W_k of the deformation. If the grains within the aggregates are only slightly strained, then the angle should be close to 45° for simple shear ($W_k = 1$). In terms of S-C terminology, both S and C are foliations in this case (Figure 15.26) and are also termed S_i (internal) and S_m (mylonitic) (Figure 15.27).

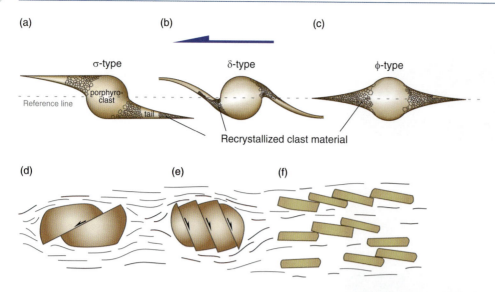

Figure 15.28 Porphyroclast systems. (a–c) Porphyroclasts with recrystallized tails. σ-type porphyroclasts have tails that do not cross the reference line, while the δ-type have tails that do. (a) and (b) show monoclinic symmetry (with the rotation axis being perpendicular to the page). The φ-type is symmetric about the reference line. (d) Fractured porphyroclast with synthetic fracture. (e) Antithetic shear fractures. (f) Tiling (imbrication) of porphyroclasts. All structures (except (c)) consistent with sinistral shear.

Mica fish

Mica porphyroclasts in mylonitic rocks tend to have tails that systematically curve away from the general orientation of the porphyroclasts, as shown in Figure 15.26. Such microstructures are known as **mica fish**, and the resulting asymmetry indicates the sense of shear. Mica fish are commonly seen to be confined by shear bands, and can be regarded as a type of S-C structure.

Foliation fish and foliation boudinage

Parts of strongly foliated mylonites sometimes back-rotate with respect to the shearing direction. The result is structures called **foliation fish** that look like mica fish but occur on a larger (for example meter) scale (Figure 14.12). These structures are also known as **asymmetric foliation boudinage**.

Boudinage

Boudinaged competent layers can be used as kinematic indicators, provided that one knows the approximate orientation of the layers before deformation started. If such a layer is located in the field of active stretching during a non-coaxial deformation, then the boudins can rotate against the shear direction, as shown in Figure 15.24c (see also Chapter 14).

Porphyroclasts

Porphyroclasts of feldspar, quartz, mica or other minerals can develop a mantle of recrystallized material that also

Figure 15.29 δ-porphyroclast indicating rotation during top-to-the-left shearing, consistent with the asymmetry of small-scale folds (right).

forms tails, as illustrated in Figure 15.28a–c. Coaxial deformations produce tail geometries that are symmetric with respect to the general mylonitic foliation (ϕ-type; Figure 15.28c). For non-coaxial deformation types, an asymmetric geometry tends to arise, and the final shape depends on W_k and the thickness of the mantle versus the core (rate of recrystallization), among other things. Asymmetric structures are classified into σ-types, where the tails have a stair-stepping geometry, and δ-types, where the tails are thinner, strongly curved and influenced by porphyroclast rotation (Figure 15.29). A characteristic difference between the two is that the tails of the σ-type are located on each side of the foliation reference line while those of the δ-type cross this line.

Many porphyroclasts show growth of quartz or other minerals in what are called porphyroclast pressure shadows or **strain shadows**. Strain shadows are asymmetric when formed in non-coaxial deformation types and show similarities with σ-type porphyroclasts (Figure 15.30).

The garnet porphyroclast in Figure 15.30 also shows an inclusion pattern. Such patterns may represent a crenulated foliation that has been overgrown by the garnet, but may also form during synkinematic growth of porphyroclasts. In the latter case the pattern indicates the sense of rotation and thus the sense of shear for non-coaxial deformations.

Folds and cleavage

Asymmetric folds and related axial planar cleavage can give information about the sense of shear in strongly deformed rocks if we know the approximate orientation of the folded layer before deformation started. In mylonite zones we often find that it is the mylonitic foliation itself that has been folded during the deformation history. Hence, the foliation must have rotated through the shear plane and into the contractional field. In such cases

the vergence of asymmetric folds indicates the sense of shear (Figure 15.29). All it takes to make mylonitic folds is for the foliation to be rotated into the field of instantaneous contraction. A rotation of a few degrees is sufficient and easily occurs where the foliation is perturbed around rigid inclusions or lenses. Alternatively, asymmetric folds can develop from pre-deformational features such as cross-beds or cross-cutting dikes that initially lie within the field of contraction. In such cases vergence must be interpreted with care, particularly if the initial orientation of the marker is unknown. Asymmetric folds can also indicate the "wrong" sense of shear in special cases, as illustrated in Figure 15.31.

Asymmetry and strong non-cylindricity (strongly curved hinge lines) are characteristic features of folds in mylonite zones (Figure 15.32 and 15.33). Sometimes the inverted limb of such asymmetric folds is displaced by small reverse or thrust faults (Figure 15.32), which helps determine the sense of shear.

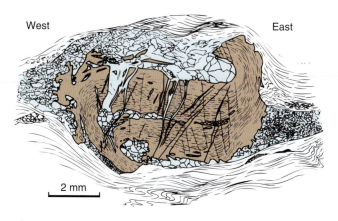

Figure 15.30 Rotated garnet with quartz tails (blue) in the strain shadows. The inclusion pattern indicates top-to-the-E sense of shear. The younger shear bands and extension fractures in the garnet indicate the opposite sense of shear. Modified from Fossen (1992).

Figure 15.32 Folds formed in non-coaxial deformation in gneisses. The vergence of the asymmetric folds indicate top-to-the-left transport. A small thrust-like structure (arrow) cuts off the inverted fold limb. The complete structure can be considered to be an S-C structure where the intrafolial fold train is caught between two C-bands.

Figure 15.31 Progressive development of folds in simple shear (strain increasing to the left). This particular initial orientation can result in a vergence apparently inconsistent with the sense of shear.

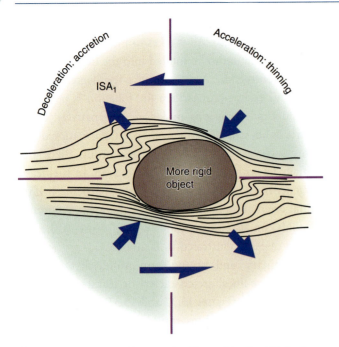

Figure 15.33 Sectors (quarters) of layer thinning/thickening around a rigid object in a mylonite zone indicating sense of shear. The structures are related to particle acceleration/deceleration and are called quarter structures.

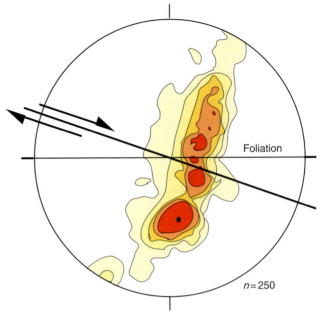

Figure 15.34 Two hundred and fifty quartz c-axes measured on the U-stage and plotted in the stereonet. An asymmetrical pattern with respect to the foliation (trending E–W in the plot), such as the one shown here, indicates the sense of shear. From Fossen (1993).

Quarter structures

The association of contraction and extension structures around tectonic lenses and rigid objects in a mylonite characterizes the sense of shear. As shown in Figure 15.33, the area around such objects can be divided into quarter sectors of accretion and stretching/thinning of layers. The structures vary according to how the foliation is rotated around the object. In the quarters of thinning, the layers have to thin, as particles experience acceleration to bypass this corner of the object. In addition to layer thinning we may find evidence of dissolution (mica concentrations). In the accretion sectors, layers thicken and fold as particles experience deceleration.

Crystallographic orientation

The orientation pattern of the optical c-axis of quartz can sometimes be used to determine sense of shear in quartz-rich mylonites. A number (c. 150–250) of orientations are measured and plotted in a stereonet with the foliation being vertical and the lineation oriented E–W in the net. The pattern, or girdle, that arises is typically asymmetric for non-coaxial deformations. The asymmetry will indicate the sense of shear, as shown in Figure 15.34. The actual girdle will depend on the crystalline slip systems

that were active and therefore the temperature during the deformation, and on the strain geometry (prolate versus oblate strain ellipsoids).

Tiling of objects

Rigid and elongated crystals in a deforming rock can be imbricated or tiled in a mylonite zone (Figure 15.28f). Such tiling requires a high density of crystals and is common in porphyritic magmatic rocks, particularly if they have not completely solidified at the onset of deformation.

Shear transfer structures

Heterogeneous rocks can experience strain partitioning during deformation in the sense that simple shear or slip is localized in weak, mica-rich layers. In some cases shear is transferred from one layer to another, particularly at the termination points of weak layers. The transfer or overlap zone can be contractional or extensional, depending on the arrangement of the weak layers and the sense of shear. Contractional structures are portrayed as fold trains and/or imbrication structures, while extensional structures are dominated by shear bands. Figure 15.35 shows how the shear direction can be determined from such structures.

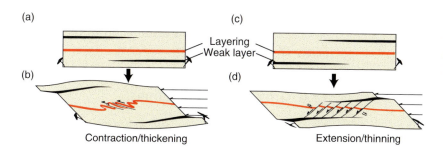

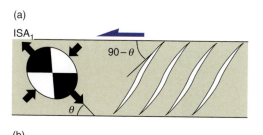

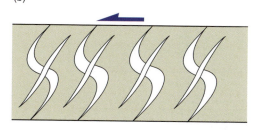

Figure 15.35 Transfer of localized shear strain from one weak layer to another can give contractional or extensional structures. Based on Rykkelid and Fossen (1992).

Microfaulted mineral grains

Mineral grains that deform in a brittle manner in a plastically deforming matrix may show sharp intragranular shear fractures. The most common example is brittle feldspar in a ductile quartz-rich matrix deforming under greenschist facies conditions. The orientation and sense of slip on such fractures relate to the sense of shear. The challenge is that such shear fractures can be antithetic as well as synthetic with respect to the shear direction (Figure 15.28d, e). Their orientation depends not only on sense of shear and W_k, but also on the shape of the grain, its orientation and that of any crystallographic cleavage or other plane of weakness that the grain may possess. Fractured mineral grains must therefore be used with care in kinematic analysis.

Fibers and veins

The orientation of extensional veins (Figure 15.36a) indicates the sense of shear in mylonites, and sometimes also W_k and the orientations of the ISA. If the veins are fibrous the fibers can give a better definition of the stretching direction.

Veins forming under non-coaxial deformation will rotate from the moment they form. This results in a sigmoidal geometry that can be used to determine the sense of shear, as illustrated in Figures 15.36b and 15.37.

Special folds can arise around veins in mylonites where the sense of shear from the fold geometry appears to oppose the sense of displacement on the vein. The geometry arises due to shear on the vein and is illustrated in Figure 15.38. This example shows that care should be exercised when determining shear sense from folds associated with veins in mylonitic rocks.

In general, populations of veins and dikes that have a variety of orientations are useful structures for reconstructing strain history and kinematics. Their appearance indicates whether they are located within the field of stretching, shortening, or stretching followed by shortening. For instance, veins that form boudinaged folds have a

Figure 15.36 En-echelon-arranged extension veins in a shear zone. The vein tips are oriented perpendicular to ISA$_1$. They are sheared into sigmoidal shapes and may be cut by younger veins.

Figure 15.37 En-echelon-arranged extension veins defining a shear zone. Note the foliation that is oriented at a high angle to each vein.

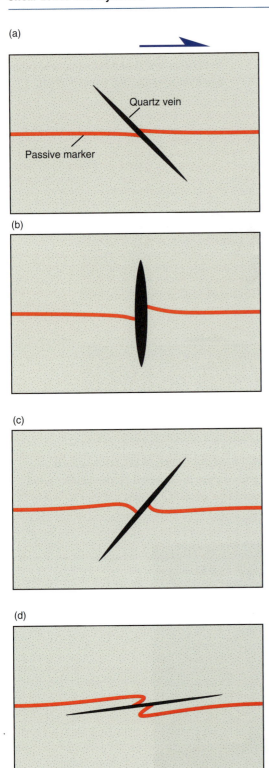

Figure 15.38 The development of an extension fracture with apparent wrong sense of displacement. (a) An extension vein forms in a shear regime. (b) The fracture opens, causing weak folding of the marker horizon. The fracture then rotates during continued shearing and the displacement along the fracture is opposite to that of the sense of shear (c, d). Based on Hudleston (1989).

history of shortening followed by extension. As discussed in Chapter 2, the sizes and distribution of these fields reflect W_k and kinematics.

15.6 Growth of shear zones

Many shear zones are simple and well-defined structures, yet our knowledge and understanding of how they form and develop is far from complete. For simplicity, it is often assumed that they form in homogeneous rocks such as granite. In practice, shear zones form at the weakest point or along the weakest layer in the rock, such as micaceous layers, partly molten zones, veins, fractures, fine-grained layers, dikes etc. This will have to be addressed separately for each shear zone.

Once a shear zone forms, we can imagine several histories of development leading to different types of shear zones, each illustrated schematically in Figure 15.39:

Type I shear zones are strain hardening shear zones where the deformation in the central part slows down as the zone thickens. The central part thus records the first part of the deformation history while the marginal parts only record the last part. Type I shear zones develop wide and plateau-type displacement profiles.

Type II shear zones are strain softening zones that quickly establish a certain thickness, but after a while the deformation localizes to the central part. Thus, the margins become inactive and the active part of the shear zone gets thinner. The result is thin shear zones with high shear strain gradients.

Type III shear zones develop a fixed thickness, and the entire zone keeps deforming without any sign of internal localization. The shear zone maintains its thickness, which equals its active thickness. This model is perhaps the least realistic one, but it may work for some types of kink bands.

Type IV shear zones have the same development as Type I zones, but the entire shear zone remains active throughout the deformation history. In other words, the zone grows in width and the strain increases from the margins toward the center. The margins only record the last increment of shearing while the central part has experienced the entire shearing history.

In general, shear zones show a positive correlation between displacement or length and thickness (Fig. 15.3). This observation indicates that most shear zones widen as deformation progresses and thus favors Type I and IV shear zones. If so, this means that the margin records the last increment of strain. In order to find out how they have developed, we could compare related shear zones

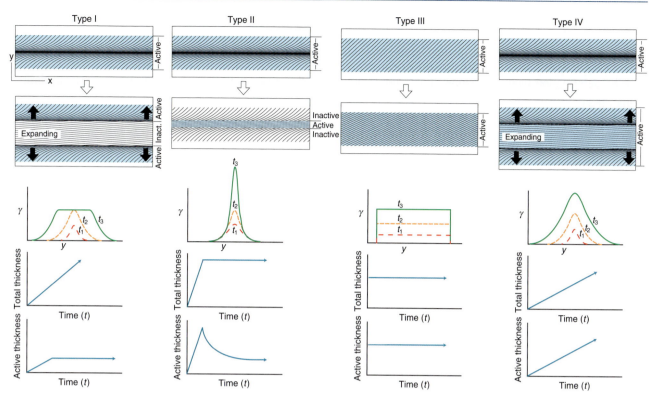

Figure 15.39 Four models for the growth of shear zones.

that show different displacements, i.e. shear zones at different stages of development.

Some shear zones are uplifted through their history, moving from the plastic regime to the plastic–ductile transitional regime and eventually into the brittle regime. This generally results in localization of shear into one or more relatively narrow zones, i.e. Type II, except that strain does not necessarily localize to the central part. Many kilometer-thick shear zones have gone through syn-kinematic exhumation and developed a late and narrow (semi-)brittle shear zone in their upper part. An example of this is the metamorphic core-complex-forming shear zones discussed in Box 17.3. Type II shear zones leave behind sheared rocks that formed at early or intermediate stages of shearing, which is very useful when unraveling the *P–T* history of a shear zone, as discussed in Chapter 21.

Shear zones can develop quite differently, depending on rock properties, fluids, deformation mechanisms and metamorphic reactions, but generally go through an early phase of widening.

Summary

Shear zones are important structures that usually contain internal structures that reflect their deformation type and deformation history, which may be important for the understanding of the tectonic development of an area. What shear zones do not directly tell us is whether they formed in an extensional, contractional or strike-slip regime. Simple shear, subsimple shear and any other type of shear zone can all form in any tectonic regime, and we have to know their orientation at the time of deformation in order to make such inferences. In other words, the shear zones have to be placed into a tectonic context. In the next three chapters we will see how shear zones and faults occur in the three main tectonic regimes. First some important points from the present chapter:

- Shear zones are zones where strain is higher than in the surrounding rock.

- Shear zones are subdivided based on ductility (ductile or brittle) and deformation mechanism (brittle/frictional or plastic).

- Ductile shear zones preserve the original continuity of passive layers.

- Plastic shear zones develop a foliation whose orientation relates to strain.

- If the deformation type (e.g. simple shear) and strain across the zone are known, these can be used to calculate the offset.

- Asymmetric structures indicate sense of shear.

- Always consider as many sense-of-shear indicators as possible to establish the shear sense.

Review questions

1. What makes a shear zone different from a fracture?

2. Can you draw the upper shear zone margin on Figure 15.9? Is it easily definable?

3. What type of data support the idea that shear zones grow in width as they accumulate displacement, and where would we look for the last increment of strain?

4. What assumption do we have to make for the last increment of strain to be representative of earlier strain increments in the zone?

5. What is meant by the term S-C structures?

6. What do you think are the most and least reliable shear sense structures described in this chapter?

7. How could you get information about the strain path (strain evolution) of a shear zone?

E-MODULE

The e-learning modules called *Shear zones* and *Kinematic indicators* are recommended for this chapter.

FURTHER READING

General
Passchier, C. W. and Trouw, R. A. J., 2006, *Microtectonics.* Berlin: Springer Verlag.

Images and descriptions
Snoke, A. W., Tullis, J. and Todd, V. R., 1998, *Fault-related Rocks: A Photographic Atlas.* Princeton, NJ: Princeton University Press.

Ideal shear zones
Ramsay, J. G., 1980, Shear zone geometry: a review. *Journal of Structural Geology* **2**: 83–99.

Ramsay, J. G. and Huber, M. I., 1983, *The Techniques of Modern Structural Geology. Vol. 1: Strain Analysis.* London: Academic Press.

Complex shear zones
Passchier, C. W., 1998, Monoclinic model shear zones. *Journal of Structural Geology* **20**, 1121–1137.

Kinematic structures
Berthé, D., Choukroune, P. and Jegouzo, P., 1979, Orthogneiss, mylonite and non-coaxial deformation of granites: the example of the

South Armorican Shear Zone. *Journal of Structural Geology* **1**: 31–42.

Dennis, A. J. and Secor, D. T., 1990, On resolving shear direction in foliated rocks deformed by simple shear. *Geological Society of America Bulletin* **102**: 1257–1267.

Lister, G. S. and Snoke, A. W., 1984, S-C mylonites. *Journal of Structural Geology* **6**: 617–638.

Passchier, C. W. and Williams, P. R., 1996, Conflicting shear sense indicators in shear zones: the problem of non-ideal sections. *Journal of Structural Geology* **18**: 1281–1284.

Platt, J. P., 1984, Secondary cleavages in ductile shear zones. *Journal of Structural Geology* **6**: 439–442.

Simpson, C., 1986, Determination of movement sense in mylonites. *Journal of Geological Education* **34**: 246–261.

Wheeler, J., 1987, The determination of true shear senses from the deflection of passive markers in shear zones. *Journal of the Geological Society* **144**: 73–77.

Strain in shear zones

Bhattacharyya, P. and Hudleston, P., 2001, Strain in ductile shear zones in the Caledonides of northern Sweden: a three-dimensional puzzle. *Journal of Structural Geology* **23**, 1549–1565.

Hudleston, P., 1999, Strain compatibility and shear zones: is there a problem? *Journal of Structural Geology* **21**: 923–932.

Simpson, C. and De Paor, D. G., 1993, Strain and kinematic analysis in general shear zones. *Journal of Structural Geology* **15**: 1–20.

Rigid objects and porphyroclasts

Bjørnerud, M., 1989, Mathematical model for folding of layering near rigid objects in shear deformation. *Journal of Structural Geology* **11**: 245–254.

Passchier, C. W. and Simpson, C., 1986, Porphyroclast systems as kinematic indicators. *Journal of Structural Geology* **8**: 831–843.

Passchier, C. W. and Sokoutis, D., 1993, Experimental modelling of manteled porphyroclasts. *Journal of Structural Geology* **15**, 895–909.

Growth of shear zones

Means, W. D., 1995, Shear zones and rock history. *Tectonophysics*, **247**: 157–160.

Sibson, R. H., 1980, Transient discontinuities in ductile shear zones. *Journal of Structural Geology* **2**, 165–171.

Soft-sediment shear zones

Lee, J. and Phillips, E., 2008, Progressive soft sediment deformation within a subglacial shear zone: a hybrid mosaic-pervasive deformation model for middle Pleistocene glaciotectonised sediments from Eastern England. *Quaternary Science Reviews* **27**, 1350–1362.

Maltman, A. J. and Bolton, A., 2003, How sediments become mobilized. In P. van Rensbergen, R. R. Hillis, A. J. Maltman and C. K. Morley (Eds.), *Subsurface Sediment Mobilization.* Special Publication **216**, London: Geological Society, pp. 9–20.

Chapter 16

Contractional regimes

Contractional faults occur in any tectonic regime, but they are most common along destructive plate boundaries and in intracratonic orogenic zones. Contractional structures received much attention from the last part of the nineteenth century up to the end of the twentieth century, when focus shifted somewhat towards extensional structures. The study of contractional faults resulted in the development of balanced cross-sections, and brought attention to the role of fault overlaps and relay structures, the relation between displacement and fault length, and the mechanical aspects of faulting. Understanding contractional faults is important not only for better understanding of orogenic processes in general, but also for improved petroleum exploration methods, because a number of the world's oil resources are located in fold and thrust belts. The fundamentals of contractional faults and related structures are covered in this chapter, with a focus on thrust structures found in orogenic belts.

16.1 Contractional faults

Contractional deformation structures form when rocks are shortened by tectonic or gravitational forces. We find contractional faults and folds in all parts of collision zones, they affect unmetamorphosed sediments in accretionary prisms associated with subduction zones, and they are common at the toe of gravitationally unstable (gliding) deltas and continental-margin sediments resting on weak mud or salt layers. Even advancing glaciers and ice sheets may generate fold-and-thrust belts, and several examples from the last ice ages exist in northern Europe and northern North America.

Consider a layered volume of rock being shortened in the direction of the layering, as shown in Figure 16.1. A number of micro- and macrostructures may result. Shortening can be accommodated by volume loss (Figure 16.1b) through the formation of dissolution seams (stylolites), solution along grain contacts or by physical compaction. A pure shear response can be envisioned where horizontal shortening is compensated by vertical thickening and where layers maintain their orientation (Figure 16.1c) or buckle (Figure 16.1d). Finally, shortening can result in contractional faults (Figure 16.1e) and genetically related fold structures, which are the main focus of this chapter.

Contractional faults and shear zones shorten the crust or some reference layer, such as bedding (Figure 16.2). When the surface of the crust is the reference, which is the case in regional analysis, contractional faults are exclusively **reverse faults** and **thrust faults**. Reverse faults are steeper than thrust faults (steeper than 30°) and do not accumulate the large displacements seen in thrusts, but there is a gradual transition between the two. Contractional faults can occur at any scale, from the microscale to regional orogenic belts and subduction zones.

When operating at outcrop scale it may sometimes be relevant to use a lithologic layering as reference. If so, normal and strike-slip faults can sometimes be seen to shorten a reference layering. This occurs where reverse faults have been rotated and overturned into apparent normal faults, or where normal faults dip at a lower angle than the reference layering, as shown in Figure 16.3.

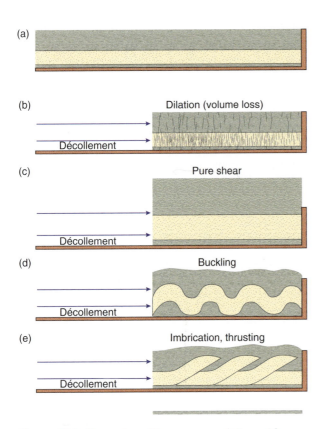

Figure 16.1 Shortening of layers can result in a wide range of strain regimes and structures: (b) dilation and cleavage formation, (c) pure shear without folding, (d) buckling, (e) imbrication.

Figure 16.2 Small thrust fault shortening Tertiary sedimentary beds in Valla de la Luna, Atacama, northern Chile.

Figure 16.3 Contractional faults that also are normal faults. They are contractional because they shorten or contract the layers that they affect.

Contractional faults are generally reverse or thrust faults, but can be other types of faults when rock layering is used for reference.

It is all a matter of reference and scale. In the following we will use the Earth's surface as our reference so that contractional faults are reverse and thrust faults only, unless otherwise stated.

16.2 Thrust faults

Nappe terminology

A **thrust** is a low-angle fault or shear zone where the hanging wall has been transported over the footwall. The movement should be predominantly dip-slip. It has been suggested that the term thrust should be reserved for horizontal displacements (heaves) in excess of 5 km, and many thrusts, including the Moine thrust in Scotland (Box 16.1) more than fulfill this requirement. Nevertheless, most geologists use the term even for outcrop-scale low-angle reverse faults, such as the one shown in Figure 16.2. For thrust faults developed in rocks that are younging upward or with upward-decreasing metamorphic grade, the following is true:

Thrust faults bring older rocks on top of younger rocks, and rocks of higher metamorphic grade on top of rocks of lower metamorphic grade.

According to this statement, stratigraphy and metamorphic grade can both be used to identify and map thrusts, and stratigraphic control in particular is very important in the mapping of thrust faults in many thrust belts. Even so, it should be noted that the premises on which the above statement is based are not necessarily fulfilled in cases where the rocks have been through an earlier phase of deformation or metamorphism.

A thrust separates the substrate from an overlying **thrust nappe**. Thrust nappes are characteristic features of contractional orogens such as the Caledonian–Appalachian orogen and the Alps. Thus, when referring to the orientations and directions of structures within thrust nappes, one commonly refers to the hinterland or foreland. The **hinterland** is the area in the central portion of the collision zone, and the **foreland** is the marginal part and thus farthest into the continent. In collisional orogens there is therefore one foreland area on each continent, separated by a common hinterland.

Although some thrust nappes occur as single units, they commonly contain a number of internal tectonic sheets, each separated by a thrust fault. The smallest units in a thrust nappe are referred to as "horses", as discussed in more detail in the next section. All of these internal thrust units are thin relative to their length and width. A collection of thrust nappes that share common lithological and/or structural features are referred to as a **nappe complex**.

A thrust nappe is bounded by a basal fault known as the **sole thrust** or **floor thrust** and an overlying **roof thrust**. In the common case of stacked nappes, the roof thrust of one nappe serves as the floor thrust for the one above. However, the shallowest of a stack of thrust nappes may be limited upward by a free erosion surface. The sole thrust, which separates the entire stack of thrust nappes from a less deformed or undeformed basement, is also called a **detachment** or **décollement** – terms also used about fundamental low-angle extensional faults and for significant faults or shear zones of uncertain sense of movement.

Nappes exposed at the surface can be discontinuous because erosion has selectively removed some parts while others have been spared. An erosional remnant of a nappe is called a **klippe** (German terminology) or **outlier**. Similarly, an erosional "hole" through a nappe that exposes the underlying rock unit or nappe, such as those seen on the map of the Scandinavian Caledonides in Figure 16.4, is called a **fenster** (also German) or **window**. Comparing rocks in nappes with those of the underlying and unthrusted basement makes it possible to determine whether the nappes belonged to the same unit or not, which has implications for the estimation of thrust displacement. There is a special terminology for these relations. Starting at the bottom, the basement is called **autochthonous**, which in Greek means something like "formed where found". Slices of basement and perhaps its sedimentary cover thrust only a few kilometers or so are called **parautochthonous**. Ideally, parautochthonous units are easily correlated with the basement rocks. Above the locally derived parautochthonous unit(s) are the **allochthonous** units. Allochthonous is also derived from Greek, where "allo" means "different" and "chthon" means "ground". Allochthonous units have been transported from areas that originally were many tens or hundreds of kilometers away. There may be many allochthons or allochthonous units in a nappe pile, and, as has been done formally in the Scandinavian Caledonides (Figure 16.4),

BOX 16.1 | THE MOINE THRUST ZONE

The Caledonian Moine thrust zone in the northwest Scottish Highlands has become a classic area of thrust tectonics. Here, a large unit of late-Proterozoic metasediments (the Moine and Dalradian rocks) were thrust northwestward above Archaean basement (the Lewisian) and its sedimentary cover. Some 100 km of overthrusting led to the formation of imbricate structures, duplexes, mylonites and other structures and rocks typically associated with major thrusts. The first ideas about thrusting arose during geologic mapping in the last half of the nineteenth century. The mapping showed that old rocks (detached basement gneisses) were resting on younger ones (Cambro-Ordovician sediments), an observation that called for the concept of thrust tectonics. Stratigraphy was key to the mapping of the Moine thrust zone. In particular, the Cambro-Ordovician stratigraphy is easily recognized in the area and has made it possible to map out very impressive duplex structures throughout the thrust zone. The recognition and balancing of duplex structures was inspired by the mapping of such structures in the Canadian Rocky Mountains in the 1970s–80s.

In addition to the Moine thrust zone, the Outer Hebrides Thrust, Great Glen Fault and Highland Boundary Fault are fundamental Scottish structures. These are Precambrian faults that have been reactivated both before, during and after the Caledonian orogeny. Reactivation of preexisting faults is a characteristic feature of orogenic belts.

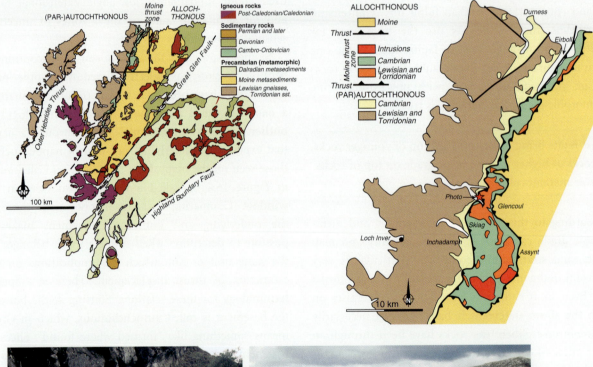

Sharp thrust contact between Precambrian gneisses and underlying Cambrian quartzite in the Moine thrust zone.

Sliver of Lewisian basement thrust above Cambrian sedimentary layers. Older rocks resting on younger ones, as seen here, are a classic feature in thrust and nappe regions.

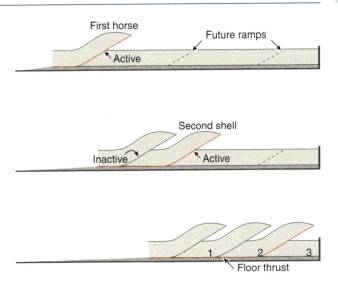

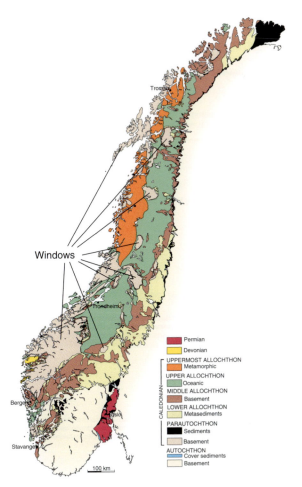

Figure 16.4 The tectonostratigraphy of the Scandinavian Caledonides. The lower and shortest transported nappes are termed the Lower allochthon, while the Uppermost allochthon contains the upper and farthest transported units. The parautochthonous units are only shortly transported, while in-situ units are called autochthonous.

Figure 16.5 The "standard" formation of an imbrication zone, known as "in sequence thrusting". The horses get younger towards the foreland (right) in this model. Deviations from this model are known as "out of sequence thrusting".

Figure 16.6 Duplex structure in dolomitic sandstones near Ny-Ålesund, Svalbard, formed during Tertiary contraction. Note S-shaped horses and floor and roof thrusts. Photo: Steffen Bergh.

they may be subdivided into the lower, middle, upper and uppermost allochthons.

Fault geometries

Contractional faults in the foreland of an orogenic zone typically form **imbrication zones** (Figure 16.5). An imbrication zone is a series of similarly oriented reverse faults that are connected through a low-angle floor thrust. If, in addition, a roof thrust bounds the zone upward, as in Figures 16.6–16.8, then the complete structure is called a **duplex structure** (e.g. the Moelven duplex in Figure 16.7). A duplex consists of **horses** that are arranged piggy-back, similar to the cards in a tilted card deck. The horses typically have an S-shaped geometry

in the vertical profile (Figure 16.9), and they tend to dip toward the hinterland. Note that horses can be folded, faulted and rotated during the thrusting history so that their primary geometries and orientations become modified.

In very low-grade and non-metamorphic sedimentary sequences deformed in the brittle regime the steep ramp-segments of imbrication zones likely form first. They form in the strongest or most competent layers, layers that fracture first during layer-parallel shortening. You can visualize this by thinking about what would happen if you compressed a sequence consisting of alternating

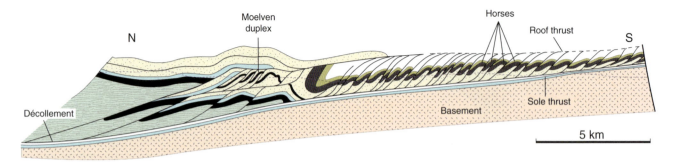

Figure 16.7 Section across the Caledonian foreland north of Oslo, Norway. The sole thrust is located in weak shales above the basement. The interpretation of a roof thrust makes the imbrication structure in the southern part a duplex structure. Based on Morley (1986).

Figure 16.8 Microscale thrust structures and related folds (thin section). These structures, including roof thrusts, floor thrusts and horses, are similar to those found at much larger scale. Tertiary thrust zone in Carboniferous shales, Isfjorden, Svalbard. Picture is 2 cm wide. Photo: Steffen Bergh.

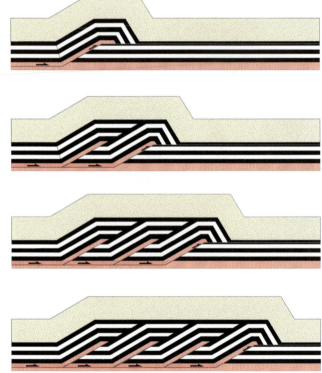

Figure 16.9 Schematic illustration of duplex development. The duplex grows toward the left (foreland), known as in-sequence thrusting or propagation. Note the characteristic double kink-fold in the frontal horse.

chocolate plates and jelly layers. The stiff chocolate would soon break and then the faults would link up along the jelly layers. That is roughly what happens in a limestone–shale sequence during layer-parallel shortening: the stiff limestone will transmit most of the stress and break. Such layers are said to act as stress gouges, and the result is steep reverse faults in limestone layers and flat thrusts in the shales.

The combination of two flat thrust segments at different stratigraphic levels connected through a steeper reverse fault (ramp) is referred to as a **flat-ramp-flat fault**, a terminology also shared with extensional faults. Ramps may also give rise to thrusts or reverse faults with opposite sense of displacement. Such **back-thrusts** form as the result of geometric complications in ramp locations (note the ramp in Figure 16.10), and seem to be favored by steep ramps.

Kinematically, an imbrication zone transfers slip from one stratigraphic layer to a higher one in the foreland direction. The lower horizontal décollement or floor thrust yields some of its displacement to each of the individual horse-bounding faults. At the top, displacement is "collected" by the roof thrust. Hence, the floor thrust terminates if all of its displacement is transferred to the roof thrust, as seen in the small-scale example in Figure 16.11.

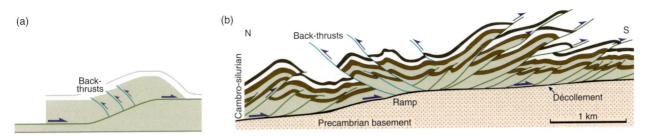

Figure 16.10 (a) Principal sketch of back-thrust formation, based on experiments and field observations. (b) Back thrusts generated above a ramp in the Caledonian sole thrust. The main thrusting direction is toward the right. North of Oslo, Norway. Based on Morley (1986).

Figure 16.11 Imbrication zone formed in the overlap zone between two mica-rich layers (black) in mylonitic gneiss. The zone formed as displacement on the lower mica-rich layer was transferred to the upper one. From Rykkelid and Fossen (1992).

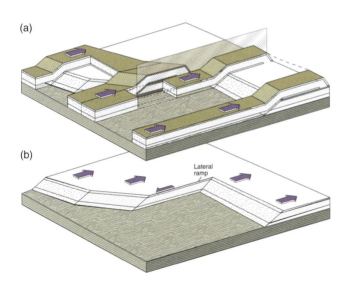

Figure 16.12 Various thrust ramps and their geometries. (a) Hanging-wall strata included, with ramps appearing as folds. (b) Hanging wall removed.

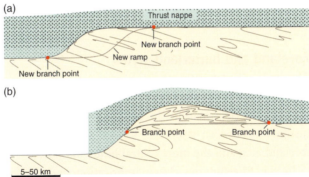

Figure 16.13 Formation of a new thrust nappe at a ramp location. Branch points are indicated.

Many ramps strike more or less perpendicular to the transport direction and are called **frontal ramps** in thrust terminology. Frontal ramps show dip-slip movements with striations in the dip direction. However, ramps may also be oblique to the transport direction (Figure 16.12). Such **oblique ramps** are oblique-slip faults formed with a combination of dip-slip and strike-slip motion. Ramps that form parallel to the movement direction of the thrust sheet are known as **lateral ramps**. Many lateral ramps are

found to be steep or vertical, and are really strike-slip faults that connect frontal ramp segments. The fact that they transfer slip from one frontal ramp to another also justifies the use of the name **transfer faults**. Another term used for this type of fault is **tear fault**.

As in any other fault system, thrust faults are connected and form three-dimensional networks. Where a fault splits into two we have a **branch point** (Figure 16.13) or, in three dimensions, a **branch line**. Branch lines enclose tectonic units (thrust nappes or horses) unless the thrust fault dies out as a blind thrust fault. The part of the branch line that delimits the unit in the front is called the **leading branch line**, while the one in the rear end is called the **trailing branch line**.

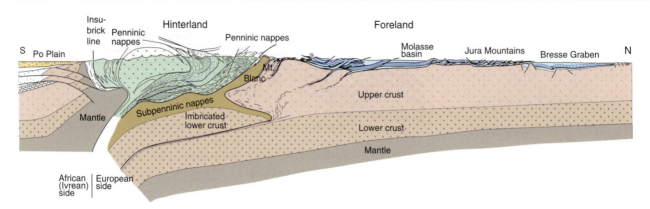

Figure 16.14 Cross-section through the Alps, showing the thin-skinned foreland deformation in the north (right) and the more pervasive and complicated deformation in the hinterland. Imbrication of the lower crust is indicated, based on seismic information. Based on Schmid and Kissling (2000).

Branch points and branch lines can occur at outcrop scale, for example in an imbrication zone, but can also be found on a regional scale. It has been suggested that the orientation of branch lines can be used to constrain the movement direction of individual nappes. This is based on the assumption that ramps are predominantly frontal and lateral; oblique ramps may be less common but also occur.

Foreland and hinterland styles

The style of deformation in a contractional regime depends on the lithologies involved and the depth at the time of deformation. In orogenic belts there is a distinction between structures formed in the marginal foreland area and the more central hinterland area.

In the foreland the classic imbrication and duplex structures described above are very common. We find these structures in the sedimentary sequence covering the underlying basement. In the Alpine example shown in Figure 16.14, the foreland-style is exemplified by the Jura Mountains, where Jurassic–Cretaceous sediments are folded and imbricated. A section through a part of the Caledonian foreland structures is illustrated in Figure 16.7, showing extensive imbrication and duplex formation. In both cases the basement is practically undeformed and separated from the shortened sedimentary cover by a basal sole thrust or décollement. This setting is known as **thin-skinned** tectonics and is characteristic of deformed foreland areas. The style of deformation is simple, as compared to more internal parts of collision zones, with a stratigraphic control that enables correlation across faults.

In the hinterland the basement is involved, and the style is therefore referred to as **thick-skinned**. Thrust

nappes in the hinterland can therefore be much thicker than those in the foreland and consist of mostly metamorphic and magmatic rocks. The interpretation of the Alpine collision zone shown in Figure 16.14 suggests that the entire lower crust is imbricated. Another characteristic of the hinterland is the occurrence of island-arc or outboard terranes, thrust onto the continental margin from a pre-collisional ocean during continent–continent collision. Hinterland nappes can vary from internally unstrained to penetratively strained. Where they are penetratively folded they may be referred to as **fold nappes**. Nappes that are internally deformed to the extent that they show a general mylonitic fabric are called **mylonite nappes**.

Thin-skinned contractional tectonics produces classic thrust structures with duplexes and related folds, while thick-skinned tectonics forms much larger nappes that may be more complexly strained.

In an orogenic zone the basement is overridden by a wedge-shaped stack of thrust nappes. The orogenic wedge thickens and lengthens over time, as fragments of the down-going crust are sliced off and incorporated within the allochthonous units and as oceanic rock units are added. The rest of the basement is transported deeper into the subduction zone and undergoes high-grade metamorphism and related deformation. Hinterland structures therefore tend to form at greater depths than foreland structures, a fact that favors plastic shear zones and plastic strain in general. There is of course brittle deformation going on at shallow levels in the hinterland zone, commonly influenced by extensional faulting.

16.3 Ramps, thrusts and folds

The temporal development of imbrication zones can vary, but the model that is regarded as normal is illustrated in Figures 16.5 and 16.9. Here, the individual sheets or horses are formed in sequence, so that successively younger faults form in the direction of thrusting. This foreland-directed progression of duplexes and imbricate zones is called **in-sequence thrusting**. The foreland-most thrust is the youngest one and carries the other horses on its back toward the foreland.

In-sequence thrusting enables the zone of contractional deformation to expand in the foreland direction, consistent with progressive broadening of active collision zones.

Thrust faults that do not follow this systematic pattern of expansion are said to be **out of sequence**. Out-of-sequence thrusting can influence the overall geometry of a duplex or imbrication zone and complicate stratigraphic relations. Even if thrusting is in-sequence the result may vary according to how much displacement each horse accumulate. In Figure 16.9 the amount of displacement is small and equal on each fault. If we increased the displacement successively on each horse, we would build a stack of horses rather than the train seen in Figure 16.9. It is clear that we can make a spectrum of imbrication geometries by varying these parameters. An example of a structure where stacking has occurred is shown in Figure 16.15.

Tectonic horses in duplexes form by the successive formation of ramps in competent layers that act as **stress guides**, meaning that they transmit and focus stress better than their adjacent layers. Hence stratigraphy controls both the locations and sizes of tectonic horses: the thicker the competent layers, the larger the horses. Furthermore, weak layers control the location of décollements.

Ramps tend to form in stiff layers, detachments in weak ones.

The largest duplexes or imbrications are found in the hinterland where basement is involved. Here, large parts of the crust, perhaps even the entire crust, can be imbricated. The identification of such large-scale duplexes relies on deep-seismic data, whereas small-scale structures can be seen in outcrop (Figure 16.6) or even in thin-section (Figure 16.8).

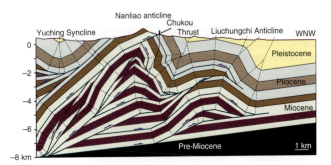

Figure 16.15 Interpretation of the Nanliao anticline, southern Taiwan, using surface information and well data. The anticline is interpreted as a stack of imbricated thrust sheets piled on top of each other. Interpretation by John Suppe (1983).

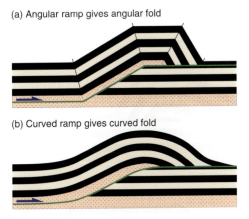

Figure 16.16 The connection between ramp and fold geometry.

Fault-bend folds

The moment a ramp is established and the hanging wall starts climbing above it, the hanging-wall layers are deformed into a **fault-bend fold**. The geometry of the fold reflects the geometry of the ramp. Angular ramps produce angular, kink-like folds, while more gently curved ramps result in less angular folds (Figure 16.16). The relation between ramp and fold geometry is simple enough that they can be modeled by means of simple computer programs. Knowing the fold geometry we can predict the ramp geometry and vice versa. Angular ramps and kink-like folds are particularly popular because they are relatively simple to construct geometrically.

A classic fault-bend fold forms where a tectonic unit is passively transported over a ramp (bend) in its sole thrust.

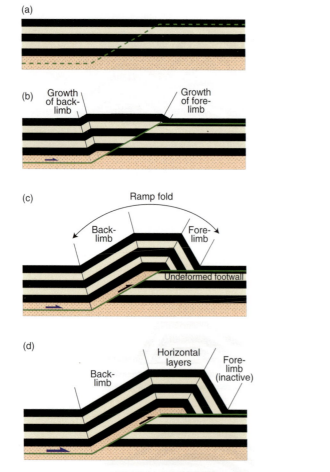

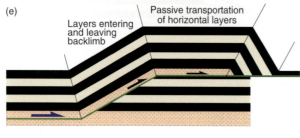

Figure 16.17 Development of a fault-bend fold. Note that layers are rotated to become hinterland-dipping in the ramp region and then rotated back to horizontal as they leave the ramp. Also note that the forelimb is passively transported toward the foreland (right) while the backlimb is stationary.

An interesting aspect of the deformation history associated with fault-bend folds is the history of deformation experienced by hanging-wall layers as they enter and pass the ramp. First the hanging-wall layers are bent upward to accommodate the shape of the ramp. Then, as they pass the ramp, they are retrodeformed to their original, usually horizontal, orientation (Figure 16.17). Thus, the layers are deformed twice over a short transport distance. In this sense the ramp is an area in which hanging-wall

layers are "processed" before being transported toward the foreland at a higher stratigraphic level. It is also interesting that while the fault-bend fold is stationary, the fold in the trailing edge of the horse or sheet is passively transported toward the foreland. The mechanism is commonly assumed to be flexural slip or shear, which preserves layer thickness and length and allows for simple kinematic constructions of cross-sections such as Figure 6.17.

Fault-propagation folds

Like normal and strike-slip faults, many reverse and thrust faults form a ductile fold zone around their tips as they form or propagate. The tip-fold zone is particularly well developed where thrust faults affect non- and low-metamorphic sedimentary rocks. The fold associated with the fault tip is a **fault-propagation fold** – a name originally applied to the particular type of fold that develops ahead of a propagating thrust, but which can also be used more generally for folds forming in front of any propagating fault tip. Fault-propagation folds differ from fault-bend folds and other folds in that they move together with the propagating fault tip. A fault-bend fold on the other hand is located at the ramp and remains stationary, with rocks passing into and out of the fold. Fault-bend folds also tend to have steeper and sometimes overturned layers in the forelimb. A seismic example is shown in Box 16.2.

Classic fault-propagation folds form in subhorizontal strata where faults propagate up-section. A simple model for the formation of such a fault-propagation fold is shown in Figure 16.18. Keeping layer thickness constant allows for simple construction of the fold geometry as a function of displacement gradient, fault dip and fault geometry. An asymmetric foreland-verging fold is the result.

> A fault-propagation fold forms above the tip line of a thrust to accommodate the deformation in the wall rock around the tip.

In general, the thrust fault will break through the fault-propagation fold if it keeps accumulating displacement. The result may be drag folds along the fault, notably in the hanging wall, but sometimes also in the footwall (Figure 16.19). Thrust faults tend to break through the steep short limb or lower syncline, creating high-angle relations between the fault and the hanging-wall layers (Figures 16.19 and 16.20).

BOX 16.2 | SEISMIC IMAGES

It may be challenging to image reverse faults and related structures on seismic images, particularly where layers and faults are steep and stratigraphy is repeated in the vertical direction. Nevertheless, contractional structures are important oil traps in many parts of the world and correct identification and interpretation may be crucial. In such cases well information is helpful, particularly if dipmeter data are available.

This example is from the Provincia Field in the Andean intramontane Magdalena basin in Colombia, and shows a detachment fold structure modified by reverse faults rooted in the detachment (fault-propagation fold). Unconformities and growth of strata show that the structure evolved during sedimentation, and its anticlinal shape defines an oil trap. Vertical lines are wells with dipmeter data that help constrain the interpretation, which was carried out by Andrés Mora, Cristina Lopez and colleagues in Ecopetrol-ICP.

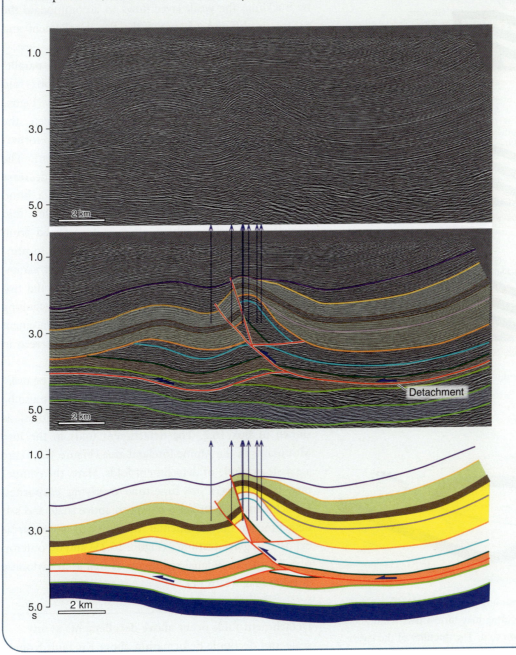

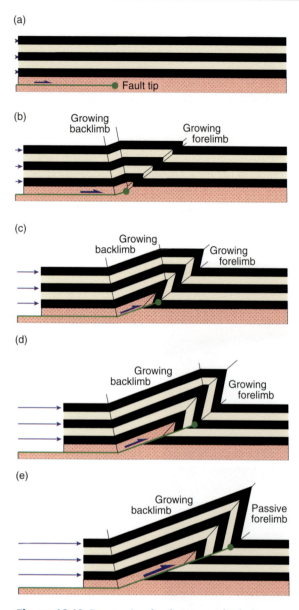

(a)

(b) Growing backlimb   Growing forelimb

(c) Growing backlimb   Growing forelimb

(d) Growing backlimb   Growing forelimb

(e) Growing backlimb   Passive forelimb

Fault tip

Figure 16.18 Progressive development of a fault-propagation fold.

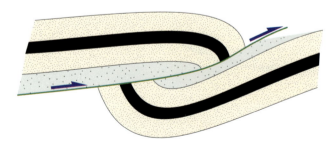

Figure 16.19 Layers commonly show drag near thrust and reverse faults. This type of drag may form where a fault cuts through a fault propagation fold. The ductility of the layers and the geometry of the ramp determine the final fold geometry.

Detachment folds

Fault-bend folds form at ramps, and fault-propagation folds form where faults propagate across the layering. However, there is another type of folds that can form where slip is solely along the layering. This type of folds, known as **detachment folds** or **décollement folds**, forms where layers above a detachment shorten more than their substrate. In fact, the substrate is commonly found to be undeformed, as portrayed in Figure 16.21a. Detachment folds tend to develop above very weak layers, such as overpressured shales or evaporates, typically concentric folds (Type 1B). As the folds form by buckling, the weak layer flows to accommodate the geometric difference between the flat décollement and the folded layers above.

Detachment folds are generally upright and parallel (constant layer thickness), sometimes with box fold geometry and oppositely dipping axial surfaces (Figure 16.22). A strong viscosity contrast between the folded layer and its surroundings promotes the formation of a series of buckle folds (a fold train). The offset along the detachment shows a gradual decrease toward the foreland, and may terminate as a blind fault. A special type of detachment-controlled fold sometimes develops where offset is transferred from one weak layer to a higher one (Figure 16.21b). In principle, this is similar to what happens at thrust ramps, but in some cases competent layer(s) between the two levels deform by buckling rather than by imbricate faulting and duplex formation. Vertical transfer of displacement may be caused by termination of weak layers, as seen in Figure 16.21b.

Classic detachment folds are upright, but can be overturned due to distributed simple shear. They may also be broken by faults to form fault-propagation folds, as shown in Box 16.2. The detachment folds in the Jura Mountains in the Alpine foreland area (Figure 16.14) are a classic example of detachment folds. Here, the competent folded layers are limestone overlying evaporites. Detachment folds are also common above shale and salt detachments on continental margins, in some areas exposed to glaciotectonics (Figure 16.23) and on outcrop scale in unconsolidated sediments as well as in mylonitic gneisses (Figure 16.11).

Detachment folds occur above décollements at any scale and decouple deformation above the fault from the commonly undeformed substrate.

Figure 16.20 Fault-propagation fold in the Tertiary thrust and fold belt in Mediumfjellet, Svalbard.

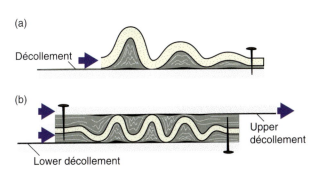

Figure 16.21 (a) Detachment folds developing above a décollement (detachment) during shortening. (b) A related type of folds forming between two décollements when displacement is transferred up-section in the direction of transport.

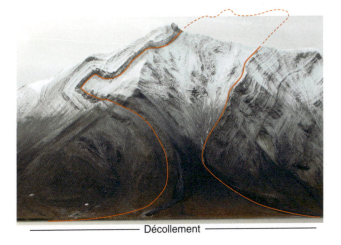

Figure 16.22 Large-scale detachment fold related to the Tertiary thrust and fold belt in Spitsbergen. The detachment is located below sea level. Photo: Mina Aase.

16.4 Orogenic wedges

The wedge model

Fold and thrust belts show an overall wedge-shaped geometry in cross-section, thinning toward the foreland. Such tectonic wedges typically occur in mountain ranges and in accretionary prisms above subduction zones. Sometimes just the low-grade foreland section is considered, in other cases the entire section from the foreland to the hinterland of a full orogenic belt is regarded. Both tend to show a wedge-shaped geometry.

The formation of tectonic wedges is commonly compared to the build-up of a wedge-shaped pile of snow or soil in front of a snowplow or bulldozer (Figure 16.24). At shallow crustal levels, where brittle deformation mechanisms dominate, the shape of the wedge depends not only on the force applied and gravity, but also on (1) the friction along the basal thrust or décollement, (2) the internal strength or frictional coefficient of the material within the wedge, and (3) any erosion at

Figure 16.23 Quaternary detachment folds formed when the last major Scandinavian ice sheet pushed south Eocene diatomite and volcanic ash deposits in northwest Jylland, Denmark. The detachment itself is not exposed.

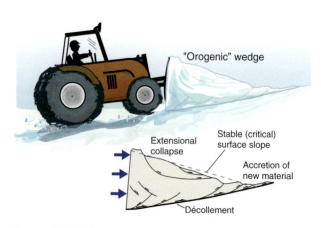

Figure 16.24 The creation and evolution of orogenic wedges are in many ways similar to the accretion of a snow or soil wedge in front of a bulldozer.

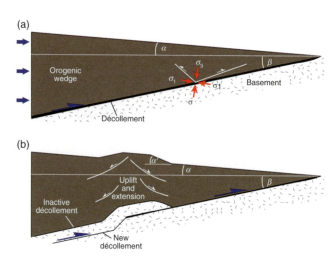

Figure 16.25 (a) Principles of an orogenic wedge. There is a close connection between the dip of the décollement, its friction, the applied force at the end, gravity and the internal strength or rheological properties of the wedge. (b) The incorporation of a basement slice (known as tectonic underplating) generates uplift and instability. Consequently, the wedge responds by thinning through normal faulting. Hence, normal faults can form in an active orogenic wedge.

the surface of the wedge (Figure 16.25a). In principle the wedge geometry is independent of the actual size of the wedge, so as the wedge grows in length by frontal imbrication it will also deform internally to maintain a stable shape.

The shape of an orogenic or accretionary wedge is controlled by the basal friction, the strength of the wedge material and erosion. Weak wedges with low basal friction develop wedges that are long and thin in cross-section.

The **basal friction** is a major controlling factor in the wedge model. The lower the basal friction, the lower and longer the wedge. In orogenic wedges the basal friction is controlled by the properties of the relatively weak basal décollement. The décollement commonly consists of lithologies rich in phyllosilicate minerals, and elevated fluid pressures may be even more important. Fluids play a particularly important role in subduction zones where the décollement zone acts as a conduit for fluids released from wet sediments and, in the metamorphic zone, recrystallization and dehydration reactions. In fact, the discovery that fluids may significantly weaken thrusts helped explain the classical mechanical problem of thrusting (see Box 16.3). A uniform frictional coefficient is commonly applied in simple models, while a gradual variation that leads to a curved slope of the wedge may be more realistic. We can relate to the snowplow analog: when the basal friction increases because of irregularities at the subsurface, more force is needed and snow or sand piles up to form a higher wedge with a steeper slope. This shows how the shape of the wedge depends on the basal friction.

The stress within the wedge must everywhere be identical to the strength of the material being deformed, i.e. the stress must be critical at every point in the wedge. This is why the orogenic wedge model is referred to as the **critical taper** or **critical wedge model**. Wherever the stress gets higher, the material will immediately deform until equilibrium is regained. In mathematical models, the Coulomb fracture criterion is used to model critical wedges, which is particularly relevant for shallow (upper crustal) wedges or parts of wedges. Such models are known as **Coulomb wedges**. Larger and deeper orogenic wedges, such as the Caledonian wedge of southwest Norway (Figure 16.26), are controlled by plastic flow laws, and simple wedge models for viscous and plastic media have been developed for these cases.

Erosion and deposition at the surface of the wedge, where material is removed, added or redistributed, will lower the surface slope and make the wedge unstable. The result is that material in the wedge rises vertically by internal redistribution of rocks and sediments. In practice, this means reverse faulting and perhaps folding so that equilibrium is achieved and the surface slope stabilizes. During this process rocks move vertically so that metamorphic rocks are brought (closer) to the surface.

The characteristic stable wedge geometry is achieved when the wedge is everywhere at the critical angle and on the verge of collapse. Material is accumulated in the front through imbrication and the formation of duplex structures, as described in previous sections of this chapter. In shallow wedges the subsurface remains undeformed. In the hinterland region of a large

---

BOX 16.3 | THRUST NAPPES: MECHANICALLY IMPOSSIBLE?

The bulldozer model was rejected for a while in favor of gravity-driven models because of mechanical considerations: How can enormous thrust nappes be pushed horizontally or up-hill for tens or hundreds of kilometers without being internally crushed? Simple mechanical calculations tell us that the nappes will crush or fold at the initial stage. Hubbert and Rubey resolved the issue in a classic work from 1959, where they emphasized the importance of pore pressure in the basal thrust zone. Overpressure close to lithostatic pressure considerably decreases the push that is needed to move the nappe. High fluid pressure (P_f) in the décollement zone decreases the effective normal stress σ_n by an amount P_f. Using the Coulomb fracture criterion the shear stress on the décollement then becomes

$$\sigma_s = C - (\sigma_n - P_f)\mu$$

Others have pointed out the importance of the fact that the nappe does not move as a rigid block. Instead, the movement occurs by the accumulation of numerous small slip events, either in the form of earthquakes or as episodes of creep. The analogy used to explain this type of movement is the way a caterpillar moves (study one if you do not know). In this way, the frictional resistance that has to be overcome becomes much smaller than by simultaneous slip on the entire basal thrust.

---

orogenic wedge, however, basement blocks may be incorporated, leading to reorganization of the décollement zone and its frictional properties, and to vertical growth of the wedge. This is illustrated schematically in Figure 16.25b, where a basement slice is ripped off and incorporated into the wedge, which locally thickens. Such local thickening creates a slope instability that again is compensated for by means of local thinning through extensional deformation. The thinning occurs in the upper part of the wedge by normal faulting, while the top-to-foreland sense of movement along the décollement zone continues.

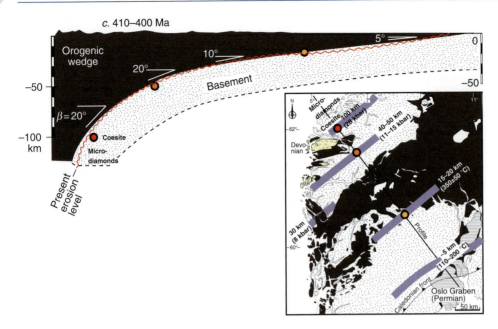

Figure 16.26 Simplified reconstruction of the orogenic wedge in southwest Norway at the end of the Caledonian orogeny, based on paleopressure and temperature data. Modified from Fossen (2000).

Gravitational models

The wedge or bulldozer model where plate motion drives the "bulldozer" is not the only model that has been suggested to explain the formation of wedge-shaped volumes of predominantly shortened rocks. Gravity has also been suggested as the dominant driving force during orogenic thrusting toward the foreland. As an orogenic belt develops, the highest mountains develop in the central part. The elevated hinterland rock volume represents a gravitational potential that may or may not drive thrust nappe motion (Figure 16.27). The Tibetan Plateau in the Himalayan mountain chain is a modern example. The first variant of the gravitational model assumed that thrust faults formed as rock units (thrust nappes) slid down from the elevated hinterland. This model is sometimes referred to as the sliding or gliding model (Figure 16.28a).

The **gliding model** was popular in the 1950s–60s, particularly in the Alps where many of the well-mapped thrust faults are dipping towards the foreland. However, seismic imaging of active orogenic belts shows that the basal décollement always dips toward the hinterland. The present foreland dip in the Alps is most likely a late modification. The gliding model is therefore not applicable to major thrusts, but can still be important on a smaller scale at shallower levels within orogenic wedges.

Gravitational collapse is a term that is now principally associated with normal faults, although it may drive reverse and thrust faults too. Gravitational collapse in contractional zones occurs when the thickened crust is too weak to sustain its own weight and collapses. Such a collapse may be related to weakening due to general

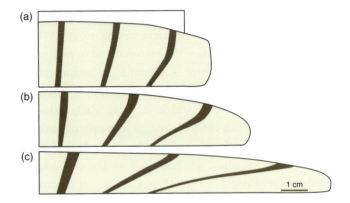

Figure 16.27 Experimental simulation of gravitational spreading. Note that most of the spreading occurs by flow of the upper part, while the basal part shows little or no movement. Together with the distortion of the originally vertical markers this indicates a significant component of simple shear. However, the change in height and length in the shear (horizontal) direction reveals a pure shear component. The deformation is therefore subsimple shear. Based on Ramberg (1981, p. 224).

heating and intrusion of warm and weak magma. Another model calls on delamination of the lower and densest portion of the lithosphere. Removal of the dense root causes uplift of the lithosphere and possibly upward collapse of the less dense remaining root, both of which generate a gravity-induced push toward the foreland (see Figure 17.21). The collapse may result in new-formed normal faults or shear zones, or reactivation of existing thrusts as low-angle extensional faults.

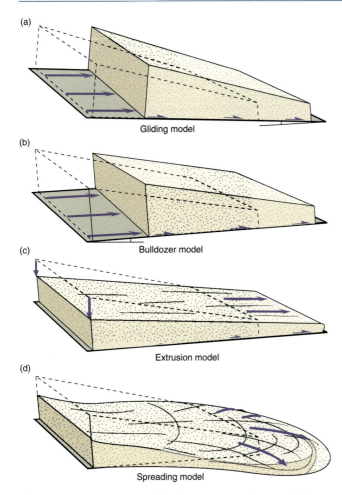

Figure 16.28 Various models for nappe translation. Gliding (a) requires a foreland-dipping décollement while (b) implies up-dip translation. (a) and (b) are rigid translations, driven by gravity (a) and a push from behind (b). (c) and (d) are also gravity driven, differing by boundary conditions (whether material is constrained on the sides or not). Combinations of these models are possible.

In principle, gravity can push thrust nappes towards the foreland. This may occur if the thick and elevated hinterland part of the orogenic wedge thins by flow towards the foreland. The process shares many similarities with that of glaciers, whose flow is also gravity driven. When the material in the wedge only extrudes toward the foreland, perpendicular to the orogenic front, the model is called an **extrusion model** (Figure 16.28c). However, if the elevated area spreads out in a more or less radial or concentric pattern, as illustrated in Figure 16.28d, it is referred to as a **spreading model**. Both of these related models can be simulated in the kitchen using dough, which is weak enough that it collapses gravitationally

and spreads out. When unconstrained, the dough flows in a radial displacement pattern (the spreading model), but constrained by two sidewalls it extrudes in a pure-shear manner (the extrusion model). These mechanisms can also be modeled more sophisticatedly in the laboratory (Figure 16.29) or numerically. Hans Ramberg explored this type of deformation by means of physical experiments in his famous laboratory in Uppsala, and later numerically on a computer, and was able to model many examples of nappe movements driven by gravity collapse alone. Some would say that his models apply fully to orogenic belts, while many would argue that the approach downplays the horizontal "push" in orogenic settings. Regardless, one of the interesting aspects of such gravitational collapse is that the entire wedge or nappe pile is thinned, compensated by lateral extension. In these models the thrust is passively formed as a consequence of spreading over a rigid basement (Figure 16.28b), and the displacement increases toward the foreland. Technically it dies out somewhere in the hinterland.

Models that explain orogeny and large-scale thrusting must consider gravity-driven collapse as well as pure push-from-behind-type shortening.

The extrusion and spreading models have implications also for the strain and deformation history. The vertical thinning and lateral extension of the wedge is a significant component of coaxial deformation throughout much of the wedge. Pure shear and subsimple shear dominate the simple extrusion model, and three-dimensional coaxial deformation characterizes radial spreading. It is evident from the model shown in Figure 16.29 that the

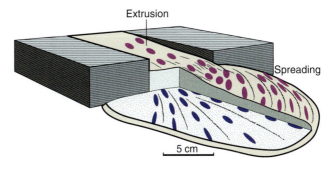

Figure 16.29 The spreading model explored in the laboratory. The deformation between the two rigid boxes is extrusion. Once the material leaves the constraining boxes, spreading generates radial strain patterns in the lower part, and concentric stretching in the upper parts of the material. Strain ellipses are indicated. Based on Merle (1989).

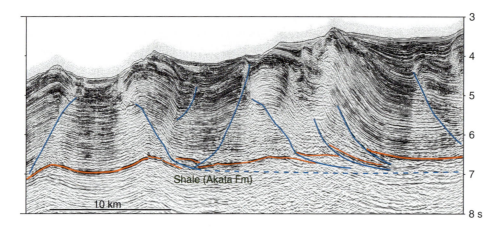

Figure 16.30 Example of contraction structures in the lower and frontal part of the Niger delta, formed by gravitational push from higher parts of the delta slope. Three-dimensional seismic data provided courtesy of Veritas DGC Ltd.

spreading model gives a radial distribution of the *X*-axis of the finite strain ellipsoid in the lower part, and a concentric pattern in the upper part. Both the shape and orientation of the strain ellipsoid should therefore be investigated when considering models of nappe transport. Curved thrust fronts could be another indication of gravity spreading.

Some nappes, such as the Jotun Nappe of the Scandinavian Caledonides, show little or no internal deformation above the intense non-coaxial deformation along their bases. They are not spreading or extruding nappes, because such nappes must be weak enough to deform internally, which requires a certain temperature and an appropriate mineralogy. Quartz-rich nappes would, for example, be more likely to collapse than feldspar-rich nappes at temperatures in the range of 300–500 °C. Allochthonous salt, described in Chapter 19, spreads gravitationally at the surface because of salt's ability to flow even at surface conditions. And large portions of continental slope deposits glide on salt or clay-rich sediments to form thrusts, imbrication structures and detachment folds in their down-slope parts. The Niger delta is an example of gravity tectonics where the wedge model works well (Figure 16.30).

Inversion tectonics

When the crust contracts it is easily influenced by preexisting structures. We have already seen how detachments form along weak lithologic layers such as shale and salt. Similarly, preexisting faults will become reactivated if they have an orientation favorable of slip.

The reuse of faults during contraction is often best expressed when normal faults are reversed. This phenomenon, and the opposite (reverse faults being reactivated as normal faults) is commonly referred to as **inversion**. Normal faults commonly exist when contraction initiates.

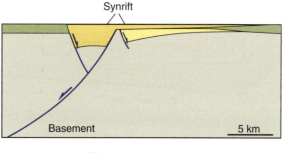

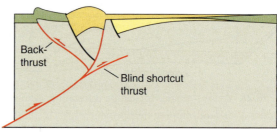

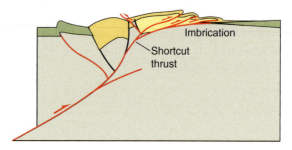

Figure 16.31 Example of inversion model where two phases of extension (Triassic and late Jurassic–early Cretaceous) (marked as synrift) were shortened during reversal of slip on the main fault and formation of new thrusts at shallow levels. Contractional faults shown in red. Based on structures mapped in northern Chile by Amilibia *et al.* (2008).

Most, if not all orogenic belts, have a pre-contractional history of rifting and continental margin normal faulting. Hence, reactivation of normal faults is to be expected during orogeny. The exception is newly deposited sediments in accretionary prisms and non-orogenic settings such as gravity-driven thrusting in the lower parts of continental slopes.

Normal faults tend to be steeper than their contractional counterparts, particularly in their upper part. This makes them less favorable for reactivation when σ_1 is switched to horizontal. Besides, steep normal faults can only accumulate limited horizontal shortening. Contraction therefore commonly results in more gently dipping **shortcut faults**, particularly in the upper and steepest part of normal faults. A schematic example based on the former South-American continental margin in northern Chile is shown in Figure 16.31. The normal fault is eventually turned into a thrust ramp with an imbrication zone forming in synrift sediments deposited during pre-contractional rifting. Back-thrusts similar to those indicated in Figure 16.10 also form. Hence, ramps in regions of thrusting may form at the locations of extensional structures.

Summary

Contractional structures are common and important structures, primarily in orogenic settings but also in the lower parts of many continental margins. The transportation of enormous sheets of the crust for up to hundreds of kilometers is very impressive, and the visual beauty and geometric–kinematic consistency of smaller-scale duplexes and fault-propagation folds are highly appreciated among geologists. Much of the terminology presented in this chapter is from areas of low-metamorphic or unmetamorphosed sedimentary rocks. Similar structures also occur in thick-skinned hinterland settings, added to plastic shear zone structures and mylonite zones discussed in the previous chapter. Important things to remember are:

- The mechanical or rheological situation is always important for the resulting structures.

- Detachments form in weak zones and can carry thick piles of nappes when assisted by high fluid pressures.

- Detachments may separate undeformed footwall rocks from deformed hanging-wall rocks, or units of different deformation style and strain.

- Ramps can form in competent units acting as stress guides, but can also form at locations of preexisting normal faults by means of reactivation.

- Allochthonous units have been transported by means of thrusting, autochthonous rocks have not.

- Thrust nappes and sheets build up foreland-thinning wedges that can be modeled by means of the critical wedge model.

- Changes in the conditions within or underneath the wedge can cause it to become unstable and thicken by thrusting and folding, or collapse by extensional faulting.

- Large-scale collapse of the thickest and central part of a wedge can drive thrusting in the foreland.

Review questions

1. What are branch lines and how could they be useful?

2. If you were looking for oil, what type of contractional structure covered in this chapter would you look for?

3. What is meant by a mylonite nappe?

4. What is the difference between a fault-propagation fold and a fault-bend fold?

5. Where could we expect to find back-thrusts?

6. What is the ideal setting for detachment folds to form?

7. How can extensional faults form in a thrust-and-fold belt?

8. What conditions determine the shape of an orogenic wedge?

9. How can those conditions change during an orogenic event?

10. Why is it likely to have preexisting normal faults in an orogenic belt?

E-MODULE

 The e-learning module called *Contraction* is recommended for this chapter.

FURTHER READING

Thrust nappes

McClay, K. and Price, N. J. (Eds.), 1981, *Thrust and Nappe Tectonics*. Special Publication **9**, London: Geological Society.

Ramberg, H., 1977, Some remarks on the mechanism of nappe movement. *Geologiske Föreningen i Stockholm Förhandlingar* **99**: 110–117.

Thrust structures

Bonini, M., Sokoutis, D., Mulugeta, G. and Katrivanos, E., 2000, Modelling hanging wall accommodation above a rigid thrust. *Journal of Structural Geology* **22**, 1165–1179.

Boyer, S. E. and Elliott, D., 1982, Thrust systems. *American Association of Petroleum Geologists Bulletin* **66**: 1196–1230.

Butler, R. W., 1982, The terminology of structures in thrust belts. *Journal of Structural Geology* **4**: 239–245.

Butler, R. W. H., 2004, The nature of "roof thrusts" in the Moine Thrust Belt, NW Scotland: implications for the structural evolution of thrust belts. *Journal of the Geological Society* **161**: 1–11.

Elliot, D., 1976, The motion of thrust sheets. *Journal of Geophysical Research* **81**: 949–963.

Mitra, G. and Wojtal, S. (Eds.), 1988, *Geometries and Mechanisms of Thrusting, with Special Reference to the Appalachians*. Special Paper **222**. Geological Society of America.

Fault-bend folding

Suppe, J., 1983, Geometry and kinematics of fault-bend folding. *American Journal of Science* **283**, 684–721.

Strain

Coward, M. P. and Kim, J. H., 1981, Strain within thrust sheets. In K. R. McClay and N. J. Price (Eds.), *Thrust and Nappe Tectonics*. Special Publication **9**, London: Geological Society, pp. 275–292.

Sanderson, D. J., 1982, Models of strain variations in nappes and thrust sheets: a review. *Tectonophysics* **88**: 201–233.

The wedge (critical taper) model

Dahlen, F. A., 1990, Critical taper model of fold-and-thrust belts and accretionary wedges. *Annual Reviews Earth Planetary Science* **18**: 55–99.

Fault-propagation folding

Erslev, E. A., 1991, Trishear fault-propagation folding. *Geology* **19**: 617–620.

Narr, W. and Suppe, J., 1994, Kinematics of basement-involved compressive structures. *American Journal of Science* **294**: 802–860.

Thrusts and fluids

Fyfe, W. and Kerrich, R., 1985, Fluids and thrusting. *Chemical Geology* **49**: 353–362.

Hubbert, M. K. and Rubey, W. W., 1959, Role of fluid pressure in mechanics of over-thrust faulting. *Geological Society of America Bulletin* **70**: 115–166.

Non-orogenic gravity-driven thrusting

Corredor, F., Shaw, J. H. and Bilotti, F., 2005, Structural styles in the deep-water fold and thrust belts of the Niger Delta. *American Association of Petroleum Geologists Bulletin* **89**, 753–780.

Thrusting and hydrocarbon accumulations

McClay, K. (Ed.), 2004, *Thrust Tectonics and Hydrocarbon Systems.* American Association of Petroleum Geologists Memoir **82**.

A few examples of thrust belts and orogenic structures

Hossack, J. R. and Cooper, M. A., 1986, Collision tectonics in the Scandinavian Caledonides. In M. P. Coward and A. C. Ries (Eds.), *Collision Tectonics.* Special Publication **19**, London: Geological Society, pp. 287–304.

Law, R. D., Butler, R. W. H., Holdsworth, R., Krabendam, M. and Strachan R. (Eds.), 2010, *Continental Tectonics and Mountain Building: The Legacy of Peach and Horne.* Special Publication **335**, London: Geological Society.

Law, R. D., Searle, M. P and Godin, L. (Eds.), 2006, *Channel Flow, Ductile Extrusion and Exhumation in Continental Collision Zones.* Special Publication **268**, London: Geological Society.

McQuarrie, N., 2004. Crustal scale geometry of the Zagros fold-thrust belt, Iran. *Journal of Structural Geology* **26**: 519–535.

Chapter 17

Extensional regimes

Traditionally, extensional structures have received less attention than their contractional counterparts. However, the tide turned in the 1980s when it was realized that many faults and shear zones traditionally thought to represent thrusts carried evidence of being low-angle extensional structures. First recognized in the Basin and Range province in the western USA, it is now clear that extensional faults and shear zones are widespread in most orogenic belts. Most would agree that the study of extensional structures has significantly changed our understanding of orogens and orogenic cycles. The current interest in extensional faults is also related to the fact that many of the world's offshore hydrocarbon resources are located in rift settings, and many hydrocarbon traps are controlled by normal faults. Also, the development of most hydrocarbon reservoirs requires a sound understanding of extensional faults and their properties and complexities.

17.1 Extensional faults

Extensional faults cause extension of the crust or of some reference layering in deformed rocks. An extensional fault is illustrated in Figure 17.1b, affecting horizontal layers. That fault shows a displacement that is close to the thickness of the layering, so that sense and amount of displacement can be easily identified. Other extensional faults have accumulated up to a hundred kilometers of offset, not quite as much as the many hundreds of kilometers estimated on some thrust faults and strike-slip faults, but still quite considerable. At this scale the crust itself is our natural choice of reference. If the distance between two points on the surface of the Earth, one on each side of the fault, is increased during deformation, then there is extension in that direction. But this can also occur across a strike-slip fault, depending on the relative positions of the two points. So, we have to evaluate extension *perpendicular* to the strike of a fault to evaluate if it is a true extension fault. This is the principal extension direction across an extensional dip-slip fault (Figure 17.1b), just as it is the principal shortening direction for contractional dip-slip faults. A perfect strike-slip fault shows no change in length perpendicular to the fault.

For smaller faults, the term extensional fault can be used for faults that extend a given reference layer, regardless of the orientation of the layering. In this sense, reverse faults can be extensional faults as long as the reference layer is extended by the faults. An example of an extensional reverse fault is shown in Figure 17.2, together with an interpretation of how it formed by rotation. It is therefore useful to specify a reference surface, for example by using the terms crustal extension and layer-parallel extension.

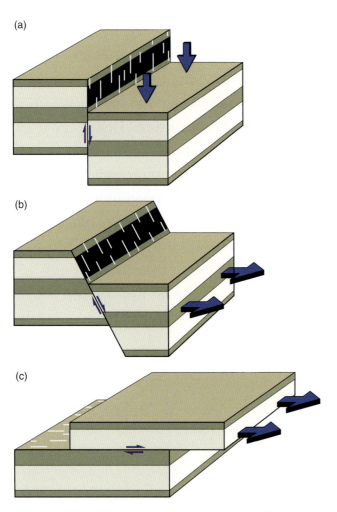

Figure 17.1 Using the surface of the Earth as a reference, extensional faults (b) are a spectrum of normal faults between vertical (a) and horizontal (c). Vertical and horizontal faults are neither extensional nor contractional.

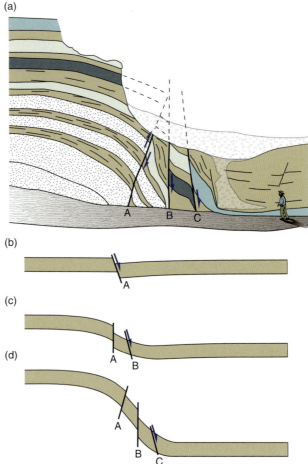

Figure 17.2 (a) Reverse faults and normal faults coexisting in a folded layer along a fault. Both the reverse and normal faults are extensional faults because they extend the layers. The reverse faults probably formed as normal faults (b) before being rotated during the folding (c, d). San Rafael Desert, Utah.

A reverse fault can also be an extensional fault if a tectonic or sedimentary layering is used for reference.

Fault dip is also important. As indicated in Figure 17.1, extensional faults cover a spectrum of fault dips between vertical (Figure 17.1a) and horizontal (Figure 17.1c). Vertical faults imply neither extension nor shortening of the crust, only vertical motions (Figure 17.1a). We can think of vertical block faulting as a large-scale analog to playing the piano keyboard – keys move vertically but the keyboard retains its length. Such vertical tectonics dominated much of the Colorado Plateau in the western USA during the Cretaceous Laramide phase. In general, faults that are oriented perpendicular to a layer do not stretch or shorten the layer. Horizontal (or layer-parallel) faults represent the other end-member: they neither shorten nor extend horizontal (or fault-parallel) layers (Figure 17.1c). Layer-parallel faults occur as flat portions (flats) of curved extensional as well as contractional faults.

Extensional faults are commonly thought to initiate with dips around 60°, based on the Coulomb fracture criterion and Anderson's theory of faulting as discussed in Section 7.3 (Figure 7.13). Field mapping and seismic interpretation show that both high- and low-angle extensional faults are common. In fact, they coexist in many extensional settings. How can we explain this finding?

The easiest explanation is that most or all rocks have an anisotropy inherited from earlier phases of deformation. Thus, a simple explanation for very steep faults is that they represent reactivated joints or strike-slip faults. Recall that joints are close to vertical because they form perpendicular to σ_3, which tends to be horizontal in the upper crust.

Arguing along the same lines, low-angle normal faults can form by reactivation of thrust faults, and many low-angle extension faults have indeed been interpreted as such. At the same time, experiments and field observations indicate that some high- and low-angle extensional faults formed under a single phase of extension, without the use of preexisting weak structures. In particular, some low-angle normal faults must have rotated from initial high-angle faults to low-angle structures, while other low-angle faults are thought to have formed directly without much rotation. We will start exploring these and related observations by means of a simple model of fault rotation known as the domino model.

17.2 Fault systems

The domino model

Sections through a rifted portion of the upper crust typically show a series of rotated fault blocks arranged more or less like domino bricks or overturned books in a partly filled bookshelf (Figure 17.3a). This analogy has given rise to the name **bookshelf tectonics** or the (**rigid**) **domino model**:

The rigid domino model describes a series of rigid fault blocks that rotate simultaneously in a uniform sense.

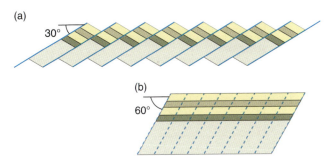

Figure 17.3 (a) Schematic illustration of rigid domino-style fault blocks. (b) Such fault blocks can be restored by rigid rotation until layering is horizontal. In this case we have applied 30° rotation and displacement removal.

BOX 17.1 | THE RIGID DOMINO MODEL

- No block-internal strain.

- Faults and layers rotate simultaneously and at the same rate.

- Faults end up with the same offset, which is constant along the faults.

- All faults have the same dip (parallel).

- Faults have equal offset.

- Layers and faults are planar.

- All blocks rotate at the same time and rate.

BOX 17.2 | THE GULLFAKS DOMINO SYSTEM

The North Sea Gullfaks oil field consists of a domino system limited by an eastern horst complex. The domino system consists of four to six main blocks, each subdivided by smaller faults. Well data show that a large number of subseismic faults and fault-related structures exists, including a large population of deformation bands. Furthermore, the dipping layering shows a systematic westward decrease in dip within each block. Therefore, a perfect rigid domino model does not work. If the blocks are back-rotated rigidly so that the layering becomes horizontal on average, the faults only obtain a 45° dip. This is lower than the ~60° fault dip expected from the mechanical considerations discussed in Chapter 6. The discrepancy can be explained by internal fault block strain.

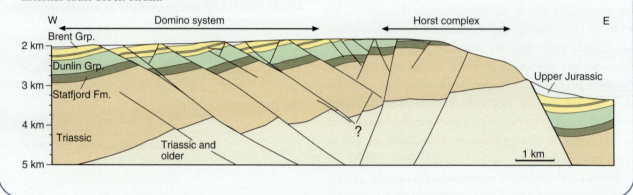

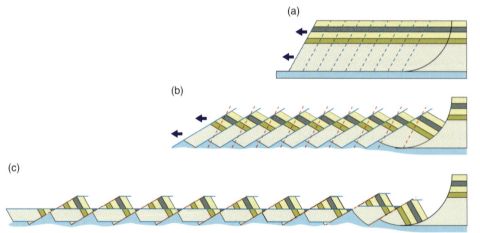

Figure 17.4 Schematic illustration of the development of a domino system. (a) The transition to the undeformed footwall is accommodated by a listric fault. (b) A new set of faults develops at high extension. (c) The resulting fault pattern can become quite complex. See Nur (1986) for a more detailed description.

The rigid domino model is easy to handle, and Figure 17.3 shows how rigid rotation of blocks and faults can restore such systems. However, the characteristics and constraints of this model (see Box 17.1) impose geometric challenges when applied to geologically realistic situations such as the example portrayed in Box 17.2. As always when applying models we have to consider boundary conditions and compatibility with the surroundings. In particular, open voids and overlaps are unacceptable. The first challenge is what is going on at each end of the domino system. This is quite elegantly resolved by introducing an appropriate listric fault, as shown in Figure 17.4. A listric fault can also be placed on the other side, accompanied by an oppositely dipping set of domino fault blocks. A graben will then be needed in the middle to connect the two sets.

The second compatibility problem exists between the base of the blocks and the substrate. This problem can be solved by introducing a mobile medium at the base of the rotating fault blocks, such as clay, salt or intruding magma. Where the blocks are large enough to reach the brittle–plastic transition in the crust, the basal space problem can be eliminated by plastic flow of the subsurface rocks.

We could also resolve the basal space problem by penetrative deformation of the basal parts of the domino blocks. However, this would represent a deviation from the ideal domino model and its assumption of rigid and therefore internally unstrained fault blocks. If this requirement is met, then restoration of a geologic section in the extension direction is easy, as shown in Chapter 20. The blocks are simply back-rotated until the layering becomes horizontal, and displacement is removed.

The soft domino model

Fault blocks seldom or never behave as rigid objects, and certainly not in rift systems with relatively weak rocks or unconsolidated sediments. Besides, we have already seen (Chapter 8) that faults occur in populations where fault sizes (width, length, displacement, area) tend to be distributed according to a power law (for example, Figure 8.12). The rigid domino model requires all faults to be of equal length and displacement, not showing any displacement gradient.

Because of these natural deviations from the rigid domino model, a **soft domino model** is defined, which permits internal strain to accumulate within blocks. This allows for variations in fault sizes, fault displacement variations and folding of layers.

The soft domino model allows for strain within domino fault blocks.

The amount of extension can no longer be found by means of simple line balancing or by rigidly back-rotating blocks. Instead, a representative model for the internal fault-block deformation must be chosen. A ductile simple shear deformation is an easy approach, as discussed in Chapter 20.

Why do domino systems form?

Extension of the crust can result either in a more or less symmetric horst-and-graben system or in a domino system of the types discussed above. The total extension and crustal thinning may be the same, but the fault arrangement depends on the way the rocks respond to the deformation, i.e. how strain is accommodated in the crust (compare Figure 17.3 with 17.5). Clearly, one of the most important factors that promote the development of asymmetric domino-style fault systems is the presence of a weak low-angle layer or structure. This can be an over-pressured formation, a mobile clay(stone), a salt layer or a preexisting fault that is prone to being reactivated.

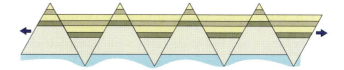

Figure 17.5 The alternative to domino-style stretching is the development of horst-graben systems. This deformation style is, ideally, symmetric and an overall pure shear strain.

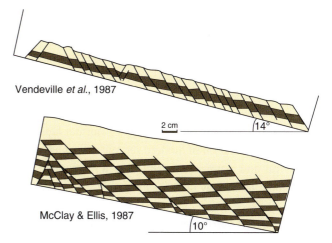

Vendeville *et al.*, 1987

2 cm   14°

McClay & Ellis, 1987   10°

Figure 17.6 Sandbox experiments where the base of the model is tilted prior to extension. The tilting may be the cause for the uniform dip and dip direction of the faults.

Conversely, the absence of such a gently dipping weak layer or décollement favors a more symmetric horst-and-graben system. The importance of a dipping (rather than horizontal) basal soft layer or décollement has been demonstrated in several physical experiments, as shown in Figure 17.6.

Multiple fault sets in domino systems

If a domino system is exposed to high extension, as in Figure 17.4, then the faults will rotate so far away from their initial and favorable orientation that a new set of faults forms (Figure 17.4b, red faults). This happens when the shear stress along the first faults decreases below the critical shear stress of the deforming rock, which depends on the strength of the host rock and on the mechanical properties (friction) of the faults. For realistic frictional values it can be shown that new faults are expected after 20–45° rotation. For an initial fault dip of 60° this means that domino faults can rotate to a dip of 40–15° while they are still active. From that point on the original domino faults become inactive, and new faults will form that cut the old ones and rotate everything further. Field evidence in favor of this model is reported from areas of high crustal extension, such as in the Basin and Range province in the western USA.

17.3 Low-angle faults and core complexes

For a long time, low-angle faults with significant displacements were almost exclusively mapped in fold-and-thrust belts, where they all were considered to be contractional faults. The traditional view that normal faults are high-angle structures, typically with dips around 60°, was well established. This somewhat simplistic distinction, which has support in rock mechanics considerations, was applied less rigidly after field mapping in the Basin and Range province in the 1970s, and later in many other parts of the world, had revealed that extensional low-angle faults are fairly common in several extensional regions (for example in the Caledonides, as shown in Fig. 17.7), even though the majority of normal faults are steeper. Interpretation of modern seismic images as well as numerical and physical modeling has also revealed that extensional faults may occur as low-angle structures.

The problem with low-angle normal faults

Low-angle extensional faults represent a mechanical challenge. In the general case where σ_1 is vertical, the formation of such faults would be mechanically infeasible for ordinary rock types. The unlikely reactivation of preexisting low-angle faults in the upper crust, combined with the observation that few low-angle faults are seismically active, calls for an alternative explanation. The most natural explanation is that low-angle faults are rotated normal faults with steeper initial dips.

Under normal or ideal circumstances, low-angle normal faults must form by rigid or soft (ductile) rotation of higher-angle faults.

Rotated normal faults

The model of fault rotation shown in Figure 17.4 already provides us with a model for how low-angle normal faults may form. The first (blue) sets of faults in this figure were rotated by a later (red) set to a horizontal orientation. However, many large-scale low-angle extensional faults are persistent and not broken up by younger faults such as those seen in Figure 17.4c (red faults). Hence, we must look for other models to explain such structures. The model that gained popularity toward the end of the 1980s involves extension at the scale of the entire crust. The isostatic effects of extension require that the base of the model is mobile – a condition that is not present in physical experiments with a fixed base, such as the examples shown in Figure 17.6.

Figure 17.8 illustrates the principle of the model. A listric normal fault forms in the upper crust, flattening along a weak detachment zone near the brittle–plastic transition. The hanging wall to large detachments is sometimes referred to as the **upper plate**, and the footwall as the **lower plate**. After a certain amount of extension a new fault forms in the hanging wall, while the first fault is inactivated. Inactivation is partly due to isostatic uplift of the thinned portion of the crust. This uplift rotates the original fault to the point where it becomes mechanically favorable to create a new fault in the hanging wall. This process repeats itself until a series of rotated domino-style blocks and related half-grabens is established. In this model, the steepest fault will be the youngest and active fault (compare the geometry of these with horses in a thrust duplex).

Note that these domino-style fault blocks are different than those of the classic domino model because they

Figure 17.7 The low-angle fault underneath the Hornelen Devonian basin in the Scandinavian Caledonides separates Devonian sandstones and conglomerates from mylonitic rocks of the Nordfjord-Sogn Detachment. Photo: Vegard V. Vetti.

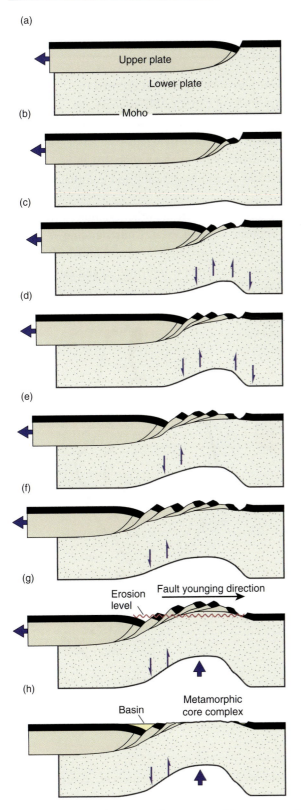

Figure 17.8 Development of a metamorphic core complex during crustal-scale extension and isostatic compensation. Note how new wedge-shaped fault blocks are successively torn off the hanging wall. Also note how isostatic compensation is accommodated by means of vertical shear. Based on Wernicke and Axen (1988).

develop at different times. The model is thus an example of how a domino-style fault block array can arise by a process that does not comply with the ideal domino model.

Rolling hinges and metamorphic core complexes

The model shown in Figure 17.8 involves simple shear deformation of the lower and middle crust. During this deformation, the upper crust is thinned and Moho is elevated. Erosion of the uplifted fault complex in the upper plate adds to this effect and eventually leads to exposure of lower plate rocks. The deformation in the upper plate is brittle, but the shearing along the horizontal detachment is initially plastic. However, this deformation takes on a more brittle character as the detachment is uplifted. Eventually, when the detachment is exposed it will appear as a core of metamorphic and mylonitic rocks overprinted by brittle structures in a window through upper plate rocks (Figure 17.8h). The lower plate rocks in this window are referred to as a **metamorphic core complex**, and such core complexes were first mapped and described in the Basin and Range region in Arizona and Nevada. Similar complexes have since been found in the entire Cordillera of western North America (Box 17.3) and many other parts of the world, including areas associated with mid-oceanic rifts.

A closer look at Figure 17.8 reveals that the vertical shear on the footwall (right) side of the figure gradually ceases. The oppositely directed shear on the hanging-wall side remains active underneath the active part of the fault system, moving in the hanging-wall direction. If we consider the crustal flexure as a fold, then the hinge zone is seen to move or roll in the hanging-wall direction during the course of the extensional development. This effect has led to the name **rolling hinge model** – a model that has been applied to many metamorphic core complexes.

> The rolling hinge model is a soft (ductile) fault rotation model where rotation migrates through the footwall as it is progressively unroofed.

Direct formation of low-angle faults

While models involving rotation of high-angle normal faults to low-angle extensional faults are both popular and realistic, some low-angle extensional faults seem to have formed with low initial dips. A characteristic feature of many such faults is that they cut high-angle faults,

BOX 17.3 | METAMORPHIC CORE COMPLEXES

A typical metamorphic core complex consists of a core of metamorphic rocks (gneisses) exposed in a window through non-metamorphic rocks, typically sedimentary rocks of considerably younger age. These two units or plates are separated by a detachment that shows evidence of significant shear offset. This detachment is brittle, overprinting mylonitic non-coaxial fabrics in the underlying gneissic lower plate. In general a core complex is controlled by a low-angle extensional detachment or shear zone that thins the upper plate (hanging wall) so that metamorphic lower-plate rocks ascend isostatically and eventually become exposed at the surface.

The core complex illustrated here is asymmetric with a single detachment. There are other examples where two detachments exist, one on each side of the dome and with opposite sense of shear. Understanding metamorphic core complexes and their development requires detailed structural field work combined with other sources of information, such as radiometric age determinations, stratigraphic information and seismic data. The map shows the locations of metamorphic core complexes and associated extension directions in the North American Cordillera.

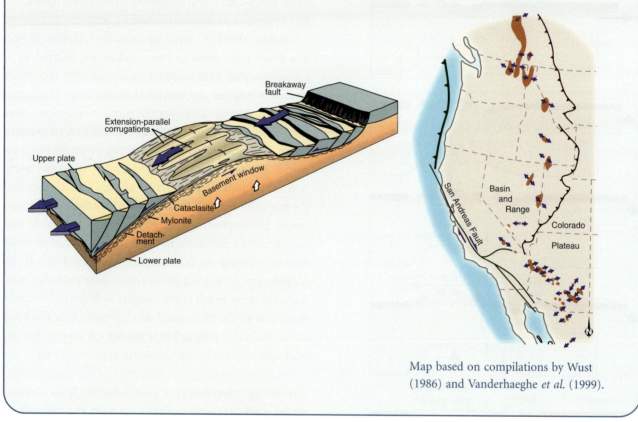

Map based on compilations by Wust (1986) and Vanderhaeghe *et al.* (1999).

contrary to what the models in Figures 17.8 and 17.4 show. This type of extensional faulting has been reproduced in experimental work, as shown in Figure 17.9 (red Fault 5 cutting Fault 1).

There must be a reason for a low-angle fault to take over the extension. An explanation could perhaps be that the high-angle faults lock up, but it is more likely to be caused by a subhorizontal zone of weakness.

In undeformed sedimentary sequences, overpressured shales or evaporite layers represent anomalously weak layers along which an extensional fault may flatten. In previously deformed rocks, preexisting faults or shear zones may create the anisotropy that causes the formation of a low-angle extensional fault. Thrust faults are low-angle structures that are easily reactivated under extension, for instance when the stress conditions change

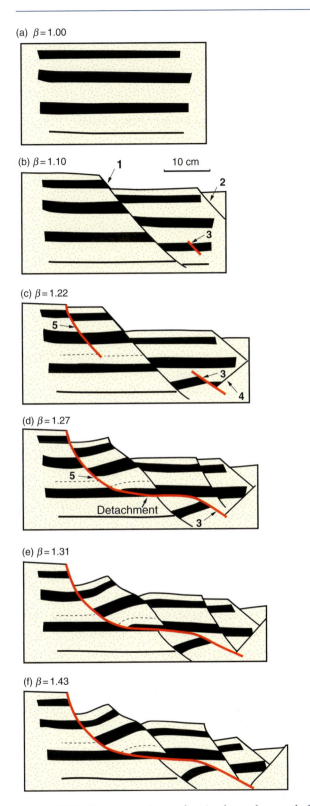

(a) $\beta = 1.00$

(b) $\beta = 1.10$

10 cm

(c) $\beta = 1.22$

(d) $\beta = 1.27$

Detachment

(e) $\beta = 1.31$

(f) $\beta = 1.43$

Figure 17.9 Plaster experiment showing how a low-angle fault can form at a relatively late stage during an extensional deformation history. Note that the late fault (red) cuts preexisting high-angle faults and that a horst is about to establish above the flat fault segment. Black pseudo-layering was painted on the side walls prior to extension. From Fossen *et al.* (2000).

at the end of an orogeny. In fact, it seems to be the rule rather than the exception that orogenic belts contain low-angle normal faults or extensional shear zones that have formed by reactivation of thrusts.

> **The direct formation of low-angle faults requires low-angle mechanically weak structures or layers in the crust, or anomalous stress orientations.**

17.4 Ramp-flat-ramp geometries

We have already seen how extensional faults can have a listric geometry. Another geometry that is particularly common for large-scale extensional faults is the combination of two ramps linked by a subhorizontal segment. Such **ramp-flat-ramp geometries** generate extensive hanging-wall strain because the hanging wall must adjust to the fault geometry during fault movements. An example of a ramp-flat-ramp geometry is shown in Figure 17.9 (red fault), where the flat develops by linkage of two steeper fault segments.

As illustrated in Figure 17.10, a series of wedge-shaped fault blocks may develop above the ramp-flat-ramp fault, where the faults either die out upward or reach the surface (also see the hanging wall to Fault 6 in Figure 17.11f). A series of such faults or fault blocks is called an extensional **imbrication zone**, similar to the use of the term in the contractional regime. A related type of extensional structure is a series of lenses that together form an **extensional duplex**. Extensional duplexes have floor and roof faults similar to contractional duplexes.

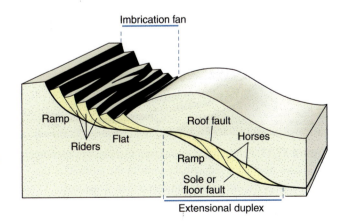

Figure 17.10 Schematic illustration of extensional imbrication and a duplex structure. Horses and riders are shown in yellow.

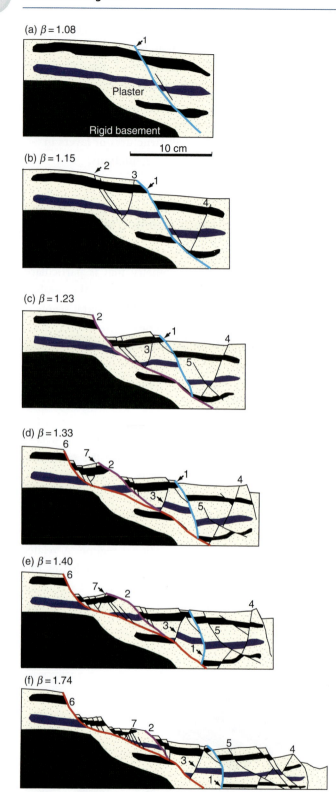

(a) $\beta = 1.08$

Plaster

Rigid basement

10 cm

(b) $\beta = 1.15$

(c) $\beta = 1.23$

(d) $\beta = 1.33$

(e) $\beta = 1.40$

(f) $\beta = 1.74$

Figure 17.11 Footwall collapse as seen in a plaster experiment. Note how new faults sequentially form in the footwall at the same time as the dip of the controlling fault is reduced.

A horst complex is sometimes seen to develop above the flat in experiments (Figure 17.11) and in rift systems such as the North Sea (Box 17.2), and domino-style fault arrays commonly develop behind such horsts as the

footwall collapses. In Figure 17.11 the development of a horst complex is repeated. The first horst is defined by Faults 1 and 3, and the second by Faults 2 and 7.

17.5 Footwall versus hanging-wall collapse

We often think of extensional imbrication zones as forming by the successive formation of faults and slices in the hanging wall. The general observation that the hanging wall tends to be more strained than the footwall supports this view. This process certainly occurs, and is referred to as **hanging-wall collapse**. An example of hanging-wall collapse is seen in Figure 17.11, where more and more faults are added to the hanging wall of Fault 6 during the last (d–f) part of the experiment. Another example is seen in Figure 17.8.

Experiments such as that shown in Figure 17.11 also tell us that **footwall collapse**, where new faults successively form in the footwall, can occur. We see this in the first stage of the experiment, where a new synthetic fault (2) forms in the footwall of Fault 1 (Figure 17.11a–c). At a later stage, Fault 6 forms yet farther into the footwall, before its hanging wall collapses into an imbrication zone.

Footwall collapse is common where huge fault blocks form and rotate in rift systems, and the elevated crests of these blocks collapse under the influence of gravity and tectonic forces. In some cases gravity alone causes collapse of the footwall. Gravitational collapse and slumping on curved fault surfaces results, creating complex stratigraphic relationships that represent challenges in petroleum exploitation. Slumping is commonly controlled by the presence of weak layers such as mud, overpressured sediments or salt, and typically develops contractional folds and faults in the toe zone in addition to extensional faults in its central and rear part (Figure 17.12).

17.6 Rifting

A rift forms where the crust is pulled apart by tectonic forces. There can be several factors leading to the formation of a rift, and two end-member models are known as active and passive rifting. In the **active rifting model**, the rift is generated by rising hot mantle material or plumes in the asthenospheric mantle, causing doming and adding tensile stresses to the domed area. The result

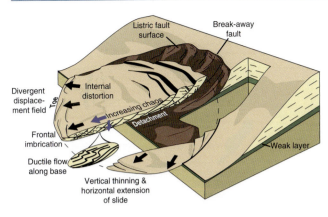

Figure 17.12 Slumping generates extensional structures in the middle and rear part, and contractional structures in the frontal part.

is a rift dominated by magmatism and not necessarily so much extension. In the **passive rifting model**, the rift forms because of far-field stresses related to plate tectonics. Passive rifts tend to form along zones of inherited weakness in the lithosphere, such as reactivated contractional structures along former orogenic zones.

Active rifting is controlled by mantle plumes while passive rifting is controlled by plate tectonic stress.

Many natural rifts tend to contain components of both of these models. In a somewhat simplified case, initial rifting may result from large-scale doming of the crust (Figure 17.13a). Steep fracture systems form at this stage, and may reach deep enough to promote magma generation and intrusions from the mantle. The following and main stage is the stretching stage, where the crust is vertically thinned and laterally extended (Figure 17.13b). Major faults and fault blocks form during this stage. Once stretching has come to a halt, the final subsidence stage is reached (Figure 17.13c). The crust cools, the basement is deepened and postrift sediments are deposited. Faults are mostly limited to those formed by differential compaction.

The extensional development of a rift is reflected by its sedimentary record. The **prerift** sequence is the sedimentary package deposited prior to extension. The **synrift** sequence is constituted of sediments deposited during the rifting. Synrift sediments show thickness and facies variations across growth faults, and hanging-wall thickening and footwall thinning or non-deposition is characteristic. The **postrift** sequence is controlled by the geometry of fault blocks and thermal subsidence after cessation of extension.

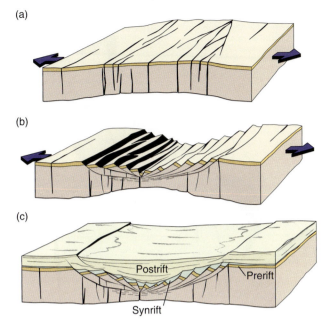

Figure 17.13 Three stages in rift development. (a) Early extension creating or rejuvenating deep-going fractures. Strain is low at this stage, and magma locally fills deep fractures as dikes. (b) The stretching phase, during which major fault complexes and arrays form. Synrift sediments not shown. (c) Postrift subsidence and sedimentation. Compactional faults in postrift sequence due to differential compaction.

Two very simple end-member models for rift growth can be envisioned based on the ground covered earlier in this chapter. One is related to the domino model, where the final width of the rift is established at the initial stage so that the domino-forming faults can accumulate displacement and rotate along with the domino blocks as extension accumulates. The other end-member model involves large-scale footwall collapse. In this model the rift expands from the rift axis so that faults get progressively younger away from the central graben. Natural rifts are of course complex with elements of several models. In general, rift development depends on several factors, including the processes and thermal structure of the mantle and the mechanical structure (distribution and orientation of preexisting weak structures) of the crust.

17.7 Half-grabens and accommodation zones

Symmetric rifts are rare, and most rifts have a master fault on one of the flanks. As we move along a rift, we may see that the location of the master fault flips from

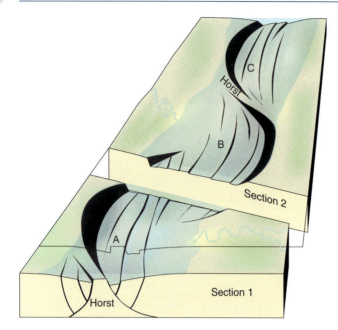

Figure 17.14 Rift system composed of interfering and overlapping half-grabens. The overlaps were called accommodation zones by Rosendahl and coworkers in the 1980s, based on observations from the East African Rift. Different types of half-graben arrangement occur. Accommodation zones may contain horsts (section 1) or grabens (section 2).

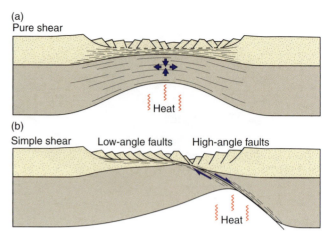

Figure 17.15 Two idealized models for crustal extension and rifting. The pure shear model is symmetric, with maximum heat underneath the middle of the rift. The simple shear model is dominated by a generally low-angle shear zone that produces an asymmetry on the rift.

The crust may be thinned symmetrically and uniformly by overall pure shear, or asymmetrically and more localized with a controlling dipping shear zone.

---

one side of the rift to the other. This is a common feature of rift systems and has been explored in detail in the Lake Tanganyika area of the East African rift system by Bruce R. Rosendahl and coworkers. Here the rift is developed as a series of oppositely dipping half-grabens. Each half-graben has a curved, half-moon shaped geometry and where one ends another, typically oppositely dipping, half-graben takes over. Depending on the arrangement of the grabens and secondary faults in their hanging walls, basinal highs (horsts) or lows (grabens) may form (Figure 17.14). The term **accommodation zone** is sometimes used specifically for this type of half-graben overlap structure. Note that (almost) symmetric sections are only obtainable from the zone between two oppositely dipping half-grabens, as shown in Figure 17.14.

17.8 Pure and simple shear models

Crustal stretching in rift zones is sometimes discussed in terms of the pure shear and simple shear models (Figures 17.15 and 17.16). The pure shear model is also called the **McKenzie model**, and the simple shear model is sometimes referred to as the **Wernicke model**, named after the authors who published the respective models in the 1970s and 1980s.

In the **pure shear model**, which is the older of the two, the total contribution of individual faults in the rift creates a symmetric thinning of the crust. The overall strain is pure shear, and horizontal extension is balanced by vertical thinning. The lower crust is thinned by plastic deformation mechanisms, while the upper crust deforms by brittle faulting.

While the pure shear model is overall symmetric by nature, the **simple shear model** results in an asymmetric rift, and in this sense is more consistent with observations discussed in the previous section. The term simple shear is used because this particular model is controlled by a dipping detachment fault or shear zone that transects the crust, and possibly the entire lithosphere. The detachment involves a localized simple shear that is significant enough that the term simple shear model is warranted. The two sides of a rift controlled by a dipping detachment are geometrically different, as is the thermal structure. In the pure shear model the highest temperature gradient is found underneath the middle of the basin, while it is offset in the simple shear model (Figure 17.15). This has consequences for uplift and subsidence patterns and therefore basin development, and different versions of the simple shear model exist that yield different results.

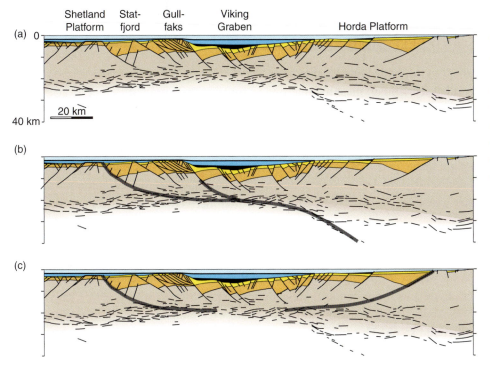

Figure 17.16 Section based on a deep seismic line across the northern North Sea. This section has been interpreted in terms of pure shear (b) as well as simple shear (c) and may be considered to contain elements of both models. Based on Odinsen *et al.* (2000).

17.9 Stretching estimates, fractals and power law relations

The amount of extension or stretching across a rift can be found in several different ways. Geologic sections based on deep seismic profiles that portray the crustal thinning beneath the rift can be used. If area conservation is assumed, i.e. no material transport in or out of the section and no assimilation of crustal material by the mantle, then extension can be estimated by restoring the section (see Chapter 20) to the point where the crustal thickness is constant and equal to that of the present rift margins.

A purely structural method of estimating the amount of extension is the summing of fault heaves along a reference horizon across the rift. If block rotation is modest, the sum should equal the total extension along the section. However, a mismatch is typically found between the strain estimate obtained from summing fault heaves and that calculated from area balancing.

In most cases the summing of heaves gives the lower stretching estimate. In some cases the two estimates differ by a factor of two. Why this discrepancy?

Balancing the crust typically give higher extension estimates than summing fault heaves across a rift.

One possibly explanation is that the lower part of the rifted crust is somehow assimilated by the mantle. Hence, the assumption of area conservation does not hold and extension is overestimated. On the other hand, assimilation of crustal material does not seem to be extensive enough to explain the discrepancy, except perhaps in active rifts with widespread magmatism.

Placing the fault displacement model under scrutiny, it becomes obvious that the contribution from subseismic faults, i.e. structures that are too small to appear on the geologic section from which fault heaves are summed, is left out. We may perhaps think that faults that are too small to appear on a seismic line or geologic section would not make much of a difference anyway. However, if the number of such small faults is large, they can sum up to a significant amount of extension. But how can we correct for fault extension if the faults are below the resolution of observation?

This problem was extensively explored in the 1980s and 1990s, and it was then realized that the distribution of fault offsets (or heaves) in many fault populations varies systematically according to a power law relation (see Box 17.4). The method was to collect fault displacement data from seismic lines, geologic maps and outcrops, and plot them in cumulative plots with logarithmic axes. Fault offset is measured for each fault along a chosen horizon

BOX 17.4 | FRACTALS AND SELF-SIMILARITY

A fractal is a geometric form that can be divided into smaller parts where each part has a form corresponding to the larger one. Fractals are called self-similar, meaning that one or more of their properties repeats at different scales. This goes for many properties of geologic structures, such as faults and folds, although there may be other aspects that are not self-similar. For example, large folds may consist of lower-order folds with corresponding shape, which again may consist of yet smaller folds of the same shape characteristics. Or a fracture set as seen from a satellite image may look like a field observation or thin-section fracture pattern from the same area. They will not be identical, but at some scales (e.g. 100 km above the surface, at the surface and under the microscope) they may look similar. This is why a scale often needs to be added to pictures of geologic structures.

Special properties or sizes of fracture populations are usually considered. In one dimension the fault offset is such a size. If the distribution of offset values from a fault population defines a straight line in a log-log plot, as shown in Figure 17.17, then the offsets are *self-similar*. This means that for a randomly chosen fault offset there is always a fixed number of faults with one-tenth of the chosen offset. So, for each fault with 1 km displacement there will be, for example, 100 faults with 100 m offset. For each fault with 100 m offset there will be 100 faults with 10 m offset and so on. The relation is described by the exponent D (Equation 17.2). In two and three dimensions, self-similarity and fractal dimensions can be applied to fault length and other geometrical aspects of faults. Fractal theory has many applications in geology and it is useful to know its mathematical basis and geologic applications.

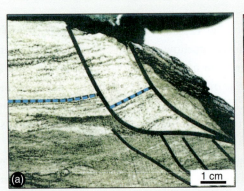 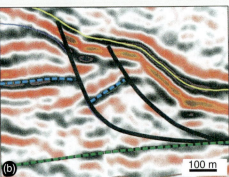

Structures on (a) centimeter scale (core) and (b) 100 m scale, both from the North Sea Statfjord Field. The two structures look very similar and illustrate the concept of self-similarity as applied to fault geometry. From Hesthammer and Fossen (1999).

on parallel seismic lines or geologic profiles and sorted on a spreadsheet. The offset values are then plotted along the horizontal axis and the cumulative number along the vertical axis. In practice, this means plotting the largest offset first with cumulative number 1, the second largest offset with cumulative number 2 and so on.

Alternatively, the cumulative number per kilometer can be plotted along the vertical axis. In the latter case the frequency is portrayed, i.e. how frequently faults with a given offset statistically occur in the given profile direction.

Many fault offset populations plot along a fairly straight segment, which implies a **power law distribution**. A power law or self-similar relation implies that the data define a more or less straight line when plotted in the log-log diagram. Mathematically this can be described by the expression

$$N = aS^{-D} \tag{17.1}$$

where S is displacement, throw or heave, N is the cumulative number of fault offsets and a is a constant. The exponent D describes the fractal dimension or slope of the straight line. Since we are working in log-log space it makes sense to rewrite the expression as

$$\log N = \log a - D \log S \qquad (17.2)$$

which is the equation for a straight line with slope $-D$. The exponent D describes the relation between the number of small and large offsets. A large D-value implies that there is a large number of small faults for each large fault. Hence, the larger the D-value, the higher the strain contribution from small faults, and the larger the error involved in extension estimates where small (subseismic) faults are left out. Common D-values from natural fault populations tend to fall between 0.6 and 0.8.

Real data seldom define a perfectly straight line in cumulative log-log diagrams, and the example shown in Figure 17.17 is no exception. However, a central straight segment may be found (between 10 and 100 m in Figure 17.17), truncated by curved segments at each end. The truncation is caused by underrepresentation of faults with very small displacements (the problem of resolution) and faults with very large displacements (the profiles do not always intersect the largest faults in an area). This effect is sometimes referred to as the censoring effect. To compensate for the lower truncation effect, the straight segment can be extended into the domain of small fault displacements (upward into the domain of small displacements in Figure 17.17). But it cannot be extended forever, and at some point this relationship is bound to break down. In porous rocks and sediments this could occur when approaching the grain size of the deformed rock, if not before.

As an example, the Mesozoic extension across the northern North Sea rift is estimated at almost 100 km by balancing the crustal area (crustal thinning). Summing fault heaves from regional seismic line interpretations indicates an extension of about 50 km. It has been shown that much or all of this difference can be accounted for by subseismic faults, if the power-law relation can be extrapolated down to small subseismic faults.

The contribution of subseismic faults may account for the discrepancy between crustal balancing and summing fault heaves shown in a regional profile across the rift.

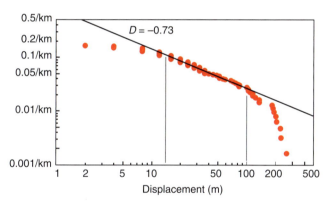

Figure 17.17 Log-log plot of fault displacement against cumulative number (normalized against length). The data define a fairly straight line in the central part. Undersampling of small and large faults explains the deviation from the straight line at each end. Data from a subpopulation of faults in the North Sea Gullfaks Field. From Fossen and Rørnes (1996).

17.10 Passive margins and oceanic rifts

If a continental rift is extended far enough, the crust will break and be replaced by oceanic crust. A **passive margin** is then established on each side of the rift, which is now located in oceanic crust. In the Viking Graben in the northern North Sea, stretching related to the late Jurassic–early Cretaceous rift ceased at around 150% ($\beta = 1.5$). Stretching beyond 1.5 generally results in initial magmatism and volcanism, and progressively more magmatism until oceanic crust initiates at a stretching factor around 3.

Little seismic activity occurs in passive margins. Fault movement in such settings is mostly gravity driven, resulting not only in slumping but also large-scale extensional fault systems soled on weak layers of salt or clay, shown as a black layer in Figure 17.18. Large-scale

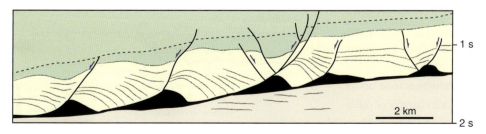

Figure 17.18 Gravity-driven synsedimentary extension above a low-angle detachment in the Kwanza Basin on the passive West African continental margin (Angola). Fault blocks are sliding on a thin and weak salt layer and are therefore detached from the substrata. The salt is flowing plastically and accommodates area problems caused by the fault block rotation. Modified from Duval et al. (1992).

examples from the margins of the Atlantic Ocean are shown in Chapter 19.

While passive continental margins gradually subside and become covered by clastic sediments, tectonic activity along oceanic rifts is usually significant. One of the main differences between continental and oceanic rifts is the much more extensive addition of magma and heat in the latter. In addition, there is no erosion apart from gravity sliding in oceanic rifts.

Hot magma and thin lithosphere cause the oceanic rift area as a whole to be a positive (elevated) structure, with a relatively narrow graben along the central axis. The potential energy represented by the relatively high rift elevation is partially released by means of listric normal faults. Low-angle detachments with normal sense of movement are described from mid-oceanic ridges, together with metamorphic core complexes that are geometrically and kinematically similar to those found in areas of continental extension. In such **oceanic metamorphic core complexes** mantle rocks are exposed in submarine windows due to tectonic unroofing by means of normal faulting. The situation is similar to that portrayed in Figure 17.8, but with no erosion.

Our knowledge of the structural geology along oceanic ridges is hampered by their inaccessibility, but new high-quality deep-ocean topographic and seismic data are giving new information about the structural processes along oceanic ridges.

17.11 Orogenic extension and orogenic collapse

Extension is by no means limited to rift zones and passive margins. Some of the most impressive extensional faults and shear zones are found in active mountain belts and in orogenic zones where plate convergence has ceased.

Orogeny is one of several stages in the Wilson cycle. In other words, orogenic belts are often built on older divergent plate boundaries or rifts, and they typically rift again at a later stage. In the early stages of a typical orogenic cycle, while an ocean still exists between two converging continents, extension occurs by means of back-arc rifting (Figure 17.19a). Stretching also occurs in the upper part of the oceanic crust where it enters the subduction zone under the island arc. This stretching is a large-scale example of outer arc extension of buckled layers discussed in Chapter 11.

During the later stage continent–continent collision, extensional faults and shear zones may form in the orogenic

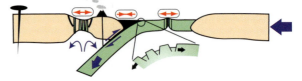

(a) Island-arc splitting, subduction and sea-floor spreading
Pre-collisional

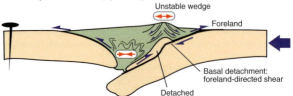

(b) Unstable orogenic wedge
Syn-collisional (syn-orogenic)

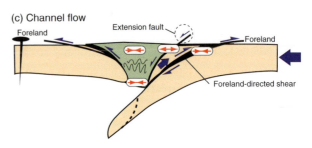

(c) Channel flow

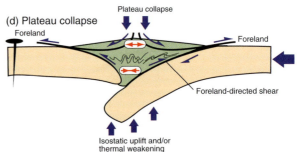

(d) Plateau collapse

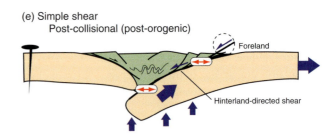

(e) Simple shear
Post-collisional (post-orogenic)

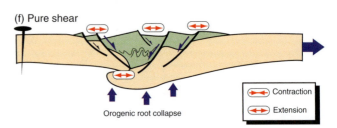

(f) Pure shear

Figure 17.19 Different types of extension connected with an orogenic cycle. Modified from Fossen (2000).

wedge where and when the wedge becomes unstable, as discussed in the previous chapter. If a large basement slice is incorporated, the wedge thickens excessively and responds to this instability by creating normal faults or shear zones (Figure 17.19b). It has been suggested, with reference to the active Himalayan orogen, that a detached and heated basement slice can have low enough density to ascend buoyantly with the formation of a thrust fault on its lower side and a normal fault on its upper side (Figure 17.19c). A hinterland-thinning geometry of the basement slice would fuel the extrusion of such slabs toward the foreland.

Figure 17.20 indicates how this model may work. The slice may be a rigid block, but may well be soft and internally flowing. The extrusion of light and hot basement material in the hinterland by means of low-viscous flows toward the foreland is commonly referred to as **channel flow**. Again, the top of the channel defines a normal shear zone, forming in an overall contractional regime.

A third model for synconvergent extension concerns changes in the thermal structure of the lower crust and the lithospheric mantle. During a continent–continent collision, crustal material is subducted and heated. Heating weakens the crust, potentially to the point where it collapses under its own weight along extensional faults and shear zones (Figure 17.19d). We can call this model **gravitational orogenic collapse**.

Extensional collapse driven by gravity occurs when the crust is too thick (weak) to support its own weight.

We usually think of gravitational orogenic collapse as a collapse of the mountainous upper part of the orogenic edifice, but it may be equally important to consider what is going on at the base of the crust. Continental subduction can force continental crust down to perhaps a hundred kilometers or so. Much of the subducted crust is lighter than its surrounding lithosphere and thus buoyant, while the lower part of the plate, the lithospheric mantle, is (for some time) colder than its surroundings and potentially denser. The same is true for any oceanic crust attached to the leading edge of the continent. In addition, phase transitions, for example eclogitization, can increase the density of rocks in the root zone. It is likely that the densest part of an orogenic root can be removed by sinking into the mantle below. This model is called the **delamination model** (Figure 17.21) and results in more rapid heating of the

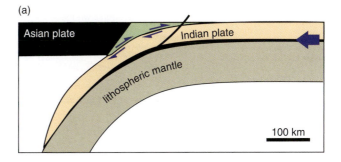

(a)

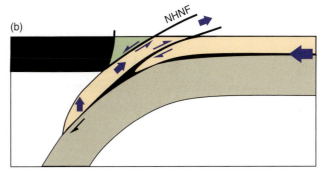

(b)

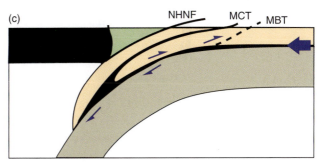

(c)

Figure 17.20 Model of the formation of a major normal fault (NHNF) in the Himalayan orogen. A slice of the continental crust is detached and uplifted by a combination of compression and buoyancy. The slice is underlain by a thrust (MCT) and overlain by a normal fault (NHNF). Based on physical modeling by Chemenda *et al.* (1995). NHNF, North Himalaya Normal Fault; MCT, Main Central Thrust; MBT, Main Boundary Thrust.

(remaining) orogenic root, partial melting and increased magmatic activity.

Delamination of a dense root releases the buoyant part of the root, which leads to orogenic uplift (Figure 17.19d). The root now collapses upward and this model can be called an **orogenic root collapse** (Figure 17.19f). This can happen by a coaxial strain mechanism where the root spreads out laterally along the base of the crust. It could probably also drive channel flow. A natural consequence of a root collapse and regional uplift is that the upper orogenic edifice also collapses gravitationally. This is happening in the Tibet Plateau today, and has been explained by root delamination, root collapse and similar models.

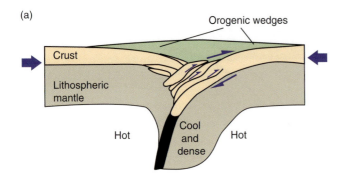

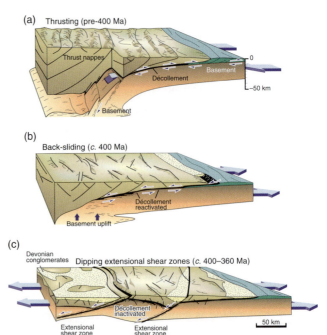

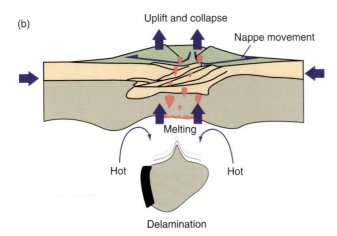

Figure 17.21 The delamination model causing orogenic collapse and orogenic root collapse. (a) Cold and dense lower root pulls down continental crust. (b) Delamination and descent of dense root by delamination, causing upward collapse of deep continental crust, uplift and collapse in the upper part of the collision zone (in the orogenic wedge) with high-level transport of rocks from the hinterland toward the foreland.

Figure 17.22 The development of low-angle extensional faults and shear zones in the southern Scandinavian Caledonides. (a) Emplacement of nappes during plate convergence. (b) Back-sliding of the orogenic wedge, causing reversal of the sense of shear along the basal detachment. (c) Formation of hinterland-dipping shear zones and faults that cut both the thrusts and the basement. Based on Fossen (2000).

17.12 Postorogenic extension

Throughout the divergent history of an orogenic belt, the sense of movement on the basal thrust is always toward the foreland, as for instance during the Caledonian example shown in Figure 17.22a. Once the sense of shear is reversed and the orogenic wedge moves toward the center of the collision zone (Figure 17.22b), the orogen kinematically enters the divergent or postorogenic stage. At this stage extensional deformation dominates at all crustal levels. One mode of postorogenic extensional deformation involves reversal of the basal thrust and higher-level thrusts within the orogenic wedge (Figures 17.19e and 17.22b). Such reactivation of basal thrust zones can cause the formation of metamorphic core complexes, as described in the previous chapter.

Another mode of extension is the formation of hinterland-dipping shear zones cutting through the crust. Such shear zones can form after the hinterland has been uplifted and orogenic thrusts rotated to orientations unfavorable for extensional reactivation (Figures 17.19f and 17.22c). Such hinterland-dipping shear zones typically affect the entire crust and transect and further rotate the reversed basal thrust. Well-developed examples of such structures have been interpreted in the Scandinavian Caledonides.

Summary

Extensional deformation structures in the form of normal faults and shear zones or detachments are extremely common in many tectonic settings and also in non-tectonic settings (e.g. gravitational sliding). They control rift structures, post-contractional stages of mountain belt cycles and are important also during orogeny. In the next chapter we will see that they also occur in strike-slip settings. Some important points and questions regarding extensional structures are presented here:

- Extensional faults are faults that extend a reference surface, typically the surface of the crust, or bedding for smaller faults.

- Several large-scale extensional fault systems can be modeled by means of the soft domino model, which allows for block-internal strain.

- Dipping detachment promotes the formation of uniformly rotated fault blocks (domino systems).

- Much like contractional faults, extensional faults can form imbrication zones, duplexes and flat-ramp-flat geometries.

- A metamorphic core complex consists of metamorphic rocks exposed in a window and separated from overlying faulted non- or very low-grade metamorphic rocks by an extensional detachment.

- There is a systematic relationship between the number of small faults relative to large faults in many extensional fault populations.

- Once established, such relationships can be used to estimate the number or density of faults beyond the resolution of a data set.

- Extensional faults and detachments are commonly found in orogens and form both during and after the collisional history.

- Synconvergent gravitational collapse of the central part of a collision zone can cause extension of the hinterland and thrusting toward the foreland.

- Post-convergent extension commonly occurs by reactivation of thrusts as extensional detachments.

Review questions

1. How can a reverse fault in some special cases also be extensional?

2. What is unrealistic about the domino fault model?

3. How can low-angle faults form if we expect normal faults to form with dips around 60°?

4. Give at least two examples of fault development that create domino-like fault blocks but contradict the domino model.

5. Explain the formation of an extensional fault set that overprints earlier extensional faults formed during the same extensional phase.

6. In what settings can metamorphic core complexes form?

7. What is meant by the rolling hinge mechanism during extensional detachment faulting?

8. Why does crustal balancing yield different estimates of extension than summing fault heaves across many rifts?

9. What may cause the entire orogen to uplift during or at the end of the collisional history?

E-MODULE

 The e-learning module called *Extension* is recommended for this chapter.

FURTHER READING

Jackson, J. and McKenzie, D., 1983, The geometrical evolution of normal fault systems. *Journal of Structural Geology* 5: 472–483.

Wernicke, B. and Burchfiel, B. C., 1982, Modes of extensional tectonics. *Journal of Structural Geology* 4: 105–115.

Orogenic collapse and extension

Andersen, T. B., Jamtveit, B., Dewey, J. F. and Swensson, E., 1991, Subduction and eduction of continental crust: major mechanisms during continent–continent collision and orogenic extensional collapse, a model based on the Norwegian Caledonides. *Terra Nova* 3: 303–310.

Braathen, A., Nordgulen, Ø., Osmundsen, P. -T., Andersen, T. B., Solli, A. and Roberts, D., 2000, Devonian, orogen-parallel, opposed extension in the Central Norwegian Caledonides. *Geology* 28: 615–618.

Dewey, J. F., 1987, Extensional collapse of orogens. *Tectonics* 7: 1123–1139.

England, P. C. and Houseman, G. A., 1988, The mechanics of the Tibetan plateau. *Royal Society of London Philosophical Transactions* Series A **326**: 301–320.

Fossen, H. and Rykkelid, E., 1992, Postcollisional extension of the Caledonide orogen in Scandinavia: structural expressions and tectonic significance. *Geology* **20**: 737–740.

Hacker, B. R. 2007, Ascent of the ultrahigh-pressure Western Gneiss Region, Norway. *Geological Society of America Special Paper* **419**, 171–184.

Houseman, G. and England, P., 1986, A dynamical model of lithosphere extension and sedimentary basin formation. *Journal of Geophysical Research* **91**: 719–729.

Wheeler, J. and Butler, R. W. H., 1994, Criteria for identifying structures related to true crustal extension in orogens. *Journal of Structural Geology* **16**: 1023–1027.

Channel flow

Godin, L., Grujic, D., Law, R. D. and Searle, M. P., 2006, Channel flow, ductile extrusion and exhumation in continental collision zones: an introduction. In R. D. Law, M. P. Searle and L. Godin (Eds.), *Channel Flow,*

Ductile Extrusion and Exhumation in Continental Collision Zones, Special Publication **268**, London: Geological Society, pp. 1–23.

Rotated normal faults and metamorphic core complexes

Brun, J. P. and Choukroune, P., 1983, Normal faulting, block tilting, and décollement in a stretched crust. *Tectonics* **2**: 345–356.

Buck, W. R., 1988, Flexural rotation of normal faults. *Tectonics* **7**: 959–973.

Davis, G. H., 1983. Shear-zone model for the origin of metamorphic core complexes. *Geology* **11**: 342–347.

Fletcher, J. M., Bartley, J. M., Martin, M. W., Glazner, A. F. and Walker, J. D., 1995, Large-magnitude continental extension: an example from the central Mojave metamorphic core complex. *Geological Society of America Bulletin* **107**: 1468–1483.

Lister, G. S. and Davis, G. A., 1989, The origin of metamorphic core complexes and detachment faults formed during Tertiary continental extension in the northern Colorado river region, U.S.A. *Journal of Structural Geology* **11**: 65–94.

Malavieille, J. and Taboada, A., 1991, Kinematic model for postorogenic Basin and Range extension. *Geology* **19**: 555–558.

Nur, A., Ron, H. and Scotti, O., 1986, Fault mechanics and the kinematics of block rotations. *Geology* **14**: 746–749.

Scott, R. J. and Lister, G. S., 1992, Detachment faults: Evidence for a low-angle origin. *Geology* **20**: 833–836.

Wernicke, B. and Axen, G. J., 1988, On the role of isostasy in the evolution of normal fault systems. *Geology* **16**: 848–851.

Extensional faults in contractional regimes

Burchfiel, B. C. *et al.*, 1992, The south Tibetan detachment system, Himalayan orogen: extension contemporaneous with and parallel to shortening in a collisional mountain belt. *Geological Society of America Special Paper* **269**.

Platt, J. P., 1986, Dynamics of orogenic wedges and the uplift of high-pressure metamorphic rocks. *Geological Society of America Bulletin* **97**: 1037–1053.

Rift systems

Angelier, J., 1985, Extension and rifting: the Zeit region, Gulf of Suez. *Journal of Structural Geology* **7**: 605–612.

Gibbs, A. D., 1984, Structural evolution of extensional basin margins. *Journal of the Geological Society* **141**: 609–620.

Roberts, A. and Yielding, G., 1994, Continental extensional tectonics. In P. L. Hancock (Ed.), *Continental Deformation*. Oxford: Pergamon Press, pp. 223–250.

The pure and simple shear models for rifting

Kusznir, N. J. and Ziegler, P. A., 1992, The mechanics of continental extension and sedimentary basin formation: a simple-shear/pure-shear flexural cantilever model. *Tectonophysics* **215**: 117–131.

McKenzie, D., 1978, Some remarks on the development of sedimentary basins. *Earth and Planetary Science Letters* **40**: 25–32.

Wernicke, B., 1985, Uniform-sense normal simple shear of the continental lithosphere. *Canadian Journal of Earth Sciences* **22**: 108–125.

Accommodation zones

Rosendahl, B. R., 1987, Architecture of continental rifts with respect to East Africa. *Annual Review of Earth and Planetary Science* **15**: 445–503.

Chapter 18

Strike-slip, transpression and transtension

Strike-slip faults constitute an important class of faults that have been studied for more than 100 years. They first received attention in California, Japan and New Zealand, where very long strike-slip faults with considerable displacement intersect the surface of the Earth. They are known for their close association with devastating earthquakes, particularly in places such as California and Turkey. Understanding such faults and the tectonic regimes in which they occur is therefore of public as well as academic interest. In this chapter we will address the basic types of strike-slip faults, their formation and tectonic settings, and also look at transpression and transtension – three-dimensional deformations that link strike-slip, extensional and contractional regimes.

18.1 Strike-slip faults

Strike-slip faults are faults where the displacement vector is parallel to the strike of the fault and thus parallel to the surface of the Earth, as shown schematically in Figure 18.1. **Strike-slip shear zones** are the deeper versions dominated by plastic deformation mechanisms, although there is a tendency to use the term strike-slip fault deliberately about both. Strike-slip faults (and shear zones) are typically steeper than other faults, and many appear as fairly straight structures in map view. Curvatures and geometric irregularities also occur along strike-slip faults, but more commonly in the vertical than in the horizontal section, perpendicular to the displacement vector. However, curvatures in map view do occur and have important implications for the structures associated with strike-slip faults. Strike-slip faults occur on all scales, and represent some of the longest and most famous faults in the world. The San Andreas Fault in California and the North Anatolian Fault in Turkey are two of our most famous and large-scale strike-slip faults, and also some of the most feared ones when it comes to earthquake hazards.

Strike-slip faults and shear zones tend to be steep and many are relatively straight in map view.

A strike-slip fault can be **sinistral** (left-lateral) or **dextral** (right-lateral) and ideally involves no vertical movement of rocks. While the extent of reverse and normal faults is strongly limited by the thickness of the crust, strike-slip faults could extend around the entire globe and, from a theoretical point of view, accumulate an infinite amount of displacement. Such Earth-spanning strike-slip faults have never been found, but the idea illustrates the fact that strike-slip faults can accumulate large displacements. For this reason, famous steep faults and shear zones, such as the Great Glen Fault in Scotland, the Tornquist Zone in northern Europe, the Billefjorden Fault in Spitsbergen, the Great Slave Lake Shear Zone in Canada, the Nordre Strømfjord shear zone in southwest Greenland and the Alpine Fault in New Zealand, have been assigned several hundred or thousand kilometers of lateral displacements, although some argue for more restricted offsets for some of them.

18.2 Transfer faults

Strike-slip faults have several different kinematic roles and are accordingly given different names. **Transfer faults** are strike-slip faults that transfer displacement from one fault to another. In general, any kind of fault that is connected to at least one other fault is involved in displacement transfer, but the term is used specifically for a particular type of strike-slip fault whose tips terminate against other faults or extension fractures. Transfer faults are therefore bounded and cannot grow freely, which has implications for their displacement–length relations.

Transfer faults transfer displacement between two extensional or contractional faults by means of strike-slip motion.

Transfer faults occur on all scales and they connect a range of structures. They can connect open or mineral-filled extension fractures (Figure 18.2), veins, dikes, normal faults of the same (Figure 18.3a) or opposite (Figure 18.4) dip directions, oblique faults, reverse faults (Figure 18.3b) and more. At larger scale, transfer faults offset continental rift axes, locally juxtaposing oppositely dipping rift faults.

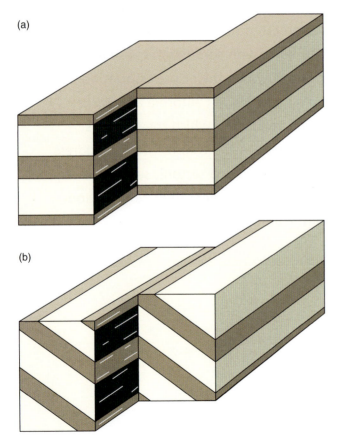

(a)

(b)

Figure 18.1 Pure strike-slip faults show no offset in any section if layers are vertical or horizontal (a) or strike parallel to the fault (b). They can therefore be difficult to identify from seismic data alone. The sense of shear shown here is sinistral (left-lateral).

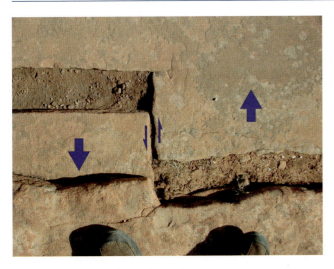

Figure 18.2 Sinistral transfer fault between two extension fractures. Path to Delicate Arch, Utah.

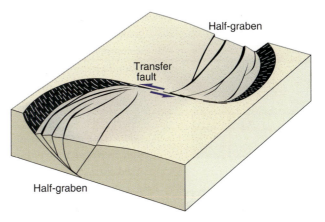

Figure 18.4 Strike-slip faults that connect half-grabens of opposite polarity are a kind of transfer fault. Such transfer faults are common in rifts such as the East African rift system, the North Sea rift and the Rio Grande rift.

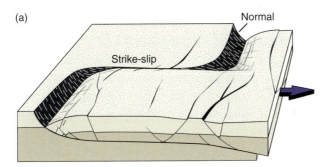

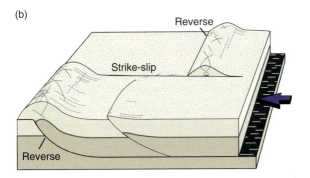

Figure 18.3 Strike-slip movements can occur along lateral ramps in both extensional and contractional settings. Such strike-slip faults are transfer faults and can attain significant offsets and little or no along-strike variation in displacement. Each end of the transfer faults is connected to extensional or contractional faults.

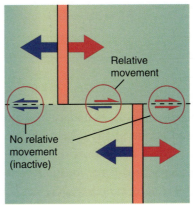

Figure 18.5 Transform fault at a mid-ocean ridge (perspective and map view). The fault is only active between the ridge segments (except for minor vertical adjustments). The offset is constant along the active part of the fault, and its length grows at a rate that is directly proportional to the spreading rate.

In mid-oceanic ridges, oceanic ridge grabens are shifted along transfer faults. When oceanic transfer faults, illustrated in Figure 18.5, first were discovered in the 1960s they were given the name **transform faults**.

Transform faults are large (kilometer-scale or longer) strike-slip faults that segment plates or form plate boundaries. The term was first used about the many transfer faults that define plate boundaries or offset mid-ocean ridges, as illustrated in Figure 18.6. In other places they connect mid-ocean ridges to destructive plate boundaries (island arcs; Figure 18.6b), or they connect two segments of a destructive plate boundary (Figure 18.6c). Transform faults that define plate boundaries can be very long, particularly those that occur in the continental crust.

The most famous example is the 1200 km long **San Andreas Fault** in California, which represents a continental

(a) Ridge–ridge

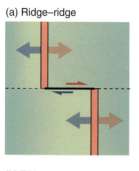

(b) Ridge–arc

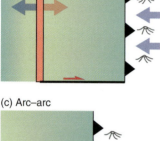

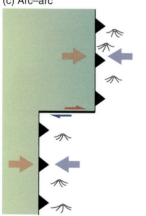

(c) Arc–arc

Figure 18.6 Transform faults are strike-slip faults connected by plate boundaries. (a) Fault between two mid-ocean spreading ridge segments. (b) Transform fault connecting a rift segment and an island arc/ subduction zone. (c) Fault displacing a destructive plate boundary.

transform fault along the boundary between the North American and Pacific plates. Large transform faults are actually fault zones rather than simple faults. The San Andreas and associated faults constitute a zone of more or less parallel faults of various lengths in a ∼100 km wide zone. Folds also occur along this zone, together with reverse and normal faults. We will come back to such structures later in this chapter. For now we emphasize the fact that, among the many faults in the zone, usually only one is active at any given time. In this sense, a fault zone such as the San Andreas is different from most active shear zones in the plastic regime where strain is accumulating in all or a significant portion of the zone.

18.3 Transcurrent faults

Transcurrent faults is a term preferentially used for strike-slip faults in continental crust that have free tips, i.e. they are not constrained by other structures. Their free tips move so that the fault length increases as displacement

accumulates. Such strike-slip faults thus follow a normal displacement–length relationship, i.e. maximum displacement increases systematically with increasing fault length (Figure 8.50). This does not mean that they never encounter any obstacles or complications during growth. In fact, transcurrent faults are free to grow, interact and link up to form longer structures, just like the extensional faults shown in Figure 8.38, but they will never have the special kinematic role that transform faults have.

Transcurrent faults have free tips and grow in length as they accumulate strike-slip displacement.

In contrast to extensional and contractional faults, transcurrent (and other strike-slip) faults do not dominate wide areas, such as the areas of normal faulting in continental rifts or the wide zone of thrust faults in a fold-and-thrust belt, but rather restrict themselves to a single zone. The role of a transcurrent fault is to shift rocks laterally, and because most faults are weaker than the host rock this is most easily done through continued shear along the existing strike-slip zone. However, as for any type of fault, the width of a strike-slip fault zone increases as it grows longer and accumulates displacement.

Free strike-slip faults (transcurrent faults) form within plates and are therefore **intraplate faults**. In contrast, transforms that occur along plate boundaries (discussed in the previous section) are **interplate faults**.

Long transcurrent faults intersect the surface of the Earth. At depth they may terminate against structures such as thrust faults, extension faults and, as shown schematically in Figure 18.7, subduction zones, or they may penetrate the brittle–plastic transition and continue downward as steep plastic shear zones.

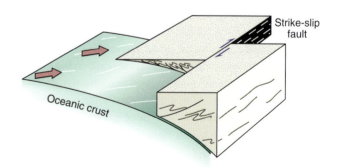

Figure 18.7 Strike-slip fault rooted in a subduction zone with an oblique subduction vector. This model has been applied to the San Andreas Fault where the oceanic plate is the Pacific Plate and the detached block is the Salina block.

18.4 Development and anatomy of strike-slip faults

Single faults (simple shear)

Strike-slip faults form when individual parts of the crust move at different rates along the surface of the Earth. Just like normal and reverse faults, strike-slip structures are complex when viewed in detail. Several secondary structures are associated with strike-slip faults, and experiments have helped us explore some of the most important ones. Riedel's clay experiments from the early 1900s are the most famous. His setup is shown in Figure 18.8, and consists of two stiff wooden blocks covered by a clay layer. The blocks were slid past one another, and stress was transferred to the overlying clay, which deformed progressively.

Riedel soon realized that the clay layer did not develop a clean, single fault but a deformation zone comprising an array of small fractures. These subsidiary fractures are classified based on their orientation and sense of slip relative to the trend of the overall strike-slip zone. The first sets of fractures to form are shear fractures. One set, known as **Riedel shear fractures** or R-fractures (or, less formally, Riedel shears), make a low angle with the overall shear zone and show the same sense of slip. Those are the fractures indicated in Figure 18.8b. In Figure 18.9a R-fractures are shown together with another set of fractures known as **P-shear fractures** (P-shears). P-fractures usually develop after the establishment of R-fractures, and their development is probably related to temporal variations in the local stress field along the shear zone as offset accumulates. A third set of shear fractures is seen in Figure 18.9a (dashed lines), identified as antithetic fractures that make a high angle to the zone. These are called **R'-shear fractures**, and are generally less well developed than R-fractures.

A strike-slip zone may develop by the linkage of various small-scale brittle structures that initiate early in the process.

In addition to the brittle R-, R'- and P-fractures, extension or **T-fractures** can occur (blue fractures in Figure 18.9b). In the shear-zone setting of Riedel's clay model, T-fractures will form perpendicular to the maximum instantaneous stretching axis (ISA$_1$, red arrows in Figure 18.9b). For large strike-slip zones, normal dip-slip faults will show more or less the same strike orientation as T-fractures. Folds can also develop in strike-slip shear zones (green structures in Figures 18.9b, c and 18.10), typically before deformation is localized into discrete faults. The axial traces of the

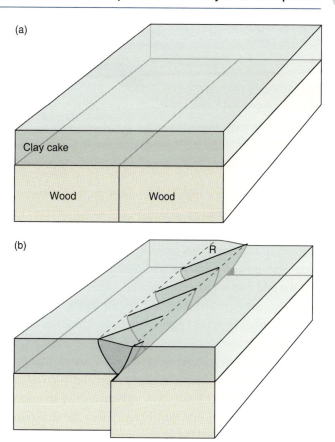

Figure 18.8 Physical model with two wooden blocks underneath a clay layer. Note the geometry of the R-fractures and the upward widening of the shear zone.

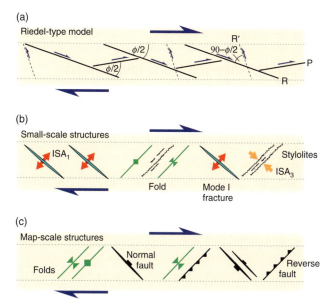

Figure 18.9 Structures formed by dextral strike-slip motion. (a) Riedel-model where R and R' are synthetic and antithetic Riedels. P-shears are secondary and connect R and R' surfaces. ϕ is the angle of internal friction. (b) Other small-scale structures that can form along a strike-slip zone. (c) Large-scale structures.

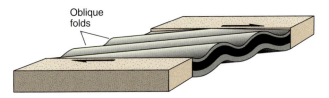

Figure 18.10 Folds formed by strike-slip movement in a wide deformation zone where the layering is horizontal. Note that the axial traces and fold axes form an acute angle with the shear zone.

folds are initially at a high ($\sim$90°) angle to the *minimum* instantaneous stretching axis (ISA$_3$, orange arrows in Figure 18.9b), provided that the layering is more or less horizontal. Dipping layers can also be folded in strike-slip zones, but in such cases the relationship between the ISA and the fold axes is more complicated. Other contractional structures, such as stylolites and reverse faults, can also form in strike-slip zones (Figures 18.9b, c). These will have approximately the same orientation as the fold axes.

A simple folding experiment can be carried out by performing a simple shear movement with your two hands and a piece of fabric. Folds form immediately at an angle to the shear direction, and their hinges reflect the direction of instantaneous stretching at the onset of folding. Subsequently, the folds rotate as the shear strain accumulates. Try it!

The development where individual shear fractures and extension fractures form and eventually link up indicates a way in which a strike-slip fault can form and grow. Other ways to form long strike-slip faults are indicated in Box 18.1, and examples of long strike-slip faults are presented in Box 18.2.

Conjugate strike-slip faults (pure shear)

Strike-slip faults may occur as simple structures or in zones of more or less parallel fault strands. However, strike-slip faults can also form conjugate sets (Figure 18.11), implying that they were active at about the same time under the same regional stress field.

Conjugate strike-slip faults fit well into both Anderson's model and the Coulomb fracture criterion. In simple terms, the acute angle between the two sets is bisected by σ_1 (red arrow in Figure 18.11), and the angle itself is determined by the internal friction of the rock. Kinematically, such faults result from pure shear in the horizontal plane, where shortening in one direction is compensated by orthogonal extension in the other. In this ideal model no extension or contraction occurs in the vertical direction.

The most famous large-scale example of conjugate strike-slip faults is the fault system on the north side of the Himalaya, as illustrated in Figure 18.12b. Here the

BOX 18.1 | HOW DO LONG STRIKE-SLIP FAULTS FORM?

Rock mechanics considerations tell us that when a shear fracture forms, the stress field in the tip regions will generate local tensile fractures, so-called wing cracks. This was first shown by cutting a straight fracture in a plastic plate and then imposing a stress field as shown in the figure. How then can strike-slip faults of up to several hundred kilometers length form?

Besides the model where faults grow by linkage of small fractures, it has been suggested that faults form from extension fractures or joints that already are long. But not as long as hundreds of kilometers. Such lengths must be achieved by linking, perhaps of reactivated joints. The faulted joint model requires a separate phase of extension and jointing prior to the strike-slip faulting, which makes strike-slip faulting a two-phase history of deformation. An alternative model is that growing strike-slip faults use other weak structures. If joints and faults are not available, dikes and steep layer interfaces or foliations may be used. Some of these structures are much longer than joints, and strike-slip faults can more easily form as long structures without the formation of wing cracks that lock up the fault.

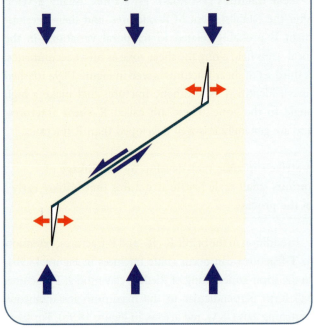

Indian continent moves north into the Eurasian continent, and some of the convergent movement is accommodated by active strike-slip faults. Strike-slip faults in this region accommodate lateral transport of material away from the

BOX 18.2 | SLIDING SIDEWAYS IN THE NORTH ATLANTIC

Several extensive and topographically outstanding faults transect the North Atlantic Caledonides from Svalbard in the north to the British Isles in the south. The most famous and conspicuous ones are perhaps the Great Glen Fault in Scotland and the Billefjorden Fault in Svalbard. These structures show many of the characteristics of strike-slip faults, including straight fault traces and steep dips. The faults themselves are largely covered and exposed parts are complicated by repeated reactivation, but they separate rock units or terranes that show different tectonometamorphic evolutions. Hence, most geologists agree that they have acted as strike-slip faults toward the end of the Caledonian orogeny.

The amount of strike-slip offset is difficult to estimate, hence very different interpretations have been forwarded. The most dramatic estimates assign about 1–2000 km of late Caledonian lateral offset to this strike-slip system. The British geologist W. Brian Harland was an early supporter of large lateral displacements, and in the 1960s he suggested a connection between the Billefjorden Fault in Svalbard and the Great Glen Fault in Scotland – an interpretation that implies a strike-slip system extending for more than 3500 km. It may even be connected with strike-slip faults in the Appalachians of Newfoundland to the southeast. Most geoscientists involved in this discussion now think that the offset is in the order of a few hundreds rather than thousands of kilometers.

Regardless, a significant sinistral strike-slip system in the central parts of the Caledonian collision zone indicates that the Caledonian orogeny was oblique and transpressional, with the strike-slip component being located to the hinterland.

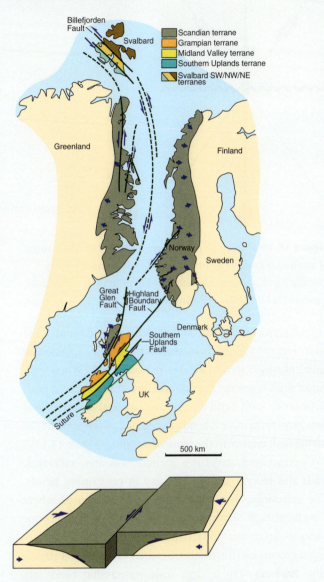

Simplified map of the North Atlantic Caledonides (Devonian reconstruction), emphasizing the strike-slip faults. The concept of partitioning into orogen-perpendicular shortening and orogen-parallel strike-slip movements is also indicated.

It is now clear that at various times and under changing stress fields, several of these steep faults have acted in a normal as well as reverse sense. This is a characteristic feature of strike-slip faults, and results in complex fault zones that may be difficult to interpret.

collision zone, and shortening perpendicular to this zone. The physical model shown in Figure 18.12a illustrates the idea. This model implies that the area north of the Himalayan collision zone is weaker than the stiff Indian plate.

Fault bends and stepovers

In sections containing the displacement vector (map view) ideal strike-slip faults are perfectly straight. However, even the simplest experimental models produce subsidiary faults

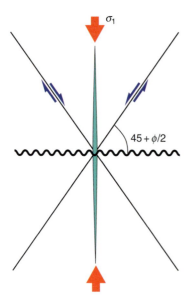

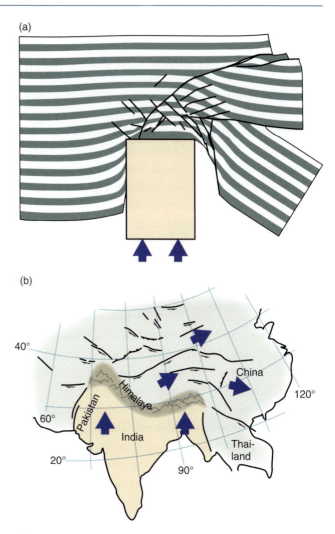

Figure 18.11 Conjugate pure shear model for the formation of strike-slip faults. The orientation of extension fractures (vertical) and stylolites (horizontal) are indicated. The model assumes that both sets are active more or less at the same time.

or fault segments that are oblique to the general trend of the fault (Figure 18.9). Such anomalies are usually explained by fault linkage, as seen in Figures 18.8 and 18.9a. When individual fault segments overlap and link in map view, a **fault stepover** or **fault bend** forms. Contractional or extensional structures form in such bends, depending on the sense of slip on the fault relative to the sense of stepping (Figure 18.13).

Contractional structures include stylolites, cleavages, folds and reverse faults, and form in **restraining bends**. The restraining bend in Figure 18.13 is located where a sinistral fault steps to the right. Subparallel reverse or oblique-slip contractional faults bounded by the two strike-slip segments can form and are called **contractional strike-slip duplexes**. On a large scale, restraining bends are recognized as areas of positive relief. After some time the contractional structures are likely to be transected by a new and straighter fault strand. The strain hardening within the restraining bend is then reduced or eliminated, although some irregularities may remain. The new fault in the restraining zone is likely to develop from P-shear fractures.

Bends along strike-slip faults are exciting structures that contain extensional or contractional structures depending on the sense of slip and stepping (right or left).

Releasing bends form where a sinistral strike-slip fault steps to the left, as in Figure 18.13, or a dextral fault steps to the right. Such bends produce extensional structures such as normal faults and extension fractures. Extension

Figure 18.12 (a) Formation of strike-slip faults in front of a stiff indenter. Two sets of faults, which can be considered conjugate, form while material is extruded sideways. The model is thought to represent an analog to what is going on in the area north of the Himalaya (b). The experiment is described by Tapponnier *et al.* (1986).

fractures are common in mesoscale releasing bends, while faults with a significant normal slip component tend to dominate larger scale examples. Series of parallel extensional faults bounded on both sides by strike-slip faults, as shown in Figure 18.13, are called **extensional strike-slip duplexes**. Normal faults generate negative structures, i.e. basins that can be filled with sediments at various scales. Figure 18.14 illustrates how Death Valley is located in a releasing bend where normal faults have lowered the otherwise mountainous region to a level close to, and locally under, sea level. Over time such a basin will widen or lengthen as new normal faults form. The Dead Sea is another famous example of such a basin, created in an overlap zone between two strike-slip transforms. Releasing-bend basins along strike-slip faults are called **pull-apart basins**.

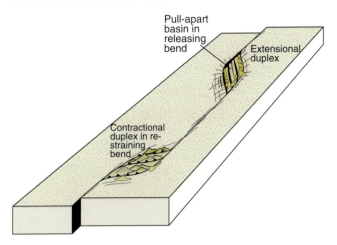

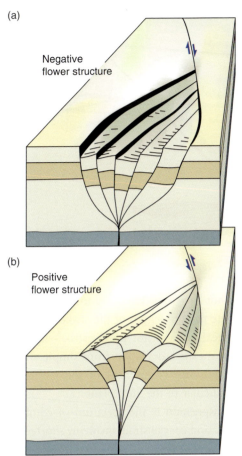

Figure 18.13 Extensional duplex (transtension) and contractional duplex (transpression) developed at bends or stepovers along a strike-slip fault system. Large-scale examples may lead to basin formation and local orogeny.

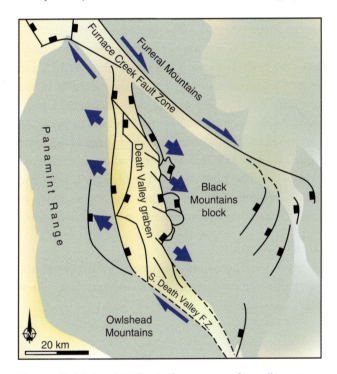

Figure 18.14 Death Valley is the type area for pull-apart basins. By type area we mean the area where it was first defined, in this case by Burchfiel and Stewart (1966). Based partly on Wright *et al.* (1974).

Figure 18.15 Principal features of (a) negative and (b) positive flower structures developed at the location of releasing and restraining bends, respectively.

where vertical movements are associated with normal faults, reverse faults or folds. A characteristic feature of such bends is their tendency to split and widen upward, as indicated in Figure 18.15. These structures are called **flower structures**. Flower structures that are associated with restraining bends are called positive, and those associated with releasing bends are called negative flower structures.

Strike-slip faults thus bifurcate and widen toward the surface, particularly in restraining and releasing bends. One reason for this may be the changes in the mechanical properties near the surface. While a fault is generally much weaker than its surroundings, this contrast decreases in less consolidated sedimentary rocks near the surface. In this setting it is therefore easier to form multiple faults.

Seismic imaging and flower structures

Reflection seismic data provide information about strike-slip faults in depth. Pure strike-slip faults can be difficult to detect from seismic data alone, not only because most are too steep to set up reflections, but also because horizontal layers or layers striking parallel to the fault show no displacement in the vertical direction (Figure 18.1). The clue is then to look for restraining and releasing bends,

18.5 Transpression and transtension

We have seen that bends in strike-slip faults can produce local components of contraction or extension. The type of deformation occurring in such bends is referred to as **transpression** and **transtension**. These modes of

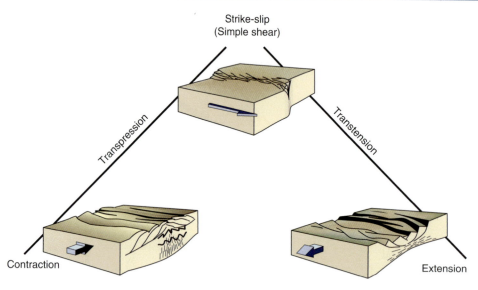

Figure 18.16 Transpression and transtension connect contraction, strike-slip and extension.

deformation do not have to be restricted to fault bends – they can dominate the full length of the strike-slip fault if the fault or shear zone is not purely strike-slip. For a shear zone this means a deviation from simple shear. It contains an additional component of shortening or extension perpendicular to the fault plane.

In general, transpression is the spectrum of combinations of strike-slip and coaxial strain involving shortening perpendicular to the zone (Figure 18.16), and transtension encompasses the combinations of strike-slip and perpendicular extension. In other words;

Transpression (transtension) is the simultaneous combination of strike-slip or simple shear motion along a structure and shortening (extension) perpendicular to it.

For a vertical shear zone of finite width, transpression and transtension can be modeled quite simply. The strike-slip component is a horizontal simple shear displacement along a vertical zone, while the shortening can be modeled as a coaxial deformation with horizontal shortening and vertical and/or lateral extension. Let us make the coaxial component a pure shear with vertical extension, as illustrated in Figure 18.17. This is the simplest mathematical model proposed for transpression, first presented by Sanderson and Marchini in 1984. Using the deformation theory from Chapter 2, the simultaneous application of the two components can be represented by the deformation matrix:

$$\mathbf{D} = \begin{bmatrix} 1 & \Gamma & 0 \\ 0 & k & 0 \\ 0 & 0 & k^{-1} \end{bmatrix} = \begin{bmatrix} 1 & \frac{\gamma(1-k)}{\ln(k^{-1})} & 0 \\ 0 & k & 0 \\ 0 & 0 & k^{-1} \end{bmatrix} \quad (18.1)$$

The matrix is three-dimensional because the pure and simple shear components act in two perpendicular planes. γ is the simple shear or strike-slip component, while the k-value determines the amount of shortening or extension across the zone. For $k = 0.7$ the zone is thinned 30% and we have transpression. For $k = 1.2$ the thickness of the zone is increased by 20% and the deformation is transtensional. As always, once we have established a deformation matrix we can find the orientation and shape of the strain ellipse, W_k and ISA, and by using incremental matrixes we can calculate rotation patterns of passive lines and planes (Appendix A). Let us use this opportunity to explore some aspects of transpression and transtension.

Strain ellipsoid

At deep levels where strain chiefly accumulates by plastic deformation mechanisms, strain markers may allow

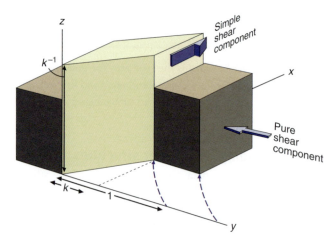

Figure 18.17 Sanderson and Marchini's (1984) simple model for transpression: homogeneous strain between two rigid blocks.

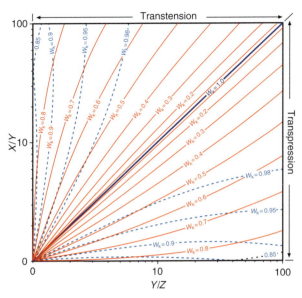

Figure 18.18 Flinn diagram where constant W_k paths are shown. Based on Fossen and Tikoff (1993).

ductile strain to be mapped within shear zones. Using the deformation matrix (Equation 18.1) it can be shown that for our model:

Transpression results in oblate ellipsoids (flattening) while transtension generates prolate ellipsoids

---

(Figure 18.18). The long axis X of the strain ellipsoid is always horizontal for transtension (Figure 18.19c, d) and vertical for transpression with a strong pure shear component ($W_k < 0.81$) (Figure 18.19a). For simple shear-dominated transpression where the pure shear component is relatively small, X is initially horizontal and later changes to vertical. A switch occurs between Y and X when the deformation path hits and bounces off the horizontal axis in the Flinn diagram (blue stippled lines in the flattening field of Figure 18.18). This occurs for deformations where Y grows faster than X, so that the two at some point become of equal length and thereafter change roles. A perfect flattening strain marks the transition.

A similar switch between Y and Z occurs for transpression, in this case through an instance of perfect constriction. We can see both of these switches for the two $W_k = 0.85$ lines in Figure 18.18, from which it is clear that high strains are generally required for the switches to take place. Other paths will emerge if the kinematic vorticity number W_k changes during the deformation history, or if other models of transpression are chosen. However, the following is a good rule of thumb:

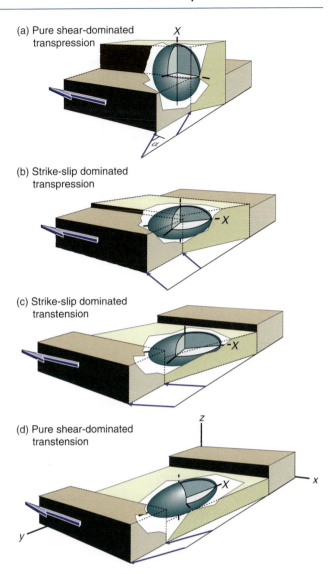

Figure 18.19 The orientation and shape of the finite strain ellipsoid for the four classes of transpression/transtension discussed in the text. From Fossen *et al.* (1994).

A pure shear-dominated transpression zone gives vertical lineations, while a strong simple shear component favors horizontal lineations.

The orientation of the longest axis of the strain ellipse in the horizontal plane, be it X or Y, can be found using the equation

$$\theta' = \tan^{-1}[(\lambda - \Gamma^2 - 1)/k\Gamma] \tag{18.2}$$

Here θ' is the angle between the longest horizontal axis and the vertical shear zone. Knowing the strain ratio R in the horizontal section, we can estimate W_k by means of Figure 18.20a.

(a)

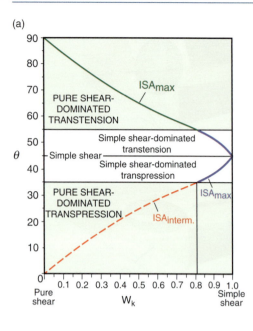

(b)

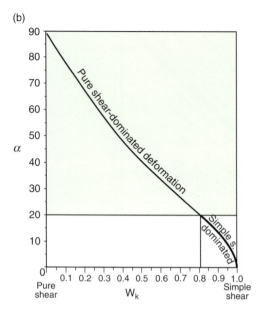

Figure 18.20 (a) The k connection between W and the angle θ between the maximum horizontal instantaneous stretching axis and the shear zone. (b) The relationship between α (the orientation of the oblique flow apophysis) and W (for both transpression and transtension). Based on Fossen and Tikoff (1993).

Linear structures

We just noted that the lineation in a zone of transpression is oblique to the shear zone by an angle θ', or vertical where there is a significant pure shear component. We sometimes have lines and planes that will rotate more or less passively during the deformation history. Using matrix premultiplication (Section 2.22) and small strain increments we can calculate how lines and planes rotate during progressive deformation. We will here focus on lines, although plane rotations can be explored in the same way. Lines of different initial orientations outline different paths for any given deformation type, as illustrated in the stereonets on Figure 18.21 (left). For simple shear (perfect strike-slip shear zone) lines rotate along great circles (Figure 18.21, $W_k = 1$). For pure shear ($W_k = 0$) a symmetric pattern emerges. For transtension the paths are controlled by the oblique flow apophysis, and lines will eventually end up parallel to this apophysis (Figure 18.21). This means that deforming linear structures will concentrate at an orientation oblique to the shear zone and the shear direction. In our model the angle α between the apophysis and the shear zone will be

$$\alpha = \tan^{-1}[(\ln k)/\gamma] \tag{18.3}$$

The angularity will be qualitatively similar but not identical to the obliqueness of the stretching lineation (Equation 18.2). Hence, using the lineation to determine the transport direction is not precisely correct in transtension zones. In practice, if we can identify a strain gradient, e.g. from the low-strained margin toward the center of the zone, then the rotation pattern can be compared to the theoretical patterns shown in Figure 18.21, and the type of transpression or transtension can be found. If strain data can be added, then Figure 18.18 can also be used.

It is important to understand that the simple model for transpression and transtension referred to above is only one of many possible models. Not all strike-slip shear zones are vertical, and lateral extrusion or other kinematic models are possible. Nevertheless, the simple model shown in Figure 18.17 illustrates how transpression can be modeled and how it can involve vertical transport of rock, which at large scale and in combination with erosion brings metamorphic rocks closer to the surface.

18.6 Strain partitioning

Natural rocks tend to be anisotropic and heterogeneous so that strain is distributed unevenly throughout a volume of deformed rocks. This could primarily be related to the pre-deformational anisotropy of a rock, and secondarily to structures that form during deformation. In particular, once a weak fault or shear zone forms it is likely to continue to localize the simple shear component while the surrounding volume must accommodate coaxial components of the deformation that is needed to balance the strain prescribed by external constraints or boundary conditions (Figure 18.22). This phenomenon is called **strain partitioning**:

Strain partitioning is the internal decomposition of the total strain across a deformed zone into zones or domains of different types of strain.

Such selective redistribution of strain within deformation zones can occur in many ways and settings, for instance in ductile shear zones at outcrop scale where simple shear is partitioned into shear bands oblique to the shear zone and back-rotation of the rock between the shear bands. However, the term strain partitioning has become very popular

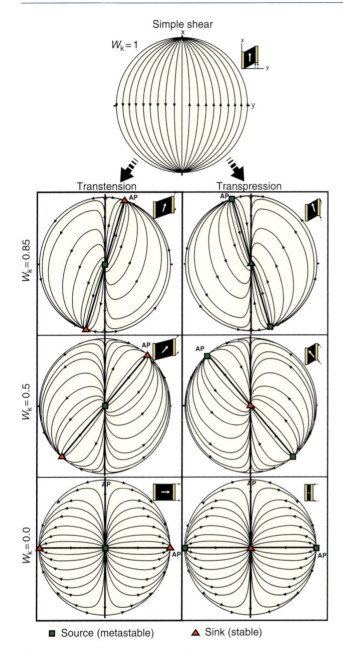

Figure 18.21 Rotation patterns for passive linear structures, shown in stereonets (equal area projection). AP: flow apophysis. From Fossen *et al.* (1994).

among geoscientists trying to understand the deformation of oblique plate margins and their accretionary prisms.

Plate margins and large-scale strain partitioning

An oblique plate margin is one where the plate motion or plate velocity vector is oblique to the margin. The deformation along such margins is one of overall transpression if the margin is obliquely convergent, and transtension if the margin is obliquely divergent. If we know the movement vector of one plate relative to the other, then we also know the orientation of the oblique horizontal flow apophysis, since the two are identical (Figure 2.21).

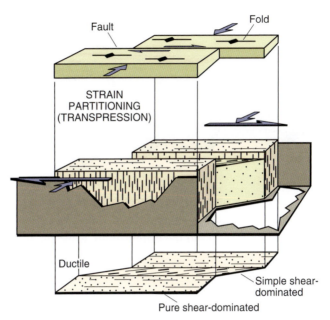

Figure 18.22 Illustration of strain partitioning in a zone of transpression. Some blocks experience mostly simple shear, while others are pure shear-dominated. At high crustal levels the simple shear zone is reduced to one or a few faults.

This very useful fact can be used to model structures in deformation zones along plate boundaries.

Strain partitioning along strike-slip dominated plate boundaries involves a balance between the amount of strike slip or simple shear on one hand, and the amount of perpendicular shortening or extension (coaxial strain) on the other. If most of the simple shear is localized along one or a few strike-slip faults, then the rest of the zone will be dominated by coaxial strain.

Inside the plate margin deformation zone, deformation may partition freely as long as it sums up to the total deformation prescribed by the plate motion vector and the orientation of the plate margin.

The San Andreas fault system, generally regarded as a strike-slip zone, is slightly transpressional because the plate vector makes a 5° angle to the plate margin in central California (Figure 18.23). From this information alone it can be deduced from Figure 18.20b that $W_k = 0.985$ and the deformation is very close to simple shear (strike-slip). This situation predicts the orientation of ISA_1 to be at 42.5° to the plate margin. However, because the simple shear component is taken up by strike-slip faults, the domains between the faults accommodate a significant amount of horizontal shortening (pure shear). This explains why ISA_1 (and σ_1) is not oriented at 42.5° but at a much lower angle to the plate margin in these domains. It also explains why

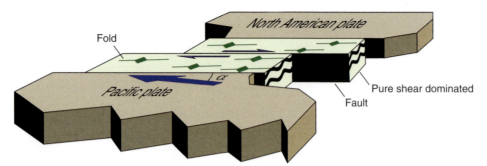

Figure 18.23 Schematic illustration of the San Andreas fault zone in California. The angle α between the two flow apophyses is only 5°, but since the simple shear is localized to the faults, pure shear dominates the volume between the faults, which is why the fold axes are subparallel to the faults and not like those in Figure 18.10.

folds with axes almost parallel to the faults exist in the pure-shear dominated domains. Even young, open folds are oriented at 6–12° to the plate margin, while the total strain in the zone predicts an initial fold orientation closer to 40°. Strain partitioning is therefore an important concern in many strike-slip zones and plate margins.

Summary

We have seen how strike-slip zones form important structures in many parts of the world, particularly along transform plate margins. Strike-slip structures are particularly interesting where geometric complications occur, or where the overall deformation deviates from simple shear (transpression and transtension). Although there are many models of transpression and transtension that can be explored, simple models based on map pattern and field detail can explain many important features and generate additional questions that contribute to further understanding of these interesting structures. Some important points and relevant questions:

- Strike-slip zones occur along transform plate boundaries as transform faults.

- They also occur inside continents as transcurrent faults.

- Transpression (transtension) occurs where there is a component of shortening (extension) across a strike-slip zone.

- Strain partitioning is found within many transpressional deformation zones.

- Simple models of transpression predict partitioning into domains of coaxial strain separated by simple shear zones or strike-slip faults.

Review questions

1. What are the characteristics of large-scale strike-slip faults or shear zones?

2. What is the role of transfer faults and in what settings do they occur?

3. What type of structures form where a strike-slip fault makes a stepover or an abrupt bend?

4. What would a profile look like across a restraining bend? Releasing bend?

5. What type of setting does Death Valley represent?

6. What could happen to strike-slip faults at depth?

7. How would you explain flattening strain in a shear zone with a clear strike-slip offset?

8. Like Death Valley, the Dead Sea is a place where you can walk on dry land below sea level, and an apparently unmotivated "hole in the ground" along a strike-slip fault. How do you think it formed?

9. How can strike-slip faults accommodate large-scale pure shear?

E-MODULE

 The e-learning module called *Strike-slip faults* is recommended for this chapter.

FURTHER READING

Strike-slip faults (general)

Sylvester, A. G., 1988, Strike-slip faults. *Geological Society of America Bulletin* **100**: 1666–1703.

Wilcox, R. E., Harding, T. P. and Seely, D. R., 1973, Basic wrench tectonics. *American Association of Petroleum Geologists Bulletin* **57**: 74–96.

Woodcock, N. H. and Schubert, C., 1994, Continental strike-slip tectonics. In Hancock, P. L. (Ed.), *Continental Deformation.* Oxford: Pergamon Press, pp. 251–263.

Restraining and releasing bends

Aydin, A. and Nur, A., 1982, Evolution of pull-apart basins and their scale independence. *Tectonics* **1**: 91–105.

Burchfiel, B. C. and Stewart, J. H., 1966, "Pull-apart" origin of the central segment of Death Valley, California. *Geological Society of America Bulletin* **77**, 439–442.

Cunningham, W. D. and Mann, P. (Eds.), 2007, *Tectonics of Strike-slip Restraining and Releasing Bends.* Special Publication **290**, London: Geological Society.

Peacock, D. C. P., 1991, Displacements and segment linkage in strike-slip fault zones. *Journal of Structural Geology* **13**: 1025–1035.

Westaway, R., 1995, Deformation around stepovers in strike-slip fault zones. *Journal of Structural Geology* **17**: 831–846.

Transpression and transtension

Oldlow, J. S., Bally, A. W. and Lallemant, G. A., 1990, Transpression, orogenic float, and lithospheric balance. *Geology* **18**: 991–994.

Sanderson, D. and Marchini, R. D., 1984, Transpression. *Journal of Structural Geology* **6**: 449–458.

Tikoff, B. and Greene, D., 1997, Stretching lineations in transpressional shear zones: an example from the Sierra Nevada Batholith, California. *Journal of Structural Geology* **19**: 29–39.

Treagus, S. H. and Treagus, J. E., 1992, Transected folds and transpression: how are they associated? *Journal of Structural Geology* **14**: 361–367.

Theoretical modeling of transpression/transtension

Fossen, H., Tikoff, T. B. and Teyssier, C. T., 1994, Strain modeling of transpressional and transtensional deformation. *Norsk Geologisk Tidsskrift* **74**: 134–145.

Jones, R. R., Holdsworth, R. E., Clegg, P., McCaffrey, K. and Travarnelli, E., 2004, Inclined transpression. *Journal of Structural Geology* **26**: 1531–1548.

Robin, P.-Y. F., 1994, Strain and vorticity patterns in ideally ductile transpression zones. *Journal of Structural Geology* **16**: 447–466.

Tikoff, B. and Teyssier, C. T., 1994, Strain modeling of displacement-field partitioning in transpressional orogens. *Journal of Structural Geology* **16**: 1575–1588.

Hydrocarbon accumulations

Harding, T. P., 1976, Predicting productive trends related to wrench faults. *World Oil,* June 1976: 64–69.

Seismic expressions

Hsiao, L.-Y., Graham, S. A. and Tilander, N., 2004, Seismic reflection imaging of a major strike-slip fault zone in a rift system: Paleogene structure and evolution of the Tan-Lu fault system, Liaodong Bay, Bohai, offshore China. *American Association of Petroleum Geologists Bulletin* **88**: 71–97.

Physical modeling

Clifton, A. E., Schlische, R. W., Withjack, M. O. and Ackermann, R. V., 2000, Influence of rift obliquity on fault-population systematics: results of experimental clay models. *Journal of Structural Geology* **22**: 1491–1509.

McClay, K. and Dooley, T., 1995, Analogue models of pull-apart basins. *Geology* **23**: 711–714.

Tikoff, B. and Peterson, K., 1998, Physical experiments of transpressional folding. *Journal of Structural Geology* **20**: 661–672.

Examples of strike-slip faulting

Storti, F., Holdsworth, R. E. and Salvini, F. (Eds.), 2003, *Intraplate Strike-slip Deformation Belts.* Special Publication **210**, London: Geological Society.

Tapponnier, P., Peltzer, G. and Armijo, R., 1986, On the mechanics of the collision between India and Asia. In M. P. Coward and A. C. Ries (Eds.), *Collision Tectonics.* Special Publication **19**, London: Geological Society, pp. 115–157.

Chapter 19

Salt tectonics

Salt as a rock has properties and behavior that are very different from most other rocks. This implies that when sedimentary sequences containing salt layers are deformed, they develop their own characteristic and often very fascinating structural styles. Salt ridges, pillows, diapirs and even salt glaciers are special structures that are of importance in many settings. Even where the salt is restricted to a thin layer, it can control the structural expression and increase the areal extent of the deforming area because of its tendency to act as a décollement. Salt-related structures are of great importance to geologists working in regions of extensional as well as contractional tectonics, and are also important because many petroleum provinces contain salt layers or are deformed by salt tectonics. This chapter deals specifically with salt structures and salt tectonics, and provides an overview of salt structure geometries, processes and tectonic settings with examples from several places around the world.

19.1 Salt tectonics and halokinesis

Salt layers form integral parts of the stratigraphic column in many sedimentary basins, including cratonic basins, rift basins, passive margin basins and foreland basins, and play a significant role when salt-bearing sedimentary sequences are exposed to deformation, be it in an extensional, contractional or strike-slip setting. The section represented by Figure 19.1 shows an example of a rift setting, where the Permian Zechstein salt layer extends for hundreds of kilometers. Figure 19.2 shows that this salt unit can be traced across much of northern Europe (from the UK to Poland), and that the originally even-thickness salt layer has reshaped into concentrations of circular or elliptical shapes in map view. The cross-section (Figure 19.1) clearly shows that the salt forms structures that deviate from the generally subhorizontal layer orientation in the basin.

The salt in this section has deformed (flowed), and we use the term **salt tectonics** when salt is involved in deformation to the extent that it significantly influences the type, geometry, localization and/or extent of deformation structures that form. This term covers any salt-related deformation and deformation structures, including salt detachment-related deformation. Another term used about the movement of subsurface salt and the formation of salt diapirs is **halokinesis**, formed by the Greek words for salt or halite (halos) and movement (kinesis).

The influence of salt during deformation depends on its thickness, extent and position in the stratigraphic column, the degree of basement reactivation and the physical properties of the overlying strata. The deformation can be local and unrelated to plate-tectonic strain, completely driven by density contrasts between the salt and its overburden. In other cases, and perhaps more

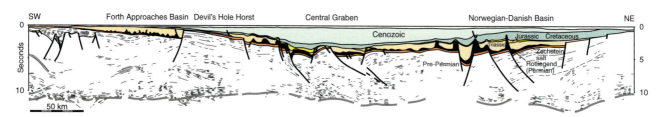

Figure 19.1 Cross-section (interpreted seismic line) through multiple salt structures of the southern North Sea. Based on Zanella *et al.* (2003). Location of profile shown in Figure 19.2.

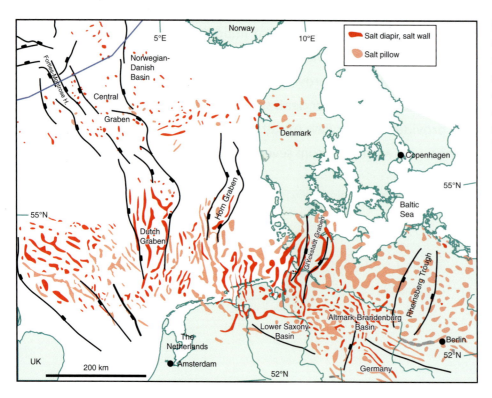

Figure 19.2 Map of salt structures and major faults of the southern North Sea basin and northern Europe. Note the close connection between the orientation of salt structures and faults. Partly after compilation by Scheck *et al.* (2003).

commonly, salt plays a more responsive role in a region under the influence of a regional tectonic stress field. Below we will look at some of the most common structures associated with salt movement, after a quick look at what makes salt so special.

19.2 Salt properties and rheology

Salt has physical and rheological properties that make it fundamentally different from most other common rocks, as summarized in Box 19.1. First of all, salt layers in sedimentary sequences may consist of pure halite, but commonly contain additional minerals, notably anhydrite, gypsum and clay minerals. Pure halite has the relatively low density of 2.160 g/cm^3, while that of impure salt is slightly higher. This makes salt less dense than most carbonate rocks, but still denser than unlithified siliciclastic sediments. So what about the density contrast that, according to many introductory textbooks, drives salt diapirism?

The answer is related to the difference in compaction between salt and porous sediments. Salt, which has practically no porosity even at very shallow burial depths, is almost incompressible and hence does not become much denser with increasing overburden. This is very different from most sedimentary rocks, where compaction is significant. Hence, as sediments compact physically and chemically during burial, their density increases with increasing burial depth. Thus, a density inversion (dense layer underlain by a less dense layer) arises once the compacting overburden becomes denser than the salt. At that point a gravitationally unstable situation is established which, under certain conditions, can result in flow of salt toward the surface.

At what depth can we expect density inversion and salt buoyancy to occur? It depends on the compaction curve, which again depends on the sediment properties and burial rates (sedimentation rates). The buoyancy of salt at various depths must therefore be calculated according to local conditions, but we can make some general estimates.

Siliciclastic sediments tend to attain densities in excess of 2.2 g/cm^3 at about 1–2 km. The exact depth depends on grain size, sorting and mineralogy, the type of sediment, and for siliciclastic sediments clay compacts faster than sand. Simple calculations can be performed by adding the density of the rock minerals (ρ_s) to that of the pore fluid (ρ_f):

$$\rho = \Phi\rho_f + (1 - \Phi)\rho_s$$

In this formula the porosity (Φ) reflects the compaction. For a siliciclastic sediment with 30% porosity, an average mineral density of 2.7 g/cm^3, and with salty pore water (density of 1.04 g/cm^3), the rock density is approximately the same as for slightly impure salt (2.2 g/cm^3). Hence, a porosity value lower than ∼30% is required for the overburden to become denser than the salt. This occurs at relatively shallow depths if the sediment is fine grained, perhaps as little as 600–700 m, while for a sandy sequence a density of 2.2 g/cm^3 may be achieved at 1500–2000 m depth. Furthermore, the average density of the overburden must exceed that of the underlying salt for a diapir to reach the surface by buoyancy alone. This requires burial to at least 1600 m, and more typically closer to 3000 m. In the Gulf of Mexico this depth is around 2300 m.

Salt layers become buoyant and gravitationally unstable when buried to the depth where the average density of the overburden exceeds that of the salt.

Rheologically, salt deforms plastically during loading even at surface conditions. Only if strain rates become high, such as those associated with earthquakes and mining (or when hit by a hammer), will salt fracture. Under most geologic conditions, salt will flow as a viscoelastic medium. This is why salt mines are being used as waste repositories.

The elastic component of salt deformation can be neglected for most purposes because of the low relaxation rates involved, and the flow can be considered **viscous**. In more detail, two deformation mechanisms dominate salt deformation; **wet diffusion** and **dislocation creep**. Wet diffusion, where material is dissolved and transported along grain boundaries by means of a thin fluid film (Chapter 10), is the dominating deformation

---

BOX 19.1 | SALT PROPERTIES

Common constituent of sedimentary basins
Mechanically very weak
Low density (2.160 g/cm^2)
High heat conductivity
Almost incompressible
Impermeable
Viscous, behaves like a fluid
Causes wide areas of deformation
Enhances horizontal detachments
Creates structural fluid traps

mechanism when there is some water present in the salt. Since wet diffusion involves transport along grain boundaries and surface area increases with decreasing grain size, wet diffusion is generally faster and thus more important in fine-grained than coarse-grained salt. Low strain rate and differential stresses also promote wet diffusion. Dry salt, with no fluids to transport matter, is different. In that case dislocation creep is the principal deformation mechanism, unless the salt simply fractures (dry salt at or near surface conditions).

Either of these deformation mechanisms is easily activated in salt, meaning that salt has very low yield strength and thus flows very easily. Together with the fact that salt is almost incompressible under most geologic conditions, this means that salt can be treated as a fluid in geologic modeling.

What makes salt so special is its low density and ability to flow like a fluid even at the surface, and particularly when loaded.

19.3 Salt diapirism, salt geometry and the flow of salt

Salt structures

Salt volumes in the crust have for a long time been known to take on a variety of geometric shapes, from elongated structures known as salt anticlines and salt pillows to more localized structures such as salt stocks. Although many of these structures are loosely referred to as diapirs they have different names, as illustrated schematically in Figure 19.3. The term **diapir** derives from the Greek words for through (*dia*) and pierce (*peran*) and is used in geology to describe a body, usually of salt, magma or water-saturated mud or sand, that gravitationally moves upward and intrudes the overburden. Hence, some structures, such as **salt pillows**

and **salt anticlines** that just bend and uplift the overlying layers, are not diapirs *sensu stricto*, because they do not intrude or pierce the overburden. However, most of these structures represent various stages that could lead to the formation of a true diapir. The process through which a diapir develops is known as **diapirism**.

A salt diapir is a mass of salt that has flowed ductilely upward and discordantly pierced the overburden.

The flow of salt from a salt layer into a salt structure is usually referred to as **salt withdrawal** or, more correctly but less commonly, **salt expulsion**. In simple terms, there are two principal types or end-members of flow. One, **Poiseuille flow**, occurs when salt flows into a salt structure during the growth of a salt anticline or diapir (Figure 19.4a). In this case flow is restricted by the viscous shear forces acting along the boundaries of the salt, an effect known as **boundary drag**. This effect causes the salt to flow faster in the central part of the salt layer than along the top and bottom. Hence, thin salt layers flow slower than thick ones, implying that flow in a thick (tens of meters

(a) Salt flowing into diapir

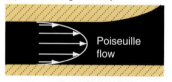

Poiseuille flow

(b) Salt beneath sliding block

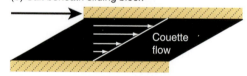

Couette flow

Figure 19.4 The two principal types of flow occurring in deforming salt layers. Arrows indicate velocity and the velocity profile is parabolic in (a) and linear in (b).

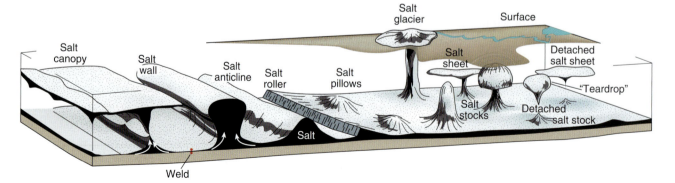

Figure 19.3 Different types of salt structures, their names and geometries. Maturity increases from the central part of the figure to the left and right.

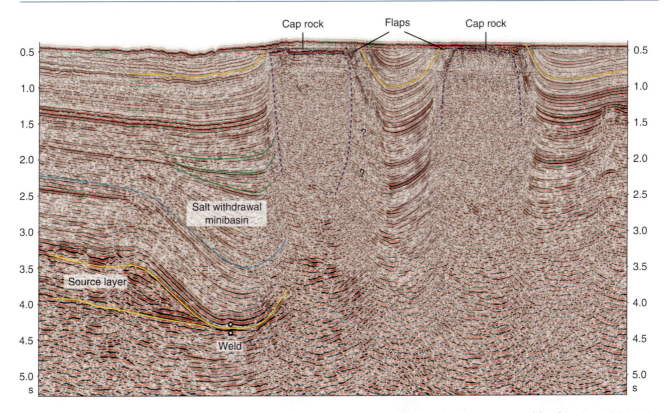

Figure 19.5 Salt structures imaged on seismic 2-D line. The two structures have surfaced (now covered by thin Quaternary cover) and developed reflective caps. Deeper parts of the salt structures are obscured by seismic noise. The source layer has been interpreted based on a salt weld structure. The growth history is recorded in the sedimentary record: thickness changes near the diapirs start above the blue horizon, after deposition of a ~1.5–2 km sedimentary load. Note minibasin above the weld. Data courtesy of the Norwegian Petroleum Directorate.

or more) salt layer is slowed down when the salt is reduced to a thickness of a few tens of meters. If the salt becomes completely exhausted the boundary layers become attached to each other, and the contact is referred to as a **salt weld** (Figure 19.3), typically indicated on geologic and seismic sections by pairs of dots (Figure 19.5).

Salt welds may pin salt anticlines and domes, and thus terminate the growth of salt structures.

Salt welds will always contain some remnant salt, and even if the remaining salt is only centimeters thick it will represent a weak zone that is prone to localized strain. The importance of a salt weld is that it terminates lateral flow of salt into adjacent salt structure(s).

The other type of flow is known as **Couette flow** (Figure 19.4b), which involves simple shearing within the salt layer as the overburden is translated relative to the substrate. This type of flow is typical for salt layers acting as décollements, but the two types of flow may well be superposed on each other. Ideally, in Couette flow there is no boundary effect of the type occurring in Poiseuille flow.

Salt diapir geometries

Salt diapirs can take on a variety of forms. In plan view they may be elongated, typically along the direction of regional faults or graben axes, as seen in the northern Europe examples of Figure 19.2. They may also be circular, a shape portrayed by many of the diapirs in the Central Graben (NW part of Figure 19.3). Some portray triangular shapes in cross-section, with the salt thinning upward. This is common where salt structures are connected with faults in the overburden (see reactive salt structures below). Salt diapirs that have plug-like shapes are known as **salt stocks** (Figure 19.3). Elongated salt structures that appear as stocks in perpendicular cross-sections are called **salt walls**. Figure 19.6 shows well-developed examples of salt walls from the Moab area (Paradox Basin) of southeast Utah.

Many diapirs have a **stem** and a wider upper part, known as the **bulb**. In extreme cases the stem may be missing, so that the salt bulb is completely detached from its source layer. Isolated salt bulbs are sometimes called **teardrop diapirs**. Such detached bodies of salt may show an asymmetric geometry that is caused by lateral movement of salt over younger sedimentary strata, in which case the salt is

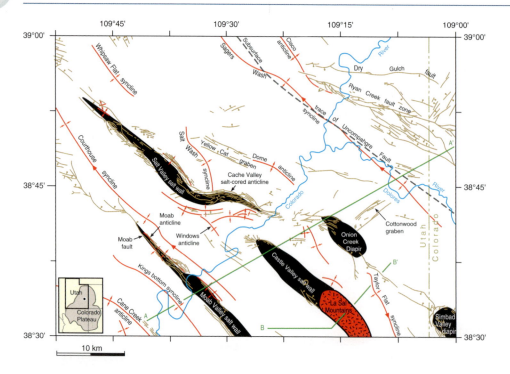

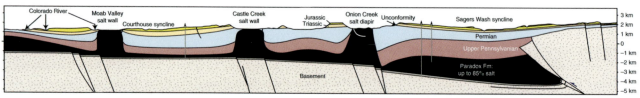

Figure 19.6 Salt structures in the Moab area of southeast Utah in map view and cross-section. Elongated salt walls dominate, but more circular diapirs also occur. The salt walls seem to follow the NW–SE trend of long-lived faults in the basement, while younger collapse-related faults occur in the overburden. Did these salt structures initiate and grow during contraction or extension? Simple questions are sometimes hard to answer. Based mainly on Doelling (2001).

said to be **allochthonous**. Diapirs can also flatten out and join at one, and less commonly at several, stratigraphic levels, forming various types of **salt stock canopies**.

Most of the information we have about the geometry of salt structures is obtained from seismic data. Salt is acoustically quite homogeneous and does not often produce seismic reflectors. Hence, fairly pure salt shows up on seismic lines as "seismically transparent" areas with a characteristic pattern of noise. The top of the salt is usually well imaged, but the walls are commonly too steep to show up on seismic lines (Figure 19.5). Salt structures and layers also absorb so much energy that sub-salt reflectors may be difficult to identify. It may therefore be difficult to distinguish between salt bulbs and salt stocks. In some cases it may even be difficult to distinguish between salt structures and impact structures, as illustrated in Box 19.2. Nevertheless, seismic imaging is continually improving, and modern 3-D surveys can give images of salt that were unachievable just a decade ago.

Modeling salt structures: the fluid approach

In the 1930s the role of the density contrast between salt and the overburden received attention, as did the ability of salt to flow even at geologically low temperatures. It led to a period during which not only the salt, but also the overlying denser sediments were modeled as fluids. It is relatively easy to physically model this situation, as demonstrated in 1934 by Lewis Lomax Nettleton, who put a layer of corn oil on top of less-dense crude oil. The result of this unstable situation was the formation of a diapir, as blobs of the light oil ascended through the heavy oil (Figure 19.7), a bit similar to so-called lava lamps, where blobs of heated wax ascend from the bottom of the lamp. In such fluid experiments, the dense fluid ("sediment") will sink into the underlying less dense fluid ("salt").

A more sophisticated modeling device, known as the **centrifuge**, has been applied since around 1960, and the person who more than anyone explored the possibilities and pushed the limitations of the centrifuge was Hans Ramberg and associated workers in his laboratory in

BOX 19.2 | SALT OR IMPACT STRUCTURE?

It can be difficult to decide whether a circular structure is salt diapir-related or an impact structure. Both mechanisms produce circular structures with faults, disturbed stratigraphy etc.

Upheaval Dome in Canyonlands National Park, Utah, is one such example. It is located in the marginal parts of the Paradox Basin, which contains salt of Carboniferous age. Salt movement in this area has caused a series of salt anticlines, notably in the Moab and Arches National Park area, but circular structures of the Upheaval Dome type are absent.

Silverpit "Crater" is a similar structure in the North Sea, identified from seismic interpretation as a depression containing circular faults (ring faults). It has also been claimed to be an impact as well as a salt structure, and the issue is not settled.

So how do we separate between impact and salt diapir structures? The best structure to look for is perhaps evidence of shocked quartz, planar deformation lamellae that are diagnostic of the high-pressure deformation related to a meteor impact event. The lamellae are bands of dislocations arranged in crystallographic directions, and can be seen under the transmission electron microscope. Such lamellae have been reported from the Upheaval Dome structure (Buchner and Kenkmann 2008), adding to the evidence in favor of an impact origin. It can be added that an impact would weaken the stratigraphy so that salt more easily could flow toward the surface, so a combination of the two models is possible.

Right: View of the central part of Upheaval Dome, Utah. Lower left: Seismic interpretation of the Silverpit structure (Stewart and Allen, 2005). Lower right: Terrain model of Upheaval Dome.

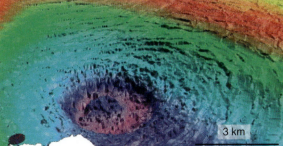

3 km

~1 km

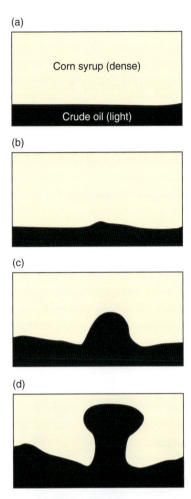

(a)

Corn syrup (dense)

Crude oil (light)

(b)

(c)

(d)

Figure 19.7 Experiment where a dense corn syrup is placed on top of a less dense oil. Experiment published by Nettleton in 1934.

Uppsala, Sweden. Ramberg and other workers used weak ductile material as rock analogs and deformed his models at up to 2000*g* in the centrifuge. Hence, in the centrifuge the role of gravity is accentuated. Also, the densities and viscosities of the experimental materials and the thicknesses of stratigraphic layers (geometry) are designed to form a downscaled version of a realistic geologic situation where relevant proportions and contrasts are maintained. This important part of a physical experiment is called **scaling**, and was discussed in Chapter 1 (Section 1.8). Note that realistic centrifuge modeling does not require that the densities and viscosities of the model layers are identical to those of the geologic layers they are meant to represent, but the density and viscosity *contrasts* between the various layers must be the same in the model as in nature.

It turned out that the centrifuge experiments produced diapirs very similar to salt diapirs observed in the upper crust. The progressive development of diapirs can be studied through such modeling. An example is

shown in Figure 19.8, where we can see development from open domes or pillows (Figure 19.8a) through tall diapirs (Figure 19.8b) to isolated or mushroom-shaped volumes of salt that intrude the overburden (Figure 19.8c) and eventually form salt canopies (Figure 19.8d). During the run, the model can be paused and layers of sediments can be added. Eventually salt spreads out at or near the surface to form salt canopies. The Dasht-e Kavir (great salt marsh) of Iran is an area where the centrifuge model has successfully reproduced salt structures, including mushroom-shaped diapirs and salt canopies (compare Figures 19.8 and 19.9). The high degree of ductility of the overburden in this area is unusual and related to disseminated salt in some of these stratigraphic layers. Their ductile nature, with the formation of folds rather than faults and fractures, is well illustrated in Figure 19.10.

In the theoretical case where the overburden behaves like a perfect fluid, the layers overlying the salt will thin without fracturing as the salt rises, and the structure is not a diapir *sensu stricto*. Such structures are uncommon, which indicates that only the mobile salt layer should be given fluid properties in models.

Modeling salt structures: the brittle approach

There is now general consensus that modeling the overburden as a fluid is a gross oversimplification. Fluids have no shear strength, whereas actual rocks and sediments do. Gravity is indeed the driving force, but for a salt diapir to initiate it has to overcome the strength of its roof. Hence, if the overburden is a strong, brittle rock (limestone or compacted siliciclastic sedimentary rock) the density contrast itself is not enough to initiate diapirism. Recall that a certain sedimentary thickness is needed for a significant density inversion to occur (perhaps 2–3 km), implying that sedimentary strata overlying salt have gone through lithification and thus mechanical strengthening. Physical considerations tell us that diapirism is very unlikely to initiate without assistance from tectonic faulting or fracturing. On the other hand, once the roof is weakened and diapirism is initiated, classic gravity-driven diapirism can more easily proceed.

Brittle fracture can reduce the strength of the roof and allow salt to ascend and form diapirs.

For this reason, the centrifuge model has been largely replaced by simple deformation boxes, where extension or contraction can be applied by moving the sidewalls, where tilting can add gravitational forces and where

(a)

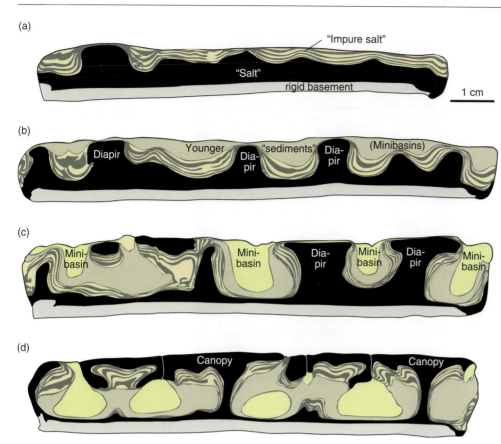

Figure 19.8 Sections through a scaled centrifuge model at different stages of development. The sections cannot be directly correlated, but give a good impression of the development of diapirs that end up forming a canopy. The model was made at the Hans Ramberg Laboratory in Uppsala to reproduce Dasht-e Kavir salt diapirs in Iran, as described by Jackson *et al.* (1990).

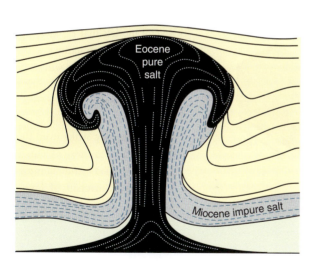

Figure 19.9 Typical geometry of mushroom-shaped diapirs found in the Dasht-e Kavir, Iran. More information about these structures are found in Jackson *et al.* (1990).

performed at the University of Rennes (France) and the University of Texas at Austin, where the model shown in Figure 19.11 was produced.

Controls on salt flow

Even though salt overlain by denser layers represents a gravitationally unstable situation, the salt will not move unless there is a gravitational or mechanical anomaly of some kind. Lateral variations in overburden thickness and/or density are known under the term **differential loading**. Many even consider variations in strength as a type of differential loading. Differential loading is an important reason why salt moves, and is related to lateral variations in the state of stress created in the salt layer. Since salt behaves like a *viscous* fluid we can talk about **pressure differences**, which salt cannot sustain over time. Hence, differential loading initiates flow of salt toward low-pressure areas. Salt movement in response to differential loading does not rely on density inversion, and therefore is fundamentally different from buoyancy. This means that differential loading is effective even at very shallow depths, causing salt to move once new sediments are being deposited unevenly on top of the salt layer.

sedimentation during diapirism can be controlled. These models, typically consisting of viscous silicone gel overlain by (brittle) sand, have successfully been used over the past few decades to reproduce a large number of seismically mappable salt structures in sedimentary basins, and much important experimental work has been

Figure 19.10 Salt structures exposed in the Dasht-e Kavir, Iran. The circular shapes of the diapirs are visible. Note the folds in the overburden, which also contain some salt. Image: NASA.

10 km

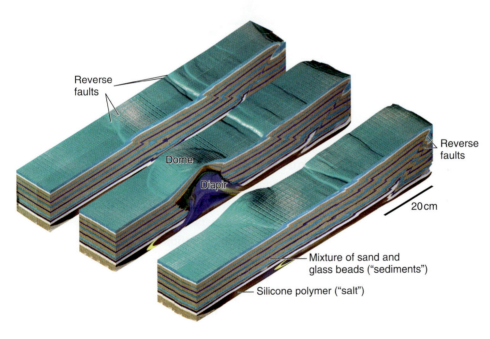

Reverse faults

Dome

Diapir

Reverse faults

20 cm

Mixture of sand and glass beads ("sediments")

Silicone polymer ("salt")

Figure 19.11 Example of physical experiment involving salt analog overlain by "sediments". This is one of several models that explore the result of contracting a sedimentary sequence with a preexisting salt diapir. During the experiment the diapir roof domed, and reverse faults developed along strike from the dome. Image courtesy of Tim Dooley, Applied Geodynamics Laboratory, University of Texas at Austin.

Differential loading occurs if, for some reason, an area is more heavily loaded (or eroded) than its circumference, in which case the salt starts to flow away from the area of maximum loading. This difference can, as illustrated in Figure 19.12a, be caused by a lateral variation in overburden thickness and/or slope, or by variations in lithology and rock density (sedimentary facies variations, local occurrence of lava etc.). Also, a difference in elevation can trigger salt flow, as indicated in Figure 19.12b.

In general, most salt structures are associated with faults and folds, and many (but not all) such structures indicate that tectonic strain played a role during salt movement. If a salt body is shortened, the weak salt is likely to flow upward and shorten in the horizontal direction, like toothpaste being squeezed out of an upright tube. In the case of regional extension, horizontal stretching or unloading causes the salt to expand laterally and thin vertically. In this simple model the boundaries

(a) Varying overburden thickness

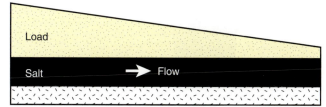

(b) Dipping salt and overburden layers

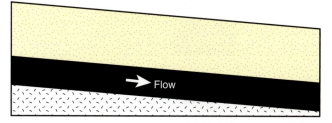

(c) Combination of (a) and (b)

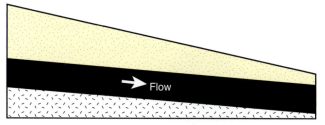

Figure 19.12 Situations where loading causes salt to flow. (a) Flow related to differential loading (e.g. a river delta), which causes differences in pressure. (b) Flow caused by elevation difference due to tilting. (c) Combination of the two, as commonly seen on continental margins. These simple models have restricted (pinned) ends so that the overburden cannot slide on the salt.

on each side of our salt body control the flow of salt, and this boundary effect can be referred to as **displacement loading**.

Vertical (sedimentary) or lateral (tectonic) loading of salt can cause salt to flow, independent of any density inversion or burial depth.

Thermal loading refers to the fact that hot salt expands and becomes more buoyant. This can accelerate salt flow toward the surface. It has also been suggested that thermal loading can result in convection (hot salt rising and cool salt sinking within the salt structure). This may not be an important process in most salt structures, but could explain some peculiarly coiled salt structures (vortex structures) found in the Dasht-e Kavir in Iran (Figure 19.9).

Structures above and around salt diapirs

Salt flow may affect the wall rock, simply because of the friction between the flowing salt and its surroundings. This can result in frictional **drag folds**, formed during salt movement. Such drag folds can form in unconsolidated or poorly consolidated sedimentary layers, but in most other cases salt will be much weaker than the wall rock, in which cases the premises for forming frictional drag are absent.

Nevertheless, rotation (folding) of layers around salt structures is common, and other explanations than friction are found. For example, wide drag-like zones can form because salt is practically incompressible while neighboring siliciclastic sediments compact. Hence, such **differential compaction** can cause apparent drag along upward-thinning salt structures. In other cases forceful intrusion of salt into the overburden occurs, as will be discussed in the next section. As a result, **flaps** of sedimentary layers can forcefully be pushed upward during diapirism (Figure 19.13a). A third folding mechanism is the effect of salt withdrawal, i.e. thinning of the source layer as salt enters the salt diapir or salt wall. Salt withdrawal causes overlying layers to subside locally, and a **rim syncline** or **minibasin** forms around the salt structure. This mechanism usually works together with differential compaction to form flanking minibasins with layers that define a downward-amplifying fold.

Seismically resolvable Class 2 folds around salt structures form from vertical shear as salt withdraws and differential compaction occurs.

A more complex structure forms as shallow diapirs grow during rapid sedimentation (passive diapirism, see below), as shown in Figure 19.14. The diapir roof and rim are then draped by thin layers of sediments, but its surface relief increases as sediments compact and the diapir rises. At some point the relief is high enough that the scarp fails gravitationally and/or the salt breaks through and becomes exposed. At the same time, the strata (flaps) along the rim of the diapir are further steepened and even overturned in some cases before becoming buried under new sediments. This development repeats, and the result can be stacking of so-called **halokinetic cycles** separated by unconformities near the salt structure.

A gentle dome structure is usually found above salt structures that have not surfaced (Figure 19.15). Again, such a dome structure could be related to differential compaction as well as to salt growth. Faults above the top

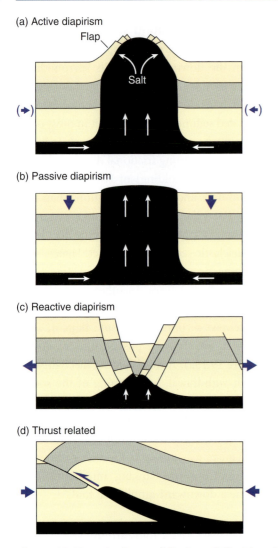

(a) Active diapirism

Flap

Salt

(b) Passive diapirism

(c) Reactive diapirism

(d) Thrust related

Figure 19.13 Main classes of diapirs and diapirism. (a) Active diapirism driven by density contrast and forceful ascent of salt (buoyancy), (b) passive evolution as sediments are deposited around the salt structure, (c) Reactive formation in response to extension, and (d) salt structuring during contraction (thrusting).

of a salt diapir are common, and in most cases these are normal faults, forming as the layers bend and stretch during the salt movement. Only in rare cases do reverse faults occur above the margins of salt diapirs, due to forceful jacking of the roof by the salt. In either case, faults above salt structures tend to reflect the map-view geometry of the salt structure, so that (ideally) circular diapirs develop circular fault patterns, while elliptical diapirs and salt walls develop faults with more linear traces. Tectonic stress or inherited fracture orientations may however influence the structural pattern above salt structures. **Concentric faults**, commonly seen above circular salt diapirs and illustrated by black faults in Figure 19.15, can in many cases be related to the collapse

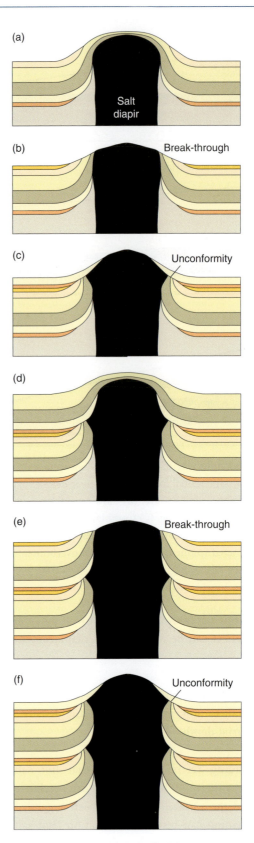

(a)   Salt diapir

(b)   Break-through

(c)   Unconformity

(d)

(e)   Break-through

(f)   Unconformity

Figure 19.14 Formation of halokinetic sequences along a vertical salt structure. Occasionally the rising salt breaks through the roof (or the roof sediments gravitationally slide away) and bends the layers. An unconformity forms as the next sedimentary sequence starts depositing. Based on Rowan *et al.* (2003).

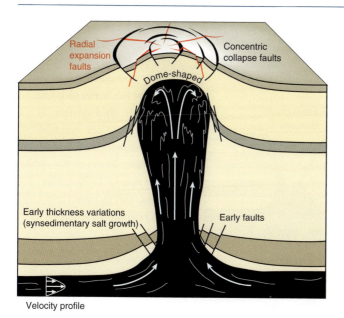

Figure 19.15 Some common subsidiary structures associated with salt domes.

of salt structures. In addition, **radial faults** (red faults in Figure 19.15) have been described from some salt domes, created by the expansion of the underlying salt. Small-scale fractures also form above salt structures, and concentric, radial and other patterns have been reported.

Within the salt itself there will be significant strain gradients, since the flow velocity decays to zero toward the adjacent rocks. The flow is probably simple along the feeding salt layer, but within the salt structure itself patterns can be more complex. This is clear from the folded pattern of bedding observed in salt mines and within exposed salt domes, and relates to the non-linear flow of salt inside diapirs.

19.4 Rising diapirs: processes

Once a diapir starts rising it will grow into a diapir with a shape that depends on the sedimentation or erosion rate, tectonic regime, strength of the overburden, gravitational loading, temperature of the salt, salt layer thickness, salt extent (salt availability) and more. In traditional models of diapirism, diapirs can force their way upward through the overburden, driven by differential, thermal or displacement loading. During such **active diapirism**, flaps of the roof are forced aside, causing significant upward rotation of layers along the upper parts of the salt walls (Figure 19.13a). As mentioned above, buoyancy forces related to gravity inversion are generally not sufficient to initiate active diapirism. Only when a diapir is well established and the overburden becomes thin can diapirism be driven by buoyancy forces.

Passive diapirism is a term used for exposed or very shallow diapirs that rise continually at a rate that more or less keeps pace with sedimentation (Figure 19.13b). It therefore requires that a diapiric structure initiates by some other mechanism. Once established, surrounding sediments subside as they compact and sink into the source salt layer, which thins as salt is fed into the diapir structure. This process, where subsidence creates a minibasin around the salt diapir where new sedimentary strata accumulate, is sometimes referred to as **downbuilding**. If this process goes on for a long time and if the source layer is thick enough, tall salt diapirs may form. If the salt rise rate is greater than the sedimentation rate, the diapir develops into an upward-widening structure. In the opposite case, where the rise rate is less than the sedimentation rate, the diapir takes on an upward-narrowing shape. It is not unusual to find passive diapirs with steep walls, indicating a balance between sedimentation rate and salt rise rate (flow rate). Depletion of the source layer may reduce or stop salt movements, causing the diapir to become buried.

We have already emphasized that weakening of the roof by means of faulting and fracturing is necessary for most salt structures to develop. The strains involved in such fracturing may be minimal. However, in some cases the overburden is more significantly affected by tectonic faulting, and salt fills in spaces between displaced fault blocks (Figure 19.13c). In this case the salt rises as a reaction to significant tectonic strain, usually extension, and the process is known as **reactive diapirism**. Experiments have shown that reactive diapirism ceases once extension comes to a halt, demonstrating its dependence on active extensional strain. Contractional deformation can also drive diapirism, as salt can be emplaced into its overburden in the hanging wall of a thrust fault (Figure 19.13d).

Diapirism may be active and driven by loading, passive as sediments build down around the diapir, or reactive in response to tectonic strain.

Most natural diapirs contain elements of two or all of these processes of diapirism. The importance of interaction between faulting and salt diapirism calls for a closer look at salt tectonics in different tectonic regimes.

19.5 Salt diapirism in the extensional regime

Many salt accumulations occur in continental extensional basins and passive continental margins. Since such

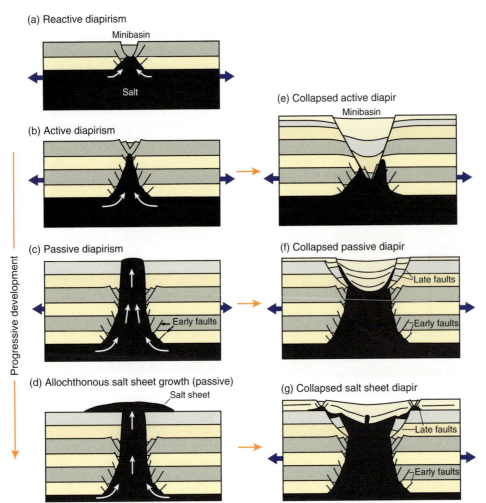

(a) Reactive diapirism

Minibasin

Salt

(b) Active diapirism

(c) Passive diapirism

Early faults

(d) Allochthonous salt sheet growth (passive)

Salt sheet

(e) Collapsed active diapir

Minibasin

(f) Collapsed passive diapir

Late faults

Early faults

(g) Collapsed salt sheet diapir

Late faults

Early faults

Progressive development

Figure 19.16 Illustration of a complete and idealized evolution of a surfacing salt diapir through stages of reactive (a), active (b), and passive (c, d) diapirism. At any stage the salt structure can collapse, giving rise to structures shown at the right-hand side (e–g). Based on Hudec and Jackson (2007) and references therein.

regions tend to experience prolonged or repeated periods of extension, many salt layers are affected by regional extensional tectonics. Specifically, many if not most salt diapir provinces initiated during regional extension, although they may later have become modified by contractional deformation, as discussed below.

Numerous physical experiments have shown that extensional faulting and fracturing weakens the overburden so that reactive diapirism can start. Graben formation in the overburden typically results in a triangle-shaped diapir that rises as extension proceeds and provides space for the salt (Figure 19.16a). The graben structure above the salt is topped by a local minibasin at this stage, although salt emplacement reduces the depth of this basin. The graben as shown in Figure 19.16a is symmetric, but can also be asymmetric with a single dominant master fault. Such asymmetric structures produce a (non-equilateral) triangular-shaped volume of salt in the footwall known as a **salt roller** (Figure 19.3). Continued evolution of a salt roller involves thickening of the salt on the footwall side of the fault.

In areas of extensional faulting and fracturing salt can start to move when the overburden above the salt diapir becomes thin enough and sufficiently weakened by extensional faulting. After an early stage of reactive salt diapirism the salt can actively (buoyantly) force itself upward if the total overburden above the source layer is, on average, denser than the salt (Figure 19.16b). There is now a significant difference in density between the salt structure and its surroundings, and the extra weight of the compacting wall sediments pressurizes the salt and causes it to flow into the salt structure. This loading can be enough to drive salt diapirism, meaning that active extension is not required at this stage. During this passive stage the roof layers are domed and stretched, resulting in steeply dipping layers or flaps along the upper margins of the salt diapir.

If the source layer is thick enough, the salt can reach the surface and act as a passive diapir (Figure 19.16c). Hence, salt can flow by reactive, active and passive diapirism during the rise of a diapir in an extensional setting. The salt may even end up flowing laterally at the surface (typically the sea floor) as an allochthonous **salt sheet**

BOX 19.3 | SALT GLACIERS

Where salt flows out of a diapir (salt plug) above sea level it flows gravitationally in the direction of highest dip. In dry regions, such as Iran, the flow occurs during periods of rainfall, when the salt is wet. Otherwise, the upper part of the salt sheet deforms brittlely by jointing. This shows the effect of fluids on deformation mechanism; wet conditions favor plastic deformation. Surface salt that flows down-dip under the influence of gravity is referred to as a salt glacier. The Zagros Mountains in Iran are known for their salt glaciers, with estimated flow rates of 0.3–16 m/year, but submarine salt glaciers also occur where the extrusion of salt is faster than the rate of dissolution of salt by the seawater. This is, for example, the case in the Gulf of Mexico, where several submarine salt glaciers have been mapped.

During the downhill flow of salt, the salt deforms by folding due to the non-constant velocity field (velocity increasing from the base upward). Recumbent folds on a variety of scales can be seen, the largest of which can be mapped out. These folds are similar (Class 2) folds, because they involve no mechanical layering and simply form because the passive layering enters the contractional field during flow.

Schematic profile through an Iranian salt glacier, based on work by Chris Talbot (1979).

The Anguru salt plug in southern Iran, where impure salt with elements of reddish-brown shales and greenish volcanic rocks flows out from the mountain in the central part of the picture, covering younger sedimentary strata (yellowish). Photo: Mahmoud Hajian.

(Figure 19.16d), as discussed in more detail in Box 19.3. Note how early faults are preserved in the deepest parts of the salt walls even at very mature stages of diapirism (Figure 19.16c, d), revealing its early extensional history.

Salt structures may go through all the stages shown in Figure 19.16a–d, but the evolution may also stop at any time during the process, or reverse in the sense of salt diapir collapse. The right-hand column of Figure 19.16 shows the result of collapse of salt diapirs after various stages of growth. The actual evolution in any given example depends on the balance between extension rate, sedimentation rate, the viscosity (temperature, purity) of the salt and salt availability (thickness of salt layer). During extension, the width of the salt diapir increases, so the flow of salt into the structure must more than balance the extension rate for the structure to grow.

If the structure widens too fast during extension, the salt structure will collapse. Clearly, if the salt source layer is exhausted and a salt weld is formed, collapse will occur during continued extension regardless of the extension rate. A history of growth and collapse, or rise and fall of a salt diapir, is very common, and reproduced nicely in laboratory experiments where extension and sedimentation rates can be controlled.

Extension can cause salt diapirs to form, and can also stretch and widen them so that they collapse vertically.

19.6 Diapirism in the contractional regime

When a sedimentary sequence overlying a salt layer is shortened during regional contraction (tectonic or purely gravitational), the sequence above the salt may start to buckle. The buckling creates low-pressure antiformal crests into which the salt flows and in this way accommodates the folding of its overburden. Long salt anticlines form in this way, and may cause significant variations in depositional thickness if the process is syn-sedimentary, with thinning or pinching out of layers toward the antiformal crests.

The formation of salt anticlines during contractional folding of the overburden is not really diapirism, since the overburden is still intact: diapirism requires piercing of the roof. Piercement can occur by means of thrust formation, as shown in Figure 19.13d, but the geometry of the resulting salt structure is very asymmetric and may not be very similar to classic diapir structures of the kinds discussed above.

While buckling and reverse faulting is one way of initiating salt diapirs during regional contraction, most large-scale diapir structures found in areas of regional contraction probably initiated during extension and have since been amplified or reworked during shortening. It is important to realize that already existing salt diapirs represent weak elements that preferentially deform if the region is subjected to horizontal shortening. When this occurs, the salt structures become narrower and salt is squeezed upward and also laterally in many cases. Several characteristic structures can result, depending on the amount of salt, structural development and sedimentation. The spectrum of structural outcome is larger if we allow contraction to form or rework salt structures. We will here look at some of the most common types of salt structures related to contraction.

Salt is so much weaker than other rocks that it will influence how regional contraction is expressed, be it by décollement folding and imbrication or by squeezing of existing diapirs into salt sheets and salt glaciers.

Teardrop diapirs

Contractional amplification of already established salt structures can result in the formation of **teardrop diapirs** (Figure 19.17), where an originally upward-widening

(a) Pre-contractional stage

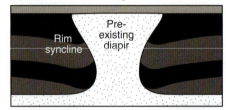

(b) Contraction, diapir rising

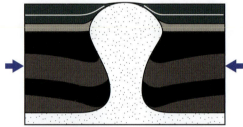

(c) Further squeezing and rise of diapir

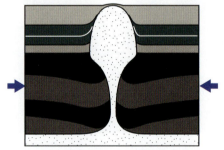

(d) Teardrop stage

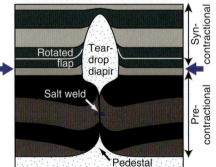

Figure 19.17 Schematic illustration of the transformation of a hourglass-shaped diapir formed during extension (a) to a teardrop diapir (d) due to contraction. Based on Hudec and Jackson (2007).

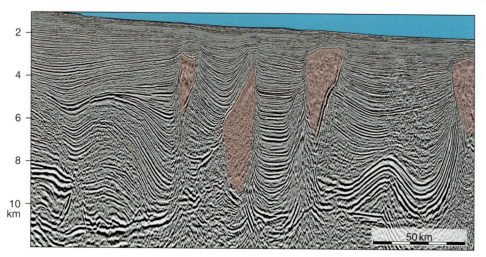

Figure 19.18 Depth-migrated seismic section from the submarine Mississippi Canyon, Gulf of Mexico, showing teardrop diapirs (detached salt stocks). The source salt layer is located at ~10–11 km depth (not interpreted). Folds formed as a result of salt withdrawal in an extensional setting. Image provided courtesy of TSG-NOPEC.

diapir is squeezed to the point where a salt weld forms in the middle of the structure. The upper and isolated part is the teardrop, and the lower root structure is called a pedestal. Salt flows upward during this process, in many cases fueled by the component of buoyancy involved, but also by the free surface at the top. In fact, buoyancy is not required for this process to occur, since the tectonic squeezing or pressurizing of the salt is sufficient. Examples of teardrop structures are commonly seen in seismic data from areas where shortening of salt structures has occurred, but they also occur in extensional regimes (Figure 19.18).

In cases where the salt weld of a teardrop structure is inclined, the weld may become activated as a reverse fault during continued shortening. The final geometry of a teardrop diapir and related structures depends on the preexisting geometry and rate of sedimentation. It also matters if the contraction is by pure shear alone, as in Figure 19.17 or if a component of shear along the basal salt layer is involved.

Salt sheets

In some cases salt can reach the surface (dry land or ocean floor) and extrude like very slow-flowing lava erupting from a volcanic feeder. Although subaerial salt sheet structures have been known for more than a century, most of the several thousand salt sheets mapped so far are submarine sheets that have been detected from seismic imaging and drilling of rifts and passive margins over the last few decades. We have already stated that this can occur without the influence of shortening, as shown in Figure 19.16d, and occurs where vertical salt movement in a surfacing diapir (feeder) is larger than the sedimentation rate. However, salt reaching and covering the land surface or ocean floor is more easily achieved,

and the structures better developed, where a contractional "squeeze" is added.

Salt that flows laterally as sheets on the surface or between stratigraphic layers is called **allochthonous**, because it overlies younger and stratigraphically higher rock or sediment layers, akin to an allochthonous thrust nappe. Salt sheets are therefore also referred to as **salt nappes**.

Salt that has been emplaced onto its stratigraphic overburden is known as allochthonous and is called a salt sheet if the width is at least five times its thickness.

Salt extrusion depends on the sedimentation rate relative to the salt flow rate. If no sediments are being deposited, the salt sheet follows the top of the upper bed, and will be confined to this stratigraphic or erosional level upon burial. This is the case with the subaerial salt structures seen in southern Iran (Box 19.3), which flow down-slope under the influence of gravity, much like a glacier. Basal irregularities and basal friction in general cause differential flow within the salt, as seen from folded markers in such **salt glaciers**. Most of the flow occurs near the base in the form of Couette flow, with the tongue of the glacier rolling over the frontal ground. However, if sediments accumulate around the margins of the salt structure, the flowing salt will periodically climb or ramp up-section away from the feeder at a rate that depends on the sedimentation rate versus the salt extrusion rate. Hence, series of ramps and flats may result (Figure 19.19a), where ramps are thought to indicate periods of more rapid sedimentation or slower salt advance. In general, the mechanism by which exposed salt sheets advance is called **extrusive advance**.

Submarine salt sheets are usually influenced by sedimentation, and many cases have been found where the

(a) Extrusive advance

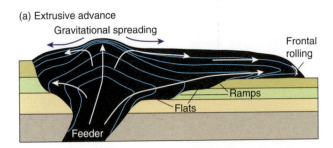

(b) Open-toed advance

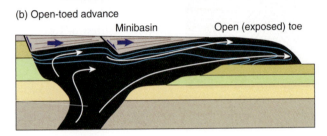

(c) Thrust advance

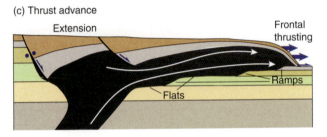

(d) Salt-wing intrusion

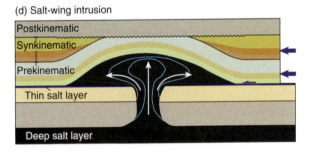

Figure 19.19 Schematic development of different allochthonous salt structures (salt sheets). Based on Hudec and Jackson (2006).

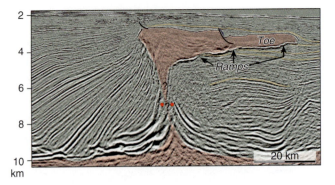

Figure 19.20 Classic salt sheet structure from the Gulf of Mexico (Mississippi Canyon). Image provided courtesy of TSG-NOPEC.

central or rear part of the salt sheet is covered by sediments, while the frontal part or toe remains exposed. Such sheets advance by extrusion through the exposed toe, and the mechanism is referred to as **open-toed advance**. Open-toed advance leads to lateral growth in a preferred direction, thereby accentuating the asymmetry of the structure. While extrusive advance is driven by salt flow up the feeder, open-toed advance involves flow resulting from sediments that accumulate on the top of the salt sheet and load the salt as the minibasin subsides into the salt (Figure 19.19b). Normal faults, as shown in Figure 19.19b and 19.20, indicate salt flows toward the toe and that there is frictional drag between the salt and the

base of the minibasin. Eventually the minibasin can reach the base of the salt layer, forming a weld that connects the feeder and the salt sheet. In this case the salt supply from the feeder is discontinued, and continued extrusion of the isolated allochthonous salt occurs by sedimentary loading or by tectonic translation if the weld starts acting as a décollement (thrust). Eventually salt sheets that spread by open-toed advance can be completely buried, as is the case with the salt sheet shown in Figure 19.20. Later erosion can sometimes reinitiate flow in such structures.

Some completely buried salt sheets advance by thrusting, a mechanism known as **thrust advance**. In this case a thrust is located in the toe zone (Figure 19.19c). Tectonically driven thrusts where salt sheets are located in the hanging wall imply the presence of a weak layer at the base of the salt or in the uppermost part of the underlying sediments, which eliminates the frictional resistance against flow that occurs during extrusive and open-toed advance. The ramps and flats observed along such thrusts are related to thrust propagation (Chapter 16) rather than variations in sedimentation rates. The ramp-flat-ramp geometries are very similar, and it may be difficult to distinguish between the basal geometries formed during extrusive and open-toed sheets on one side and thrusts on the other.

In some cases several salt sheets form in the same region, and if they are close enough the salt can coalesce to form a continuous unit, or **salt canopy**, that can cover a large area. Once formed, the canopy can act as any other salt layer, with the creation of secondary diapirs piercing even higher stratigraphic units.

Salt intrusions

The above types of salt sheets form at or near the surface. However, it is possible, although not very common, for

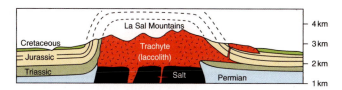

Figure 19.21 The rare occurrence of a salt diapir (salt wall) intruded by magma. The intrusion is a laccolith structure similar to salt wing intrusions described in the text. From the Moab area, Utah. Based on Doelling (2001).

salt to intrude into a stratigraphic level at depth. Such structures are called **salt (wing) intrusions** and are known from the southern North Sea, where the main salt layer (Permian Zechstein salt) has intruded into a higher and relatively thin Triassic evaporite layer. The resulting structures resemble magmatic laccolith intrusions (Figure 19.21). It appears that the presence of a higher weak evaporite layer into which the salt can intrude is required for its formation, and it also relies on regional contraction that can pressurize the main salt and form décollement folds above the upper evaporite layer. Décollement folding creates space into which salt can flow. Usually salt flows laterally, but in the case of salt intrusions it flows up-section along faults.

19.7 Diapirism in strike-slip settings

Any faulting or fracturing weakens the overburden and eases the formation of salt diapirs. However, it is in the releasing (extensional) stepovers in strike-slip systems that reactive salt diapirs tend to form. In these locations the overburden is weakened by extensive faulting and fracturing, and space is accommodated by horizontal extension. Salt diapirs in releasing stepovers develop much as described for regional extension (above): they may rise and subsequently collapse if the rate of salt flow into the diapir is slow relative to the extension rate. In restraining (contractional) stepovers or bends salt may act as local décollements along which thrusts localize.

Releasing bends are ideal locations for reactive diapirism, and the locations where one would expect diapirs to form along strike-slip faults.

19.8 Salt collapse by karstification

Salt structures whose tops are at or close to the surface are exposed to dissolution by meteoric water. This results not only in the formation of a cap rock of insoluble minerals

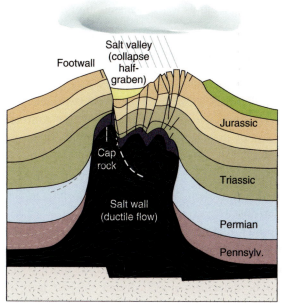

Figure 19.22 Generalized section across the salt walls of the Paradox Basin in southeast Utah. The collapse structure at the top of the salt is in part generated by reactivation of a regional system of Tertiary joints during salt dissolution, although most of the faulting is related to cataclastic deformation bands that must have formed earlier and at greater depths.

and rock layers, but can also lead to karstification, similar to karst formation in limestones and marbles. In both cases, karstification can lead to gravity collapse of overlying rocks and sediments, but the solubility of salt is higher (∼360 g/l) and thus the potential of gravity collapse greater.

Collapse of karst structures can generate fault systems above the salt structure long after salt movement comes to a halt. Such fault structures can be difficult to distinguish from crestal grabens and circular fault systems that form during extensional collapse of a salt structure during regional extension, and there can be an effect of both at the same time. The salt ridges in the Paradox Basin of southeast Utah are one of several places where salt dissolution has been interpreted to have caused relatively recent collapse of salt walls, probably by reactivation of already established faults at the crest (Figure 19.22). The presence of a cap rock at the top of the salt, i.e. a residual rock consisting mostly of gypsum and clay, shows that salt dissolution has occurred. These salt structures were exposed relatively recently as a result of the late Cenozoic and Quaternary uplift of the Colorado Plateau. Collapse of the upper part of salt structures may be related to subsurface dissolution, resulting in reactivation of existing faults as well as generation of new ones.

19.9 Salt décollements

One of the most important roles of salt is to act as a mechanically weak décollement, be it in extensional settings such as passive continental margins, or contractional settings such as orogenic wedges (fold-and-thrust belts). While salt structures such as diapirs only form when a thick salt layer is present, décollements can form in meter-thick or even thinner salt layers as long as the salt is laterally extensive.

The presence of salt can make a great difference in the way strain is accommodated and distributed. In general, the weak and viscous nature of salt enables much wider areas to be deformed, provided that the salt layer is extensive. It also allows for, and usually causes, **decoupling** of the substrate and the overburden. This means that the style of deformation in the overburden is different than that of the substrate. Décollement folding, as discussed in Chapter 16, is one example relevant to salt décollements. Completely different fault patterns above and underneath the salt layer are also typical, because faults do not extend through salt layers unless the layer is thin or the faults are very large.

In orogenic wedges salt acts as décollements, separating strongly imbricated and folded rock units from less deformed or undeformed basement. Hence, salt promotes thin-skinned tectonics in collision zones such as the foreland portions of the Alps and the Pyrenees (Figure 19.23) and in the Zagros fold-and-thrust-belt of Iran (Figure 20.3). Salt units in such settings localize tectonic strain and in this sense preserve adjacent (overlying) rock layers from strain, even though they are transported tens or hundreds of kilometers. Another consequence of the low friction of salt décollements is that the above-lying wedge of deformed rocks becomes long and thin in cross-section,

and thus extends farther into the foreland than would be the case without a salt décollement. Thus, salt décollements can make orogenic belts wider.

Salt layers in orogenic wedges easily form décollements that generate a belt of thin-skinned deformation and lower the slope of the orogenic wedge.

The role of salt in orogenic wedges is most pronounced in the brittle part of the crust, where the rheologic and mechanical contrast between salt and its surroundings is large. Deeper down, other rock units weaken as plastic deformation mechanisms become more generally activated, and at metamorphic depths thin limestone or phyllite units may play some of the same role that salt does at shallower depths.

Salt décollements may also form in response to gravity. The most extreme expressions of salt décollement deformation occur on some passive continental margins, such as offshore west Africa and Brazil. In such areas, ocean-dipping salt layers allow for gravity-driven sliding of the overburden. The dip is not impressive, perhaps only a few degrees, but sufficient due to the low strength of salt. In addition there is commonly loading of the up-dip section because of its proximity to river delta systems. The overall strain pattern is the same as for any gravity sliding process: extension in the rear and contraction in the front (Figures 19.24 and 19.25, also see Figure 17.12). Small-scale analogs can be found in several every-day situations, such as the gravity sliding of snow covering the roof of a heated house. The heat from the house generates a thin water film on which the snow slowly slides down-dip, causing extension in the upper part and contraction (folding) in the lower part (Figure 19.26).

In the upper part of continental margin salt décollements, extreme extension can establish and pull apart fault blocks above the salt. So-called **rafts** form where extensional faults completely disconnect the fault blocks

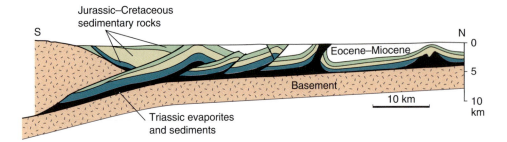

Figure 19.23 Example of the role of salt as a décollement zone decoupling the basement from its contracted sedimentary cover in the foreland region of the Pyrenean orogenic wedge. Modified from Hayward and Graham (1989).

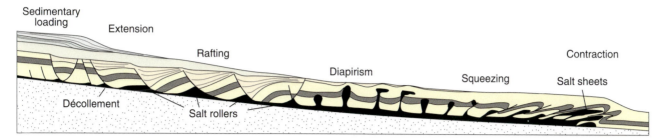

Figure 19.24 Schematic illustration of structures commonly found on passive continental margins with salt detachment.

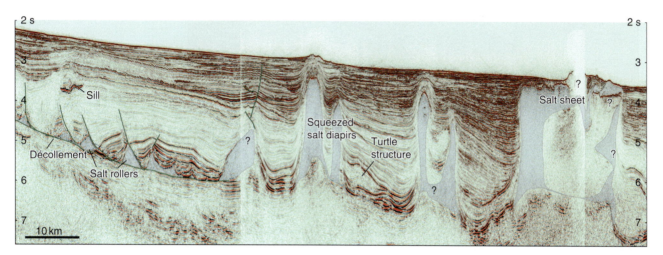

Figure 19.25 Seismic line from the Espirito Santo Basin off the coast of Brazil, showing extensional faults and rafts in the upper (left) part, salt diapirs in the central part and allochthonous salt in the right-hand, down-dip part. A décollement floors the structures, particularly in the up-dip part of the section. Image provided courtesy of CGGVeritas.

Figure 19.26 The slow movement of snow down a slippery roof creates extensional structures in the upper part and, if pinned, contractional structures at the bottom. In this case the windows and fence cause the pinning. View from my office window in Bergen.

above the salt décollement (Figures 19.24 and 19.25). Rafting in the sense that fault blocks ride as separated blocks on a salt layer is well described from the passive margin of Angola.

At some point down the décollement, extension gives way to contraction, and contractional structures occur. This occurs because the slope becomes shallower and the décollement flattens out, and also because salt layers on continental margins terminate at some point oceanward. Figure 19.27 illustrates an example where salt termination causes pinning of the décollement in the left side of the section and contractional structures in the hanging wall. Friction increases dramatically at the termination point, and sediments that come sliding down the décollement accumulate or pile up through the formation of impressive contractional structures. These structures include décollement folds, squeezed salt diapirs, and thrust or reverse faults (Figures 19.25 and 19.27). Salt sheets, which commonly form in contraction, are typically found in the lower and contractional part of salt décollements, and if enough salt is present, salt canopies may form.

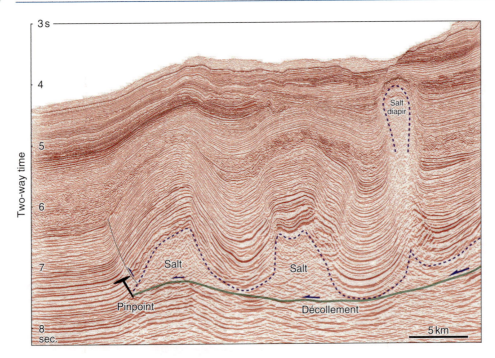

Figure 19.27 The down-dip termination of salt on the West African passive margin west of Senegal. The termination point pins the décollement, and the result is contraction (folding) of the overburden near the toe. Image provided courtesy of Ophir Energy, Rocksource ASA and AGC.

Summary

Salt is a special type of rock that deforms viscously by means of plastic deformation mechanisms under most geologic conditions. A buried salt layer can be considered as a fluid, overpressured by the weight of the overburden. The weak nature of salt causes it to localize regional deformation, its incompressibility causes flanking minibasins as surrounding sediments compact, and its low density makes it buoyant when buried beneath a critical depth. Sedimentation rates through time, sediment type, tectonic strain and salt flow rate are some of the many important variables that make every case of salt tectonics special. These are some important points and questions that relate to this chapter:

- Salt is commonly less dense than its overburden (shallow depth if covered by limestone, a couple of kilometers down if covered by siliciclastic sediments), but weakening of its roof is usually required to initiate diapirism.

- Many salt diapirs initiate due to differential loading or in response to localized faulting of the overburden.

- Diapirs are amplified and modified by passive growth, or by regional extension or contraction.

- High extension rates can widen and lower diapirs so that their upper part collapses with the formation of a capping minibasin.

- Contraction can squeeze diapirs, sometimes to the surface where they can extrude to form salt sheets and salt glaciers.

Review questions

1. Why does diapirism not initiate as a result of density inversion alone?

2. What is meant by reactive diapirism and in what tectonic regime(s) can it occur?

3. What is it about classic centrifuge models that in some way makes them unrealistic?

4. What is the difference between active and passive diapirism and how can we distinguish between them?

5. What is meant by the expression "downbuilding"?

6. Why do we not see diapirs in the upper part of a gravity-driven décollement setting such as shown in Figure 19.24?

7. What determines whether a salt wall or salt diapir forms?

8. What is the difference between a salt sheet and a salt canopy?

9. What is the effect of a basal salt layer in an orogenic wedge?

10. Can you do a rough interpretation of Figure 1.6 and identify the stratigraphic level of the salt, pencil in salt structures and interpret the folds?

E-MODULE

The e-learning module called *Salt tectonics* is recommended for this chapter.

FURTHER READING

General

Hudec, M. R. and Jackson, J. A., 2007, Terra infirma: understanding salt tectonics. *Earth-Science Reviews* **82**, 1–28.

Diapirism

Stewart, S. A., 2006, Implications of passive salt diapir kinematics for reservoir segmentation by radial and concentric faults. *Marine and Petroleum Geology* **23**, 843–853.

Vendeville, B. C. and Jackson, M. P. A., 1992, The rise of diapirs during thin-skinned extension. *Marine and Petroleum Geology* **9**, 331–353.

Vendeville, B. C. and Jackson, M. P. A., 1992, The fall of diapirs during thin-skinned extension. *Marine and Petroleum Geology* **9**, 354–371.

Allochthonous salt sheets

Hudec, M. R. and Jackson, M. P. A., 2006, Advance of allochthonous salt sheets in passive margins and orogens.

American Association of Petroleum Geologists Bulletin **90**, 1535–1564.

Passive margins (salt décollement tectonics)

Brun, J. P. and Fort, X., 2004, Compressional salt tectonics (Angolan margin). *Tectonophysics* **382**, 129–150.

Jackson, J. and Hudec, M. R., 2005, Stratigraphic record of translation down ramps in a passive-margin salt detachment. *Journal of Structural Geology* **27**, 889–911.

Salt dissolution processes

Gutiérrez, F., 2004, Origin of the salt valleys in the Canyonlands section of the Colorado Plateau Evaporite: dissolution collapse versus tectonic subsidence. *Geomorphology* **57**, 423–435.

Chapter 20

Balancing and restoration

Restoring a geologic cross-section or map to its original, pre-deformational state is an important part of making a structural interpretation. We want to be able to restore our deformed section to a geologically feasible undeformed section. For simplicity, we usually assume that either length or area (or volume in three-dimensional analyses) is preserved. If preserved, the section is balanced, meaning that length, area or volume of the restored section "balances" that of the strained section (the interpretation). Such exercises were first performed in areas of contraction, and are now routinely applied to extensional areas. Balancing puts important constraints on geologic interpretations, although there is no guarantee that a balanced section is correct. In this chapter we look at the basic premises and methods for balancing and restoration, mostly in sections and map view, and point out some of their usefulness and shortcomings.

20.1 Basic concepts and definitions

The uncertainties contained in any geologic data set always give room for many different interpretations. The principle known as Occam's Razor suggests that we should favor simple explanations and models, and this also holds true in the business of balancing geologic sections: If we were to take into account every little detail we would meet technical difficulties or simply find ourselves using our time very inefficiently. Even though a simple approach to restoration may give obvious inconsistencies or errors, it may still give valuable information about the deformation in question.

Let us now discuss what balancing and restoration actually means. **Balancing** adjusts a geologic interpretation so that it not only seems geologically reasonable in its present state, but also is restorable to its pre-deformational state according to some assumptions about the deformation. Thus, balancing is a method that adds realism to our sections and maps. A balanced section must be **admissible**, meaning that the structures that it contains are geologically reasonable both with respect to each other and to the tectonic setting, and **restorable** (retrodeformable).

Restoration involves taking a section or map and working back in time to undeform or **retrodeform** it. In terms of the deformation theory discussed in Chapter 2, this is the same as applying the reciprocal or inverse deformation matrix $\mathbf{D}^{-1}$, except that the deformation does not have to be, and in general is not, a linear transformation. We must decide whether the deformation can be explained by rotation, translation, simple shear, flexural flow or some combination of these. By applying reciprocal versions of these deformations the deformed section or map should be restored. Realistic restoration requires compatibility between different elements of the section, notably that layers within the section remain coherent. This means that the restored section has no or a minimum of overlaps and gaps, that fault offsets (except for those from earlier phases) are removed and that sedimentary layers are unfolded and rotated to planar and horizontal layers. Hence, when restoring a previously undeformed stratigraphic sequence:

We do not want to see overlaps, gaps, fault offsets, curved layers and non-horizontal layers in the restored state.

In practice we will not be able to perfectly restore sections or maps, at least not if the deformation is complex, but these are our basic goals during restoration. In other words, we want our restored section or map to look as realistic and likely as possible. Only when we have demonstrated that a geologic section can be restored to a reasonable pre-deformational state does the section balance.

A geologic section is not proven balanced until an acceptable restored version is presented.

Technically, we are not necessarily concerned with the deformation history or sequence of restorational actions; only the deformed and undeformed stages are compared. We can thus isolate different components of the deformation, such as rigid rotation, fault offset (block translations) and internal deformation of fault blocks (ductile strain, also called distortion). It is also possible to start with an undeformed model and deform it until reaching something that looks like the interpreted section. This is not called balancing, but **forward modeling**.

There are several reasons why balancing and restoration are increasingly used. They help ensure that the interpretation is realistic and provide support for strain estimates, for example by determining the amount of extension or shortening along a cross-section. In the 1960s, Clarence Dahlstrom and others applied this tool to reconstruct sections across the Canadian Rocky Mountains of Alberta prior to contraction, and calculated the amount of shortening involved. Later the same principles were used in areas of extension, such as the Basin and Range province and the North Sea basin. The Scottish geologist Alan Gibbs was one of the first to apply the principles of restoration to cross-sections from the North Sea rift.

A balanced section or map is not necessarily correct, but is likely to be more correct than a section that cannot be balanced.

Although the balancing of sections is most common, restoration and balancing can be done in one, two and three dimensions. One-dimensional restoration is known as line restoration, two-dimensional restoration is most commonly applied to cross-sections but can also be applied to maps, while three-dimensional restoration and balancing takes into account possible movements in all three dimensions.

20.2 Restoration of geologic sections

The simplest form of restoration is to reconstruct how a mapped straight line was oriented and located before the deformation started. Such simple one-dimensional

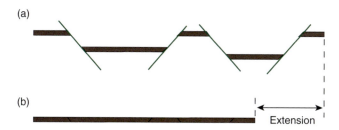

Figure 20.1 The concept of one-dimensional restoration where the marker is horizontal. In this case the line segments can be moved along the fault traces until they form a continuous layer. The extension is found by comparing the undeformed and the deformed states.

considerations are typically done to stratigraphic markers observed or interpreted in cross-section, and the principle is illustrated in Figure 20.1 for a horizontal marker horizon. Taking into account fault geometry or two or more stratigraphic markers turns section balancing into two dimensions.

Systematic section balancing was first applied in the Canadian Rockies in Alberta in the 1950s and 1960s, where petroleum-oriented structural geologists balanced thrusted and folded stratigraphic markers and reconstructed their pre-deformational lengths. The basic premises were that bed length and bed thickness measured normal to bedding remain constant. This works well for the Alberta sections, because much of the deformation in this area is localized to weak shale layers. This localization caused detachment faulting and folding by the flexural slip mechanism, which preserves layer thickness. The stratigraphic markers were mapped through traditional fieldwork, while modern mapping also relies on remote data, including seismic line interpretation with supporting well data. One-dimensional restoration can be done in any direction, and the outcome is the amount of extension or contraction in that direction. In general section restoration, however, the principal premises are that we study sections that contain the displacement vector or the maximum and/or minimum principal axes of the strain ellipsoid, and that the strain is plane (Box 20.1). These premises are sufficiently fulfilled in many (but far from all) contractional and extensional regimes that line balancing becomes meaningful.

Section balancing generally requires plane strain and orientation in the main displacement direction.

If strain is not plane, so that there is a strain component perpendicular to the section studied, material

---

BOX 20.1 | CONDITIONS FOR BALANCING OF CROSS-SECTIONS

- Geologically sound interpretation.
- Plane strain deformation.
- The section must contain the tectonic transport direction.
- The choices of deformation (vertical shear, rigid rotation etc.) must be reasonable and based on the general knowledge of deformation in the given tectonic setting.
- The result must be geologically reasonable, based on independent observations and experience.

---

will move into and out of the section, implying that lengths and areas will change. It is possible to account for such strain if the bulk strain can be estimated, but then we are into the more complex case of three-dimensional restoration. In this section we will restrict ourselves to plane strain cases. An important implication of this is that if we use one-dimensional restoration in a region of non-plane strain, the resulting interpretation will be incorrect.

Rigid block restoration

The simplest case of section restoration is where fault blocks behave as rigid blocks during deformation, so that only rigid rotation and translation is involved. Previously undeformed sedimentary layers then end up as straight lines in the section (planar in three dimensions), while fault traces can be curved. The domino system in Figure 20.2a is an example where both offset (translation) and rotation (anticlockwise) are involved. Each block can be restored by rotating the section so that the layers become horizontal, before removing the fault offsets. This can readily be done by means of a pair of scissors, or by using a computer drawing program. Before starting, pin one of the two ends and use that **pinpoint** as the point of reference. The outcome is not only the extension, but also the initial dip of the faults and the amount of block rotation. This is a simple example of constant length restoration. When using two or more horizons we see that rigid block restoration also preserves area.

Constant length and area

Assuming that the marker is restored to its initial length means that it has been extended or shortened only through the discrete formation of observable separations or overlaps. This is the basis for **constant length restoration** (or constant length balancing), which is commonly and conveniently applied during section restoration, very useful when a quick restoration is needed. It is frequently used to restore fold-and-thrust belts, not only in the Canadian Rocky Mountains, but in almost any region dominated by thin-skinned tectonics. Figure 20.3 shows an example from the Zagros fold-and-thrust belt of Iran, which is one of several areas controlled by a weak

salt décollement, and therefore well-suited for constant length restoration.

Constant length restoration works well when our layers are straight and horizontal (Figures 20.1 and 20.2a). In many cases deformed layers are curved, and the concept of constant length becomes questionable.

When deformed line segments are curved, we have a component of ductile strain, and rigid body rotation cannot restore the line.

It is still possible that the lengths of the lines are preserved during folding – it all depends on the way that strain accumulates. If the line lengths should change, area may still be preserved, and in that sense **constant area restoration** can be said to be somewhat more robust and applicable. Area balancing makes sense: if we shrink a section in one direction we somehow have to increase it by the same amount in the other direction. Using Figure 20.4b as an example, areas A and B have to be equal. This is true even if layers change length and thickness, as long as there is no compaction and no movement in or out of our section. Area balancing also has implications for depth of detachment estimates. For example, the area C in Figure 20.4c equals area D in the same figure. If we know the extension (horizontal displacement of the hanging wall) we can easily estimate the depth at which the listric fault flattens out. The same is the case for the fault-propagation fold shown in Figure 20.4b.

Flexural slip

Preservation of line length and bed thickness makes restoration simple. Flexural slip preserves both, as folding of layers by flexural slip implies slip parallel to bedding only.

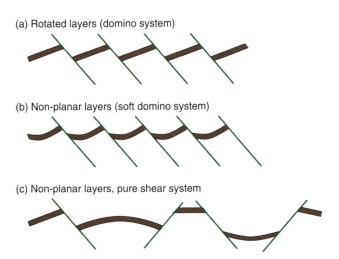

(a) Rotated layers (domino system)

(b) Non-planar layers (soft domino system)

(c) Non-planar layers, pure shear system

Figure 20.2 (a) Rotated layers can be restored by rigid rotation and removal of fault displacement if layers are unfolded. (b) and (c) Folded layers must be restored by a penetrative (ductile) deformation such as vertical or inclined shear.

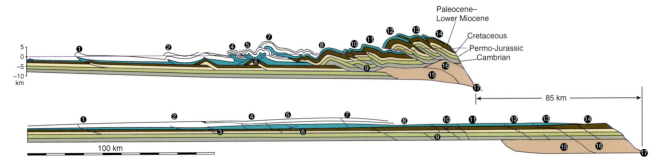

Figure 20.3 Cross-section through the Zagros fold-and-thrust belt. The section was balanced using the so-called sinuous bed method, which involves measuring the lengths of the top and bottom of each formation between faults and matching ramp and flat lengths on a restored section with those on the deformed section while maintaining bed thickness (constant area). Numbers are added to help correlate the two sections. Modified from McQuarrie (2004).

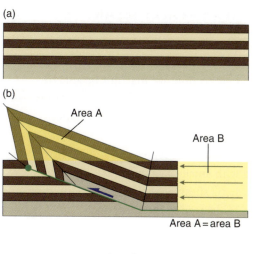

(a)

(b)

Area A

Area B

Area A = area B

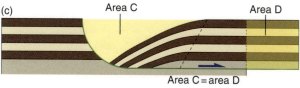

(c)

Area C

Area D

Area C = area D

Figure 20.4 Constant area balancing of sections in the transport direction. (a) Undeformed section, (b) fault-propagation fold, (c) listric extensional fault. Note that bed length changes if internal strain is shear but not if the mechanism is flexural slip or flexural shear.

If the strain is distributed we have flexural shear, still with the shear plane along bedding.

In a contractional regime layer-parallel simple shear is very common, as slip or shear easily localize to mechanically weak layer interfaces. Even if layers start to fold, layer-parallel slip or shear can create flexural slip or flexural shear folds – i.e. fold mechanisms that preserve both layer length and area.

Flexural slip and flexural shear preserve layer length and bed thickness, and therefore also area.

Flexural slip is convenient when modeling or balancing contractional structures such as fault-bend folds and fault-propagation folds. Fault-bend folds can be explored by moving a paperback book or pile of papers over a ramp. Drawing squares with circles on the side of the paper pile shows how strain accumulates without changing the length of the papers.

Restoration can be done by means of a ruler or thread by measuring the length of each marker in the deformed section. The markers are moved to their original cut-off points (locations where they were cut

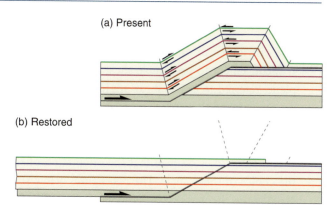

(a) Present

(b) Restored

Figure 20.5 Constant length and thickness was assumed during this restoration. Each hanging-wall line was straightened by rotation of individual segments and moved left to match the stratigraphy in the footwall, consistent with flexural slip restoration.

off at the ramp) and straightened while bed thickness is maintained, as done in Figure 20.5. If bed thickness changes in the deformed section, the flexural slip method alone will not give the correct restored section. A similar method can be used to restore flexural slip folds, such as décollement folds.

In extensional regimes, layer-parallel shear can also occur, but less commonly. This is so because slip most likely initiates at 20–30° to σ_1 (Chapter 7), which is vertical for extension and horizontal in the contractional regime. Another mechanism is therefore commonly used during restoration of extended sections, namely simple shear across the layering.

Shear

Distributed simple shear is a concept that implies that small-scale deformation can be treated as ductile deformation, where layer continuity is preserved. The actual small-scale deformation is not really important in this context, and on a seismic section it is simply referred to as subseismic (below the resolution of seismic reflection profiles) or ductile. The goal is to apply simple shear to account for the ductile deformation in the fault blocks, as exemplified in Box 20.2. As we know from Chapter 15, simple shear acting across layers rotates the layers and changes their length, and heterogeneous simple shear deflects or folds the layers. However, simple shear preserves area, so the assumption of constant area in extensional regimes is more realistic than that of preservation of bed length.

BOX 20.2 | WHAT DOES VERTICAL AND INCLINED SHEAR REALLY MEAN?

The distributed or ductile deformation of layers in a deformed section can be modeled in many ways. It is convenient to use simple shear, and the variables are shear strain and the inclination of the shear plane, referred to as the **shear angle**. At the scale of a seismic or geologic section, we may not be able to see the deformation structures that make a layer look non-planar. The structures may be subseismic faults, deformation bands, extension fractures or microscale reorganization structures. Hence, the deformation is ductile at our scale of observation, and the effects are modeled by simple shear.

It is sometimes claimed that the orientation of small faults in the hanging wall represents a guide to the choice of shear angle. The two figures below show hanging walls with small-scale faults that are mostly synthetic to the main fault. However, their arrangement calls for antithetic shear, as shown by yellow arrows. These examples illustrate the difficulty involved in using small faults to determine the shear angle.

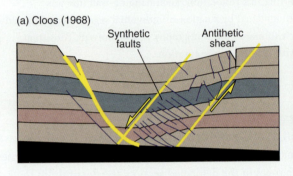

(a) Cloos (1968)

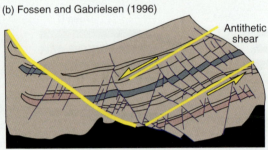

(b) Fossen and Gabrielsen (1996)

The classic use of (simple) shear in extensional deformation is hanging-wall deformation above non-planar normal faults, particularly listric extensional faults of the kind shown in Figure 20.6. Constant-area deformation of the hanging wall of a listric fault was first modeled by means of **vertical shear**. This technique is sometimes referred to as the **Chevron construction**, named for the oil company that first utilized the method (not to be confused with chevron folds mentioned in Chapter 11). Vertical shear involves no extension or shortening in the horizontal direction, but individual layers will be extended, rotated and thinned.

It was soon realized that hanging-wall shear deformation could deviate from vertical shear. Both **antithetic shear** (shear plane dipping against the main fault) and **synthetic shear** were evaluated. The differences between the two are illustrated in Figure 20.6, which shows that antithetic shear affects a larger part of the hanging wall and implies less fault offset than vertical shear. Choosing the right shear angle is not always easy, and we usually have to try a few different options to make a good choice. It seems that antithetic shear with a shear angle of around 60° works well in many deformed hanging walls above listric faults, while synthetic shear produces unrealistically

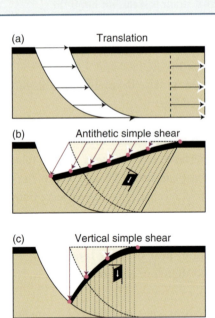
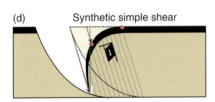

Figure 20.6 Deformation of the hanging wall above a listric fault. (a) Pure translation. (b–d) Antithetic, vertical and synthetic shear. Note the different hanging-wall geometries and the fact that the shear only affects the left part of the hanging wall.

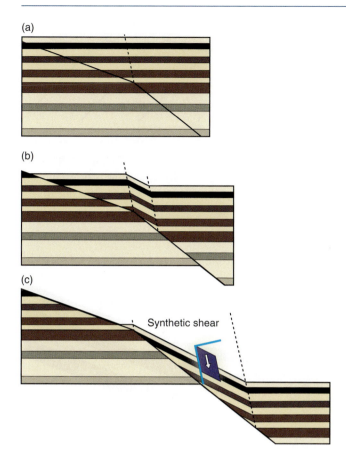

Figure 20.7 Local synthetic shear in the hanging wall to a normal fault. The simple shear is related to the downward steepening of the fault.

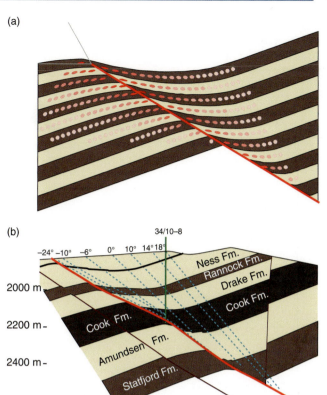

Figure 20.8 (a) Trishear applied to a gently dipping normal fault that has propagated up-section. A hanging-wall syncline forms. (b) Section from the Gullfaks Field, North Sea, constrained by seismic and well data. Dip isogons (blue) are based on dipmeter data and seismic interpretation. Note the geometric similarities.

steep hanging-wall layers in many cases (Figure 20.6d). Vertical shear may be more realistic when considering large parts of the crust.

Synthetic shear works in cases where faults steepen downward, either gradually (antilistric faults) or abruptly as in Figure 20.7. Synthetic shear also applies in some cases where a hanging-wall syncline is developed. However, in cases where the syncline widens upward, **trishear** (Box 8.3) may be a more realistic alternative. An example is shown in Figure 20.8, where forward trishear modeling (a) reproduces the present section (b) quite well.

Trishear has no fixed shear angle, but involves a mobile triangular deformation zone. The triangular zone is attached to the fault tip and represents a ductile process zone ahead of the fault. Trishear is an interesting model that explains local drag structures and folding of layers around faults, but is not applicable to entire regional sections, as is vertical or inclined shear. However, it is possible to combine general vertical shear throughout a cross-section and apply local trishear to specific faults within that section.

Other models

Other assumptions that have been used to model ductile deformation of the hanging wall of non-planar faults include **constant displacement** along the master fault and **constant fault heave**. We have already seen in Chapter 8 how displacement varies along faults, so although these assumptions may work geometrically, they are geologically unsound except for very special cases. However, if the displacement on a given fault is much longer than the section studied, the variation in displacement may be relatively small, and constant displacement may be an acceptable approximation. In general, the testing of different assumptions during section restoration provides alternative explanations, and their differences highlight the uncertainties involved.

The effect of compaction

Constant area implies that area is conserved even though the shape of a given area may change during deformation. However, compaction (vertical shortening) of sediments

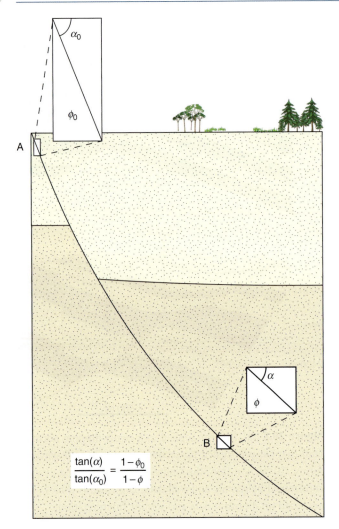

$$\frac{\tan(\alpha)}{\tan(\alpha_0)} = \frac{1-\phi_0}{1-\phi}$$

Figure 20.9 The effect of compaction on the geometry of synsedimentary faults (growth faults). The fault dip lowers with depth because of the downward-increasing compaction. Knowledge of the original porosity can be used to calculate the original fault dip at any point along the fault. ϕ_0 is typically around 0.4 (40%) for sandstone and 65–70% for clay. Note that there may be other reasons why some faults flatten with depth.

may occur during as well as after tectonic deformation, decreasing the area seen in vertical sections. The effect of compaction on faults and folds is significant where sedimentary sequences are deformed at shallow depths, prior to burial and lithification. Growth faults in delta settings are a realistic example, relevant to places such as the Gulf of Mexico. The primary effect is shown schematically in Figure 20.9, where the fault dip can be seen to decrease downward, solely as a result of compaction.

Compaction also has a geometric effect on newly formed faults in a sedimentary sequence. This is illustrated in Figure 20.10, where the upper parts of the initially straight faults become flattened and the faults slightly antilistric. This effect is due to differential compaction,

Figure 20.10 Posttectonic compaction is largest in the upper part of a sedimentary sequence. The reduction of the dip of the faults will therefore be greater in the shallow part. An initially straight fault will therefore flatten towards the top due to differential compaction.

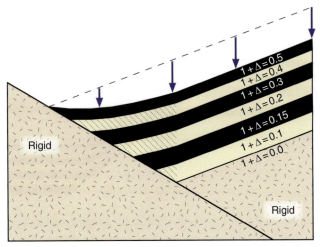

Figure 20.11 If the hanging wall compacts more than the footwall, a compaction syncline forms. In the example shown here the hanging wall is assumed to be rigid while hanging-wall layers compact from 50% in the topmost layer down to 10% in the lower layer above the rigid basement.

where shallow layers compact more than the already compacted deeper layers.

Another compaction-related effect relates to the fact that sand compacts less than mud and clay. Clay has an initial porosity close to 70%, while that of sand is around 40%. So, if a fault forms in a recently deposited sand–clay sequence, subsequent compaction will give the fault a lower dip in the more compacted clay layers than in sand layers.

There is also the effect of differential compaction on each side of a fault. If the offset on a fault becomes large, say some hundred meters or more, then the hanging-wall layers will compact more than those in the adjacent footwall, simply because the footwall layers have already been compacted to a larger extent or are crystalline rocks. For dipping extensional faults, the result is a hanging-wall syncline, as shown in Figure 20.11.

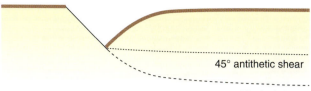

Figure 20.12 Example of a curved hanging-wall layer. The shape of the layer reflects the fault geometry at depth, but the choice of shear angle influences the estimated fault geometry. From White *et al.* (1986).

Most balancing programs utilize compaction curves for various lithologies that remove the effect of compaction during restoration.

Finding fault geometry from hanging-wall layers

At shallow levels in rift systems it is common to see reflectors in seismic sections offset by a fault, while the fault geometry at depth remains unclear. However, the direct relationship between fault geometry and layer geometry provides important insight into fault geometry at depth. Where the hanging-wall reflectors are curved, the fault is also likely to be curved. Hence, we can use the shape of the layers to calculate the shape of the fault and, in the case of a rollover structure, to estimate the depth to the detachment.

There is a direct relationship between fault geometry and hanging-wall strain for non-planar faults.

We do, however, need to choose a model for the hanging-wall strain, and the result will be different for synthetic, vertical and antithetic shear angles. Vertical shear gives a deeper detachment than antithetic shear, as illustrated in Figure 20.12, but the amount of extension involved will be greater for antithetic shear. If we have information from both hanging-wall and fault geometries, we can find the shear angle that best balances the section. For many cases, a 60° antithetic shear angle is found, supporting the antithetic shear model.

20.3 Restoration in map view

A mapped horizon that has been affected by extensional faulting is portrayed as a series of isolated fault blocks, by contractional faulting as overlapping fault blocks, or a combination of both. If we consider the case of extensional faulting shown in Figure 20.13a, the separation between the blocks reflects the amount of extension. The map pattern with all the different fault blocks is reminiscent of a jigsaw puzzle, and the restoration involves putting the pieces back to their pre-deformational positions (Figure 20.13b). The goal is to restore the puzzle to the state where the number of openings and overlaps is minimized, either manually or by computer. If the blocks involve no internal deformation and the layering is horizontal, reconstruction is, at least in principle, relatively simple.

For the common situation where layers are dipping, rigid-body rotation of the blocks may be necessary to obtain horizontal layering. If layers are non-planar, then the surface should be unfolded or additional errors will be introduced. The choice of strain model may not be easy – should we project the material points on the folded surface to the horizontal plane by means of oblique shear, vertical shear or some other transformation? If constant surface area is assumed, a flexural slip fold model is implied.

Even if such uncertainties and inaccuracies cannot be dealt with to the full extent, the outcome of map restoration is often quite useful. The displacement field emerges by connecting the locations of points before and after deformation, as shown in Figure 20.13c for our example. The orientation of displacement vectors allows the distinction between plane and non-plane strain (parallel displacement vectors versus diverging vectors, respectively) and an assessment of the influence of gravity spreading during deformation. Additionally, non-plane strain means that area will not be conserved in any cross-section through the deformed volume, and cross-sections cannot be correctly balanced: If we create an interpretation that does balance, it must be wrong! If the strain is plane, then sections chosen for cross-section balancing must be oriented parallel to the displacement vectors. Map restoration allows us to more accurately choose this direction.

Some important outcomes from map restoration include the following. Relative movement of points on each side of faults will give the local displacement vector of the fault (Figure 20.13d) or, if layers are not horizontal, the horizontal component of the displacement vectors. The nature of slip (dip-slip, oblique-slip, strike-slip etc.) on faults is thereby revealed and the number of rotation of blocks about the vertical axis also emerges from map restoration (Figure 20.13e). The amount of strain in any horizontal direction will also be apparent. Finally, the gap and/or overlap areas provides an idea about the consistency of the restoration and reliability of the interpretation.

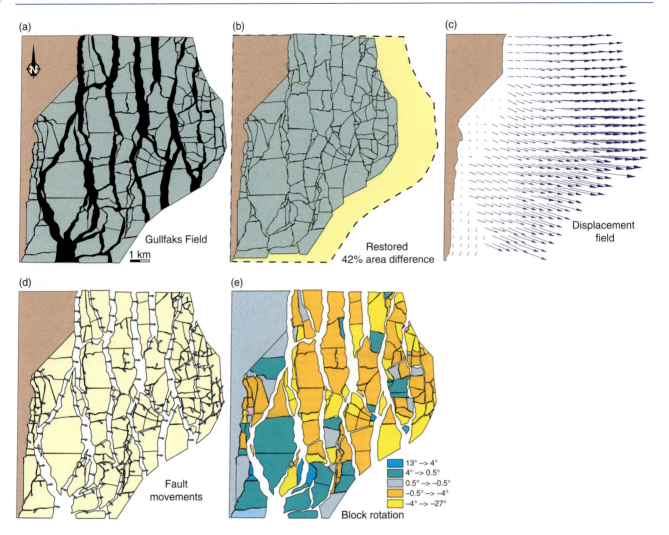

Figure 20.13 Balancing of the Statfjord Formation map of the Gullfaks Field. (a) Present situation. (b) Restored version. (c) Displacement field relative to westernmost block. (d) Fault slip. (e) Rotation about vertical axis. From Rouby *et al.* (1996).

20.4 Restoration in three dimensions

Map restoration is sometimes described as being three-dimensional. However, true three-dimensional restoration involves volume, and means simultaneous map-view and cross-sectional restoration. Fault blocks are considered as three-dimensional objects that can be moved around and deformed internally, but not independently of each other. Three-dimensional restoration is inherently complex, and an in-depth treatment is beyond the scope of this book.

During three-dimensional restoration each surface is mathematically treated as a mesh of triangles or other polygons that can be deformed. Using vertical shear, it is like holding a handful of pencils that connect the different surfaces, where each pencil is free to move slightly differently than its neighboring pencils. An advantage of three-dimensional restoration is the opportunity to account for non-plane strain. However, the many choices

involved and the time required to set up and run three-dimensional restoration models makes simpler forms of restoration more attractive. They may not be as accurate, but they may provide very important information within a reasonable time frame.

20.5 Backstripping

The restoration techniques discussed above are purely kinematic and do not take into consideration the elastic or isostatic response of the crust. This should be done when large-scale restoration is performed. **Backstripping** is a kind of isostatic restoration where the focus is on the subsidence history of a basin by successively removing sedimentary sequences and balancing isostasy. This type of restoration is applied to rift basins to determine the magnitude of lithospheric extension during rifting by estimating the post-rift subsidence rate. The procedure

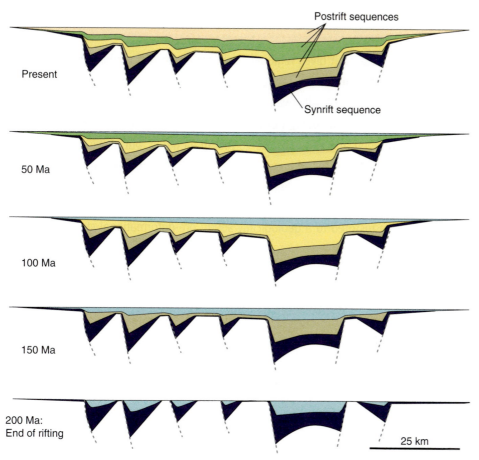

Figure 20.14 Schematic illustration of backstripping of a section through a rift, taking into consideration flexural response (isostasy), compaction and thermal subsidence. Sedimentary units are successively removed to the base of the postrift succession, and the sections can be tested against paleobathymetric markers, such as eroded fault crests. Based on Roberts *et al.* (1993).

involves stepwise removal of gradually older stratigraphy, correcting for compactional effects that can be calculated from established compaction–depth curves, and adjusting for subsidence caused by sediment loading. Just like the geometric balancing discussed elsewhere in this chapter, it can be done in one, two or, less commonly, three dimensions. One-dimensional backstripping assumes Airy isostasy while two- and three-dimensional backstripping relies on flexural isostasy, where isostatic compensation takes into account lateral variations in loading. Paleobathymetric estimates are generally needed to constrain previous stages of basin bathymetry.

We will not go into backstripping in any detail in this book (see suggested reading below for a more detailed treatment), but in general terms, modeling a cross-section in a rift setting involves:

Removal of the water layer and computation of the flexural isostatic response.

Removal of the youngest stratigraphic unit.

Decompaction of the remaining stratigraphy.

Calculation of the flexural isostatic response to removing the sediment load.

Adding thermal uplift from an estimate of β and rift age.

This procedure is repeated for every stratigraphic unit to the base of the postrift sequence, producing a series of restored cross-sections. A schematic illustration is shown in Figure 20.14.

Summary

Restoration of sections and maps can be useful because it gives regional strain estimates and explores how well different strain mechanisms can explain observed structures. If a section does not restore, both the interpretation and the assumptions used during restoration should be critically evaluated. If a section or map is fully restored to a geologically viable state the interpretation is balanced and sound, but not necessarily correct. Restoration is always compromising the complexities of natural deformation and will never be correct in all detail, but is still quite useful as long as this is kept in mind. Restoration can be done by means of paper and scissors, a drawing program on a computer or special programs designed especially for restoration purposes. Important points to remember:

- A balanced interpretation is one that has been restored to a geologically sound pre-deformational state.

- A sound restored (undeformed) state generally implies horizontal sedimentary layers and no overlaps or open spaces between fault blocks.

- If basic assumptions made during restoration are sound, a balanced geologic interpretation is considered to be a more likely interpretation.

- There is always more than one restorable interpretation.

- The conditions and assumptions used during restoration must be critically evaluated, using all available information and experience.

Review questions

1. What are the two most basic conditions that must be fulfilled for section balancing to make sense?

2. What is the difference between restoration and forward modeling?

3. What is meant by ductile strain in restoration?

4. What is the most common model for ductile strain in section restoration?

5. How could we restore a folded layer?

6. What information can map-view restoration give?

E-MODULE

 The e-learning module called *Balancing and restoration* is recommended for this chapter.

FURTHER READING

General

Rowland, S. M. and Duebendorfer, E. M., 2007, *Structural Analysis and Synthesis*. Oxford: Blackwell Science.

Contraction

Dahlstrom, C. D. A., 1969, Balanced cross sections. *Canadian Journal of Earth Sciences* **6**: 743–757.

Hossack, J. R., 1979, The use of balanced cross-sections in the calculation of orogenic contraction: a review. *Journal of the Geological Society* **136**: 705–711.

Mount, V. S., Suppe, J. and Hook, S. C., 1990, A forward modeling strategy for balancing cross sections. *American Association of Petroleum Geologists Bulletin* **74**: 521–531.

Suppe, J. 1983, Geometry and kinematics of fault-bend folding. *American Journal of Science* **283**, 684–721.

Woodward, N. B., Gray, D. R. and Spears, D. B., 1986, Including strain data in balanced cross-sections. *Journal of Structural Geology* **8**: 313–324.

Woodward, N. B., Boyer, S. E. and Suppe, J., 1989, *Balanced Geological Cross-sections: An Essential Technique in Geological Research and Exploration*. American Geophysical Union Short Course in Geology, **6**.

Extension

de Matos, R. M. D., 1993, Geometry of the hanging wall above a system of listric normal faults – a numerical solution. *American Association of Petroleum Geologists Bulletin* **77**: 1839–1859.

Gibbs, A. D., 1983, Balanced cross-section construction from seismic sections in areas of extensional tectonics. *Journal of Structural Geology* **5**: 153–160.

Morris, A. P. and Ferrill, D. A., 1999, Constant-thickness deformation above curved normal faults. *Journal of Structural Geology* **21**: 67–83.

Nunns, A., 1991, Structural restoration of seismic and geologic sections in extensional regimes. *American Association of Petroleum Geologists Bulletin* **75**: 278–297.

Westaway, R. and Kusznir, N., 1993, Fault and bed "rotation" during continental extension: block rotation or vertical shear? *Journal of Structural Geology* **15**: 753–770.

Withjack, M. O. and Peterson, E. T., 1993, Prediction of normal-fault geometries – a sensitivity analysis. *American Association of Petroleum Geologists Bulletin* **77**: 1860–1873.

Salt structures

Hossack, J., 1993, Geometric rules of section balancing for salt structures. In M. P. A. Jackson, D. G. Roberts and S. Snelson (Eds.), *Salt Tectonics: A Global Perspective*. Memoir **65**. American Association of Petroleum Geologists, pp. 29–40.

Rowan, M. G., 1993, A systematic technique for sequential restoration of salt structures. *Tectonophysics* **228**: 331–348.

Map and three-dimensional restoration

Rouby, D., Fossen, H. and Cobbold, P., 1996, Extension, displacement, and block rotation in the larger Gullfaks area, northern North Sea: determined from map view restoration. *American Association of Petroleum Geologists Bulletin* **80**: 875–890.

Rouby, D., Xiao, H. and Suppe, J., 2000, 3-D Restoration of complexly folded and faulted surfaces using multiple unfolding mechanisms. *American Association of Petroleum Geologists Bulletin* **84**: 805–829.

Backstripping

Roberts, A. M., Yielding, G. and Badley, M. E., 1993, Tectonic and bathymetric controls on stratigraphic sequences within evolving half-graben. In G. D. Williams and A. Dobb (Eds.), *Tectonics and Seismic Sequence Stratigraphy*. Special Publication **71**, London: Geological Society, pp. 81–121.

Watts, A. B., 2001, *Isostasy and Flexure of the Lithosphere*. Cambridge: Cambridge University Press.

Chapter 21

A glimpse of a larger picture

As structural geologists we need to make objective observations and analyses based on our knowledge of structural geology. We then have to put our local observations together to come up with or evaluate a larger, regional model. We have to put our observations together in time, to construct a history of deformation, or perhaps come up with a model that incorporates sedimentary information, intrusive relations or metamorphic data. Combining structural observations with other information is always necessary, and in this chapter we will have a very brief look at a few relevant examples, particularly the separation of deformation into phases, metamorphic petrology, P–T–t paths and depositional patterns. The treatment will be brief, and is meant to point out some important principles and directions rather than discussing examples and methods in detail.

21.1 Synthesizing

In this book we have looked at different classes of structures, such as faults, folds and foliations, in separate chapters. At this point we should have a basic knowledge of individual types of structures stored in our minds, ready to be activated when needed. We should be able to identify domino fault blocks, cataclastic deformation bands, subgrains, deformation bands, similar folds and many other structures discussed in this book. When examining an outcrop or a seismic section, our brain automatically searches for familiar geometries and patterns. By combining observations from many outcrops, we pose the following kinds of questions: Do observed folds form a consistent pattern with respect to style and vergence? Are the folds observed in an area or outcrop kinematically consistent so that they can be explained by means of a single deformation phase, or do we need to call for a more complicated deformation history? If faults affect a folded layer, such as in the two cases portrayed in Figure 21.1, did the folding and faulting occur during the same phase of deformation, or do we have a sequence of structures that formed at different times under different stress fields and/or physical conditions? We may have to measure up the various structures and plot the orientation data in stereonets in search of an answer. Perhaps strain analysis and kinematic analysis are needed

(a)

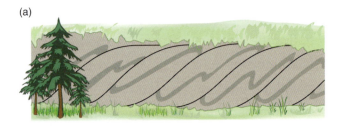

(b)

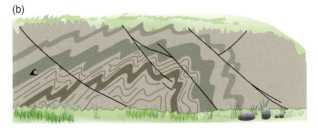

Figure 21.1 Illustration of faults and folds in outcrop. In (a), the fold pattern systematically fits that of the reverse faults, making it natural to interpret folds and faults as forming during the same deformation phase. In (b) there is no kinematic connection between the normal faults and the folds, and we interpret this as an expression of two deformation phases.

too. This may still not be enough, and we may find that we need to bring in information from related disciplines, such as metamorphic petrology, geochronology and stratigraphy.

21.2 Deformation phases

Structural geologists commonly search for evidence of two or more **deformation phases**. This was particularly the focus during the 1960s to 1980s, commonly together with metamorphic petrology and radiometric dating, and is also of fundamental importance to structural geologists of today. There are several ways to define a deformation phase, and here is one definition:

A deformation phase is a time period during which structures formed continuously within a region, with a common expression that can be linked to a particular stress or strain field or kinematic pattern.

Clearly the P–T conditions, stress field and kinematics may change continuously during deformation, in which case the concept of discrete deformation phases is less appropriate, but in many cases the concept of deformation phases is useful. Structures formed during a single deformation phase are expected to share certain features. For example, folds formed during a single phase may show a consistent vergence, and they may develop a cleavage with a mineral assemblage consistent with a certain metamorphic condition. In the brittle regime, extension fractures formed during a single phase would be expected to show the same orientation (except where influenced by local stress anomalies) and to be healed by the same minerals. Fault kinematics was discussed in Chapter 9, and a consistent pattern is expected for faults and fractures formed in the same regional stress field.

A very important principle when considering the deformational history of a region is that of **overprinting relations**. Overprinting relations implies that two or more structures are found in the same outcrop or sample so that their relative age can be determined. A brittle normal fault offsetting a plastic shear zone is good evidence for two phases of deformation. In the ductile regime, fold interference patterns (Chapter 11) can reveal characteristics of the two different phases of deformation.

Structural style and orientation is not necessarily such a good criterion, because the style of structures is influenced by local rheologic variations and properties as well as variations in strain magnitude. Hence, open and tight folds may form during the same phase, and concentric décollement folds could form above a detachment while tight shear folds could form within the detachment itself. Furthermore, extension fractures could develop in a competent limestone bed as deformation bands form in adjacent sandstone and slip surfaces in adjacent shale layers. Hence, localities that show overprinting relations are invaluable in determining relative age.

Sorting out the age relations between structures is not always easy, even if the structures all appear in the same outcrop or area. Faults of different orientations may well form at the same time, for instance the faults shown in Figure 8.4.

21.3 Progressive deformation

Many rocks have undergone two or more phases of deformation, some that can be related to tectonic events such as collision of an island arc complex with a continental margin or a continent–continent collision. Such tectonic events can sometimes produce several phases each. Deformation that represents the combined effect of two or more phases is called **polyphasal**. From about 1960 through the 1980s the concept of deformation phases was commonly applied at outcrop scale or even hand-sample scale. Most modern geologists would say that a deformation phase should be of more regional character than that, requiring the same relation to be found in a number of outcrops within a region. There is no clear definition as to how regional a set of structures must be before we can speak about a separate deformation phase. For example, a bend in a strike-slip fault could produce a set of deformation structures in the bend area that overprint earlier structures (Figure 18.13). How large should the bend be before we apply the term separate deformation phase to this set of structures? Probably quite large (a square kilometer or more?), but as often the case in structural geology, the term can be used in somewhat different ways.

Another relevant example where the term polyphasal deformation may be challenged is where folding and cleavage formation occur in shear zones and mylonite zones. Folds can form sporadically in active mylonite zones, depending on local geometric effects such as tectonic lenses. Quarter structures discussed in Chapter 15 (see Figure 15.33) are a good example, as are sheath-fold formations of the type portrayed in Figure 11.39. Folds and foliations that form sporadically in a shear zone during progressive shearing can produce refolded

folds and overprinting foliations that are not considered evidence of polyphasal deformation. Instead, we think of such structures as formed during **progressive deformation**. However, if the shear zone were inactive for a substantial time interval, and were then reactivated at a different crustal depth and/or stress regime, we would talk about a new phase of deformation.

Polyphasal deformation implies discrete deformation phases, while progressive deformation involves more continuous and gradual development at a local or regional scale.

Note that any deformation phase involves progressive deformation, since structures develop over time.

21.4 Metamorphic textures

Metamorphic conditions are important, as they strongly influence the type of structure that will form and the microscale deformation mechanisms and processes at work during deformation. **Prograde metamorphism** involves temperature–pressure increase, while **retrograde metamorphism** describes the opposite case. Metamorphic or P–T paths can be found by analyzing metamorphic assemblages, and in deformed rocks these assemblages can often be associated with deformation structures. Prograde metamorphic mineral parageneses are commonly overprinted by retrograde deformation structures and assemblages formed during exhumation.

Porphyroblasts

Sometimes metamorphic minerals grow to become crystals that are much larger than the general grain size of the rock. Such large metamorphic crystals are called **porphyroblasts** and are common in mica-bearing schists and gneisses. The porphyroblasts shown in Figure 21.2 show inclusions that outline an internal foliation, which is strongly curved in Figure 21.2a and much straighter in Figure 21.2b. Such internal foliations represent remnants of foliations at the time of growth, reflecting metamorphic conditions and fabric orientation and geometry at that time. Hence, porphyroblasts provide important small windows into earlier stages of the deformation history. Figure 21.3 illustrates how an early foliation can be preserved within porphyroblasts.

But what is the implication of a straight versus curved porphyroblast inclusion pattern? Porphyroblasts may grow prior to, during or after a given deformation phase,

(a) (b)

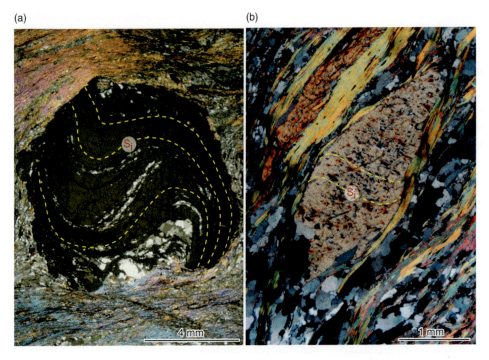

4 mm 1 mm

Figure 21.2 (a) Garnet porphyroblast in micaschist, showing sigmoidal pattern. The inclusion pattern is an earlier foliation now disconnected and at high angle to the external foliation. (b) Amphibole porphyroclast with (mostly) straight inclusion trails. Straight trails may indicate pre- or intertectonic growth, but the curvature close to the boundary may indicate that deformation initiated during the last part of the growth history.

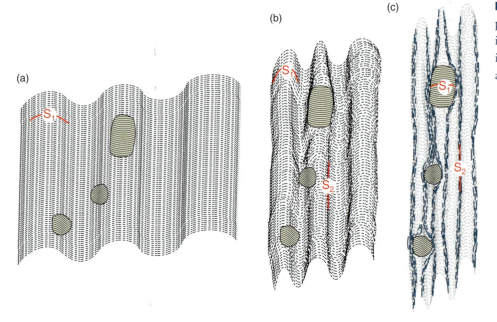

(a) (b) (c)

Figure 21.3 Development of porphyroblasts with an oblique internal foliation (S_1) that was inherited from the rock fabric at the time of growth (a).

and are respectively termed **pretectonic**, **syntectonic** and **posttectonic**. Some also use the term **intertectonic** in cases where growth can be shown to have occurred between two phases of deformation, although most would simply regard this as pretectonic. Syntectonic inclusion trails tend to be curved because, once nucleated, porphyroblasts represent rigid objects that easily rotate during further growth. If you sketch this out step by step you will see that a curved inclusion pattern results.

According to this finding, the example shown in Figure 21.2a can be interpreted as syntectonic. Pretectonic (intertectonic) inclusion patterns tend to be straight, hence the straight central part of the inclusion pattern seen in Figure 21.2b may well be pretectonic, while the outer curvature suggests rotation toward the end of the growth history. It should be added that pretectonic (intertectonic) inclusion trails may also be curved, which is the case if an earlier deformation phase produced a curved (crenulation) cleavage – a scenario outlined in Figure 21.3.

Posttectonic porphyroblasts are simpler to identify because they simply overgrow the present fabric, which can be traced continuously through the porphyroblasts.

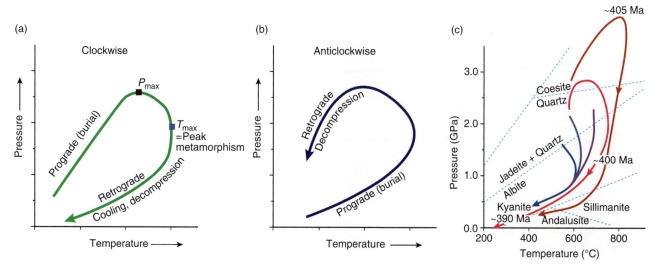

Figure 21.4 Clockwise (a) and anticlockwise (b) *P–T* paths. (c) *P–T–t* paths for different locations in subducted Caledonian continental crust in southwest Norway, relevant to Figures 16.26 and 17.22. Based on compilation by Labrousse *et al.* (2004). Ages relate to the two deepest paths. Well-known phase transitions are shown for reference.

P–T paths

Rocks undergoing tectonometamorphism during an orogenic event experience a change in metamorphic conditions that can be portrayed as a path in pressure–temperature (*P–T*) space, referred to as ***P–T* paths** (Figure 21.4). The rock must have, and usually has, a memory of different stages or phases of its tectonometamorphic development for us to estimate its *P–T* path. For example, many porphyroblasts contain minerals as inclusions that grew during *P–T* conditions that were different from those reflected by the matrix mineralogy. It is sometimes possible to estimate the *P–T* conditions of both stages by means of microprobe analyses of minerals that are in equilibrium.

The same can be done for mineral parageneses preserved within wall rocks or tectonic lenses in shear zones, as shown in Figure 21.5 where a Proterozoic paragenesis is preserved in the wall rock to Early Paleozoic eclogite shear zones (a and b) and in lenses in wider eclogite shear zones (c). Pseudomorphs of metamorphic minerals that grew at higher *P–T* conditions and retrogressed at a later stage also provide information about former (peak?) metamorphic conditions. For example, quartz pseudomorphs after coesite are found in some ultra-high-pressure terranes. Also, chemical zoning of porphyroblasts and reaction textures may give information on the *P–T* development. This type of information is the foundation of the field of **thermobarometry**, and based on laboratory experiments and calibrations it gives us the opportunity to plot the tectonometamorphic development in *P–T* diagrams of the type shown in Figure 21.4 (a and b).

Without really going into the field of metamorphic petrology here, we will point out that *P–T* diagrams are important because they reflect metamorphic conditions during or between deformation phases, and when metamorphic mineral assemblages can be connected to structures such as cleavage or discrete shear zones we have a tight link with structural geology and tectonics.

P–T diagrams can be constructed if we can retrieve information about different stages during the *P–T* history and thus rely on heterogeneous strain and incomplete metamorphic resetting.

Also the shape of *P–T* paths is of interest. Several *P–T* paths exhibit either clockwise or anticlockwise (counterclockwise) paths in diagrams where temperature is plotted along the horizontal axis and pressure along the vertical axis. A **clockwise path** is illustrated in Figure 21.4a, and is characteristic for cases where relatively cold continental crust is rapidly subducted, and therefore reaches peak pressure before the rock warms up to achieve the temperature peak of the *P–T* path. Subduction of the Baltic shield during the Caledonian continent–continent collision is a well-constrained example (see Figure 16.26), and related clockwise *P–T* paths are shown in Figure 21.4c. Most clockwise *P–T* paths imply that rocks undergo heating during the first part of the exhumation history. Note that a *P–T* path and its shape depend on the rate of burial and exhumation and the particle path followed in the orogenic wedge, which again is controlled by both

(a)   (b)

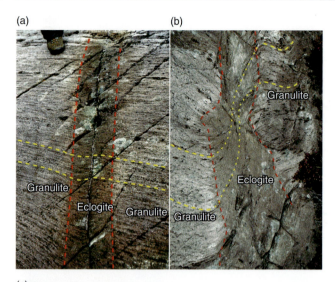

(c)

Figure 21.5 Close connection between metamorphism and deformation structures during Caledonian eclogitization of Grenvillan-age granulites of the Bergen Arcs, southwest Norwegian Caledonides. Selective eclogitization along a fracture (a), along a plastic shear zone (b) and in a wide shear zone (c). (a–c) represent increasing strain and metamorphism, and the selective eclogitization is explained by fluid infiltration along fractures and shear zones. Calculated P–T conditions and U/Pb ages are different for the granulites and eclogites, as demonstrated by Austrheim (1987), Bingen *et al.* (2004) and Raimbourg *et al.* (2005).

regional tectonics and local structural development. Figure 21.4b shows an **anticlockwise path**, and such paths may be envisioned for any tectonic environment in which heating precedes crustal thickening, for example where rifting precedes orogeny or where intrusion of hot magma into the lower crust precedes thickening.

A prerequisite for constructing a P–T path is that enough information about the metamorphic history can be found in the rocks. Mineral parageneses must be formed at several points along the path and must survive the later part of the tectonometamorphic history. Minerals in a dry rock may not alter or recrystallize, even if the conditions change dramatically. However, once fluids enter the rock, metamorphic reactions and recalibration readily occur. This means that a rock may have seen metamorphic conditions not recorded by its mineral assemblage. Fluids are closely associated with shear zones, hence a shearing event may produce shear zones or fractures along which fluids enter, so that minerals representative for the given P–T conditions form preferentially along the shear zone. This close connection between deformation and metamorphism is important and provides heterogeneity that helps us obtain information about the conditions during deformation and pre-deformational conditions. Commonly, shear zones of different ages contain different metamorphic mineral assemblages and thus valuable thermobarometric information that helps us construct the P–T path.

21.5 Radiometric dating and *P–T–t* paths

The direction (clockwise or anticlockwise) of a P–T path can easily be estimated based on simple overprinting relations seen in the field or thin section, but it is useful to add absolute time constraints by means of radiometric data, as done in Figure 21.4c. In simple terms, different methods date either the time of metamorphic equilibrium or the time of cooling through a closure interval independent of any metamorphic reactions. Direct dating of metamorphism is commonly done by means of U-Pb dating of zircon, monazite or sphene that has grown during metamorphism. The method gives the age of zircon growth, and laser methods sometimes give the growth history of zoned zircons.

A useful structural synthesis heavily relies on any dates that can constrain the tectonothermal history of rocks.

Most $^{40}Ar/^{39}Ar$ analyses of amphibole and white mica are regarded as giving cooling ages, indicating the time the mineral passed the temperature at which argon diffusion closes. This temperature is regarded to be around 500 °C for hornblende and 350 °C for muscovite. In addition, it is possible that white micas growing during shearing around or slightly below the 350 °C retention temperature retain their argon from the time they crystallize, so that their $^{40}Ar/^{39}Ar$ ages reflect the time of growth. If so, they date the tectonometamorphic event represented by the shear zone they occur in. The $^{40}Ar/^{39}Ar$ method is also applied to K-feldspar, which in this context acts like a multidomainal mineral that closes to argon over an interval from 350 to 150 °C and therefore can be used to model time–temperature paths over this interval.

The lowermost end of a P–T path can be constrained by apatite fission track analyses, where an annealing temperature for apatite of 150 °C is used. By combining these and other radiometric methods not mentioned here, and perhaps independent stratigraphic evidence, the P–T path can be constrained in absolute time, and the result is a **P–T–t path**. It should be noted that radiometric evidence may be difficult to interpret, and that even published results and conclusions should be subject to critical evaluation.

Dating of intrusive rocks that intrude during or between tectonometamorphic events are useful, and dating of such rocks may give crystallization ages that are more accurate and reliable than other geochronologic information. Several methods and techniques other than those mentioned here are in use, and new techniques that allow better precision and smaller error limits have evolved over the last few decades.

Microtextures and deformation mechanisms

Thermobarometric information and radiometric constraints on tectonothermal events should be linked to microscale deformation processes. Using our knowledge of how rocks and minerals deform under different P–T conditions helps us identify sets of structures that share the same deformation mechanisms. All together, this can help us recognize, map out and group different types of structures within a region.

21.6 Tectonics and sedimentation

Sediments and metasediments are important in metamorphic regions because they were deposited at the surface. If a (meta)sedimentary unit shows an intact or modified primary basal contact, we know that its substrate was at the surface at the time of deposition. In cases where we know the age of the sediments, we have a very useful piece of information that can be added to P–T–t paths. For instance, the path shown in Figure 21.4c from the southwest Norway Caledonides is constrained by early Middle Devonian sediments (~397 Ma), and predated by ~425 Ma sediments that were deposited on oceanic crust now found as ophiolitic fragments in the collision zone.

The connection between sedimentation and tectonics is even more obvious when deformation occurs at or near the surface. We have already seen how local basins or minibasins can form around and between growing salt diapirs as salt withdraws (Figures 19.5 and 19.18). Depositional patterns around salt diapirs, with their thickness changes, unconformities and migration patterns of depositional centers, reflect the growth history of salt structures and formation and movement of related faults. The

formation of kinematic sequences illustrated in Figure 19.14 is one type of such relations. Deciphering of stratigraphic patterns can be done by successive restoration or backstripping, which may give information that otherwise would be difficult to obtain.

The syntectonic sedimentary record preserved in rifts, orogens and strike-slip settings reflects the tectonic history of the area with regard to timing of fault movements, salt growth or collapse, exhumation, metamorphic events or other local or regional tectonic events.

Sedimentation patterns associated with faults are very common in any tectonic environment that involves some component of vertical movement. Pull-apart basin fill (Chapter 18) reflects the timing of fault movement and rate of subsidence, which again is related to geometry and slip rate. In contractional regimes we find clastic wedges in front of propagating thrust nappes and growing orogenic wedges. Such foreland basins reflect the surface relief created by the allochthonous units and their composition. An approaching orogenic front may generate increased clastic input to the foreland basin, more rapid sedimentation, coarser grain size and a mineralogy and clast type that may be exotic to the original foreland area. For example, the sudden appearance of chromite in the Caledonian foreland sequence near Oslo has been interpreted as evidence for ophiolitic allochthons approaching the basin from the hinterland. At a later point the orogenic front reached this area and the layers were involved in thin-skinned shortening. The entire sequence, which is referred to as synorogenic, thus contains important information about the orogeny long before it was involved in orogenic deformation.

In rift basins the sedimentary layers are grouped into pre-, syn-, and postrift sequences. More locally, a major normal fault, such as the one shown in Figure 21.6, will have a syntectonic wedge of clastic fill that generally shows thinning and grain-size fining away from the fault, and an increase in dip downward – features that are illustrated in both Figures 21.7 and 21.8. The down-dip increase in dip is more pronounced in Figure 21.7 because of the strongly listric fault geometry. Listric faults rotate layers very efficiently as the fault slips and new layers are added. The model illustrated in Figure 21.7 is used to explain how a >20 km thick sequence of conglomerates and sandstones that has never been buried to more than about 10 kilometers could accumulate during the collapse of the Caledonian orogen in southwest Norway. This type

Figure 21.6 Half-graben with rotated hanging-wall layers showing an internal unconformity and stratigraphic thinning away from the fault, characteristic of syntectonic sedimentation. Caleta Herradura, Chile. The cliff is 50–70 m high.

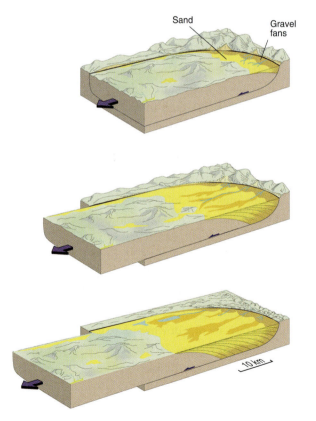

Figure 21.7 Listric fault model that allows for accumulation of sediments with a stratigraphic thickness of tens of kilometers. This model explains stratigraphic relations in Devonian basins that formed during extensional collapse of the Caledonian orogen. From Fossen *et al.* (2008).

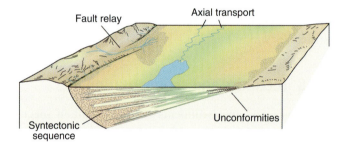

Figure 21.8 Syntectonic sediments in hanging walls to normal faults form wedge-shaped volumes with the coarsest material deposited along the fault scarps. Fault relays attract coarse-grained fluvial systems that generate large relay-zone fans.

of mechanism can, in principle, accumulate an infinite amount of stratigraphic thickness without burying the sediments deeper than the depth of the detachment and illustrates well how fault geometry and tectonics may control sedimentation and stratigraphy.

Fault relay zones also influence depositional patterns, and **relay zone fans** are likely to be considerably larger than other fans developing along active faults, as illustrated in Figure 21.8. This type of interplay between tectonism and depositional patterns is of interest to petroleum geologists looking for coarse-grained clastic wedges along fault systems (stratigraphic traps). Fault slip rate is another important factor, since high slip rates influence depositional patterns more than faults with low slip rates.

Summary

The most important aspect of structural geology is making careful observations, collecting and recording thorough and detailed notes, and treating those data as objectively as possible. The more examples of deformed rocks we study, in the field, in seismic data and through experiments, the better we will understand the evolution of structures under various crustal conditions. We should be aware of other relevant information, some of which has been mentioned in this chapter:

- Overprinting relations are essential when reconstructing the tectonic history of a region.

- Structures can be forming continuously and progressively over a long time period, or in discrete phases separated by periods of tectonic quiescence.

- Metamorphic minerals are useful because they may indicate the P–T conditions at the time of formation, and some may be dated radiometrically.

- Polyphasally deformed and metamorphosed rocks commonly have a memory of more than one tectonometamorphic event due to localized strain and incomplete metamorphic alteration.

- Structures forming close to the surface, notably faults and salt structures, may influence topography and thus sedimentary patterns.

Review questions

1. How can metamorphic petrology help evaluating whether a set of overprinting structures formed during progressive deformation or as separate deformational events?

2. Can you list other methods and criteria that can be used for this purpose?

3. How can we distinguish between pretectonic (intertectonic), syntectonic and posttectonic porphyroblasts?

4. In what tectonic environment can we expect clockwise P–T paths to form?

5. Make a P–T–t diagram based on this information from the eclogite province of the Bergen Arcs (for references, see Bingen *et al.*, 2004): (1) Granulite facies: almost 1 GPa and 800–850 °C at ~930 Ma (U-Pb ages of zircon). (2) Eclogite facies: 1.8–2.1 GPa and ~700 °C at 423 ± 4 Ma (U-Pb zircon rim crystallization age). (3) Amphibolite facies retrogressive shearing: 0.8–1.2 GPa and ~690 °C at 409 ± 8 Ma (Rb/Sr). Other ages: Dikes predating amphibolite facies shearing: 422 ± 6 Ma to 428 ± 6 Ma (Rb/Sr) and 418 ± 9 (U-Pb).

6. What characterizes pre-, syn- and posttectonic sedimentary sequences in a half-graben setting?

FURTHER READING

Barker, A. J., 1998, *Introduction to Metamorphic Textures and Microstructures*, 2nd edition. Cheltenham: Stanley Thornes.

Best, M. G., 2003, *Igneous and Metamorphic Petrology*, 2nd edition. Oxford: Blackwell.

Gawthorpe, R. and Leeder, M. R., 2000, Tectono-sedimentary evolution of active extensional basins. *Basin Research* **12**: 195–218.

Passchier, C. W. and Trouw, R. A. J., 2006, *Microtectonics*. Berlin: Springer.

Spry, A., 1969, *Metamorphic Textures*. Oxford: Pergamon Press.

Appendix A More about the deformation matrix

The deformation matrix not only represents a precise definition of deformation, but also contains a wealth of information about the deformation. With a little knowledge of linear algebra it is possible to extract the strain ellipse (ellipsoid), its orientation, rotation and strain of lines and planes, and for steady-state deformation flow parameters such as flow apophyses and instantaneous stretching axes can be found. The methods and formulas given here can be inserted into spreadsheets or can be explored using spreadsheets already prepared and posted on the webpage for this book.

A.1 The deformation matrix and the strain ellipsoid

The theory presented in this book is based on the decomposition of deformation into simple shear (γ) and pure shear or coaxial (k) components that may or may not imply a change in area or volume, as defined in Chapter 2.

In two dimensions (plane strain) the deformation matrix transforms a point or vector (x, y) to a new position (x', y'):

$$\begin{bmatrix} x' \\ y' \end{bmatrix} = \begin{bmatrix} D_{11} & D_{12} \\ D_{21} & D_{22} \end{bmatrix} = \begin{bmatrix} x \\ y \end{bmatrix} \tag{A.1}$$

or

$$\mathbf{x'} = \mathbf{D}\mathbf{x} \tag{A.2}$$

This is known in linear algebra as a linear transformation, implying homogeneous deformation. To utilize the deformation matrix we must know the principles of deformation and vector operations. Consult an elementary linear algebra textbook for details about theory; we will here constrain ourselves to methods and equations that help us get information about the deformation. It could, however, be useful to have some understanding of eigenvectors and eigenvalues.

Any non-zero vector $\mathbf{x}$ and a corresponding number λ in the equation

$$\mathbf{A}\mathbf{e} = \lambda\mathbf{e} \tag{A.3}$$

are an eigenvector and an eigenvalue, respectively. For reasons not explained here, the matrix $\mathbf{A}$ that we want to analyze for eigenvectors and eigenvalues is not $\mathbf{D}$ but the matrix product $\mathbf{DD}^{\mathrm{T}}$:

$$\mathbf{DD}^{\mathrm{T}}\mathbf{e} = \lambda\mathbf{e} \tag{A.4}$$

It can be shown that there are only two eigenvectors with corresponding eigenvalues for a 2×2 matrix and three for a 3×3 matrix. What happens to an eigenvector during the transformation (deformation)? Equation A.3 shows that it only extends or shortens (depending on the eigenvalue λ). No shear strain (or shear stress if we consider the stress matrix) is present, and the eigenvectors represent the orientations of the principal strain axes (or stress axes). The eigenvalues represent the quadratic stretches, hence the square root of an eigenvalue is the length of a principal strain axis.

Eigenvectors and eigenvalues are easily calculated by means of computer programs, and two-dimensional and simple three-dimensional matrices can be handled in a spreadsheet. Let us do plane strain, allowing for sub-simple shear with or without additional volume change, as an example:

$$\mathbf{D} = \begin{bmatrix} k_x & \frac{\gamma(k_x - k_y)}{\ln(k_x/k_y)} \\ 0 & k_y \end{bmatrix} = \begin{bmatrix} k_x & \Gamma \\ 0 & k_y \end{bmatrix} \tag{A.5}$$

First form the symmetric matrix $\mathbf{DD}^{\mathrm{T}}$:

$$\begin{bmatrix} k_x & \Gamma \\ 0 & k_y \end{bmatrix} \begin{bmatrix} k_x & 0 \\ \Gamma & k_y \end{bmatrix} = \begin{bmatrix} k_x^2 + \Gamma^2 & k_y\Gamma \\ k_y\Gamma & k_y^2 \end{bmatrix} \tag{A.6}$$

Now we want to find the eigenvectors (three for three dimensions: $\lambda_1 > \lambda_2 > \lambda_3$) and eigenvectors ($\mathbf{e}_1 > \mathbf{e}_2 > \mathbf{e}_3$). For two dimensions this is not very difficult to do by hand if you have some knowledge of linear algebra, but we will just give the result here:

$$\lambda = \frac{\Gamma^2 + k_x^2 + k_y^2 \pm \sqrt{(\Gamma^2 + k_x^2 + k_y^2)^2 - 4k_x^2 k_y^2}}{2} \tag{A.7}$$

$$\mathbf{e} = \begin{bmatrix} -k_y\Gamma \\ \Gamma^2 + k_x^2 - \lambda) \\ 1 \end{bmatrix} \qquad (A.8)$$

Note that Equation A.7 gives two solutions and therefore two eigenvectors. Also note that some of the equations deduced from the deformation matrix collapse for perfect simple shear and pure shear.

A.2 Change in area or volume

The area or volume change (Δ) involved in homogeneous deformation can be found by calculating the determinant of the matrix $\mathbf{D}$, denoted det $\mathbf{D}$ (see Box 2.1). The determinant is identical to the product of the eigenvalues of $\mathbf{DD}^T$. The volume change (area change in two dimensions implies volume change if there is no strain along the third dimension) becomes det $\mathbf{D} - 1$ times 100%. For our plane strain matrix (Equation A.5) we get

$$\det \mathbf{D} = \begin{vmatrix} k_x & \Gamma \\ 0 & k_y \end{vmatrix} = k_x k_y \qquad (A.9)$$

We can see that in this case there is no volume change if $k_x = 1/k_y$, since the determinant then becomes 1.

A.3 Orientation of the strain ellipsoid

The angle θ' between the largest principal strain axis and the shear direction is

$$\theta' = \cos^{-1}(e_{11}) \qquad (A.10)$$

where e_{11} is the first component of the normalized longest eigenvector of $\mathbf{DD}^T$ (the longest eigenvector corresponds to the largest eigenvalue, denoted λ_1). Note that the normalized form of a vector $\mathbf{x}$ is $\mathbf{x}/(\mathbf{x}^T\mathbf{x})^{1/2}$. For the two-dimensional example in Section A.1 the angle can be found from this equation:

$$\theta' = \tan^{-1}\left(\frac{\Gamma^2 + k_x^2 - \lambda_1}{-k_y\Gamma}\right) \qquad (A.11)$$

A.4 Extension and rotation of lines

We can study the change in orientation of any line from its initial orientation, given by the unit vector $\mathbf{l}$ (made up of the direction cosines of the line) to the new direction $\mathbf{l}'$ by the transformation

$$\mathbf{l}' = \mathbf{D}\mathbf{l} \qquad (A.12)$$

The angle of rotation (ϕ) of this line can be found from the formula

$$\cos\phi = \frac{\mathbf{l}^T\mathbf{l}'}{\sqrt{\mathbf{l}'^T\mathbf{l}'}} \qquad (A.13)$$

and the quadratic extension (λ) of the line is simply

$$\lambda = \mathbf{l}'^T\mathbf{l}' \qquad (A.14)$$

The angle β between the largest principal strain axis ($\mathbf{e}_1$) and the line is

$$\beta = \cos^{-1}(\mathbf{e}_1^T\mathbf{l}') \qquad (A.15)$$

where $\mathbf{e}_1$ is the normalized eigenvector corresponding to the largest eigenvalue (λ_1) of $\mathbf{DD}^T$. The new vector $\mathbf{l}'$ has, in general, a length different from unity, but may be normalized to reveal the new direction cosines with respect to the coordinate axes.

A.5 Rotation of planes

Planes are treated by means of their poles (normals). If $\mathbf{p}$ is the pole to a plane prior to deformation, the new orientation of the plane is given by $\mathbf{p}'$, where

$$\mathbf{p}' = \mathbf{p}\mathbf{D}^{-1} \qquad (A.16)$$

The rotation of $\mathbf{p}$ equals the rotation of the plane, and can be found by using the equation

$$\cos\phi = \frac{\mathbf{p}^T\mathbf{p}'}{\sqrt{\mathbf{p}'^T\mathbf{p}'}} \qquad (A.17)$$

A.6 ISA

In terms of the deformation matrix the principal stretching axes can be found by estimating the eigenvectors of $\mathbf{DD}^T$ for a very (strictly speaking an infinitely) small amount of strain. For our deformation matrix above (Equation A.5) the angle θ between the fastest stretching direction (ISA$_1$) and the shear direction or x-axis of our coordinate system is

$$\theta = \tan^{-1}\left\{ -\frac{2}{\gamma}\left(\ln k_x - \frac{\ln(k_x + k_y)}{2} \pm \frac{\sqrt{\ln(k_x + k_y)^2 + \gamma^2}}{2} \right) \right\} \qquad (A.18)$$

A.7 Flow apophyses

Flow apophyses for a steady (interval of) deformation can be extracted from the deformation matrix. We already know that pure shear has flow apophyses parallel to the coordinate axes (strain ellipse axes), two for two-dimensional considerations and three perpendicular apophyses in three dimensions. For the deformation matrix in Equation A.5 the simple shear component in the x-direction causes one of the flow apophyses to be oblique. The apophyses become

$$\begin{bmatrix} 1 \\ 0 \end{bmatrix}, \quad \begin{bmatrix} \dfrac{\gamma}{\ln(k_x/k_y)} \\ 1 \end{bmatrix} \tag{A.19}$$

These two apophyses correspond to AP_1 and AP_2 in Figure 2.19. The third apophysis (AP_3) is perpendicular to the x-y plane that we are looking at in this account. The acute angle α between the two flow apophyses is

$$\alpha = \tan^{-1}\left(\frac{\ln(k_1/k_2)}{\gamma}\right) \tag{A.20}$$

The relationship between flow apophyses and ISA_1 for plane strain is given by

$$\alpha = 90 - 2\theta \tag{A.21}$$

A.8 Kinematic vorticity number (W_k)

For a deformation that is stable over the time that our strain accumulates, meaning that flow apophyses and ISA remain constant, the kinematic vorticity number W_k can be calculated from the deformation matrix. For our plane strain example it becomes

$$W_k = \frac{\gamma}{\sqrt{2(\ln k_x)^2 + 2(\ln k_y)^2 + \gamma^2}} \tag{A.21}$$

or

$$W_k = \cos\left\{\tan^{-1}\left(\frac{2\ln k}{\gamma}\right)\right\} \tag{A.22}$$

It can also be shown that

$$W_k = \cos\alpha = \cos(90 - 2\theta) \tag{A.23}$$

where α is defined in the previous section.

A.9 Polar decomposition of D

Non-coaxial deformations involve rotation of the strain ellipsoid during deformation and for this reason are referred to as rotational deformations, as opposed to non-rotational or coaxial deformations such as pure shear. If we want to extract and quantify the rotational component of the deformation represented by a deformation matrix D we can decompose (split) D into a strain matrix S and a rotational matrix R so that

$$D = SR \tag{A.24}$$

The symmetric matrix S only contains the pure strain component, describing the shape of the strain ellipse/ellipsoid while R contains the rotation component.

The rotation matrix R

The rotation matrix R is defined as

$$R = \begin{bmatrix} \cos\omega & -\sin\omega \\ \sin\omega & \cos\omega \end{bmatrix} \tag{A.25}$$

The angle ω can be written in terms of the deformation components, and for our plane strain case

$$\tan\omega = \frac{D_{12} - D_{21}}{D_{11} + D_{22}} = \frac{\Gamma}{(k_x + k_y)}$$
$$= \frac{\gamma(k_x - k_y)}{\ln(k_x - k_y)(k_x + k_y)} \tag{A.26}$$

Thus the expression

$$\omega = \tan^{-1}\left(\frac{\Gamma}{k_x + k_y}\right) \tag{A.27}$$

can be entered into the rotation matrix (Equation A.25).

The strain matrix S

Reorganizing Equation A.24 gives us the following expression for the matrix S:

$$S = DR^{-1} = \begin{bmatrix} D_{11} & D_{12} \\ D_{21} & D_{22} \end{bmatrix} \begin{bmatrix} \cos\omega & \sin\omega \\ -\sin\omega & \cos\omega \end{bmatrix}$$
$$= \begin{bmatrix} D_{11}\cos\omega - D_{12}\sin\omega & D_{11}\sin\omega + D_{12}\cos\omega \\ D_{21}\cos\omega - D_{22}\sin\omega & D_{21}\sin\omega + D_{22}\cos\omega \end{bmatrix} \tag{A.28}$$

The length of the long and short axes of the finite strain ellipse, and the diagonal elements of matrix S, are given completely in terms of deformation components. For our plane strain example we then have

$$S = \begin{bmatrix} k_x\cos\omega - \Gamma\sin\omega & k_x\sin\omega + \Gamma\cos\omega \\ -k_y\sin\omega & k_y\cos\omega \end{bmatrix} \tag{A.29}$$

If we wish we can take this a step farther. S is always symmetric and positive definite for geologically realistic transformations, and therefore orthogonally diagonalizable. This simply means that S can be transformed into a diagonal matrix by means of a new matrix P whose columns are the eigenvectors of S:

$$S = PS_d P^{-1}$$

The diagonal S_d is now a diagonal matrix with the principal quadratic strains along the diagonal:

$$S_d = \begin{bmatrix} \lambda_1 & 0 \\ 0 & \lambda_2 \end{bmatrix} \tag{A.30}$$

$$S = \begin{bmatrix} e_{11} & e_{21} \\ e_{12} & e_{22} \end{bmatrix} \begin{bmatrix} \lambda_1 & 0 \\ 0 & \lambda_2 \end{bmatrix} \begin{bmatrix} e_{11} & e_{12} \\ e_{21} & e_{22} \end{bmatrix} \tag{A.31}$$

Interpretation

We can now write the total transformation as

$$D = PS_d P^{-1} R \tag{A.32}$$

or

$$D = \begin{bmatrix} e_{11} & e_{21} \\ e_{12} & e_{22} \end{bmatrix} \begin{bmatrix} \lambda_1 & 0 \\ 0 & \lambda_2 \end{bmatrix} \begin{bmatrix} e_{11} & e_{12} \\ e_{21} & e_{22} \end{bmatrix} \begin{bmatrix} \cos \omega & -\sin \omega \\ \sin \omega & \cos \omega \end{bmatrix} \tag{A1.33}$$

We can interpret this decomposition in the following way: First the matrix **R** rotates everything through an angle ω. This describes the rotational component of the deformation. Then what becomes the long axis of the strain ellipse (ellipsoid) is rotated by P^{-1} to become parallel to the x-axis of the coordinate system where it is strained by **S**, then rotated back in position by **P**. When relating this to Equation A.32, remember that matrix multiplication is not commutative and works backwards, i.e. the last matrix is applied first.

Appendix B Stereographic projection

In geology, three-dimensional orientation data, such as bedding, foliations and lineations, are commonly projected into two dimensions by means of stereographic projections. Once plotted, the data can be analyzed and compared, and structural relations and features can conveniently be evaluated. Stereographic nets have therefore been used for the entire period of modern structural geology, first by means of manual projection methods, and more recently by means of plotting programs available for personal computers.

B.1 Stereographic projection (equiangular)

Stereographic projection is about representing planar and linear features in a two-dimensional diagram. The orientation of a plane is represented by imagining the plane to pass through the center of a sphere (Figure B.1a). The line of intersection between the plane and the sphere will then represent a circle, and this circle is formally known as a *great circle*. Except for the field of crystallography, where upper-hemisphere projection is used, geologists use the lower part of the hemisphere for stereographic projections, as shown in Figure B.1b. We would like to project the plane onto the horizontal plane that runs through the center of the sphere. Hence, this plane will be our **projection plane**, and it will intersect the sphere along a horizontal circle called the **primitive circle**.

To perform the projection we connect points on the lower half of our great circle to the topmost point of the sphere or the zenith (red lines in Figure B.1c). A circle-shaped projection (part of a circle) then occurs on our horizontal projection plane, and this projection is a **stereographic projection** of the plane. If the plane is horizontal it will coincide with the primitive circle, and if vertical it will be represented by a straight line. Stereographic projections of planes are formally called cyclographic traces, but are almost always referred to as **great circles** because of their close connection with great circles as defined above.

Once we understand how the stereographic projection of a plane is done it also becomes obvious how **lines** are projected, because a line is just a subset of a plane. Lines thus project as points, while planes project as great circles. A great circle (as any circle) can be considered to consist of points, each of which represents a line within the plane. Hence, a line contained in a plane, such as a slickenline or mineral lineation, will therefore appear as a point on the great circle corresponding to that plane.

In Figure B.2 we have also projected the line that is normal to a given plane, represented by the **pole** to the plane. The projection is found by orienting the line through the center and connecting its intersection with the lower hemisphere with the zenith (red line in Figure B.2a). The intersection of this (red) line with the projection plane is the pole to the plane. Hence, planes can be represented in two ways, as great circle projections and as poles. Note that horizontal lines plot along the primitive circle (completely horizontal poles are represented by two opposite symbols) and vertical lines plot in the center.

For stereographic projections to be practical, we have to establish a grid of known surfaces for reference. We have already equipped the primitive circle with geographical directions (north, south, east and west), and we can compare the sphere with a globe with longitudes and latitudes. In three dimensions this is illustrated in Figure B.3, looking from the south pole toward the north pole: longitudes and latitudes are the lines of intersection between great circles (the original meaning) and so-called small circles. If we now project the small and great circles onto the horizontal projection plane, typically for every 2 and 10 degree interval, we will get what is called a stereographic net or **stereonet**.

The longitudes are planes that intersect in a common line (the N–S line), and thus appear as great circles in the stereonet. The projections of the latitudes, which are not planes but cones coaxial with the N–S line, are usually referred to as **small circles** (also their projections onto the stereonet). The net that emerges from the particular projection described above is called the **Wulff net**.

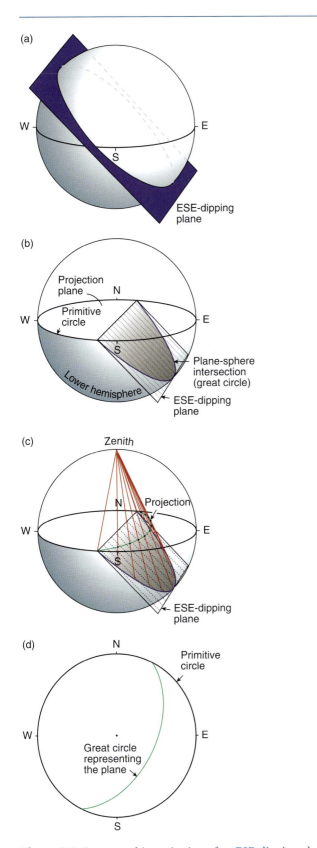

(a)

W — | — E

S

ESE-dipping plane

(b)

Projection plane

N

Primitive circle

W — | — E

S

Lower hemisphere

Plane-sphere intersection (great circle)

ESE-dipping plane

(c)

Zenith

N   Projection

W — | — E

S

ESE-dipping plane

(d)

N

Primitive circle

W — | — E

Great circle representing the plane

S

Figure B.1 Stereographic projection of an ESE-dipping plane.

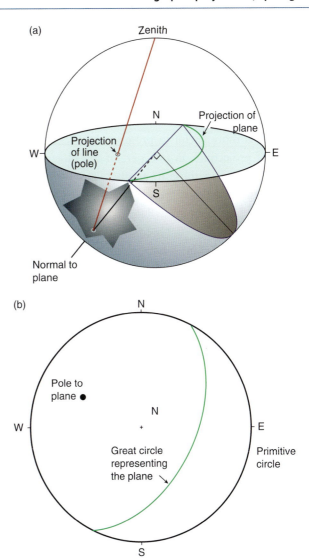

(a)

Zenith

N   Projection of plane

Projection of line (pole)

W — | — E

S

Normal to plane

(b)

N

Pole to plane ●

N

W — + — E

Great circle representing the plane

Primitive circle

S

Figure B.2 Stereographic projection of a line, in this case the normal or pole to a plane.

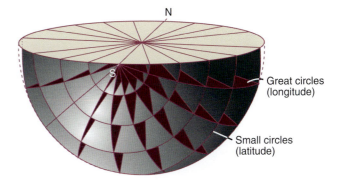

N

S

Great circles (longitude)

Small circles (latitude)

Figure B.3 The stereonet. To compare this with longitude and latitudes of our planet, remember that the north–south axis is horizontal in the stereonet, and vertical in common views of the Earth.

B.2 Equal area projection

The Wulff net makes it possible to work with angular relations (it preserves angles between planes across the net), which can be useful in some cases, for instance for crystallographic purposes. However, for most structural purposes it is more useful to preserve area, so that the densities of projections in one part of the plot can be directly compared to those of another. The method of plotting is the same, but because the projection is not stereographic but equal area (Figure B.4), the positions of planes and lines in the plot become somewhat different. The net is called a **Schmidt net** or simply an **equal area net** (Figure B.5 shows the equal area projection). Multiple data plotted in an equal area net can be contoured with respect to density, which can be useful when evaluating concentrations of structural data around certain geographic directions. Contouring is typically done for

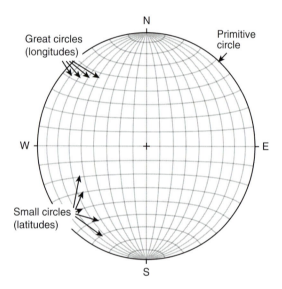

Figure B.4 The equal area net.

crystallographic axes such as quartz c-axes (Figure 15.34), and the contour values exhibit the number of points as the percentage within a given 1% area of the stereonet. Contouring is easily done by means of one of the many computer programs available for personal computers.

B.3 Plotting planes

Planes can be represented in a stereonet in two different ways: by means of great circles or poles (Figure B.2). Figure B.5 gives a demonstration of how to plot both by hand, and we will start with a great circle representation.

A plane striking 030 (or N 30° E) and dipping 30° to the SE is plotted as an example. Tracing paper is placed over a pre-made net (an equal area net was chosen), and the centers are attached by means of a thumbtack. The primitive circle and north (N) are marked on the transparent overlay. We then mark off the strike value of our plane, which is 030 (Figure B.6a), and then rotate the overlay so that this mark occurs above the N direction of the underlying stereoplot (Figure B.6b). For our example, this involves rotating the overlay 30° anticlockwise. We then count the dip value from the primitive circle inwards, and trace the great circle that it falls on (Figure B.6b). When N on the tracing paper is rotated back to its original orientation (Figure B.6c) we have a great circle that represents our plane. The shallower its dip, the closer it comes to the primitive circle, which itself represents a horizontal plane.

The procedure is quite similar if we want to plot poles. All we do differently is to count the dip from the center of the plot in the direction opposite to that of the dip, which in our example is to the left. When done correctly, there will be 90° between the great circle and the pole of the same plane (see Figure B.6b). The pole thus falls on the opposite side of the diagram from that of the

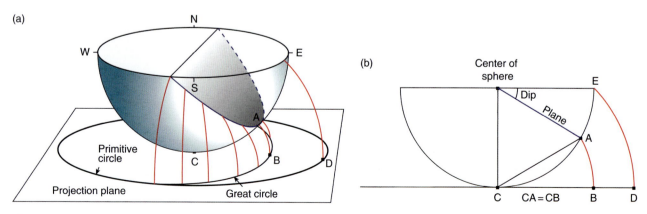

Figure B.5 The equal area projection. A plane is projected onto the projection plane, which in this case has been made tangential to the lower pole of the sphere. The projection is illustrated in 3-D (a) and along a profile through C and A (b).

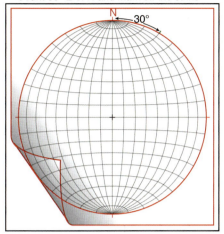

(a) Mark primitive circle and N, E, S and W.
Count 30° E of N and mark off the strike

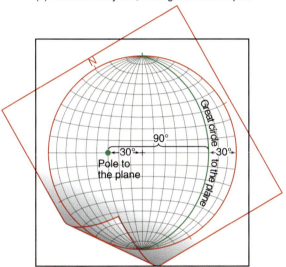

(b) Rotate overlay 30°, mark great circle or pole

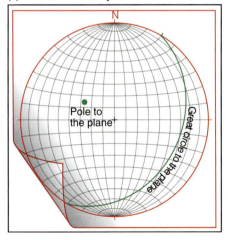

(c) Back-rotate overlay 30°

Figure B.6 Plotting the plane N 030° E, 30 NE in the stereonet (equal area projection).

corresponding great circle. Poles are generally preferred in structural analyses that involve large amounts of orientation data, and particularly if grouping of structural orientations is an issue (which commonly is the case).

B.4 Plotting lines

Plotting a line orientation is similar (but different) to plotting a plane orientation. For example, a line plunging 40° toward 030 (NE) is considered. As for the plane, we mark off the trend (030) (Figure B.7a), rotate the overlay either 30° anticlockwise, as for the plane (Figure B.7b), or until it reaches the E direction (a clockwise rotation of 60° for our example). Now count the plunge value along the straight line toward the center, starting at the primitive circle, and mark off the pole (Figure B.7b). Back-rotate the overlay, and the task is completed (Figure B.7c).

B.5 Pitch (rake)

When doing fault analyses it is useful to plot both the slip plane and its lineation(s) in the same plot. In this case the lineation will lie on the great circle that is representing the slip plane. The angle between the horizontal direction and the lineation is called the **rake** or **pitch**, and is plotted by rotating the great circle of the plane to a N–S orientation and then counting the number of degrees from the horizontal (N or S), i.e. the pitch value measured in the field (Figure B.8). Users of the right-hand rule will always measure the pitch clockwise from the strike value, so that the angle could be up to 180°. The right-hand rule has been used in Figure B.8. Others measure the acute angle and count from the appropriate strike direction, in which case the pitch will not exceed 90°.

B.6 Fitting a plane to lines

If two or more lines are known to lie in a common plane, the plane is found by plotting the lines in a net. The lines are then rotated until they fall on a common great circle, which represents the plane we are looking for.

B.7 Line of intersection

The line of intersection between two planes is perhaps most easily seen by plotting the great circles of the two planes, in which case the line of intersection is represented by the point where the two great circles cross. When plotting poles to planes, the line of intersection is the pole to the great circle that fits (contains) the two poles.

(a) Mark primitive circle and N, E, S and W.
Count 30° E of N and mark off the trend

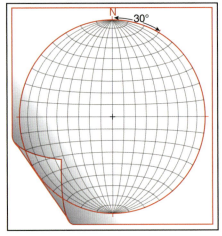

(b) Rotate overlay 30°, count plunge value toward
center and mark off pole

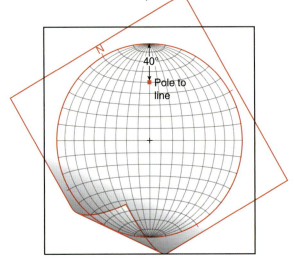

(c) Back-rotate overlay

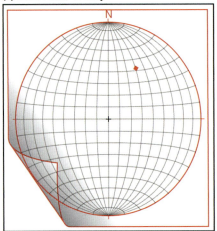

Figure B.7 Plotting the line plunging 40° towards N 030° E in the stereonet (equal area projection).

B.8 Angle between planes and lines

The **angle between two planes** is found by plotting the planes as poles and then rotating the tracing paper until the two points fall on a great circle. The angle between the planes is then found by counting the degrees between the two points on the great circle (Figure B.8, where the angle of two sets of fractures is considered). The **angle between two lines** is found in a similar manner, where the two lines are fitted to a great circle (the plane containing both lines) and the distance between them (in degrees) represents the angle (Figure B.8, lineations).

B.9 Orientation from apparent dips

Finding the orientation of a planar structure from observations of **apparent dips** can sometimes be useful. If two or more apparent dips are measured on two arbitrarily oriented planes, and each of those two planes is represented by a great circle and a point representing the apparent dip (measured at the outcrop), then the great circle that best fits the points represents the plane we want to find. An inverse problem would be to determine apparent dips of a known planar fabric or structure as exposed on selected surfaces. We then plot the planar fabric as a great circle, and the apparent dip will be defined by the point of intersection between the planar fabric and the surface of interest.

B.10 Rotation of planes and lines

Rotation of planar and linear structures can be done by moving them along a great circle, the pole of which represents the rotation axis. Rotation about a horizontal axis is easy: just rotate the tracing paper so that the rotation axis falls along the N–S direction, and then rotate poles by counting degrees along the small circles.

Rotating along an inclined axis is a bit more cumbersome. It involves rotating everything so that the axis of rotation becomes horizontal, then performing the rotation as above, and finally, back-rotating so that the axis of rotation achieves its original orientation.

B.11 Rose diagram

Sometimes only the strike component of planes is measurable or of interest, in which case the data can be represented in the form of a rose diagram. A rose diagram is the principal circle subdivided into sectors, where the

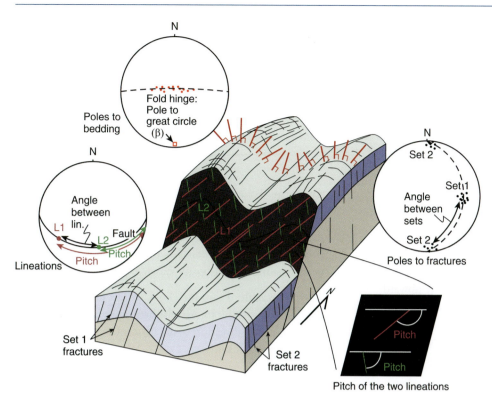

Figure B.8 A constructed, but realistic, situation where various structural elements in a deformed rock sequence are represented in stereonets. Plotting bedding orientations reveals the *b*-axis (local orientation of the hinge line). The angle between the fracture sets can be found by counting degrees along the great circle that fits both data sets. Fault data are plotted separately, showing the fault plane as a great circle and lineations as dots on that great circle.

number of measurements recorded within each sector is represented by the length of the respective petal. This is a visually attractive way of representing the orientation of fractures and lineaments as they appear on the surface of the Earth, and can also be used to represent the trend distribution of linear structures (Figure 1.12).

B.12 Plotting programs

All of these operations and more can be done more quickly by means of stereographic plotting programs, such as the one generously made available to the structural community by Richard Allmendinger (1998). However, understanding the underlying principles is the key to success when using such programs. Several plotting programs also have statistical add-ons that are quite useful.

FURTHER READING

Lisle, R. J., 2004, *Stereographic Projection Techniques for Geologists and Civil Engineers.* Cambridge: Cambridge University Press.

Glossary

Accommodation zone: Zone between two overlapping fault segments where offset is transferred from one fault segment to the other. Used specifically for the zone connecting two oppositely dipping half-grabens (Rosendahl *et al.*, 1986).

Active diapirism: Diapirs forcing their way upward through the overburden, driven by differential, thermal or displacement loading. Weakening of the overburden by means of fracturing is generally involved.

Active folding: Folding of layers by layer-parallel shortening controlled by contrasts in viscosity between layers (buckling).

Active markers: Markers or structures that react mechanically to the stress field by having a different viscosity from their matrix. Such markers may buckle or boudinage, and do not give a representative image of the general state of strain in the rock.

Active rifting: Rifting as a response to upwelling of the asthenosphere, which generates tensile stresses that lead to normal faulting and stretching in the lithosphere.

Allochthonous: Tectonic unit that has been transported too far for direct correlation with the substrate. Derived from Greek: "allo" means "different" and "chthon" means "ground". Typically used for nappes that have moved tens of kilometers or more.

Allochthonous salt: Salt that has been detached from its source layer, usually by contractional deformation.

Angular shear: Change in angle for a pair of lines that were orthogonal before deformation. More specifically, the angular shear along a reference line is the change in angle of a line that was perpendicular to the reference line before deformation.

Anisotropic volume change: Volume change that is created by shortening or extension in one or two directions only.

Anticlinal: Fold where rock layers get younger away from the axial surface of the fold.

Anticrack: Engineering term for closing fracture, i.e. fractures that show compactional displacement.

Antiform: Fold where the limbs dip down and away from the hinge zone.

Antiformal syncline: A syncline (strata get older away from its axial surface) that has the shape of an antiform, i.e. a syncline turned upside down.

Antithetic fault: From the Greek word "antithetos", placed in opposition to. An antithetic fault is a fault dipping in the opposite direction to an adjacent master fault or dominating fault set.

Antithetic shear: Shear acting antithetic to the sense of displacement of a reference fault. Used as a section restoration technique.

Aperture: The distance between the two walls of a fracture.

Area change: Change in area due to deformation. Implies volume change unless compensated for in the third dimension.

Aseismic slip: Stable sliding, as opposed to stick-slip.

Aspect ratio: Long dimension divided by short dimension of an ellipse or rectangle.

Asperity: Irregularity along a fracture surface.

Autochthonous: Lithologic unit in or along an orogenic belt that has not been tectonically transported. The Greek word "auto" means "the same" in this connection.

Axial plane: A planar axial surface, not necessarily parallel to the bisecting surface.

Axial plane cleavage: Cleavage that is subparallel to the axial surface of a fold. The cleavage must have formed during the process of folding. Also called axial planar cleavage.

Axial surface: The theoretical surface connecting the hinge lines of consecutive surfaces in a fold structure.

Axial trace: The theoretical line that connects hinge points across a fold.

Axially symmetric extension: Extension in one principal direction (X-axis of the strain ellipsoid) and equal shortening in the other two (Y and Z). Implies perfect constrictional strain. Equal to uniform extension.

Axially symmetric shortening: Shortening in one principal direction (Z) and equal extension in the other two (Y and X). Implies perfect flattening strain. Equal to uniform shortening.

Back-thrust: Thrust displacing the hanging wall toward the hinterland, i.e. opposite to the general thrusting direction.

Backstripping: Isostatic basin restoration where focus is on the subsidence history of a basin by successively removing sedimentary sequences and balancing isostasy.

Backward modeling: Starting with the present state and modeling (restoring) back to the pre-deformational stage.

Balancing: The construction or interpretation of a geologic profile or 3D model that can be reconstructed by means of geologically realistic processes to a geologically sound undeformed state.

Basin: In fold terminology, this is a dome turned upside-down.

Bending: Folding mechanism that occurs where forces are applied at a high angle to the layering.

Bisecting surface: Surface that divides a fold into two parts. When the bisecting surface is vertical, the limbs should have the same dip.

Blastomylonite: A mylonite that has recrystallized posttectonically. Grains show no preferred orientation (equant grains) and little or no internal strain.

Blind fault: Fault that terminates without reaching another fault or the surface. Traditionally used in thrust fault terminology (blind thrust).

Bluntness: The angularity or curvature of folds as observed in cross-sections perpendicular to the hinge line.

Body forces: Forces that affect the entire volume of a rock, the inside as well as the outside.

Bookshelf tectonics: Popular name for the development of rotated fault blocks forming according to the domino model.

Borehole breakouts: Stress method that uses the geometry of a borehole to estimate the maximum horizontal stress.

Boudinage: The process leading to the formation of boudins.

Boudins: Structures forming during systematic segmentation of preexisting layers. Classic boudins form by extension of layers that are more competent than the matrix. See also *foliation boudinage*.

Boundary drag: The restriction of flow in a layer by the viscous shear forces acting along the boundaries. Particularly relevant for salt and viscous magma.

Box fold: Fold with two axial planes and two hinge zones that formed simultaneously. Reminiscent of a (bottomless) box in cross-section.

Branch line: A line of intersection between two intersecting faults. Used for any type of fault (normal, reverse or strike-slip).

Branch point: Point in a section or map where two fault traces join.

Breached relay ramp: Relay ramp that has been cut by a fault, transforming it from a soft-linked into a hard-linked overlap structure.

Breccia: Cohesive or non-cohesive fault rocks consisting of randomly oriented fragments resulting from brittle fracturing. Breccia fragments must constitute more than 30% of the rock.

Brittle deformation mode: Deformation by means of brittle deformation mechanisms (fracturing, frictional sliding, cataclastic flow).

Brittle strain, brittle deformation: Deformation by fracturing (discontinuous deformation).

Brittle shear zone: Shear zones dominated by brittle deformation mechanisms. Also called frictional shear zones. Also used for shear zones that disrupt originally continuous markers.

Buckle folds: Folds that form by buckling. They show a certain regularity with regard to wavelength and amplitude as a function of layer thickness and the viscosity contrast between layer(s) and the matrix.

Buckling: A folding mechanism that occurs when layers that are more competent (higher viscosity) than the matrix are compressed parallel to the layering. As stress increases the layer becomes unstable and buckles through the amplification of minute irregularities along the layer interfaces.

Bulb: Upper thick part of a teardrop diapir.

Byerlee's law: Relation between critical shear stress on a fracture and the related normal stress across it. The normal stress reflects the depth in the crust, hence this law models critical shear strength through the frictional upper crust.

Characteristic earthquake model: Each slip event is equal to the others in terms of slip distribution and rupture length.

Cataclasis: Brittle crushing of grains (grain size reduction), accompanied by frictional sliding and rotation. Derived from a Greek word for crushing.

Cataclasite: Cohesive and fine-grained fault rock. Cataclasites are subdivided into those that have 10–50% matrix (protocataclasite), 50–90% matrix (cataclasite) and >90% matrix (ultracataclasite).

Cataclastic deformation band: Deformation band where cataclasis is an important deformation mechanism.

Cataclastic flow: Flow of rock during deformation by means of cataclasis, but at a scale that makes the deformation continuous and distributed over a zone.

Centrifuge: Spinning device used for physical modeling where gravity can be scaled.

Channel flow: Large-scale flow of relatively low-viscosity heated rocks in a "channel" from the continent–continent collision zone of an orogen. The kinematics is that of extrusion, with a thrust below and normal movement above the channel.

Chemical compaction: Compaction by means of wet diffusion, i.e. dissolution at grain contacts and/or stylolite formation.

Chevron fold: Fold with angular hinge and where the axial surface forms more or less perpendicular to σ_1.

Chevron method: The vertical shear method used during restoration or balancing of sections.

Chocolate tablet boudinage: Boudinage in two directions (in the XY-plane), forming more or less square or rectangular boudins in three dimensions.

Christmas-tree folds: Secondary folds superimposed on a larger and preexisting upright fold, usually by means of gravity collapse. The Christmas-tree pattern emerges when the primary fold is an upright antiform.

Clay injection: Injection of clay along a fault, normally because a tensile fracture opens due to local overpressure.

Clay smear potential (CSP): relationship between the thickness of a faulted clay layer and the distance from the clay layer along the fault in a sequence of sandstone with one or more clay layers. CSP is used in fault seal analysis.

Clay smear: Smearing or, less commonly, injection of clay along the fault core.

Cleavage: A tectonic foliation formed at low-grade metamorphic conditions and related to folding. A cleaved rock breaks more easily along the cleavage.

Cleavage refraction: A change in cleavage orientation across an interface between layers of contrasting competence.

Climb: The process by which edge dislocations jump to another slip plane to get around obstacles in the lattice.

Coaxial deformation: Lines along ISA do not rotate during the deformation; $W_k = 0$. The principal strain axes (X, Y and Z) remain stationary throughout the deformation history.

Coble creep: See *grain boundary diffusion.*

Coefficient of sliding friction: The shear stress required to activate slip on a fracture divided by the normal stress across the fracture.

Cohesion: The solidness of a medium. A cohesive rock does not fall apart very easily, while a non-cohesive medium easily disintegrates. Cementation increases cohesion in sedimentary rocks and fault gouges.

Cohesive strength: The intersection of the fracture criterion or envelope with the vertical axis of the Mohr diagram. It is the theoretical value of the shear stress that it takes for a rock to fail in shear along a plane across which there is no normal stress. In many cases the cohesive strength of a rock is approximately twice its tensile strength ($C = 2T$).

Compaction: Shortening in one direction while the other two directions are unstrained.

Compaction band: Deformation band involving compaction without shear. Compaction bands are found in highly porous sandstones.

Compaction cleavage: Cleavage formed by lithostatic compaction of sediments into sedimentary rocks. Best developed in mudrocks.

Compaction curve: Graph expressing changes in porosity with burial depth for a given sediment(ary rock).

Compactional shear zones: Shear zones with a component of compaction across the walls.

Competency: A relative expression that compares the mechanical strength or resistance to flow of a layer or object to that of its adjacent layers or matrix. Competent objects are more resistant to flow than their matrix.

Compression: Expression used extensively for compressional stresses. Contraction or shortening should be used for strain.

Compressive strength: The amount of compression that a rock can withstand before fracturing, usually many times its tensile strength (eight times according to Griffith).

Concentric faults: Circular faults related to collapse of underlying rocks, for example due to salt collapse and collapse of karst structures.

Concentric folds: Folds whose well-rounded arcs approximate half-circles so that limbs and hinges are inseparable (or they can be said to consist of hinges only).

Conjugate faults: Two intersecting faults that formed under the same stress field. Such faults show opposite sense of shear and make 30° to σ_1.

Conjugate folds: See *box folds*.

Constant area restoration: Restoration of map or cross-section where area is preserved.

Constant displacement restoration: Displacement is considered to be constant along the fault(s).

Constant-horizontal-stress reference state: Reference state of stress assuming that the lithosphere has no shear strength at a certain depth and that it behaves like a fluid below this compensation depth.

Constant length restoration: Restoration of one or several markers in a cross-section where the length of the marker is the same before and after restoration.

Constitutive laws: Laws or equations describing the relationship between stress and strain.

Continuous cleavage: Cleavage where the distance between individual cleavage domains is indistinguishable in hand sample, i.e. less than 1 mm.

Contraction: Reduction in length. Synonymous with shortening.

Contractional fault: Fault that shortens a reference horizon, which may be lithologic layering or the surface of the Earth.

Contraction fracture: Closing fracture, usually stylolitic. Also known as anticrack.

Corrugations: Cylindrical undulations on shear zones, faults or slip surfaces. Occur from micro- to map scale. Large-scale examples may be related to fault growth by segment linkage or shear zone-parallel folding; outcrop-scale also to frictional carving of fault walls.

Couette flow: Simple shear-type flow caused by translation of the overburden relative to the substrate.

Coulomb envelope: The linear failure envelope predicted by the Coulomb fracture criterion.

Coulomb material: Material that conforms to the Coulomb fracture criterion.

Coulomb wedge: Model of orogenic wedge where the Coulomb fracture criterion is applied to the wedge material.

Creep: Generally used for slow geologic processes. More specifically used for the (slow) way that permanent plastic deformation accumulates at long-term constant stress by various microscale or atomic-scale deformation mechanisms (diffusion creep, dislocation creep etc.).

Creep mechanisms: Deformation mechanisms at work in a crystal or crystal aggregate that respond to sustaining stress by the gradual accumulation of plastic strain. These are separated into diffusion creep (grain boundary diffusion, volume diffusion and pressure solution) and dislocation creep (dislocation climb, dislocation glide and recrystallization).

Crenulation cleavage: Cleavage formed by microfolding at low metamorphic conditions of phyllosilicate-rich and well-foliated rocks.

Crenulation lineation: Lineation formed by crenulation of phyllosilicate-rich layers. Closely related to intersection lineations.

Critical taper model: See *critical wedge model*.

Critical tensile strength: The tensile stress when the material is at the verge of failure.

Critical wedge model: Model of allochthonous units in a collision zone (subduction zone, foreland fold-and-thrust belt) where the units take on a wedge-shaped geometry controlled by basal friction, internal strength and erosion/deposition. Also called critical taper model.

Critically stressed: Stressed to the limit of its strength, so that the material is at the verge of failure.

Cross-slip: Process allowing dislocations to change slip planes in order to bypass obstacles.

Cylindrical fold: Fold with straight hinge line, so that an imaginary cylinder can be fitted to the hinge zone.

Damage zone: Zone of brittle deformation structures (fractures, deformation bands and/or stylolites) around a fault. The zone has a density of such structures that is higher than the surrounding rocks.

Décollement: Large-scale detachment, i.e. fault or shear zone that is located along a weak layer in the crust or in a stratigraphic sequence (e.g. salt or shale). The term is used in both extensional and contractional settings.

Décollement folds: Folds formed above a décollement or detachment, where sub-décollement layers are undisturbed by the folding. Identical to detachment folds.

Decoupling: Expression used in cases where the upper part of a section deforms in a different style than the lower part, or if the structures are not directly connected between the two levels. The two parts are separated by a décollement or very weak layer (typically salt).

Deformation: The change of the shape, position and/or orientation as a result of external forces. The deformation is found by comparing the undeformed and the deformed states and positions.

Deformation bands (I): Millimeter-thick zones of strain localization formed by grain reorganization and/or grain crushing. Shear bands with some compaction across the band form the most common type.

Deformation bands (II): Microscopic zones in a mineral grain with similar optical orientation (extinction), forming between dislocation walls.

Deformation gradient tensor: See *deformation matrix*

Deformation matrix: Transformation matrix that relates the undeformed and the deformed states of a deformation. A deformation matrix describes a linear transformation and therefore homogeneous deformation. It represents a complete description of the deformation (but not the deformation history).

Deformation mechanisms: Mechanisms at the microscale that are active during deformation, including cataclasis, frictional grain boundary sliding, dry and wet diffusion, dislocation glide, dislocation climb, plastic grain boundary sliding, twinning and kinking. Recrystallization is a process that generally occurs by means of two or more mechanisms. The term deformation mechanism can also be used in a more general way.

Deformation phase: A time period during which structures form continuously within an area or region. The structures would show a common expression that can be linked to a particular stress or strain field or kinematic pattern, although the style may vary (for example; both open and tight folds may be related to the same phase because strain can be heterogeneous, but they share the same shortening direction).

Deformation twins: The result of mechanical twinning. Deformation twins are common in calcite crystals.

Delamination model: Model where the dense root of an orogen detaches and sinks into the underlying mantle.

Detachment: Low-angle or horizontal fault or shear zone separating an upper plate (hanging wall) from a lower plate (footwall). Detachments are typically reactivated weak layers or structures.

Detachment folds: Folds formed in competent layers above a detachment during shear or slip on the detachment.

Deviatoric stress: The difference between the total stress and the mean stress. Closely related to tectonic stress.

Dextral: Right-lateral, moving right relative to a point of reference.

Diapir: A body, usually of salt, magma or water-saturated mud or sand that gravitationally moves upward and intrudes the overburden.

Diapirism: The process associated with the formation of diapirs.

Differential compaction: Two or more areas experiencing different amounts of compaction, for example across a steep fault where new sediments preferentially accumulate on the downthrown fault block.

Differential loading: Unevenly distributed load, for instance delta lobes generating locally thick layers of sediments.

Differential stress: The difference between the largest and smallest principal stresses, i.e. the diameter of the Mohr circle.

Diffusion: The movement of vacancies (holes) in an atomic lattice. Volume diffusion occurs within the lattice, while grain boundary diffusion (Coble creep) occurs along the grain boundaries.

Dilation (US), dilatation (UK): Volume change, usually implying volume loss (negative dilation being volume gain). Isotropic dilation involves the same amount of extension in all directions. A common example of anisotropic dilation is uniaxial strain. In the context of shear zones dilation usually implies compaction across the zone.

Dilational shear zones: Shear zones with a component of dilation (negative compaction) across the walls.

Dilation band: Deformation band where displacement is dilational (volume increase) without shear. Dilation bands show an increase in porosity and are relatively uncommon as compared to other types of deformation bands.

Dip isogons: Theoretical lines connecting points of equal dip on the upper and lower boundaries of a folded layer oriented in an upright position (vertical bisecting surface).

Dip separation: Apparent fault displacement as observed in a vertical section in the dip direction of the fault. Dip separation equals true displacement for dip-slip faults.

Dipmeter log: Well log showing dip and azimuth of planar features based on interpretations of resistivity measurements along the wellbore. Measurements are done by running a dipmeter tool through the wellbore, and the planar features represent bedding, deformation bands or fractures.

Dip-slip fault: Fault with the slip vector oriented along the dip direction of the fault surface, i.e. a perfectly reverse or normal fault.

Disaggregation band: A deformation band formed by non-destructive granular flow (rotation and frictional sliding of grains). Commonly forms in sand and poorly consolidated sandstones.

Discrete crenulation cleavage: Crenulation cleavage with a sharp discontinuity between QF- and M-domains, in contrast to *zonal crenulation cleavage.*

Disharmonic folds: Folds in multilayered rocks that change shape and wavelength along the axial trace.

Disjunctive cleavage: Domainal cleavage that is independent of previous foliation(s). Typical for very low grade metasediments. May be divided into stylolitic, anastomosing, rough and smooth, according to the morphology of the domains. Disjunctive cleavage contrasts with crenulation cleavage, where a preexisting foliation is reworked by microfolding and solution.

Dislocation: An atomic-scale line defect within a crystal lattice. Dislocations can move by means of glide and climb mechanisms and dislocation formation and motion causes plastic deformation. Edge and screw dislocations are the principal types of dislocations in naturally deformed rocks.

Dislocation creep: Strain accumulation through the movement of dislocations through the crystal, where mechanisms called climb and cross-slip are used to bypass obstacles in the lattice.

Dislocation glide: The self-healing (and caterpillar-style) process by which edge dislocations move (see Figure 10.12).

Dislocation walls: Concentration of dislocations forming walls within a crystal. Dislocation walls mark the boundary between deformation bands and subgrains.

Displacement: The difference between the location of a point before and after deformation. For faults, displacement is the relative motion of two originally adjacent points on each side of the fault.

Displacement field: The field of vectors describing the distortion of points in a deformed medium, i.e. vectors connecting the pre- and post-deformational positions of particles.

Displacement–length ratio: The ratio between the maximum displacement of a fault and its length. Usually measured on a cross-section or a map, which introduces a sectional uncertainty.

Displacement loading: Loading by changing the lateral boundary conditions, for instance by compressing or extending a volume of rocks or sediments.

Displacement vector: Vector connecting the positions of a material point (e.g. a sand grain) before and after deformation.

Distortion: Strain.

Domainal cleavage: Cleavage composed of domains of different minerals, usually micaceous M-domains and quartzofeldspathic QF-domains. When individual domains are visible in hand sample, the domainal cleavage is a spaced cleavage.

Dome: Bowl-shaped geometry with layers dipping in every direction from a summit.

Domino (fault) model: Model where parallel normal faults define fault blocks that rotate like domino bricks during deformation. Also called bookshelf mechanism. In the classic domino model blocks are by definition rigid, but the soft domino model allows for internal block deformation.

Domino faults: Set of parallel normal faults separated by rotated fault blocks (domino blocks) where bedding is dipping antithetic to the faults.

Doubly plunging fold: Fold plunging in two directions because of curved hinge line.

Downbuilding: Term used for the process where sediments accumulate along a passive salt diapir. The sediments build downward with regard to the surface of the Earth.

Drag (folds): Zone of folding on one or both sides of a fault or salt structure. The folding must be related to the fault formation and/or growth. Originally a genetic term implying that the folding is controlled by frictional resistance along a fault. Now used as a purely descriptive term.

Ductile deformation: Continuous deformation at the scale of observation, resulting from any deformation mechanism (brittle or plastic). Some geologists restrict the term to crystal-plastic deformation.

Ductile shear zone: Shear zone with no internal shear-related discontinuities.

Duplex: Tectonic unit consisting of a series of horses that are arranged in a piggy-back fashion between a sole and a roof thrust. Also used for similar structures in extensional and strike-slip settings (extensional and strike-slip duplexes).

Dynamic analysis: The analysis that explores the relationship between stress and strain.

Dynamic recrystallization: Synkinematic recrystallization, i.e. continual crystallization during deformation. Revealed in shear zones by slightly non-equant grains that define a new fabric at an angle to the foliation.

Edge dislocation: Type of linear defect in a crystal lattice. See Figure 10.11.

Effective stress: The total stress minus the pore pressure in a porous rock or sediment.

Elastic deformation (strain): Deformation (strain) that disappears when the applied stress is removed.

Elastic material: Material that deforms elastically.

Elongation: $e = (l - l_0)/l_0$, where l_0 and l are the lengths of the line before and after the deformation, respectively.

Enveloping surface: Surface enveloping or tangenting a series of geometric features, such as a series of fold hinges or Mohr circles.

Extension: A measure of how much longer a line or object has become due to deformation.

Extension fracture: Fracture formed by extension perpendicular to the fracture walls. The amount of extension can be minute, as for joints, or can be larger, as for veins.

Extensional duplex: Duplex forming along an extensional fault, where individual riders are separated by extensional faults and bound by a roof fault and a floor fault.

Extensional fault: Fault that extends (increases the length of) a reference horizon, which may be lithologic layering or the surface of the Earth.

Extrusion: Strain model where rock moves in a uniform direction with no strain along Y (plane strain). Applied to outcrop-scale shear zones, and also to orogens (particularly the Himalayan orogen) as a model where a unit (thrust nappe) in the orogenic wedge translates toward the foreland faster than the overlying unit. The result is an allochthonous unit bound by a thrust below and a normal fault above. See also *channel flow*.

Extrusive advance: The gravity-controlled mechanism by which an exposed salt sheet (salt glacier) advances.

Fabric: The configuration of planar and/or linear objects in a penetratively deformed rock. An L-fabric is composed of linear features, while an S-fabric consists of planar elements.

Failure envelope: The curve enveloping a series of Mohr circles representing different differential and mean stress values (different positions along the x-axis of the Mohr diagram). Each circle touches the envelope. The failure envelope describes the stress conditions at failure for different stress conditions in a given medium (rock).

Far-field stress: Equivalent to *remote stress*.

Fault: Surface or narrow tabular zone with displacement parallel to the surface (zone). Generally used for brittle structures (structures dominated by brittle deformation mechanisms).

Fault bend: Bend in the fault trace or fault surface. Although the term does not imply a particular evolution, many fault bends probably represent hard-linked structures formed by fault linkage.

Fault-bend fold: Fold forming in the hanging wall in response to a bend or kink in the fault surface. Traditionally a fold forming above a thrust fault ramp.

Fault core: Central high-strain zone of a fault where most of the displacement is taken up. Enveloped by the fault damage zone. The fault core can consist of non-cohesive rock flour or strongly sheared phyllosilicate-rich rock called fault gouge, or cohesive cataclasite, and can contain lenses of wall rock. The core thickness can vary from less than 1 mm for small (meter-scale) faults to around 10 m for large (kilometer-displacement) faults.

Fault cut: Stratigraphic section missing in a well due to omission by faulting. The fault cut is estimated based on stratigraphic information from nearby wells or outcrops.

Fault cut line: The line of intersection between a fault surface and another (usually stratigraphic) surface cut by the fault. There are two such lines, known as hanging-wall and footwall cut (or cutoff) lines.

Fault damage zone: See *damage zone*.

Fault gouge: Fine-grained and clay-rich non-cohesive rock located in the core to faults, formed by crushing and chemical alteration of the host rock.

Fault grooves: Linear tracks or grooves carved out on a slip surface by asperities on one of the walls.

Fault juxtaposition diagram: Diagram illustrating the lithologic contact relations along a particular fault, e.g. sand–sand, sand–shale etc. Closely related to triangle diagram, which is a general juxtaposition diagram for a synthetic fault (gradually changing displacement).

Fault linkage: The process where two adjacent fault segments interfere and connect, forming a hard-linked or soft-linked structure.

Fault plane solution: Stereographic projection containing information about the first motion caused by an earthquake based on seismic observations from a number of seismographs. It consists of two orthogonal planes separating compressional from tensional motion. One of these planes represents the fault orientation. The beach-ball style plots also give us the sense of slip (normal, reverse etc.) and approximate locations of the principal stress axes.

Fault-propagation fold: Fold forming ahead of a propagating fault tip. Traditionally used for thrust faults, but can be used for any type of fault (normal, strike-slip or reverse).

Fault stepover: Link between two more or less parallel faults that are not aligned. The faults must be close to each other so that their stress fields can interfere. The two faults can be soft-linked or hard-linked.

Fault strain: Strain calculated for an area or volume affected by numerous faults. As a concept, two- and three-dimensional strain applies only to ductile (continuous) deformation, but since ductility is scale dependent, it can, as an approximation, be applied to fractured rocks with distributed fractures.

Fault trace: Intersection between a fault and any given surface, such as the surface of the Earth, a stratigraphic interface or a cross-section.

Fault zone: A series of closely spaced subparallel faults or slip surfaces, forming a zone. The thickness of the zone must be small relative to its length. Recent use: The central and most strained part of a fault, coinciding with or somewhat wider than the fault core.

Fenster: Erosional exposure of the rock unit underlying a nappe (window).

Fiber lineation: See *mineral fiber lineation.*

Fissure: Fluid-filled extensional fracture.

Flaps: Folded layers along the upper walls of salt diapirs, dipping away from the diapir (unless they have been inverted). Flaps form as a moving salt diapir breaks through its overburden, lifts, rotates and shoulders aside roof layers.

Flat-ramp-flat fault: Fault with a subhorizontal segment connected with steeper segments on each side. Used about extensional as well as reverse/thrust faults.

Flexural flow: Folding by layer-parallel simple shear. Also called flexural shear.

Flexural folding: Folding by means of flexural slip, flexural shear or orthogonal flexure.

Flexural shear: Fold mechanism where layers are deformed by layer-parallel simple shear. The shear strain is zero at the hinge point, increasing toward the inflection points. Sense of shear is away from the hinge zone, i.e. opposite on the two limbs. Also called flexural flow.

Flexural slip: Slip along bedding interfaces during folding. As for flexural shear, slip increases away from the hinge line, being opposite on the two limbs. Typical in folded layers of high contrasts in strength.

Floor thrust: Thrust fault defining the base of a duplex structure or the basal thrust of a nappe or nappe complex, i.e. a sole thrust.

Flow: A term used for rocks in the perspective of geologic time: Given enough time and appropriate physical conditions (temperature, pressure, fluid availability), rocks flow by means of plastic or brittle deformation mechanisms. A distinction can be made between cataclastic flow and plastic flow.

Flow apophyses: Apophyses separating domains of different particle motion during flow (deformation).

Flow laws: Mathematical models describing the relation between deformation rate, stress and deformation mechanism for a given rock.

Flow parameters: Parameters describing the deformation at any instant or interval of the deformation history. For steady-state deformation, they are representative of the entire deformation history. Important flow parameters are the velocity field, flow apophyses, ISA and vorticity.

Flow pattern: The pattern outlined by the particle paths during flow (distributed deformation).

Flower structure: The upward-splitting and widening pattern of strike-slip faults as seen in cross-section.

Fold axis: The straight hinge line of a cylindrical fold.

Fold hinge: See *hinge.*

Fold limb: The two parts of the fold that are separated by the hinge zone, i.e. by the area of maximum curvature.

Fold nappe: Nappe that is internally folded throughout and appears to have originated by shearing of an inverted fold limb.

Foliation: Usually a tectonic planar structure formed in the plastic regime. Foliations are characterized by flattening across the structure. Also used for primary structures such as bedding or magmatic layering, in which case the term primary (in contrast to secondary or tectonic) foliation should be used.

Foliation boudinage: The formation of boudins in strongly foliated metamorphic rocks, where the boudins are separated by en-echelon arranged shear fractures, small shear zones or extension fractures.

Foliation fish: Volume in a strongly foliated rock that is back-rotated relative to the rest of the rock, displaying a fish-like geometry.

Footwall: The surface underneath a non-vertical fault.

Footwall collapse: The formation of one or more secondary faults in the footwall to a fault. Expression most commonly used for normal faults.

Footwall uplift: Uplift of the footwall of a normal fault, which typically is in the order of 10% of the fault throw.

Forced folds: Folds formed as basement blocks move along preexisting faults, forcing covering sedimentary layers to fold into monoclinal structures.

Foreland: The peripheral or frontal part of a thrust region or orogenic belt, dominated by thin-skinned tectonics and very low to non-metamorphic conditions.

Forward modeling: Modeling a process or the development of a cross-section etc. from the beginning to the end or present state, i.e. opposite to backward modeling.

Four-way dip closure: Dome structure.

Fracture: A sharp planar discontinuity. An ideal fracture is narrow (thinner than 1 mm), involves a displacement discontinuity as well as being a mechanical discontinuity, and is weak so that the rock preferentially breaks along the fracture. It also conducts fluids. A distinction is drawn between extension fractures and shear fractures, and sometimes also contraction fractures.

Fracture cleavage: A dense array of fractures that mimics a cleavage. While ordinary cleavage involves shortening across the cleavage, fracture cleavage involves shear along or extension across the structure. The term is a bit confusing and is best omitted.

Fracture toughness: A material's resistance to continued growth of an existing fracture. High fracture toughness implies high resistance against fracture propagation and therefore low propagation rates.

Frictional regime: The regime where the physical conditions favor the brittle deformation mechanism in the crust, i.e. the upper part of the crust. Identical to the brittle regime, but emphasizes the dependence on deformation mechanism (and not deformation style).

Frictional shear zone: Shear zones dominated by brittle deformation mechanisms. Also called brittle shear zones.

Frictional sliding: Sliding on a fracture that has a certain friction without activation of plastic deformation mechanisms.

Frontal ramp: Ramp oriented perpendicular to the transport direction. Used traditionally about thrust ramps, but now also about ramps along extensional faults.

General shear: Deformation that is more complex than simple shear, usually involving three-dimensional strain. Used synonymously with subsimple shear by some geologists.

General shear zone: Shear zone that deviates from the ideal shear zone model.

Geometric striae: Linear irregularities on a slip surface, e.g. surface corrugations.

Gliding model: Orogenic model popular in the 1950s and 1960s in which nappes were thought to gravitationally glide from elevated parts of the orogen.

Gneissic banding: Banding or layering where individual bands consist of different minerals, often representing transposed and flattened dikes or other primary structures.

Graben: German for grave. A depression bounded by two more or less strike-parallel but oppositely dipping normal faults or vertical faults.

Grain boundary diffusion: Diffusion of crystal vacancies along grain boundaries – a plastic deformation mechanism. Also called Coble creep.

Grain boundary migration: Recrystallization of crystals by means of migration of grain boundaries.

Grain boundary sliding: Plastic deformation mechanism where grains slide as a result of diffusion (not to be confused with frictional grain boundary sliding, which is a brittle mechanism).

Granular flow: Particles flowing by frictional sliding and rolling (translation and rigid rotation). Typical for deformation of loose sand or soil. Also called particulate flow.

Gravitational orogenic collapse: Orogen collapsing under its own weight, occurring when gravitational forces created by the weight of the elevated top of the orogen exceed the strength of the orogenic edifice.

Griffith fracture criterion: Equation 7.6, which is non-linear (curved) in the Mohr diagram.

Groove lineation: Lineation defined by fault grooves.

Growth fault: A shallow normal fault that has moved during deposition of sediments on the hanging-wall side. The hanging-wall strata thicken toward the fault and may also be more coarse-grained close to the fault. The fault displacement increases downwards as fault dip decreases.

Hackles, hackle marks: Plumose curvilinear patterns defined by gentle relief on extension fractures, radiating from the nucleation point of the fracture or fanning away from a curvilinear axis. Identical to plumose structures.

Half-graben: Structural depression controlled by one master normal fault and typically also antithetic (oppositely dipping) minor normal faults in the rotated hanging-wall layers. An asymmetric graben is a graben structure somewhere between a perfect half-graben and a symmetric graben.

Halokinesis: The study of salt tectonics, i.e. the formation of salt diapirs and related structures due to flow of salt in the subsurface.

Halokinetic cycle: Cyclic sedimentary sequences along salt diapirs where cycles are separated by unconformities and caused by non-steady rise of the diapir. Halokinetic cycles are vertically stacked along the walls of diapirs.

Hanging wall: The rock volume above a dipping fault.

Hanging-wall collapse: The formation of one or more secondary faults in the hanging wall to a fault. Expression most commonly used about normal faults.

Hard link: Expression used in fault overlap zones where the overlapping faults are connected by at least one fault that is mappable at the scale of observation.

Harmonic folds: Folds that are repeated with similar shape along the axial trace.

Heave: The horizontal component of the dip separation of a fault. Equal to the horizontal component of the true displacement vector for dip-slip faults.

Heterogeneous deformation: Deformation varies within the area or volume in question. Also called inhomogeneous deformation.

Heterogeneous strain: The state of strain (strain ellipse or ellipsoid) varies within the area or volume in question. Also called inhomogeneous strain.

High-angle fault: Fault dipping more than $30°$.

Hinge: The area of maximum curvature of a folded surface, i.e. the zone that connects the fold limbs.

Hinge line: The line of maximum curvature, i.e. the line defined by consecutive hinge points on a folded surface. Linear and known as the fold axis for cylindrical folds.

Hinge point: Point of maximum curvature on a folded surface.

Hinge zone: Zone of maximum curvature on a fold.

Hinterland: The central or internal zone of an orogen, as opposed to the foreland. The hinterland is characterized by basement involvement and locally high metamorphic grade.

Homogeneous deformation: Deformation is everywhere the same within the area or volume in question.

Homogeneous strain: The state of strain is everywhere the same within the area or volume in question, meaning that the strain in the entire area or volume can be represented by a single strain ellipse or ellipsoid.

Homologous temperature: The ratio of a material's temperature T to its melting temperature T_m using the Kelvin scale.

Horizontal separation: The fault-related separation of layers observed on a horizontal surface across the fault.

Horse: The smallest tectonic unit in thrust terminology: a tectonic sheet bounded by thrust faults on each side and occurring in trains in duplex structures. S-shaped geometry common. Now also used in normal fault terminology (horses in extensional duplexes).

Horsetail fractures: Fractures splaying off the tip of a larger fracture.

Horst: Elongated area that is stratigraphically elevated relative to rocks on each side. A horst is bounded by normal faults that are vertical or dipping away from the horst.

Hybrid fracture: Combination of shear (Mode II) and opening (Mode I) fractures.

Hydraulic fracturing: Increasing the fluid pressure in an interval in a well until the formation pressure is exceeded and the rock fractures. Used to stimulate hydrocarbon producers or water injectors.

Hydrostatic stress: State of stress where the stress is the same in all directions (spherical stress ellipsoid). Occurs in fluids only, including magma.

Ideal shear zone: Ductile simple-shear zone with or without additional compaction or dilation across its planar and parallel walls.

Imbrication zone: A series of reverse faults dipping in the same direction and soling out on a floor thrust, but not necessarily bounded by a roof thrust. Also used for similar arrangement of normal faults.

Incremental strain ellipse: Strain ellipse for a small part of the total deformation history. The incremental strain ellipses sum up to the finite strain ellipse.

In-sequence thrusting: Thrusts getting younger toward the foreland.

Inhomogeneous deformation/strain: See *heterogeneous deformation/strain*.

Instantaneous stretching axes (ISA): Directions of maximum ongoing stretching (ISA_1), minimum stretching (maximum shortening; ISA_3) and an intermediate axis perpendicular to the other two (ISA_2). These axes are defined at any instant during the deformation.

Interlimb angle: The internal angle between the two limbs of a fold.

Internal rotation component of strain: The difference (in degrees) between the orientation of a material line situated along the maximum principal strain axis and the orientation of the same line prior to deformation.

Interplate faults: Faults defining a plate boundary.

Intersection lineation: Lineation formed by the intersection between two planar structures, such as bedding and a cleavage.

Intertectonic: Between two phases of deformation.

Intraplate faults: Faults occurring within tectonic plates, i.e. not defining part of a plate boundary.

Inverse deformation: The deformation (transformation) that takes a deformed object back to its undeformed state. Also called reciprocal deformation or strain.

Inversion: (1) Turning stratigraphy upside-down by means of recumbent folding (inverted limb). (2) Reactivating a normal fault as a reverse fault. (3) Turning a basin into a high and vice versa (related to the previous definition).

ISA: See *instantaneous stretching axes*.

Isochoric: Having constant volume or area.

Isotropic medium: A medium that has identical mechanical properties in all directions, so that it reacts identically to stress regardless of its orientation.

Isotropic volume change: Volume change caused by shrinking or expanding a volume by the same amount in all directions. Also called volumetric strain.

Joint: Extensional fracture, often laterally extensive (up to hundreds of meters) with very small (microscopic) displacement.

Juxtaposition: Description of the lithologic contact relations on a fault. Commonly expressed in a juxtaposition or triangle diagram.

Juxtaposition seal: Seal where sand is in complete contact with shale across the fault.

Kinematics: From Greek "kinema", meaning motion. The description of how rock masses or objects in rocks move as a result of deformation.

Kinematic axes: A kinematic framework defined by three orthogonal axes, *a*, *b* and *c*, with the *a*-axis representing the transport direction and *c* being perpendicular to the shear plane.

Kinematic indicator: Any structure indicating the sense of shear or transport during a deformation event. Examples include shear bands in mylonites, rotated porphyroclasts, drag folds and Riedel shears associated with faults.

Kinematic vorticity number, W_k: dimensionless number representing the ratio between the rate of rotation of the strain ellipsoid and the rate of strain accumulation. $W_k = 0$ for pure shear and 1 for simple shear.

Kink bands: More or less synonymous with kink folds, emphasizing the fact that one limb of such asymmetric folds commonly defines bands of anomalous dips.

Kink folds: Small (typical centimeter-scale) angular folds with straight hinges, thought to form with axial planes at an acute angle to σ_1.

Klippe: An erosional remnant of a thrust nappe.

L-fabric: Linear fabric.

L-tectonite: Strongly deformed rock dominated by a linear fabric.

LS-tectonite: Strongly deformed rock dominated by both linear and planar fabric elements.

Lateral ramps: Ramps that form parallel to the transport direction of an allochthonous unit.

Leading branch line: The frontal branch line.

Line defect: Linear defect in the crystal lattice, known as a dislocation.

Lineament: Straight or gently curved line feature on the surface of the Earth (or another planet), identified and mapped by means of remote sensing imagery. A lineament is likely to represent a geologic structure or lithologic contact.

Linear fabric: A lineation that penetrates the rock.

Lineation: Linear structure formed by means of tectonic strain, e.g. rotated amphibole needles in an amphibolite, stretched aggregates of quartz and feldspar in a granitic gneiss, or striations on a fault surface. The linear objects are pervasive (metamorphic rocks) or limited to a fracture surface (brittle regime).

Listric (fault): From Greek "listros", meaning spoon-shaped. Geometric term that describes the downward flattening geometry of some faults. Faults that steepen by depth are sometimes called antilistric.

Lithostatic pressure: The product of the density and height of the overlying rock layers, multiplied by g, the acceleration due to gravity.

Lithostatic reference state: Reference state of stress in the crust where the crust is considered a medium without shear strength (i.e. a fluid) and where stress in any direction is the product of density, depth and g.

Low-angle fault: Fault dipping less than $30°$.

Lower plate: The footwall to a large-scale detachment extensional fault or shear zone.

M-domains: Microscopic cleavage domains dominated by mica (M) and sometimes also other phyllosilicates and opaque phases. M-domains are separated by QF-domains.

M-folds: Symmetric folds, typically folds occurring in the hinge zone of a larger fold.

Master fault: The largest fault in an area.

McKenzie model: The pure shear model for rifting, where the lithosphere is stretched symmetrically in an overall pure-shear style.

Mean stress: The arithmetic mean (average) of the principal stresses.

Mechanical stratigraphy: Stratigraphy based on the mechanical properties of layers rather than lithology or sedimentary characteristics.

Mechanical twinning: The formation of twins in response to a directed stress.

Metamorphic core complex: Exposed portion (window) of the metamorphic rocks of the lower plate (footwall) of a low-angle normal fault (detachment). According to the original use of this term, the upper plate should be exposed to brittle deformation while the lower one contains mylonitic rocks formed in the plastic regime.

Mica fish: Mica grains with pointed and oppositely bent tails, typically delimited by shear bands. Characteristic of micaceous mylonites (or phyllonites).

Microstructures: Structures that range in size from the atomic scale to the scale of grain aggregates, observable under the optical or electron microscope.

Microtectonics: The study of small-scale deformation structures that yield information about strain, kinematics and the deformation history.

Mineral fiber lineation: Lineation formed by growth of fibrous or elongated minerals, such as quartz, serpentine and actinolite.

Mineral lineation: Lineation deformed by parallel-arranged minerals on a surface or throughout a given volume of rocks.

Minibasin: Small basin forming atop, between or adjacent to salt diapirs.

Missing section: See *fault cut*.

Mode I fracture: Opening mode or extension fracture.

Mode II fracture: Shear fracture, where the movement is into and out of the plane of observation. Also known as the sliding mode.

Mode III fracture: Shear fracture with movement parallel to the edge, i.e. the plane of observation is parallel to the slip vector. Also called the tearing mode.

Mohr circle: Circle in the Mohr diagram that describes the normal and shear stress acting on planes of all possible orientations through a point in the rock.

Mohr diagram: Diagram where the horizontal and vertical axes represent the normal (σ_n) and shear (σ_s) stresses that act on planes through a point.

Mohr failure envelope: The failure envelope of a material found experimentally, regardless of whether it obeys the Coulomb fracture criterion or not.

Monocline: Sub-cylindrical fold with only one inclined limb (the other limb is the regionally horizontal layering).

Mullion: Linear deformation structures at the interface between a competent and an incompetent layer, where cusp shapes point into the more competent rock.

Mylonite: Well-foliated tectonic rock formed by intense plastic deformation, usually at middle crustal levels and deeper. Normally characterized by grain size reduction, although grain size increase can occur when fine-grained rocks become mylonitized. Subordinate brittle deformation can occur. Subdivided into protomylonite, blastomylonite,

mylonite and ultramylonite depending on the degree of recrystallization.

Mylonite nappe: Nappe dominated by mylonitic rocks.

Mylonitic foliation: The foliation formed during mylonitization: usually a strong and compositional foliation defined by parallel minerals and mineral aggregates, lenses and parallel layers reflecting primary structures such as dikes and bedding.

Mylonitization: The process that transforms a rock to a mylonite. This occurs predominantly by means of plastic deformation mechanisms, commonly with subordinate brittle microfracturing.

Nabarro–Herring creep: See *volume diffusion*.

Nappe: Allochthonous unit, usually a thrust nappe, but the expression has also been used for extensional allochthonous units that are resting on extensional detachments.

Nappe complex: A collection of thrust nappes that share common lithological and/or structural features and form a single unit.

Neck: Term used to describe the narrow connection between elements in pinch-and-swell structures.

Necking: The formation of pinch-and-swell structures.

Neutral point: Point in a fold hinge of no stretching or shortening.

Neutral surface: Theoretical surface in a fold hinge zone separating layer-parallel extension in the outer part from layer-parallel shortening in the inner part of the hinge.

Newtonian fluid: A material that deforms so that shear stress and shear strain are linearly related (Equation 6.22). Also called linear or perfectly viscous material.

Non-coaxial deformation: Lines along the principal strain axes do not have the same orientation before and after the deformation.

Non-coaxial deformation history: Lines along ISA as well as the principal strain axes rotate during the deformation history: $W_k \neq 0$.

Non-steady-state deformation: The particle paths and flow parameters such as ISA and W_k vary during the deformation history.

Normal drag: Rotation of layers in the walls of a fault so that the curvature is consistent with sense of offset on the fault.

Normal fault: Fault where the hanging wall has moved down relative to the footwall. Normal faults are extensional with respect to a horizontal layer or the surface of the Earth.

Normal stress: Stress or stress component acting perpendicular to the surface of reference.

Oblate: Disk-shaped geometry, used to describe the strain ellipsoid. For perfect oblate objects the maximum and intermediate axes (X and Y) are of the same length: $X = Y \gg Z$.

Oblique ramp: Ramps oriented obliquely to the transport direction (in thrust systems or extensional fault systems).

Oblique-slip fault: Fault where the displacement vector is dipping at a lower angle than the dip of the fault, i.e. a mixture of strike-slip and either normal or reverse movement.

Oceanic metamorphic core complex: Metamorphic core complex formed in an oceanic rift setting, where the metamorphic core consists of serpentinites.

Open-toed advance: Salt sheet advance by means of extrusion at the toe while the rest of the sheet is being covered by sediments.

Orogenic root collapse: Model where the orogenic root is less dense than its surroundings so that it collapses upward and spreads out laterally.

Orogenic wedge: The wedge-shaped (as seen in cross-section) area of allochthons in an orogenic zone or mountain range: thickest in the hinterland and thinning toward the foreland.

Orthogonal flexure: Fold mechanism where lines originally orthogonal to the layering remain orthogonal, producing a neutral surface separating outer-arc stretching from inner-arc shortening. Also called tangential longitudinal strain.

Outlier: An erosional remnant of a thrust nappe. Identical to klippe.

Out of sequence thrusting: Thrusts not getting systematically younger (or older) toward the foreland.

Overcoring: Strain relaxation method where the expansion of a drill core is measured and related to stress.

Overlap zone: Zone between overlapping fault segments. Can be hard linked, where the overlapping folds are physically connected, or soft linked, where the strain in the overlap zone is ductile (continuous deformation). A relay ramp is a common expression of the latter.

Overlapping faults: Two faults with approximately the same strike orientation but laterally offset with respect to each other so that the fault tips are misaligned. The tips have grown past each other by a length that is much smaller than the fault length, and the two faults must be close enough that their elastic strain fields overlap.

Overpressure: The pore pressure in a stratigraphic unit exceeds the hydrostatic pressure. This can occur in a highly porous and permeable unit (e.g. sandstone) that is captured between two impermeable layers (e.g. shale) so that the pore fluid cannot escape. Further lithostatic loading (burial) will then generate an overpressure, which reduces the effective stress and counteracts physical compaction.

Overprinting relations: Relative age relations between structures that can be observed in the field, from aerial photos, satellite images or seismic sections, for example one fault cutting another, or a fold being refolded.

P-shear fractures, P-shears: Sets of subsidiary slip surfaces arranged oblique to the zone. When the section of observation (which for a strike-slip fault should be horizontal) is considered as a profile, P-shears define a thrust-type sense of movement.

P–T path: A path in a *P–T* diagram (diagram with pressure and temperature along the axes), with an arrow on the path showing the development. The path is based on identification of metamorphic parageneses indicating pressure and temperature at different times.

P–T–t path: A path in *P–T* space that contains points that have been dated, usually radiometrically.

Paleopiezometer: Use of recrystallized grain size to estimate the level of differential stress during recrystallization (or subgrain formation), based on experimental or theoretical data.

Parallel folds: Folds showing constant layer thickness (Class 1B).

Parasitic folds: Folds occurring on the limbs or hinge of a larger fold that formed during a single process of folding.

Parautochthonous: Almost autochthonous rocks in an orogen, with only short transportation. The rocks can easily be correlated with the autochthonous.

Particle path: The path traced out by particles in a deforming medium during progressive deformation.

Particulate flow: Identical to *granular flow*.

Passive diapirism: Diapirism where salt is at or near the surface and rises due to loading by sediments continually accumulating and downbuilding around the diapir, displacing salt into the diapir.

Passive folding: Folding where the layering has no mechanical influence (no competency contrast): the layers merely act as markers.

Passive marker: Structure or object that has no mechanical or rheological contrast to its surroundings so that it deforms identically to its surrounding rock. This implies that it will not buckle or boudinage.

Passive rifting: Rifting forming in response to, and driven by, far-field forces related to plate interactions. Passive rifts form along weak zones in the lithosphere.

Pencil cleavage: Two differently oriented cleavages in shales causing the shale to break into pencil-like fragments. Typically a combination of compaction and tectonic strain.

Perfect plastic deformation: Time-independent plastic deformation where strain rate has no influence on the stress–strain curve. Perfect plastic deformation does not show any work (strain) hardening or softening.

Perfect plastic material: Incompressible material that accumulates permanent strain at a constant stress level and where this level of stress (the yield stress) cannot be increased even if we try: the yield stress is independent of strain rate.

Periodic folds: Folds along a single layer showing constant wavelength–thickness ratio (L/h).

Permanent strain: Strain that remains after the removal of the stress field that caused the strain.

Phyllitic cleavage: Continuous cleavage formed under lower-middle greenschist facies conditions in phyllitic rocks.

Phyllonite: Micaeous mylonite, typically strongly sheared low-grade metapelite.

Phyllosilicate deformation bands: Deformation bands where phyllosilicates have reoriented to form a local fabric along the band. Common in phyllosilicate-bearing sands and sandstones. Also called phyllosilicate framework structures.

Pinch-and-swell structures: Necking of (usually) competent layers without the formation of clearly separated boudins.

Pinning: Expression used about the role of non-recrystallizing minerals as obstacles in a rock that undergoes recrystallization.

Pinpoint: Fixed point of reference during restoration or balancing.

Pitch: The angle between the strike of a slip surface and the slip lineation, measured with a protractor on the slip surface. Also called rake.

Planar fabric: Foliation, cleavage.

Plane strain: Strain where the intermediate strain axis (Y) remains unaffected by the deformation ($Y = 1$), shortening in Z being compensated by extension in X. There is therefore no particle motion perpendicular to the *XZ*-plane. Plane strain produces strain ellipsoids that plot along the diagonal (*k* = 1) of the Flinn diagram.

Plastic behavior: Deformation that only occurs at a given yield stress. The deformation is permanent, and could be strain hardening or softening.

Plastic deformation: Ductile deformation resulting from plastic deformation mechanisms. Plastic deformation produces a non-recoverable change in shape (permanent strain) without failure by rupture. To some scientists (engineers in particular) plastic deformation is simply the accumulation of permanent deformation without fracturing in response to stress. Soil mechanics people therefore regard clay as a (quasi-)plastic material.

Plastic deformation mechanisms: Dislocation creep (glide + climb), twinning and diffusion. In a strict meaning of the term, diffusion is separated from both plastic and brittle deformation mechanisms.

Plastic shear zone: Shear zone dominated by plastic deformation mechanisms.

Plasticity: Deformation mechanisms where atomic bonds are broken while material coherency or cohesion is maintained. Note however that soil mechanicists relate plasticity merely to rheology (the accumulation of permanent strain).

Plumose structures: Subtle relief pattern on joints that resembles the structure of a feather, indicating the growth direction of joints. The pattern points toward the nucleation point of the joint. Commonly found on joints in fine-grained rocks. Also called hackle marks.

Poiseuille flow: Flow in a layer where the flow is fastest in the middle and decreases toward the margins due to friction.

Poisson's ratio: The ratio between strain imposed on a sample in one direction and the resulting strain perpendicular to this direction. If they balance there is no volume change, the material is incompressible, and Poisson's ratio is 0.5.

Polyclinal folds: Folds where axial surfaces show variable or contrasting dips.

Polyphasal: Having several phases; used about deformation that has occurred as discrete events, where the individual events or phases are named D_1, D_2, . . .

Pore pressure: The pressure in the fluid (water, oil or gas) filling the pore space in a porous rock.

Porphyroclast: Larger relict mineral grains, typically feldspar or other "resistant" minerals, in a strongly sheared finer-grained mylonitic matrix. Commonly with asymmetric tails that reflect sense of shear.

Postkinematic: After deformation.

Postrift: After rifting; typically used about the sequence of sediments deposited during the phase of thermal subsidence after the cessation of rifting and related extensional faulting and stretching.

Posttectonic: Something that happened after the deformation phase in question.

Power-law distribution: Data defining a straight line in a log-log diagram.

Pressure difference: Used here for difference in pressure from one location to another, typically resulting from differential loading. Creates an unstable situation in viscous rocks (fluids) that drives flow, for instance in a salt layer or in the asthenospheric mantle.

Pressure solution: Wet diffusion along grain boundaries. The only type of diffusion that can occur at submetamorphic temperatures. Commonly called dissolution or solution when found in sedimentary rocks, because temperature and chemical factors seem to be more important than pressure during compaction of sediments.

Pressure-solution cleavage: Tectonic cleavage defined by subparallel solution seams. Common in limestones.

Pretectonic: Prior to the deformation phase in question.

Primary fabric: Sedimentary and magmatic fabric formed as the rock was deposited or crystallized.

Primary foliation: Sedimentary and magmatic foliation formed as the rock was deposited or crystallized (e.g. bedding).

Principal planes of stress: The planes in a stress field that do not contain shear stresses. These planes are perpendicular to the principal stress axes.

Principal strain axes: The two (three) orthogonal axes of the strain ellipse (ellipsoid) that represent the directions and amounts of maximum extension (*X*) and shortening (*Z*). The strain ellipsoid has a third, intermediate axis (*Y*) that is orthogonal to the other two. Eigenvectors of the deformation matrix.

Principal stress axes, principal stresses: The three mutually orthogonal axes (two in two dimensions) defining the directions of the principal stresses and therefore the stress ellipsoid. Eigenvectors of the stress tensor. Values of the principal stresses are the lengths of these axes and the eigenvalues of the stress tensor.

Principal stretches: The lengths X, Y and Z of the principal strain axes.

Process zone: The zone ahead of a fracture where the formation of numerous microcracks softens the rock. The process zone propagates ahead of the fracture. Also called the frictional breakdown zone.

Prograde metamorphism: A metamorphic development toward higher metamorphic grade.

Progressive deformation: Ongoing and evolving deformation, producing a deformation history.

Prolate: Elongate three-dimensional shape reminiscent of a cigar. A perfectly prolate object has a major principal strain that is much longer than the other two axes, which are of equal length: $X \gg Y = Z$.

Protolith: The original rock with structures or textures that have not been significantly altered by the deformation in question, e.g. a protolithic lens in a shear zone.

Pseudotachylyte (US), pseudotachylite (UK): Glass or devitrified glass formed as a result of frictional melting during faulting.

Pull-apart basin: Basin formed in an extensional overlap or releasing bend of a strike-slip fault or fault zone.

Pulverization: Process occurring during faulting where rock is crushed to submicrometer particle sizes so that the fractured particles remain in place such that the original (primary) textures of the rock are clearly evident in hand sample. Pulverization is thought to be associated with supershear rupture velocity quakes (very fast rupture rates), which have been observed along strike-slip faults.

Pure shear: Plane strain coaxial deformation where particles move symmetrically around the principal axes of the strain ellipse in the XY-plane. Flow apophyses are orthogonal and $W_k = 0$.

Pure shear model of rifting: See *McKenzie model*.

Pure shear zone: Shear zone deformed by pure shear.

QF-domains: Microscopic cleavage domains dominated by quartz (Q) and feldspar (F), separated by mica-rich M-domains.

R′-fractures: Fractures in a shear setting that are antithetic to Riedel shears.

R-shears: See *Riedel shears*.

Radial faults: Faults forming a radial pattern, ideally radiating from (or projecting towards) a common center point. Typically found in the overburden of domes, notably salt domes.

Raft: Fault block in an extensional fault system atop a décollement, where the fault block is separated from pre-faulting strata in its neighboring fault blocks. The process is called rafting.

Rake: See *pitch*.

Ramp: The steep and relatively short segment of a thrust fault as it climbs to a higher level in the stratigraphy. Now also used for steep segments connecting two low-angle normal fault segments. See also *relay ramp*.

Ramp-flat-ramp geometry: A fault with a subhorizontal central segment bounded by steeper segments at each end.

Random fabric: Fabric where fabric elements show no preferred orientation.

Reactive diapirism: The process of salt rising as a reaction to significant tectonic brittle strain, usually extension.

Reciprocal deformation: The deformation (transformation) that takes a deformed object back to its undeformed state. Also called inverse deformation.

Recovery: Removal or rearrangement of dislocations in a crystal to produce domains of fewer dislocations and hence lower energy. Leads to subgrain formation.

Recrystallization: Process occurring during deformation (and diagenesis) by means of grain boundary migration and/or the formation of new grain boundaries by the localization of dislocations and point defects. Recrystallization that occurs during deformation is known as dynamic, whereas temperature-controlled recrystallization is static. Unstable minerals can also recrystallize into new minerals with different compositions during changing metamorphic conditions.

Recumbent fold: Fold with subhorizontal axial surface and horizontal hinge line.

Reduced stress tensor: The tensor containing information about the orientation of the principal stress axes and the relative (but not absolute) length of the axes.

Relay: The zone between overlapping fault segments.

Relay ramp: Folded area in a relay formed by flexing of layers between the fault tips. Usually used for subhorizontal layers that are given a ramp-like geometry in the overlap zone. The folding is due to strain transfer between the two faults.

Releasing bend: A bend along a strike-slip fault that has generated local extension.

Releasing overlap zones: Overlap zones where the fault arrangement and sense of displacement cause stretching within the overlap zone. Also called extensional overlap zones.

Remote stress: The stress field that is regionally present away from anomalies set up by local structures such as weak faults, i.e. remote with respect to some structure(s) in question. Also called far-field stress.

Repeated section: Stratigraphic section that is repeated because of faulting, i.e. the opposite of missing section. Whether one observe repeated or missing sections in a well depends on the orientations of the fault and the well. A vertical well gives repeated sections across reverse or thrust faults.

Residual stress: Stress that has been locked into a rock so that it is preserved after the external stress field has been changed or removed.

Restoration: Expression used for the reconstruction of a geologic section, map or 3-D model to its pre-deformational situation.

Restraining bends: Bends on strike-slip faults that generate local contraction or transpression.

Restraining overlap zones: Overlap zones with shortening in the displacement direction. Also called contractional overlap zones.

Retrodeform: Undeform, transform back to the pre-deformational stage.

Retrograde metamorphism: A metamorphic development toward lower metamorphic grade.

Reverse drag: Rotation of layers in the hanging wall of a fault so that the curvature is inconsistent with sense of offset on the fault. Rollover structures, which are related to fault geometry, are the most common example of reverse drag.

Reverse fault: Fault where the hanging wall has moved up relative to the footwall, implying shortening of horizontal layers.

Rheologic stratigraphy: The rheological stratification of the lithosphere or part of the lithosphere, where a subdivision can be made into layers of contrasting rheological properties.

Rheology: The study of flow (rheo in Greek) of any rock and other material that deforms as a continuum under the influence of stress. Elasticity, viscosity and combinations of these are different rheological behaviors.

Ribs and rib marks: Elliptical (conchoidal) structures occurring on extension fractures, centered on the nucleation point of the fracture. They are ridges or furrows where the fracture has a slightly anomalous orientation and are perpendicular to hackle marks.

Riedel shear fractures, Riedel shears: Sets of subsidiary slip surfaces arranged en echelon, each Riedel or R-shear being oblique to the zone or main slip surface. When the section of observation (which for a strike-slip fault should be horizontal) is considered as a profile, R-shears appear extensional. An antithetic set (R′ or R prime) also occurs, although less commonly than R-shears.

Rim syncline: Syncline along the rim of a map-scale structure. Typically used about folds encircling and outlining salt diapirs.

Rodding: Elongated mineral aggregates (lineation) forming during deformation.

Rolling hinge model: Dynamic model where the upper part of an initially steep normal fault flattens as offset accumulates. The hinge of the curved fault moves or "rolls" toward the hanging wall due to isostatic adjustments during faulting. The model results in a metamorphic core complex.

Rollover: The fold structure defined by the steepening of otherwise horizontal hanging-wall layers toward a normal fault. Normally related to a downward-flattening (listric) fault.

Roof thrust: Low-angle fault that defines the upper limitation of a duplex structure.

Rotational deformation: Expression used for non-coaxial deformations.

S-C mylonites: Mylonites showing two sets of more or less simultaneous surfaces (S and C) with an oblique angle that indicates the sense of shear.

S-fabric: Planar fabric.

S-fold: Asymmetric fold where the short limbs appear to have been rotated anticlockwise with respect to their long limbs.

S-tectonite: Strongly deformed rock dominated by a planar fabric.

Salt anticline: Salt-cored anticline, usually formed by salt movement (flow into the anticlinal structure) and bending of overlying strata (no intrusion of salt).

Salt (stock) canopy: Salt from several upward-widening diapirs that have merged at a higher stratigraphic level than their source salt layer.

Salt expulsion: Another name for salt withdrawal.

Salt glacier: Sheet or tongue of salt flowing from a surfacing salt diapir where salt is extruded, usually due to contractional strains.

Salt (wing) intrusion: Salt intruding a higher stratigraphic level below the surface. Sometimes called salt wing intrusion.

Salt nappe: Equivalent to salt sheet.

Salt pillow: A concentration of salt caused by flow of salt into a sink where the salt has not intruded the overburden.

Salt roller: Asymmetric triangular-shaped volume of salt forming in the footwall of a normal fault.

Salt sheet: Sheet of allochthonous salt formed by the lateral flow of salt from one or more diapirs, generally at or near the surface.

Salt stock: Salt diapir with more or less cylindrical geometry.

Salt tectonics: Term used when salt is involved in deformation to the extent that it influences the type, geometry, localization and/or extent of deformation structures that form. Does not imply a direct relationship with plate tectonic stress.

Salt wall: Elongated salt diapirs, where the length is many times the width of the structure.

Salt weld: Point or area where a salt layer becomes completely exhausted (removed by flow) so that the boundary layers become attached to each other.

Salt withdrawal: Salt flowing from a local area into an adjacent salt structure. Salt withdrawal can result in local sedimentary basins (so-called minibasins) above the area of withdrawal, and generate salt welds when all the salt has withdrawn.

Scale model: A model that has been properly scaled down (or up) from some natural example.

Scaling: Changing natural physical quantities to those appropriate for a given laboratory setup. Parallel to the scaling of lengths, such as the thickness of the crust, it is necessary to scale strain rate and quantities such as temperature, viscosity, stress, gravity or confining stress, cohesion and grain size.

Schist: Metamorphic deformed rock showing a well-developed schistosity. Depending on the mineralogy, schists are named mica schists, quartz schists, greenschists etc.

Schistosity: Tectonic foliation defined by coarse-grained platy minerals in rocks deformed under upper greenschist and amphibolite facies conditions.

Screw dislocation: See Figure 10.11.

Sealing fault: Impermeable or very low permeability fault in porous and permeable rocks that prevents fluids from crossing the fault. Shale or clay smearing as well as a membrane of non-permeable fault gouge may seal faults, even if there is sand–sand contact along the fault surface.

Secondary foliation: Foliation formed after the rock was deposited or crystallized.

Seismogenic zone: The zone of frequent earthquakes in the crust, which is the middle and lower part of the brittle crust, where the crust is strongest.

Self-juxtaposed seal: Reservoir against reservoir across the fault. Sealing due to fault rock development, including cataclasis and cementation in or along the fault.

Semi-ductile: Both ductile and brittle deformation style, i.e. originally continuous layers are partly deflected and partly affected by discontinuities.

Shale gouge ratio (SGR): Ratio used in fault sealing analysis, relating the amount of shale that has passed a point at a fault surface to the local fault throw.

Shale smear factor (SSF): Ratio between fault throw and the thickness of the shale source layer. SSF is used in the evaluation of the sealing properties of faults.

Sharp discontinuity: Term used about fractures (including stylolites), i.e. structures that are very thin relative to their length and height. The term discontinuity relates to abrupt changes in displacement (rate) and also implies that the structures represent mechanical anomalies characterized by reduced shear strength, tensile strength and stiffness, and also high fluid conductivity relative to the host rock. See also *tabular discontinuities*.

Shear bands: Small-scale (millimeter- or centimeter-scale) shear zones, usually in well-foliated mylonitic rocks. Single sets of shear bands make an oblique angle to the foliation that indicates the sense of shear in sheared rocks.

Shear fold: Term sometimes used for similar folds.

Shear fracture: Fracture with detectable wall-parallel displacement. Different from fault in that it only consists of a single fracture, while faults are composed of a number of linked fractures.

Shear sense indicators: See *kinematic indicators*.

Shear strain (γ): The tangent value of the change in angle ψ experienced by two originally perpendicular lines.

Shear strain rate: The rate at which shear strain accumulates.

Shear stress: Stress acting parallel to a plane of reference.

Shear zone: Tabular strain zone dominated by ductile (continuous) deformation, typically dominated by simple shear.

Shear zone walls: The two bounding sides of a shear zone. Also called shear zone margins.

Sheath folds: Highly non-cylindrical folds formed in zones of high shear strains.

Shortcut fault: Fault or fault segment that establishes a new and shorter or less painful trace, for instance during fault reactivation or due to geometric complications such as a kink in the fault plane. In the latter case a shortcut fault would cut across the kink and make the fault straighter.

Shortening: See *contraction*.

Similar fold: Fold where the inner and outer arc have exactly the same shape, implying that the distance between the arcs parallel to the axial trace is everywhere the same. Similar folds are Class 2 folds according to Ramsay's dip isogon-based classification.

Simple shear: Non-coaxial plane strain deformation where particles move along straight lines and $W_k = 1$.

Simple shear model of rifting: See *Wernicke model*.

Simple shear zone: Shear zone deformed by simple shear.

Sinistral: Left-lateral, moving left relative to the point of reference. Used about relative sense of movement on faults, indicating that the opposite fault block is offset or moves to the left.

Slaty cleavage: Tectonic cleavage in slate that forms by reorientation and pressure solution (diffusion) of minerals.

Slickenlines: Lineations found on *slickensides*, providing information about the direction of slip on faults.

Slickensides: The finely polished surface displayed by faults, formed by extensive grain crushing (milling) and/or synkinematic mineral growth.

Slickolites: The peaks or "teeth" on tectonic stylolites.

Slip: (1) Shear motion localized to a surface (slip surface). (2) Movement of a dislocation front within a crystallographic plane.

Slip plane: (1) Planar slip surface. (2) Crystal lattice plane along which dislocations move.

Slip surface: Well-defined surface or narrow (<1 mm) zone along which slip has occurred. In principal, the amount of slip can be from a few centimeters to several kilometers, but large displacements tend to produce a zone (not surface) of brittle shearing known as the fault core.

A slip plane and a fault are equivalent terms to some geologists.

Smear: Zone or membrane of fine-grained and commonly impermeable rock or sediment smeared along a fault. The smear comes from clay, shale or other fine-grained rock in the rock strata affected by faulting.

Soft domino model: The domino model modified to allow for ductile strain within fault blocks.

Soft link: Expression used for fault overlap zones where there is no fault connecting the two overlapping fault segments. Instead, a relay ramp may be present.

Soil mechanics: Engineering mechanics applied to soils.

Sole thrust: Low-angle thrust fault that marks the base of a tectonic unit.

Spaced cleavage: Cleavage where the width of the cleavage domains is at least 1 mm, i.e. distinguishable in a rock sample.

Splay faults: Smaller faults horse-tailing from a fault near its tip point.

Spreading: Gravitationally driven translation of rocks with a radial displacement or velocity pattern, similar to what happens when putting soft dough on a table.

Static recrystallization: Recrystallization of a rock or mineral grain (shortly) after deformation. Characterized by equigranular texture unstrained grains (no pronounced undulatory extinction under the microscope etc.).

Steady-state deformation: Deformation where particle paths and flow parameters such as ISA and W_k are constant throughout the deformation history.

Steady-state flow (or creep): Deformation of a material at constant strain rate. See also *steady-state deformation*.

Stem: Thin part of a teardrop diapir.

Stickolites: Tubular structures perpendicular to pressure solution seams in limestones.

Stick-slip: Sudden (seismic) slip events separated by periods of no motion.

Strain: Change in length (1-D) or shape (2-D or 3-D) due to deformation. Primarily defined for continuous (ductile) deformation, but is also used in areas of faulting: see *fault strain*.

Strain axes: See *principal strain axes*.

Strain compatibility: Fulfilled when there is no slip along an interface separating differently strained layers.

Strain ellipse: The ellipse resulting from the deformation of a unit sphere passively present in the deforming medium. The strain ellipse has two principal axes that define the directions of maximum, minimum and intermediate strains (principal strain axes).

Strain energy: The energy stored within a crystal as dislocations accumulate during straining. Proportional to dislocation density.

Strain geometry: Shape of the strain ellipsoid or Flinn k-value, commonly displayed in the Flinn diagram.

Strain hardening: The effect in which the stress level must be increased in order to maintain a fixed strain rate. Best constrained in the laboratory during rock mechanics experiments. In a shear zone or deformation band setting, strain hardening means that it is easier to deform the wall rock than to continue the deformation in the zone or band. Strain hardening is also referred to as deformation hardening or work hardening.

Strain invariants: The principal strain axes and their lengths (eigenvectors and eigenvalues of the deformation matrix), since they are independent of our choice of coordinate system.

Strain markers: Objects reflecting the state of strain in a deformed medium, i.e. lines or objects of known undeformed length, shape and/or orientation.

Strain partitioning: The physical decomposition of strain into different components from micro- to mesoscale. For example: Transpression can be partitioned into zones of pure and simple shear, e.g. along obliquely convergent plate boundaries. At the centimeter-scale, simple shear can be partitioned into oblique shear bands and domains experiencing back-rotation.

Strain rate: The rate or speed at which strain accumulates. Two different types of strain rate are in common use: elongation rate and shear strain rate.

Strain shadows: Sectors on each side of a porphyroclast or other rigid object in a shear zone or mylonite zone where minerals may crystallize into tails. May be symmetric or asymmetric, in the latter case serving as sense-of-shear indicators. Related to sectors of instantaneous stretching. Also called pressure shadows and pressure fringes.

Strain softening: The effect in which the stress level must be decreased in order to maintain a fixed strain rate, i.e. opposite to strain hardening.

Stratigraphic separation: The missing or repeated section seen in a well that penetrates a fault.

Strength: The amount of stress that a rock can sustain before it fails or yields. Differential stress ($\sigma_1 - \sigma_3$) is commonly implied in geology. Unconfined rock samples can be tested in the laboratory under axial compression, axial tension or shear, which will give different values (compressive strength, tensile strength and shear strength). However, addition of a confining pressure increases the strength of a rock, or the differential stress that is required for fracture or permanent strain to accumulate.

Stress: Stress on a surface is the force over the area on which it is applied. Typically decomposed into a normal and a shear component. The complete state of stress at a point is given by the stress tensor and illustrated by the stress ellipsoid.

Stress ellipsoid: Ellipsoid that describes the state of stress at a point. A point represents the intersection of an infinite number of planes, and the normal stress to any given plane at that point is the distance from the point to the ellipse, measured perpendicular to the plane in question. The axes of the ellipse are the principal stresses (maximum, minimum and intermediate normal stresses, designated σ_1, σ_2 and σ_3), oriented perpendicular to the principal planes of stress. A stress ellipsoid only exists when all three principal stresses are compressive or tensile. Stress ellipse in two dimensions.

Stress guide: Competent layer along which stress is preferentially transmitted. Such layers could be limestones in sedimentary rocks, or the lower part of the brittle upper crust at lithospheric scale.

Stress tensor (matrix): Tensor describing the state of stress at a point.

Stretching (s): $s = 1 + e$, where e is elongation (extension). Equal to the beta factor reported in regional estimates of the stretching of a basin.

Stretching lineation: Lineation formed by tectonic stretching of objects, such as mineral aggregates and conglomerate clasts.

Striations, striae: Linear scratches and grooves on a slip surface, formed by frictional movement of the hanging wall against the footwall of a fault.

Strike-slip duplex: Duplex formed along a strike-slip fault, similar to classic thrust- or extension-related duplexes turned on their side. Hence, there are two types: extensional and contractional strike-slip duplexes.

Strike-slip fault: Fault where the displacement is horizontal. Strike-slip faults are dextral (right-lateral) or sinistral (left-lateral).

Structural geology: The study of deformation structures in the lithosphere in order to understand their geometry, distribution and formation.

Structural style: The distinctive appearance of a set of structures. For example, folds having similar opening angle, vergence and/or metamorphic mineral association share a structural style that makes them distinct from other folds in the same rocks, or similar to folds in another area or unit. Tectonic setting, depth–temperature conditions and lithologic characteristics all influence the resulting structural style.

Stylolitic cleavage: Cleavage consisting of stylolites (pressure solution seams). Also called pressure-solution cleavage.

Subgrains: Part of a mineral grain that has a uniform optical, and therefore also crystallographic, orientation that differs from that of the rest of the grain. At the crystal lattice scale, subgrains are accumulations of dislocations in the form of dislocation walls.

Subseismic faults: Faults that are too small to be imaged on a given seismic line or survey. Their size range is defined upward by the resolution of the seismic data in each case, and other data, such as cores, outcrop data or dipmeter data, must be used to identify them.

Subsimple shear: Planar deformation intermediate between simple and pure shear, i.e. $0 < W_k < 1$ and the flow apophyses make an acute angle.

Subsimple shear zone: Shear zone where the deformation is subsimple shear.

Superplasticity: Microscale deformation process dominated by grain boundary sliding and favored by fine grain size. No longer widely used.

Surface forces: Forces that act on a surface (traction).

Surface lineations: Lineations limited to a surface (fault or fracture), i.e. not penetrative.

Synclinal: Fold where rock layers get older away from the axial surface of the fold.

Synform: Fold where the limbs dip down and toward the hinge zone.

Synformal anticline: An anticline (strata get younger away from its axial surface) that has the shape of a synform, i.e. an anticline turned upside down.

Synkinematic: Formed or occurring during deformation.

Synrift: Formed during rifting, typically used about sedimentary sequences that show thickening toward the hanging-wall side of extensional faults.

Syntectonic: During the deformation phase in question.

Synthetic fault: A small fault dipping in the same direction as an adjacent main fault.

Synthetic shear: Shear acting synthetically to the sense of displacement of a reference fault. Applied during restoration of cross-sections.

T-fractures: Extension fractures occurring in a brittle shear zone or fault. They initially indicate the instantaneous stretching direction, but easily rotate during continued deformation. Typically associated with R- and P-fractures in strike-slip fault zones.

Tabular discontinuity: Term used for strain localization structures with a measurable thickness in outcrop or hand specimen (i.e. thicker than sharp discontinuities) and a continuous change in strength or displacement across them. Deformation bands are examples of tabular discontinuities. Shear zones can be considered tabular discontinuities.

Tangent-lineation diagrams: Plots of fault data and their kinematics in stereonet that reveal paleostress orientations and possibly stress ellipsoid shape.

Tangential longitudinal strain: Fold mechanism where lines that were orthogonal to the folded layer before folding remain so after the folding is over. The mechanism implies outer-arc extension and inner-arc contraction.

Tear fault: See *transfer fault*.

Teardrop diapir: Allochthonous salt shaped like an inverted teardrop, formed by welding of the stem of an upward-widening diapir.

Tectonics: From the Greek word "tektos" – to build. The knowledge of how the lithosphere is being built as rocks or sediments yield after a period of stress build-up. Examples are plate interactions (plate tectonics), salt movements (salt tectonics), the effect of glaciers (glaciotectonics) and collapsing orogens or fault scarps (gravity tectonics).

Tectonic fabric: Fabric resulting from a tectonic process, as opposed to primary fabric.

Tectonite: Strongly strained rock, usually a mylonite. A distinction is made between L-tectonites (strong lineation), S-tectonites (strong foliation) and LS-tectonites (pronounced foliation and lineation).

Tensile strength: The intercept of a fracture criterion or envelope with the negative horizontal axis. In general: the amount of tensile stress that a medium can withstand before failing.

Tension: Pull that can (but does not have to) result in extension.

Tensile fracture, tensile crack: Extension fractures with small openings perpendicular to the walls, i.e. Mode I or opening mode fractures. Most geologists find it useful to reserve the term tension for stress, in which case the term tensile fractures should be restricted to extension fractures formed under tension.

Tensor: An nth order tensor is an object that has m^n components, m being the dimension. Zero-order tensors are scalars, first-order tensors are vectors, while second-and higher-order tensors are matrices. Tensors have properties that are independent of coordinate systems. For instance, the deformation matrix (second- or third-order tensor in section or space) will describe the strain ellipsoid and dilation regardless of the related coordinate system.

Thermal loading: The (increased) buoyancy of salt caused by heating (salt expands and thus becomes lighter when heated).

Thermobarometry: The quantitative determination of the temperature and pressure at which a metamorphic (or igneous) rock reached chemical equilibrium. Metamorphic index minerals or mineral parageneses are used to estimate P and T, based on known reactions and stability fields.

Thick-skinned deformation: Orogenic deformation involving the basement and not only the overlying sedimentary cover.

Thin-skinned deformation: Orogenic deformation not involving the basement, only the overlying sedimentary cover. Basement nappes may still occur in the orogenic wedge, but these basement rocks must stem from an area far away from the area in question.

Throw: The vertical component of the dip separation of faults.

Thrust advance: The motion of salt sheets by thrusting (shear strain localization along the base of the salt sheet).

Thrust (fault): Low-angle reverse fault with measureable displacement (often >10 km). More informally used for any low-angle reverse fault, regardless of the amount of displacement.

Thrust nappe: Map-scale rock unit that has been transported on a thrust fault for tens of kilometers or more.

Tightness (of folds): Fold geometry depending on the fold's interlimb angle. From gentle via open and tight to isoclinal.

Traction: Stress on a surface.

Trailing branch line: The rearmost branch line of a nappe or nappe complex.

Transcurrent faults: Strike-slip faults with unconstrained tip.

Transected fold: Fold whose axial plane cleavage is oblique with respect to the axial surface of the fold.

Transecting cleavage: The cleavage transecting the fold hinge of a transected fold.

Transfer fault: Fault that transfers offset from one fault to another.

Transform fault: Strike-slip fault that defines a plate boundary. Transform faults transfer movement between mid-ocean ridge segments or island-arc bounding faults.

Translation: Rigid movement without any rotation or strain. The displacement field consists of parallel and equally long displacement vectors.

Transmissibility: The ability of a fault to transmit fluids in a hydrocarbon or water reservoir.

Transposed layering: Layering formed by tectonic flattening of originally cross-cutting elements (dikes, beds, cross-beds, magmatic banding, tectonic foliation etc.) into a composite foliation of subparallel elements. The process involves high strain and the foliation reflects the flattening (XY) plane of the strain ellipsoid. Commonly seen in gneisses formed in the lower crust.

Transpression: Strike-slip zone with an additional and simultaneous shortening across the zone. Usually a three-dimensional deformation where the strain ellipses plot off the $k = 1$ diagonal of the Flinn diagram.

Transtension: Strike-slip zone with additional and simultaneous extension across the zone, i.e. opposite to transpression.

Triangle diagram: Diagram showing the juxtaposition of beds for a continuously increasing offset. The input for the construction of a triangle diagram is the local stratigraphy. Factors such as SGR-values can be calculated and incorporated into triangle diagrams.

Trishear: Model for the deformation ahead of a propagating fault tip, where shear fans out into an upward-widening zone of heterogeneous ductile isochoric deformation expressed by (fault propagation) folding of horizontal layers. See Box 8.3.

Undulatory (or undulose) extinction: Uneven extinction of mineral grains seen under the optical microscope with crossed polarizers on rotation of the stage.

Uniaxial contraction/shortening: Contraction in one direction and no strain in the plane perpendicular to this direction. Compaction of sediments is an example.

Uniaxial extension: Extension in one direction and no strain in the plane perpendicular to this direction. Uniaxial extension implies volume increase.

Uniaxial strain reference state: Model of the state of stress in the crust, where stress arises as a consequence of vertical compaction only (i.e. uniaxial compaction). Stresses arise because rocks cannot expand or contract in the horizontal plane and the stress level increases downward with the weight of the overburden.

Uniform extension: A state of strain where stretching in X is compensated for by equal shortening in the plane orthogonal to X.

Uniform flattening: A state of strain where shortening in Z is compensated for by identical stretching in all directions perpendicular to Z.

Uniform slip model: The slip at a given point is the same in each slip event, while the slipping area may vary.

Upper plate: The hanging wall to a large-scale extensional detachment fault or shear zone.

Upright fold: Fold with vertical axial surface and horizontal hinge line.

Variable slip model: Both the amount of slip and the rupture length vary from event to event.

Vein: Extension fracture filled with mineral(s).

Velocity field: A vector field describing the velocities of particles at any given moment during the deformation history.

Vergence: Term related to the geometry of asymmetric folds. The vergence direction is the direction of apparent shear if the fold is considered to be the result of simple shear.

Vertical shear: Simple shear with a vertical shear plane, i.e. particles moving along vertical lines. Commonly used to restore vertical sections.

Viscosity: A measure of the resistance of a fluid to deforming under shear stress or, less formally, of a medium's resistance to flow. "Thick" fluids therefore have a higher viscosity than "thin" or runny fluids. Over geologic time, rocks in the middle and lower crust can be considered as fluids with very high viscosity.

Viscous material: Material showing a linear relationship between shear stress and shear strain. Viscous materials also show a linear relationship between shear stress and shear strain rate.

Volume change: See *dilation*.

Volume diffusion: Diffusion creep where vacancies migrate through the crystal lattice. Also known as Nabarro–Herring creep.

Volumetric strain: Isotropic volume change.

Vorticity: The local angular rate of rotation in a fluid, or a measure of how fast a fluid rotates or circulates. Applied to rocks deforming (flowing) in the plastic regime. Mathematically, vorticity is the curl of the velocity, and therefore a vector whose axis is along the axis of rotation. It can be visualized by considering a tiny portion of the fluid or flowing rock freezing into an undeformable sphere. The vorticity vector is the axis of rotation through that sphere and the angular velocity (an expression of rotation rate) is half the vorticity. If the sphere does not rotate there is no vorticity.

Wall rock: The rock on each side of a shear zone or a fault.

Wallace–Bott hypothesis: Slip on a given fracture occurs along the greatest resolved shear stress.

Wernicke model: The simple-shear model for rifting, where the lithosphere is stretched asymmetrically with a controlling low-angle simple-shear zone in the central part of the rift.

Wet diffusion: The type of diffusion occurring by transport of ions in water films between grain contacts, generally at shallow depths.

Window: Erosional exposure of the rock unit underlying a nappe. Identical to fenster.

Wing crack: Tensile fracture forming at the tip of a shear fracture during fracture growth or reactivation, oriented oblique to the host fracture.

Work hardening/softening: More or less the same as strain hardening/softening, but strictly relates to the work (energy) involved rather than the level of stress.

Yield point: The point on a stress–strain curve that marks the transition from elastic to permanent deformation.

Yield stress: The critical stress that it takes for a rock to flow (yield).

Young's modulus: The ratio between stress and strain for an elastic material, which describes how much stress it takes to achieve a certain strain. Also called the elastic modulus.

Z-folds: Folds whose short limbs appear to have been rotated clockwise with respect to their long limbs.

Zonal crenulation cleavage: Crenulation cleavage through which the earlier foliation can be traced continuously.

References

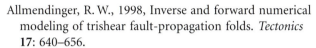

Allmendinger, R. W., 1998, Inverse and forward numerical modeling of trishear fault-propagation folds. *Tectonics* **17**: 640–656.

Amilibia, A., Sabat, F., McClay, K. R., Munoz, J. A., Roca, E. and Chong, G., 2008, The role of inherited tectono-sedimentary architecture in the development of the central Andean mountain belt: Insights from the Cordillera de Domeyko. *Journal of Structural Geology* **30**: 1520–1539.

Anderson, E. M., 1951, *The Dynamics of Faulting*. Edinburgh: Oliver and Boyd.

Angelier, J., 1994, Fault slip analysis and palaeostress reconstruction. In P. L. Hancock (Ed.), *Continental Deformation*. Oxford: Pergamon Press, pp. 53–100.

Austrheim, H., 1987, Eclogitization of lower crust granulites by fluid migration through shear zones. *Earth and Planetary Science Letters* **81**: 221–232.

Bhattacharyya, P. and Hudleston, P., 2001, Strain in ductile shear zones in the Caledonides of northern Sweden: a three-dimensional puzzle. *Journal of Structural Geology* **23**: 1549–1565.

Bingen, B., Austrheim, H., Whitehouse, M. J. and Davis, W. J., 2004, Trace element signature and U–Pb geochronology of eclogite-facies zircon, Bergen Arcs, Caledonides of W Norway. *Contributions to Mineralogy and Petrology* **147**: 671–683.

Breddin, H., 1956, Die tektonische Deformation der Fossilien im Rheinischen Schiefergebirge. *Zeitschrift Deutsche Geologische Gesellschaft* **106**: 227–305.

Buchner, E. and Kenkmann, T., 2008, Upheaval Dome, Utah, USA: Impact origin confirmed. *Geology* **36**: 227–230.

Burchfiel, B. C. and Stewart, J. H., 1966, "Pull-apart" origin of the central segment of Death Valley, California. *Geological Society of America Bulletin* **77**: 439–442.

Chemenda, A. I., Mattauer, M., Malavieille, J. and Bokun, A. N., 1995, A mechanism for syn-collisional rock exhumation and associated normal faulting: results from physical modelling. *Earth and Planetary Science Letters* **132**: 225–232.

Childs, C., Nicol, A., Walsh, J. J. and Watterson, J., 2003, The growth and propagation of synsedimentary faults. *Journal of Structural Geology* **25**: 633–648.

Choukroune, P., Gapais, D. and Merle, O., 1987, Shear criteria and structural symmetry. *Journal of Structural Geology* **9**: 525–530.

Cloos, E., 1968, Experimental analysis of Gulf Coast fracture patterns. *American Association of Petroleum Geologists Bulletin* **52**: 420–441.

Cruikshank, K. M., Zhao, G. and Johnson, A., 1991, Duplex structures connecting fault segments in Entrada Sandstone. *Journal of Structural Geology* **13**: 1185–1196.

Currie, J. B., Patnode, A. W. and Trump, R. P., 1962, Development of folds in sedimentary strata. *Geological Society of America Bulletin* **73**: 655–674.

Darby, D., Hazeldine, R. S. and Couples, G. D., 1996, Pressure cells and pressure seals in the UK Central Graben. *Marine and Petroleum Geology* **13**: 865–878.

Doelling, H. H., 2001, Geologic map of the Moab and eastern part of the San Rafael Desert 30′ × 60′ quadrangles, Grand and Emery Counties, Utah, and Mesa County, Colorado. *Utah Geological Survey Map 180*.

Durham, W. B. and Goetze, C., 1977, Plastic flow of oriented single crystals of olivine. I. Mechanical data. *Journal of Geophysical Research* **82**: 5737–5754.

Duval, B., Cramez, C. and Jackson, M. P. A., 1992, Raft tectonics in the Kwanza Basin, Angola. *Marine and Petroleum Geology* **9**: 389–390.

Dyer, R., 1988, Using joint interactions to estimate paleostress ratios. *Journal of Structural Geology* **10**: 685–699.

Engelder, T., 1985, Loading paths to joint propagation during a tectonic cycle: an example from the Appalachian Plateau, U.S.A. *Journal of Structural Geology* **7**: 459–476.

Engelder, T., 1993, *Stress Regimes in the Lithosphere*. Princeton: Princeton University Press.

Erslev, E. A., 1991, Trishear fault-propagation folding. *Geology* **19**: 617–620.

Fleuty, M. J., 1964, The description of folds. *London: Proceedings of the Geologists' Association* **75**: 461–492.

Fossen, H., 1992, The role of extensional tectonics in the Caledonides of South Norway. *Journal of Structural Geology* **14**: 1033–1046.

Fossen, H., 1993, Structural evolution of the Bergsdalen Nappes, Southwest Norway. *Norges Geologiske Undersøkelse Bulletin* **424**: 23–50.

Fossen, H., 1998, Advances in understanding the post-Caledonian structural evolution of the Bergen area, West Norway. *Norsk Geologisk Tidsskrift* **78**: 33–46.

Fossen, H., 2000, Extensional tectonics in the Caledonides: synorogenic or postorogenic? *Tectonics* **19**: 213–224.

Fossen, H. and Gabrielsen, R. H., 1996, Experimental modeling of extensional fault systems by use of plaster. *Journal of Structural Geology* **18**: 673–687.

Fossen, H. and Hesthammer, J., 2000, Possible absence of small faults in the Gullfaks Field, northern North Sea: implications for downscaling of faults in some porous sandstones. *Journal of Structural Geology* **22**: 851–863.

Fossen, H. and Rørnes, A., 1996, Properties of fault populations in the Gullfaks Field, northern North Sea. *Journal of Structural Geology* **18**: 179–190.

Fossen, H. and Rykkelid, E., 1990, Shear zone structures in the Øygarden Complex, western Norway. *Tectonophysics* **174**: 385–397.

Fossen, H. and Tikoff, B., 1993, The deformation matrix for simultaneous simple shearing, pure shearing, and volume change, and its application to transpression/transtension tectonics. *Journal of Structural Geology* **15**: 413–422.

Fossen, H. and Tikoff, B., 1997, Forward modeling of non steady-state deformations and the "minimum strain path". *Journal of Structural Geology* **19**: 987–996.

Fossen, H. and Tikoff, B., 1998, Extended models of transpression and transtension, and application to tectonic settings. In R. E. Holdsworth, R. A. Strachan and J. F. Dewey (Eds.), *Continental Transpressional and Transtensional Tectonics*. Special Publication 135, London: Geological Society, pp. 15–33.

Fossen, H., Tikoff, T. B. and Teyssier, C. T., 1994, Strain modeling of transpressional and transtensional deformation. *Norsk Geologisk Tidsskrift* **74**: 134–145.

Fossen, H., Odinsen, T., Færseth, R. B. and Gabrielsen, R. H., 2000, Detachments and low-angle faults in the northern North Sea rift system. In Nøttvedt, A. (Ed.), *Dynamics of the Norwegian Margins*, Special Publication **167**, London: Geological Society, pp. 105–131.

Fossen, H., Schultz, R., Shipton, Z. and Mair, K., 2007, Deformation bands in sandstone – a review. *Journal of the Geological Society, London* **164**: 755–769.

Fossen, H., Dallman, W. and Andersen, T. B., 2008, The mountain chain rebounds and founders. The building up of the Caledonides; about 500–405 Ma. In I. Ramberg, I. Bryhni and A. Nøttvedt (Eds.), *The Making of a Land: Geology of Norway*. Trondheim: Norsk Geologisk Forening, pp. 178–231.

Gallagher, J. J., Friedman, M., Handin, J. and Sowers, G. M., 1974, Experimental studies relating to microfracture in sandstone. *Tectonophysics* **21**: 203–247.

Gartrell, A. P., 1997, Evolution of rift basins and low-angle detachments in multilayer analog models. *Geology* **25**: 615–618.

Gleason, G. C. and Tullis, J., 1995, A flow law for dislocation creep of quartz aggregates determined with the molten salt cell. *Tectonophysics* **247**: 1–23.

Griffith, A. A., 1924, The theory of rupture. In C. B. Biezeno and J. M. Burgers (Eds.), *First International Congress on Applied Mechanics*. Delft: J. Waltman, pp. 55–63.

Griggs, D. and Handin, J., 1960, Observations on fracture and a hypothesis of earthquakes. In D. Griggs and J. Handin, (Eds.), *Rock Deformation*. Memoir **79**, Boulder: Geological Society of America.

Gueguen, E., Dogliono, C. and Fernandez, M., 1997, Lithospheric boudinage in the western Mediterranean back-arc basin. *Terra Nova* **9**: 184–187.

Hanmer, S., 1986, Asymmetrical pull-aparts and foliation fish as kinematic indicators. *Journal of Structural Geology* **8**: 111–122.

Hayward, A. B. and Graham, R. H., 1989, Some geometrical characteristics of inversion. In M. A. Cooper and G. D. Williams (Eds.), *Inversion Tectonics*. Special Publication **44**, London: Geological Society, pp. 17–39.

He, B., Xu, Y.-G. and Paterson, S., 2009, Magmatic diapirism of the Fangshan pluton, southwest of Beijing, China. *Journal of Structural Geology* **31**: 615–626.

Heard, H. C., 1960, Transition from brittle fracture to ductile flow in Solenhofen limestone as a function of temperature, confining pressure and interstitial fluid pressure. *Geological Society of America Bulletin* **79**: 193–226.

Heard, H. C. and Raleigh, C. B., 1972, Steady-state flow in marble at 500° to 800° C. *Geological Society of America Bulletin* **83**: 935–956.

Hesthammer, J. and Fossen, H., 1998, The use of dipmeter data to constrain the structural geology of the Gullfaks Field, northern North Sea. *Marine and Petroleum Geology* **15**: 549–573.

Hesthammer, J. and Fossen, H., 1999, Evolution and geometries of gravitational collapse structures with examples from the Statfjord Field, northern North Sea. *Marine and Petroleum Geology* **16**: 259–281.

Hobbs, B. E., McLaren, A. C. and Paterson, M. S., 1972, Plasticity of single crystals of synthetic quartz. In H. C. Heard, I. Y. Borg, N. I. Carter and C. B. Raleigh (Eds.), *Flow and Fracture of Rocks*. American Geophysical Union Monograph **16**: pp. 29–53.

Hobbs, B. E., Means, W. D. and Williams, P. F., 1976, *An Outline of Structural Geology*. New York: J. Wiley and Sons.

Hodgson, R. A., 1961, Classification of structures on joint surfaces. *American Journal of Science* **259**: 493–502.

Holst, T. B. and Fossen, H., 1987, Strain distribution in a fold in the West Norwegian Caledonides. *Journal of Structural Geology* **9**: 915–924.

Hossack, J., 1968, Pebble deformation and thrusting in the Bygdin area (Southern Norway). *Tectonophysics* **5**: 315–339.

Høyland Kleppe, K.-J., 2003, Strukturgelogisk analyse av forkastninger og deres innvirkning på kommunikasjon i østlige deler av Gullfaksfeltet, nordlige Nordsjøen. Unpublished Cand. scient. thesis, University of Bergen.

Hubbert, M. K. and Rubey, W. W., 1959, Role of pore fluid pressure in the mechanics of overthrust faulting. I: Mechanics of fluid-filled porous solids and its application to overthrust faulting. *Geological Society of America Bulletin* **70**: 115–205.

Hudec, M. R. and Jackson, M. P. A., 2006, Advance of allochthonous salt sheets in passive margins and orogens.

American Association of Petroleum Geologists Bulletin **90**: 1535–1564.

Hudec, M. R. and Jackson, M. P. A., 2007, Terra infirma: understanding salt tectonics. *Earth-Science Reviews* **82**: 1–28.

Hudleston, P. J., 1973, Fold morphology and some geometric implications of theories of fold development. *Tectonophysics* **16**: 1–46.

Hudleston, P. J., 1989, Extracting information from folds in rocks. *Journal of Geological Education* **34**: 237–245.

Hudleston, P. J. and Holst, T. B., 1984, Strain analysis and fold shape in a limestone layer and implications for layer rheology. *Tectonophysics* **106**: 321–347.

Jackson, J. A., Cornelius, R. R., Craig, C. H., Gansser, A., Stöcklin, J. and Talbot, C. J., 1990, *Salt Diapirs of the Great Kavir, Central Iran*. Memoir **177**, Boulder: Geological Society of America.

Kreemer, C., Holt, W. E. and Haines, A. J., 2003, An integrated global model of present-day plate motions and plate boundary deformation. *Geophysical Journal International* **154**: 8–34.

Labrousse, L., Jolivet, L., Andersen, T. B., Agard, P., Hébert, R., Maluski, H. and Schärer, U., 2004, Pressure-temperature-time deformation history of the exhumation of ultra-high pressure rocks in the Western Gneiss Region, Norway. In D. L. Whitney, C. Teyssier and C. S. Siddoway (Eds.), *Gneiss Domes in Orogeny*. Special Paper **380**, Boulder: Geological Society of America, pp. 155–183.

Lisle, J., 1984, Strain discontinuities within the Seve-Köli Nappe Complex, Scandinavian Caledonides. *Journal of Structural Geology* **6**: 101–110.

Lister, G. S. and Snoke, A. W., 1984, S-C mylonites. *Journal of Structural Geology* **6**: 617–638.

Mair, K., Main, I. and Elphick, S., 2000, Sequential growth of deformation bands in the laboratory. *Journal of Structural Geology* **22**: 25–42.

Marone, C. and Scholz, C. H. 1988. The depth of seismic faulting and the upper transition from stable to unstable slip regimes. *Geophysical Research Letters* **15**: 621–624.

McClay, K. R. and Ellis, P. G., 1987, Analogue models of extensional fault geometries. In M. P. Coward, J. F. Dewey and P. L. Hancock, P. L. (Eds.), *Continental Extensional Tectonics*. Special Publication **28**, London: Geological Society, pp. 109–125.

McQuarrie, N., 2004, Crustal scale geometry of the Zagros fold–thrust belt, Iran. *Journal of Structural Geology* **26**: 519–535.

Merle, O., 1989, Strain models within spreading nappes. *Tectonophysics* **165**: 57–71.

Morley, C. K., 1986, Vertical strain variations in the Osa-Røa thrust sheet, North-western Oslo Fjord, Norway. *Journal of Structural Geology* **8**: 621–632.

Myrvang, A., 2001, *Bergmekanikk*. Trondheim, Institutt for geologi og bergteknikk, NTNU.

Nettleton, L. L., 1934, Fluid mechanics of salt domes. *American Association of Petroleum Geologists Bulletin* **18**: 1175–1204.

Odinsen, T., Christiansson, P., Gabrielsen, R. H., Faleide, J. I. and Berge, A., 2000, The geometries and deep structure of the northern North Sea. In A. Nøttvedt (Ed.), *Dynamics of the Norwegian Margin*. Special Publication **167**, London: Geological Society, pp. 41–57.

Olson, J. and Pollard, D. D., 1989, Inferring paleostresses from natural fracture patterns: a new method. *Geology* **17**: 345–348.

Paterson, M. S., 1958, Experimental deformation and faulting in Wombeyan marble. *Geological Society of America Bulletin* **69**: 465–476.

Petit, J.-P., 1987, Criteria for the sense of movement on fault surfaces in brittle rocks. *Journal of Structural Geology* **9**: 597–608.

Platt, J. P. and Vissers, R. L. M., 1980, Extensional structures in anisotropic rocks. *Journal of the Geological Society, London* **2**: 397–410.

Raimbourg, H., Jolivet, L., Labrousse, L., Leroy, Y. and Avigad, D., 2005, Kinematics of syneclogite deformation in the Bergen Arcs, Norway: implications for exhumation mechanisms. In D. Gapais, J. P. Brun and P. R. Cobbold (Eds.), *Deformation Mechanisms, Rheology and Tectonics*. Special Publication **243**, London: Geological Society, pp. 175–192.

Ramberg, H., 1981, *Gravity, Deformation and the Earth's Crust*, 2nd edition. London: Academic Press.

Ramsay, J. G., 1967, *Folding and Fracturing of Rocks*. New York: McGraw-Hill.

Ramsay, J. G. and Huber, M. I., 1983, *The Techniques of Modern Structural Geology. Vol. 1: Strain Analysis*. London: Academic Press.

Ramsay, J. G. and Huber, M. I., 1987, *The Techniques of Modern Structural Geology. Vol. 2: Folds And Fractures*. London: Academic Press.

Ramsay, J. G. and Woods, D. S., 1973, The geometric effects of volume change during deformation processes. *Tectonophysics* **16**: 263–277.

Reusch, H., 1888, *Bømmeløen og Karmøen med omgivelser*. Oslo: Kristiania.

Roberts, A. M., Yielding, G., Kusznir, N. J., Walker, I. M. and Dorn-Lopez, D., 1993, Mesozoic extension in the North Sea: constraints from flexural backstripping, forward modelling and fault populations. In J. R. Parker (Ed.), *Petroleum Geology of Northern Europe*. London: Geological Society, pp. 1123–1136.

Roberts, D. and Strömgård, K.-E., 1972, A comparison of natural and experimental strain patterns around fold hinge zones. *Tectonophysics* **14**: 105–120.

Rouby, D., Fossen, H. and Cobbold, P., 1996, Extension, displacement, and block rotation in the larger Gullfaks area, northern North Sea: determined from map view restoration. *American Association of Petroleum Geologists Bulletin* **80**: 875–890.

Rowan, M. G., Lawton, T. F., Giles, K. A. and Ratliff, R. A., 2003, Near-salt deformation in La Popa basin, Mexico, and the northern Gulf of Mexico: A general model for passive diapirism. *American Association of Petroleum Geologists Bulletin* **87**: 733–756.

Rutter, E. H., 1976, The kinetics of rock deformation by pressure solution. *Philosophical Transactions of the Royal Society of London* A **283**: 203–219.

Rykkelid, E., 1987, Geologisk utvikling i Møkster (Selbjørnområdet) i Sunnhordland. Unpublished Cand. scient. thesis, University of Bergen.

Rykkelid, E. and Fossen, H., 1992, Composite fabrics in mid-crustal gneisses: observations from the Øygarden Complex, West Norway Caledonides. *Journal of Structural Geology* **14**: 1–9.

Rykkelid, E. and Fossen, H., 2002, Layer rotation around vertical fault overlap zones: observations from seismic data, field examples, and physical experiments. *Marine and Petroleum Geology* **19**: 181–192.

Sanderson, D. and Marchini, R. D., 1984, Transpression. *Journal of Structural Geology* **6**: 449–458.

Scheck, M., Bayer, U. and Lewerenz, B., 2003, Salt redistribution during extension and inversion inferred from 3D backstripping. *Tectonophysics* **373**: 55–73.

Schmid, S. M. and Kissling, E., 2000, The arc of the western Alps in the light of geophysical data on deep crustal structure. *Tectonics* **19**: 62–85.

Scholz, C. H., 1990, *The Mechanics of Earthquakes and Faulting*. Cambridge: Cambridge University Press.

Schultz, R. A. and Fossen, H., 2002, Displacement-length scaling in three dimensions: the importance of aspect ratio and application to deformation bands. *Journal of Structural Geology* **24**: 1389–1411.

Stewart, S. A. and Allen, P. J., 2005, 3D seismic reflection mapping of the Silverpit multi-ringed crater, North Sea. *Geological Society of America Bulletin* **117**: 354–368.

Stipp, M. and Tullis, J., 2003, The recrystallized grain size piezometer for quartz. *Journal of Geophysical Research* **30**: doi:10:1029/2003GL018444.

Suppe, J., 1983, Geometry and kinematics of fault-bend folding. *American Journal of Science* **283**: 684–721.

Talbot, C. J., 1979, Fold trains in a glacier of salt in southern Iran. *Journal of Structural Geology* **1**: 5–18.

Tapponnier, P., Peltzer, G. and Armijo, R., 1986, On the mechanics of the collision between India and Asia. In M. P. Coward and A. C. Ries (Eds.), *Collision Tectonics*. Special Publication **19**, London: Geological Society, pp. 115–157.

Tikoff, B. and Fossen, H., 1999, Three-dimensional reference deformations and strain facies. *Journal of Structural Geology* **21**: 1497–1512.

Trudgill, B. and Cartwright, J., 1994, Relay-ramp forms and normal-fault linkages, Canyonlands National Park, Utah. *Geological Society of America Bulletin* **106**: 1143–1157.

Twiss, R. J., 1977, Theory and applicability of a recrystallized grain size paleopiezometer. *Pure and Applied Geophysics* **115**, 227–244.

Twiss, R. J. and Moores, E. M., 2007, *Structural Geology*, 2nd edition. New York: H. W. Freeman and Company.

Vanderhaeghe, O., Burg, J.-P. and Teyssier, C., 1999, Exhumation of migmatites in two collapsed orogens: Canadian Cordillera and French Variscides. In U. Ring, M. T. Brandon, G. S. Lister and S. D. Willett (Eds.), *Exhumation Processes: Normal Faulting, Ductile Flow and Erosion*. Special Publication **154**, London: Geological Society, pp. 181–204.

Vendeville, B., Cobbold, P. R., Davy, P., Brun, J. P. and Choukroune, P., 1987, Physical models of extensional tectonics at various scales. In M. P. Coward, J. F. Dewey and P. L. Hancock (Eds.), *Continental Extensional Tectonics*. Special Publication **28**, London: Geological Society, pp. 95–107.

Wernicke, B. and Axen, G. J., 1988, On the role of isostasy in the evolution of normal fault systems. *Geology* **16**: 848–851.

White, N. J., Jackson, J. A. and McKenzie, D. P., 1986, The relationship between the geometry of normal faults and that of the sedimentary layers in the hanging walls. *Journal of Structural Geology* **8**: 897–909.

Wright, L. A., Otton, J. K. and Troxel, B. W., 1974, Turtleback surfaces of Death Valley viewed as phenomena of extensional tectonics. *Geology* **2**: 53–54.

Wust, S. L., 1986, Regional correlation of extension directions in the Cordilleran metamorphic core complexes. *Geology* **14**: 828–830.

Zanella, E., Coward, M. P. and McGrandle, A., 2003, Crustal structure. In D. Evans, C. Graham, A. Armour and P. Bathurst (Eds.), *The Millennium Atlas, Petroleum Geology of the Central and Northern North Sea*. London: Geological Society, pp. 35–43.

Cover and chapter image captions

Book cover   Folded Mesozoic layers in the Windgällen area, Canton Uri, Switzerland. The folds in this area are basement cored and mark the basal part of the Alpine orogenic edifice in the area. 46° 48′ 10″ N, 8° 47′ 50″ E.

Frontispiece
Small thrust fault in the rock formation known as ketobe knob on the northwest side of San Rafael Swell, Utah. This very well exposed thrust probably formed during the cretaceous sevier orogeny. Earthy-red and grayish layers belong to the Jurassic Entrada and Curtis Formations, respectively.   *page* ii

Chapter 1
Aerial photograph (digital orthophoto) of the Bighorn Basin, Wyoming, showing the Sheep Mountain Anticline and associated fold structures. The Sheep Mountain Anticline formed during the Laramide compression. 44° 39′ N, 108° 12′ W.   0

Chapter 2
Folded gneiss foliation in coastal exposures west of Bergen, Norway (Turøy). The gneissic banding is a result of Caledonian reworking of Proterozoic gneisses. 60° 27′ 16.125″ N, 4° 55′ 16.980″ E.   20

Chapter 3
Moderately deformed Neoproterozoic quartz conglomerate, where strain is exposed in sections containing the principal strain axes. The deformation occurred during the development of the extensional Raft River detachment in the Raft River Mountains, NE Utah, USA. 41° 57′ 26.6402″ N, 113° 19′ 32.208″ W.   54

Chapter 4
Balanced rock, Arches National Park, Utah. The massive "egg" of Entrada Sandstone at the top of less-resistant silty sandstones of the Dewey Bridge Member is proudly balancing stress and rock strength. 38° 42′ 03.40″ N, 109° 36′ 50.40″ W.   68

Chapter 5
Dead Horse Point, southern Utah, overlooking fractured rock formations of Permian to Jurassic ages. 38° 28′ 9.012″ N, 109° 44′ 21.120″ W.   78

Chapter 6
Archaic gneisses from Finnmark, northern Norway near the Russian Border, showing boudinaged layers that tell us about significant rheologic differences at the time of deep-crustal deformation. 69° 47′ 26.376″ N, 30° 47′ 34.800″ E.   96

Chapter 7
Checkerboard Mesa, Zion National Park. The horizontal fractures follow bedding while the vertical ones are perpendicular to bedding. These are fractures related to cooling and stress release during exhumation, enhanced by weathering. 37° 13′ 36″ N, 112° 52′ 47″ W.   118

Chapter 8
Caleta Heradurra near Antofagasta in northern Chile exposes a normal fault separating metamorphic basement (right) from gently rotated Miocene–Pliocene shallow marine sediments in the hanging wall (left). The overall geometry is that of a half-graben. 23° 12′ 6.012″ S, 70° 34′ 4.188″ W.   150

Chapter 9
Slickensided fault surface in meta-anorthosite, Bergen Arcs, southwest Norway. Shiny polished slip surfaces show well-defined striations that make them fit into a field of NW–SE stretching after the collapse of the Caledonian orogen in this area. 60° 33′ 26.388″ N, 5° 16′ 10.200″ E.   188

Chapter 10

Sheared Red Butte Granite (25 Ma) from the southern Grouse Creek Mountains, NE Utah. Dynamically recrystallized quartz exhibits a left-dipping foliation indicating dextral (top-to-the-west) sense of shear. Large zoned feldspar crystals are seen. Fabrics related to the formation of the Albion Mountains–Raft River Mountains–Grouse Creek Mountains metamorphic core complex. Thin section collection of James Evans, Utah State University. 41° 40′ 14″ N, 113° 46′ 41.5″ W.   202

Chapter 11

Folded limestones in Provo Canyon, Utah. The fold is known as the Bridal Veil Falls fold and formed during E–W shortening associated with the Cretaceous Sevier orogeny, occurring in the Charleston thrust sheet. 40° 19′ 51.140″ N, 111° 36′ 34.440″ W.   218

Chapter 12

Slaty cleavage showing marked refraction across lithology interfaces in late-Proterozoic turbidites of the Kongsfjord Formation. The cleavage is close to vertical in the fine-grained sandy layers, while it becomes left-dipping in the central shale layer. Photo taken in midnight sun in Kongsfjord, Finnmark, Norway. 70° 43′ 39.612″ N, 29° 19′ 44.400″ E.   242

Chapter 13

Lineations on foliation surfaces of strongly sheared quartz schist of the Raft River detachment, NW Utah, USA. 41° 52′ 20.322″ N, 113° 18′ 10.806″ W.   258

Chapter 14

Foliation boudinage in Proterozoic gneisses west of Bergen, Norway. The asymmetry is consistent with the top-to-hinterland (W) sense of shear recorded in these rocks during the extensional collapse of the Caledonian mountain chain in the Devonian. 60° 28′ 56.732″ N, 4° 55′ 46.381″ E.   270

Chapter 15

Picture of a ~40 cm tall section of a large Caledonian shear zone in Proterozoic gneisses west of Bergen, Norway. The small folds indicate top-to-the-right sense of shear and their limbs get attenuated as they overturn and tighten. 60° 28′ 52.887″ N, 4° 55′ 55.068″ E.   284

Chapter 16

The Glarus thrust near Flims, Switzerland. The thrust is very sharp, marked by a light gray-yellowish, meter-thick limestone called the Lochseiten limestone, acting as a lubricant at the ~15 km depth that thrusting occurred at. The hanging-wall rocks (Verrucano, 25–300 Ma) are older than the limestone (100–150 Ma) and the footwall flysch unit (35–50 Ma), consistent with thrust tectonics that puts older on younger rocks. 46° 54′ 3.08″ N, 9° 14′ 2.950″ E.   310

Chapter 17

Normal-faulted glacial (Quaternary) sand and clay layers exposed in sea cliffs near Nørre Lyngby, Jylland, Denmark. Note smearing of some of the clay layers. The "faults" are, according to modern terminology, disaggregation deformation bands and clay smears. 57° 24′ 49.29″ N, 9° 44′ 44.19″ E.   332

Chapter 18

Aerial view of the San Andreas fault in the Carrizo Plain, Central California, looking south. Note the sudden change in course of the stream channel, consistent with 130 meters of dextral (right-lateral) movement accumulated by earthquakes on this famous strike-slip fault over the last ~3700 years. 35° 08′ 00″ N, 119° 40′ 25″ W.   354

Chapter 19

Satellite image (Landsat 7) of the Dasht-e Kavir, the salt desert of Iran. The top of salt diapirs, including some salt glaciers (greenish) pierce banded salt-rich sediments that show large flow folds. Image by NASA.   370

Chapter 20

Folded deep-water sediments (chert-shale layers), deformed during emplacement of the Semail ophiolite. Hawasina nappes, Nizwa Road. 22° 53′ 56.998″ N, 57° 43′ 18.596″ E.   394

Chapter 21

Folded mylonites of the Bergsdalen Nappes near Kvamskogen, southwest Norway. While the gneissic layering was produced mainly during Caledonian thrusting, the folds formed after the kinematic reversal of the thrust zone. This reversal is related to the switch from contraction to extension, and placed the foliation in the contractional field of the instantaneous strain ellipse. 60° 23′ 15.30″ N, 6° 03′ 03.24″ E.   408

Index